Nano-microbiology for Sustainable Development

Yugal Kishore Mohanta • Bishwambhar Mishra
Ramesh Namdeo Pudake
Editors

Nano-microbiology for Sustainable Development

Springer

Editors
Yugal Kishore Mohanta
Department of Applied Biology
University of Science and Technology
Meghalaya (USTM), School of Biological
Sciences
Baridua, Meghalaya, India

Bishwambhar Mishra
Department of Biotechnology
Chaitanya Bharathi Institute of Technology
Hyderabad, Telangana, India

Ramesh Namdeo Pudake
Amity Institute of Nanotechnology
Amity University Uttar Pradesh
Noida, Uttar Pradesh, India

ISBN 978-3-031-78844-4 ISBN 978-3-031-78845-1 (eBook)
https://doi.org/10.1007/978-3-031-78845-1

This Springer imprint is published by the registered company Springer Nature Switzerland AG
The registered company address is: Gewerbestrasse 11, 6330 Cham, Switzerland

Preface

The increasing global issues of climate change, food insecurity, environmental degradation, and resource depletion necessitate inventive and sustainable solutions. The integration of nanotechnology and microbiology—two very promising scientific domains—has arisen as an effective strategy to tackle these intricate challenges. This book, "Nano-Microbiology for Sustainable Development," examines the fascinating possibilities of microbial nanotechnology in addressing the United Nations Sustainable Development Goals (SDGs), encompassing environmental sustainability, human health, and economic development. Microorganisms, despite their diminutive size, significantly contribute to the maintenance of ecosystem functions vital for life on Earth. Their capacity to create nanomaterials, decompose contaminants, and improve agricultural output has attracted growing interest from researchers globally. Notwithstanding this increasing interest, approximately 90% of microbial diversity remains unexamined, constituting a reservoir of unexploited solutions. This book seeks to integrate microbiology with nanotechnology, showcasing innovative interdisciplinary research and practical applications that can substantially aid in the attainment of the Sustainable Development Goals (SDGs).

The initial chapters explore the role of microbial nanotechnology in advancing the Sustainable Development Goals, emphasizing its capacity to provide sustainable solutions for global issues. The investigation commences with the synthesis of nanoparticles facilitated by algae, highlighting algae's contribution to the generation of eco-friendly nanomaterials for many applications in pollution management and agriculture. The subsequent sections highlight the significance of fungi in the production of agriculturally pertinent nanoparticles, concentrating on nanoparticles derived from fungi that can improve sustainable agricultural operations. The book examines the biorefining of agro-industrial waste into useful nanomaterials, transforming trash into sustainable resources for agriculture and fostering a circular economy. An essential component of sustainability is food security, which the book explores through the application of nanobiotechnology for food preservation and shelf-life extension. This method utilizes eco-friendly nanoparticles to diminish food deterioration and prolong shelf life, thus enhancing global food sustainability. A crucial emphasis is the sustainable manufacture of antibacterial nanoparticles, which provide viable solutions for health and sanitation while preserving environmental integrity.

The book also explores nano-biosensors for the detection of microbial pathogens in sustainable agriculture, a vital development for safeguarding food safety and preserving crop health. The many environmental applications of microbial nanotechnology are analyzed, especially its role as a mechanism for environmental sustainability and its potential to aid in e-waste management by facilitating the recovery of valuable materials and diminishing the ecological impact of electronic trash. Due to global disparities in technological advancement, particular emphasis is placed on the influence of microbial nanobiotechnology on the economics of developing nations. The chapters examine how the utilization of microbial nanotechnology can create economic prospects in agriculture, waste management, and sustainable industry, assisting developing countries in tackling their environmental and economic issues.

Water, a vital resource for humanity, is examined through sustainable water resource management with microbes-assisted nanotechnology. This section elucidates the mechanisms by which bacteria can filter water and eliminate pollutants, thereby advancing the Sustainable Development Goal of universal access to clean water. The book further examines the use of microbe-derived nanomaterials for the adsorption of harmful pollutants, illustrating the significance of these technologies in reducing environmental contamination. Finally, it investigates the potential of microbial nanostructures in soil remediation, highlighting their significance in restoring soil health and productivity for sustainable agriculture.

Incorporating insights from a wide range of scientific disciplines, this book aims to serve as a comprehensive resource for researchers, industry professionals, and policymakers. It brings to light the emerging and transformative field of microbial nanotechnology and its vast potential to contribute to sustainable development on multiple fronts, from environmental conservation and resource management to economic growth and global health.

We extend our heartfelt gratitude to all contributors whose expertise and dedication have made this book "Nano-Microbiology for Sustainable Development" possible. To the authors, researchers, and scientists who shared their groundbreaking work on microbial nanotechnology, thank you for your invaluable insights. Special thanks to the reviewers for their critical feedback, which greatly enhanced the quality of this book. We are also grateful to the institutions and organizations that supported this work, fostering research and innovation in sustainable development. Lastly, we thank our readers, whose curiosity and commitment inspire progress in this vital field.

We hope this book will inspire further research and collaboration, driving the continued exploration and application of microbial nanotechnology to build a more sustainable and equitable world. "Nano-Microbiology for Sustainable Development" offers a glimpse into a future where science, technology, and nature harmonize to address humanity's most pressing challenges. We invite you to explore this exciting frontier with us.

Baridua, Meghalaya, India — Yugal Kishore Mohanta
Hyderabad, Telangana, India — Bishwambhar Mishra
Noida, Uttar Pradesh, India — Ramesh Namdeo Pudake

Contents

Editors and Contributors

About the Editors

Yugal Kishore Mohanta has more than 5 years of post-PhD teaching and research experience and currently holds the position of Assistant Professor at the Department of Applied Biology within the School of Biological Sciences, University of Science and Technology, Meghalaya (USTM), India. Dr. Mohanta's current area of research pertains to nanobiotechnology and sustainable development goals. Dr. Mohanta has international research experience as both a PhD visiting scholar and a postdoctoral fellow at KU Leuven, Belgium, and the University of Nizwa, Sultanate of Oman, respectively. Dr. Mohanta has authored 100 research publications in peer-reviewed journals, which are indexed in the Science Citation Index Expanded (SCIE). These papers have high impact factors and have garnered a total of 3000 citations having an h-index of 33 and an i10 index of 61. Additionally, he has served as a reviewer for over 30 international peer-reviewed journals from Springer, Elsevier, Wiley, Frontiers, and Multidisciplinary Digital Publishing Institute (MDPI) and holds the position of associate editor for four SCIE-indexed journals. Additionally, Dr. Mohanta has made contributions to 16 book chapters, edited books for CRC Press and Elsevier-Academic Press, and authored numerous articles on scientific topics for national science publications. Dr. Mohanta currently holds the position of group leader at the Nano-biotechnology and Translational Knowledge Laboratory and serves as the principal investigator for the Science and Engineering Research Board-Department of Science and Technology (SERB-DST) project, a start-up research grant funded by the Government of India in New Delhi. Dr. Mohanta is also working as the Chief Coordinator of the Microbiologist Society in India for the North-East States of India.

Bishwambhar Mishra brings over a decade of post-PhD experience in teaching and research and is currently an Assistant Professor in the Department of Biotechnology at Chaitanya Bharathi Institute of Technology in Hyderabad, India. His primary research areas include industrial microbiology and nanobiotechnology, with a focus on their applications. His work is aimed at harnessing the potential of microorganisms and nanotechnology to address pressing global challenges such as environmental sustainability, food security, and resource management. Dr. Mishra has made significant contributions to the scientific community, authoring 60 research publications in high-impact, SCIE-indexed journals, with his work garnering 1,047 citations, an h-index of 18, and an i10-index of 33. These publications reflect the high caliber of his research, spanning various interdisciplinary domains. In addition to his extensive publication record, he has authored 20 book chapters that further advance knowledge in the field. Dr. Mishra has served as a reviewer for over 27 prestigious international journals, including those published by renowned publishers such as Springer, Elsevier, Wiley, Frontiers, and MDPI. His expertise is further recognized through his role as Topical Editor and Editorial board member of SCIE-indexed journals, where he plays a critical role in shaping scientific discourse and guiding future research.

Ramesh Namdeo Pudake is an Assistant Professor in Amity University, Uttar Pradesh, one of the top ranked private universities in India. Dr. Pudake received PhD from China Agricultural University, Beijing, PRC. After his PhD, he engaged in research in a range of topics in biotechnology. He has also worked in the Department of Agronomy at Iowa State University Ames, IA, USA, on functional genomics. Currently, he is focusing on the research on different applications of nanotechnology in agriculture. Dr. Pudake has published more than 45 research publications, 8 books, and 25 book chapters. He also has one scholarship award from the Chinese Government for pursuing his doctoral studies. He is an expert reviewer and guest editor for several journals of repute.

Contributors

Nisha Agrawal Strain Engineering, Colipi GmbH, Hamburg, Federal Republic of Germany

Bhavna Alke LAQV/Requimte, Department of Chemistry, NOVA School of Science and Technology, FCT NOVA, NOVA University Lisbon, Caparica, Portugal

Rajshree Amrit School of Biosciences and Technology, Vellore Institute of Technology, Vellore, Tamil Nadu, India

Sarita Aryal Medical Oncology, Dana Farber Cancer Institute, Harvard Medical School, Boston, MA, USA

Archana Ayyagari Department of Microbiology, Swami Shraddhanand College, University of Delhi, New Delhi, India

Rameshwari Ashish Banjara Department of Chemistry, Rajeev Gandhi Govt. Postgraduate College, Ambikapur, Chhattisgarh, India

Bharmjeet Singh International Centre for Genetic Engineering and Biotechnology, New Delhi, India

Arunima Biswas Department of Microbiology, Raidighi College, Raidighi, West Bengal, India

Renald Blundell Centre for Molecular Medicine, Biomedical Sciences Building, University of Malta, Msida, Malta

Akashi Bora Nano-biotechnology and Translational Knowledge Laboratory, Department of Applied Biology, School of Biological Sciences, University of Science and Technology Meghalaya, Ri-Bhoi, Meghalaya, India

Anuj Kumar Borah Department of Biochemistry, The Assam Royal Global University, Guwahati, Assam, India

Tanmayi Boreda Department of Biotechnology, Chaitanya Bharathi Institute of Technology, Hyderabad, Telangana, India

Rusmita Buragohain Nano-biotechnology and Translational Knowledge Laboratory, Department of Applied Biology, University of Science and Technology Meghalaya (USTM), Ri-Bhoi, Meghalaya, India

Arnab Chakraborty School of Biosciences and Technology, Vellore Institute of Technology, Vellore, Tamil Nadu, India

Palak Chaudhary University Institute of Biotechnology, Chandigarh University, Mohali, Punjab, India

Nitin Chauhan Department of Zoology, Sri Aurobindo College, University of Delhi, New Delhi, India

Jaeyeop Choi Smart Gym-Based Translational Research Center for Active Senior's Healthcare, Pukyong National University, Busan, Republic of Korea

Mautushi Das Department of Botany, School of Biological Sciences, University of Science and Technology Meghalaya, Ri-Bhoi, Meghalaya, India

Harisankar Dey Department of Botany, Ramakrishna Mission Vivekananda Centenary College (Autonomous), Kolkata, West Bengal, India

Vu Hoang Minh Doan Smart Gym-Based Translational Research Center for Active Senior's Healthcare, Pukyong National University, Busan, Republic of Korea

Chandan Kumar Gautam Department of Microbiological Sciences, North Dakota State University, Fargo, ND, USA

T. Gourav Department of Biotechnology, Chaitanya Bharathi Institute of Technology, Hyderabad, Telangana, India

Udensi Great Molecular Biosciences (Micro Track), University of Kansas (KU), Lawrence, KS, USA

Rishikesh Kumar Gupta Amity Institute of Neuropsychology and Neurosciences, Amity University Uttar Pradesh, Noida, India

Mahima Kalyanam Department of Biotechnology, Chaitanya Bharathi Institute of Technology, Hyderabad, Telangana, India

Yashasvi Kambhampati Department of Biotechnology, Chaitanya Bharathi Institute of Technology, Hyderabad, Telangana, India

Ashish Kumar Department of Biotechnology, Guru Ghasidas Vishwavidyalaya (A Central University), Bilaspur, Chhattisgarh, India

Manisha Kumari University Institute of Biotechnology, Chandigarh University, Mohali, Punjab, India

Priyanka Kumari Department of Forensic Science, Sam Higginbottom University of Agriculture, Technology and Sciences, Prayagraj, Uttar Pradesh, India

Regional Forensic Science Laboratory, Muzaffarpur, Bihar, India

Jay Kumar Department of Ophthalmology, Visual and Anatomical Sciences, Wayne State University, Detroit, MI, USA

Neelam Amit Kungwani Marine Biotechnology, Gujarat Biotechnology University, Gandhinagar, Gujarat, India

Kusumlata Department of Biotechnology, Guru Ghasidas Vishwavidyalaya (A Central University), Bilaspur, Chhattisgarh, India

Byeongil Lee Digital Healthcare Research Center, Pukyong National University, Busan, Republic of Korea
Industry 4.0 Convergence Bionics Engineering, Department of Biomedical Engineering, Pukyong National University, Busan, Republic of Korea
Smart Gym-Based Translational Research Center for Active Senior's Healthcare, Pukyong National University, Busan, Republic of Korea
Department of Smart Healthcare, Pukyong National University, Busan, Republic of Korea

C. R. Meera Department of Microbiology, St Mary's College (Autonomous), Thrissur, Kerala, India

Arti Mishra Umeå Plant Science Center, Department of Plant Physiology, Umeå University, Umeå, Sweden

Bishwambhar Mishra Department of Biotechnology, Chaitanya Bharathi Institute of Technology, Hyderabad, Telangana, India

Sudip Mondal Digital Healthcare Research Center, Pukyong National University, Busan, Republic of Korea

Sphoorthy Nadimpalli Department of Biotechnology, Chaitanya Bharathi Institute of Technology, Hyderabad, Telangana, India
Department of Chemistry, Graduate School of Science, Kyushu University, Fukuoka, Japan

C. Nagendranatha Reddy Department of Biotechnology, Chaitanya Bharathi Institute of Technology, Hyderabad, Telangana, India

Nishithendu Bikash Nandi Department of Chemistry, National Institute of Technology Agartala, Jirania, Tripura, India

Mahesh Narayan Department of Chemistry and Biochemistry, The University of Texas at El Paso (UTEP), El Paso, TX, USA

Pinku Chandra Nath Department of Food Technology, School of Applied and Life Sciences, Uttaranchal University, Dehradun, Uttarakhand, India

Amilia Nongbet Department of Botany, School of Biological Sciences, University of Science and Technology Meghalaya, Ri-Bhoi, Meghalaya, India

Mohamad Zuber Nurgat Department of Biotechnology, St Xavier's College (Empowered Autonomous Institution), Mumbai, Maharashtra, India

Junghwan Oh Digital Healthcare Research Center, Pukyong National University, Busan, Republic of Korea
Industry 4.0 Convergence Bionics Engineering, Department of Biomedical Engineering, Pukyong National University, Busan, Republic of Korea
Smart Gym-Based Translational Research Center for Active Senior's Healthcare, Pukyong National University, Busan, Republic of Korea
Ohlabs Corp., Busan, Republic of Korea

Sanjana Reddy Pailla Department of Biotechnology, Chaitanya Bharathi Institute of Technology, Hyderabad, Telangana, India

Bibhu Prasad Panda Environmental Sciences, Department of Chemistry, ITER, Siksha' O' Anusandhan (Deemed to be University), Bhubaneswar, Odisha, India

Jibanjyoti Panda Nano-biotechnology and Translational Knowledge Laboratory, Department of Applied Biology, University of Science and Technology Meghalaya (USTM), Ri-Bhoi, Meghalaya, India

Shrishty Pandey Amity Institute of Microbial Technology, Amity University Uttar Pradesh, Noida, Uttar Pradesh, India

Praneeta Panigrahi University Institute of Biotechnology, Chandigarh University, Mohali, Punjab, India

Thi Tuong Vy Phan Center for Advanced Chemistry, Institute of Research and Development, Duy Tan University, Danang, VietNam
Department of Environmental and Chemical Engineering, Duy Tan University, Danang, Vietnam

Ramesh Namdeo Pudake Amity Institute of Nanotechnology, Amity University Uttar Pradesh, Noida, Uttar Pradesh, India

Manjit Kumar Ray Department of Applied Biology, School of Biological Sciences, University of Science and Technology Meghalaya, Ri-Bhoi, Meghalaya, India

Puja Roy Nano-biotechnology and Translational Knowledge Laboratory, Department of Applied Biology, University of Science and Technology Meghalaya (USTM), Ri-Bhoi, Meghalaya, India

Khashayar Sarabandi Department of Food Chemistry, Research Institute of Food Science and Technology (RIFST), Mashhad, Iran

Pankaj Losan Sarma Department of Microbiology, The Assam Royal Global University, Guwahati, Assam, India

Ashmita Sethi University Institute of Biotechnology, Chandigarh University, Mohali, Punjab, India

Anirudh Sharma Department of Biotechnology, Jaypee Institute of Information Technology, Noida, Uttar Pradesh, India

Riya Sharma Amity Institute of Microbial Technology, Amity University Uttar Pradesh, Noida, Uttar Pradesh, India

Gajendra B. Singh University Institute of Biotechnology, Chandigarh University, Mohali, Punjab, India

Niraj Singh Department of Microbiology, The Assam Royal Global University, Guwahati, Assam, India

Sowjanya Sivakumar School of Biosciences and Technology, Vellore Institute of Technology, Vellore, Tamil Nadu, India
VIT School of Agricultural Innovations and Advanced Learning, Vellore Institute of Technology, Vellore, Tamil Nadu, India

Laxmikarthika V. Srinivasan School of Biosciences and Technology, Vellore Institute of Technology, Vellore, Tamil Nadu, India

Kirthikha Shanmuga Sunder Department of Biotechnology, Chaitanya Bharathi Institute of Technology, Hyderabad, Telangana, India

Usman T. Syed Division of Biological and Environmental Science and Engineering, King Abdullah University of Science and Technology (KAUST), Thuwal, Saudi Arabia

Guy Raymond Feuya Tchouya Department of Chemistry, Faculty of Science, Université des Sciences et Techniques de Masuku, Franceville, Gabon

Thi Thuy Truong Industry 4.0 Convergence Bionics Engineering, Department of Biomedical Engineering, Pukyong National University, Busan, Republic of Korea

Jaagriti Tyagi Amity Institute of Microbial Technology, Amity University Uttar Pradesh, Noida, Uttar Pradesh, India

Navya Tyagi Dr DY Patil Biotechnology and Bioinformatics Institute, Pune, Maharashtra, India

Shareefraza J. Ukkund Department of Biotechnology, P. A. College of Engineering (Affiliated to Visvesvaraya Technological University, Belagavi), Mangalore, Karnataka, India

Vanga Sri Varsha Department of Biotechnology, Chaitanya Bharathi Institute of Technology, Hyderabad, Telangana, India

Vijaya Laxmi Govindugari Department of Biotechnology, Chaitanya Bharathi Institute of Technology, Hyderabad, Telangana, India

Mrunal S. Wagh Research and Development, Organica Biotech Private Limited, Mumbai, Maharashtra, India

Rajasri Yadavalli Department of Biotechnology, Chaitanya Bharathi Institute of Technology, Hyderabad, Telangana, India

Lyudmila Zaynitdinova Laboratory of Biodiversity of Microorganisms, Institute of Microbiology, Tashkent, Uzbekistan

Chapter 1
Contribution of Microbial Nanotechnology to Sustainable Development Goals

Bharmjeet Singh, Nitin Chauhan, Nisha Agrawal, and Archana Ayyagari

Abstract The Sustainable Development Goals (SDGs) of the United Nations (UN) stand out today as the most needed guiding light in a tormented world facing complex global challenges. These provide an effective roadmap towards a healthier, happier, and progressive society. Microorganisms, being the smallest of living forms, have held the key for ages to a plethora of global game-changing innovations in the quest of sustainable development. In conjunction with nanotechnology, these microbes powerfully open up novel ways to address several UN-SDGs. This chapter sets off a thrilling excursion that reveals the possible contributions of microbial nanotechnology towards humankind via SDG goals. Microbial nanotechnology refers to the process of synthesizing metals, metal oxide, and other relevant nanoparticles using microorganisms, including bacteria, fungi, viruses, and algae. These microbial-assisted nanomaterials have shown incredible success in electronics, catalysis, food sector, health sciences, environment protection, catalysis, and optics. The biosynthesis approach using microorganisms has proved to be eco-friendly, more sustainable, and quicker for producing the desired nanoparticles in contrast to traditional methods, *i.e.*, physical and chemical methods. This chapter furnishes explicit information on how microbes may help in the progress of various SDGs. Also, it covers different strategies involving microbial cells for nanoparticle synthesis, and their applications in healthcare, agriculture, food, energy, and bioremediation. Furthermore, it highlights the importance of conducting ethical research and

B. Singh
International Centre for Genetic Engineering and Biotechnology, New Delhi, India

N. Chauhan
Department of Zoology, Sri Aurobindo College, University of Delhi, New Delhi, India

N. Agrawal
Strain Engineering, Colipi GmbH, Hamburg, Federal Republic of Germany

A. Ayyagari (✉)
Department of Microbiology, Swami Shraddhanand College, University of Delhi, New Delhi, India
e-mail: aarchana@ss.du.ac.in

Y. K. Mohanta et al. (eds.), *Nano-microbiology for Sustainable Development*,
https://doi.org/10.1007/978-3-031-78845-1_1

development to ensure that nanotechnology delivers outcomes that align perfectly with the larger global vision. However, like any other breakthrough, hurdles and barriers in this path would be expected. This chapter openly outlines these challenges which must be taken into consideration to maximize the potential of microbial nanotechnology.

Keywords SDGs · Microbes · Nanotechnology · Food · Health · Agriculture · Energy · Bioremediation

1.1 Introduction

The sustainability of the environment is seriously threatened by our existing lifestyle, which includes massive usage of chemicals, a greater reliance on non-renewable energy sources, and the unchecked production of waste products in industrial operations. The world urgently requires more emphasis to use green technology, cleaner manufacturing practices, and sustainable policies to preserve the earth's biosphere for the forthcoming generations. In an attempt to work together towards this goal, 193 nations adopted the 17 Sustainable Development Goals (SDG), an initiative supported by the UN to promote global economic development that is sustainable (Costanza et al., 2016). The SDGs seek to provide solutions that can promote societal and economic advancement without compromising the environment. Microorganisms, being ubiquitous, have the potential to tackle the challenges in agriculture, food security, energy, and healthcare in the best sustainable way (Doolam et al., 2024; Mishra et al., 2024). As per recent advances and development of microbial nanotechnology which deals with the synthesis, manipulation and use of nanoparticles, it is now possible to provide innovative solutions to global challenges.

For all the identified SDGs, microbial nanotechnology has already shown promising outcomes (Fig. 1.1). In the field of healthcare, revolutionary developments and improvements have been reported in drug delivery, diagnostics, imagining technologies, regenerative medicines, and personalized medicine. Moreover, for environmental protection microbial nanotechnology has delivered innovative technologies/methods, especially for the remediation of soil and water pollution (Gobinath et al., 2021). Further, for energy production, various novel nano-devices/designs for improving the charge/discharge rates, cycle stability, and energy storage capacity performance has been developed *i.e.*, Microbial fuel cells, Solar cells, batteries and supercapacitors etc. In the food sector, nano-based sensors are helping in the identification of hazardous preservatives, heavy metals, antibiotics, toxins, and pathogenic microorganisms in food. Likewise, nanobiosensors are also helping the farmers to get the real-time monitoring of various parameters that are critical for crop growth. The soil health, and environmental factors such pH, salinity, moisture content, quantity of soil nutrients, environmental stress, and pesticide residue,

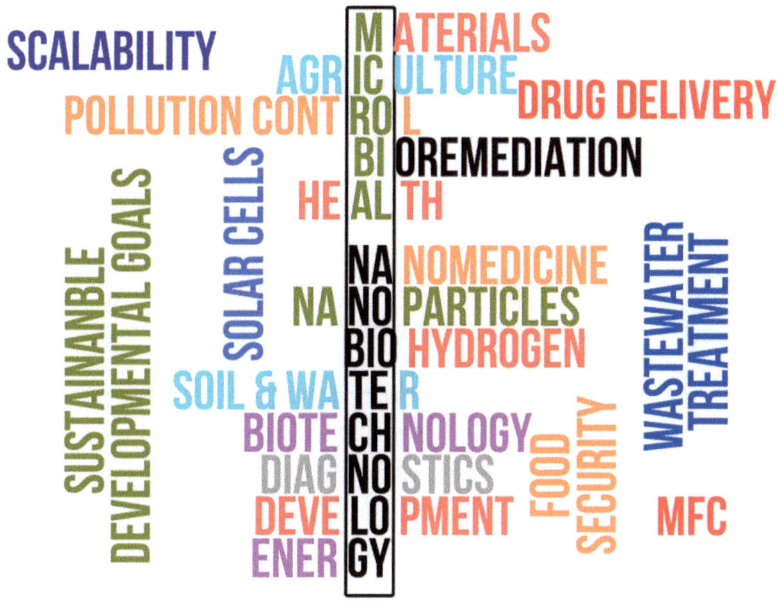

Fig. 1.1 A few of the numerous areas wherein microbial nanotechnology comes extremely applicable

among others can be monitored with biosensors. With so much advancement in the last decade, still the research and use of microbial-assisted technologies is low. The strong interdependence and collaboration between Industries, researchers, and policymakers is the need of the hour, and can't ever be overemphasized in utilizing the latest and most powerful tool of microbial nanotechnology to address the UN sustainable development goals in a more effective way. This chapter provides useful information and an understanding of microbial-assisted technologies and their appropriate applications to tackle global issues.

1.2 Microbial Nanotechnology in Food Security (SDG 2)

The current global scenario faces food security as one of the most pressing challenges. This encompasses access, availability, utilization, and stability of food worldwide. This has relevance to the United Nations' Sustainable Development Goal 2 (SDG 2) which aims to end hunger, achieve food security, improve nutrition, and promote sustainable agriculture by 2030 (Fig. 1.2). Microbial nanotechnology has immense potential in terms of enhancing food production, protecting crops, and improving food quality and safety.

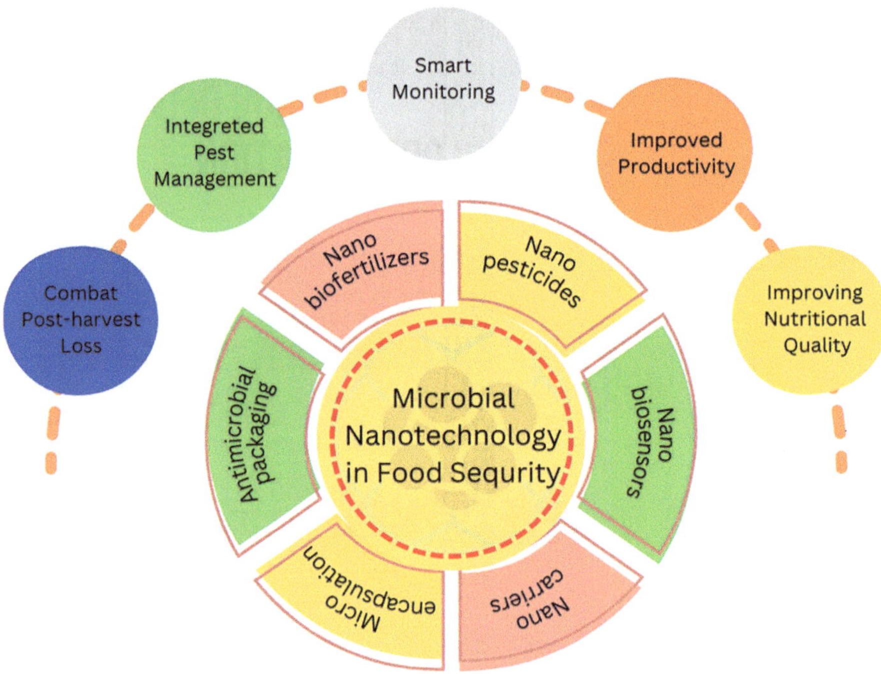

Fig. 1.2 Schematic representation of the ways in which microbial nanotechnology contributes towards food security (SDG 2)

1.2.1 Enhancing Crop Production

Microbial nanotechnology holds great promise to enhance crop yields by way of developing nano-biofertilizers and nanopesticides. Beneficial microbes such as prokaryotic nitrogen fixers and mycorrhizal fungi are popularly used as nano-fertilizers, which serve to improve the facilitation of nutrient uptake, thereby contributing well to better plant growth. Accurate quantities of these live miniature benefactors in an optimal condition help reduce the need of chemical fertilizers, thereby minimizing adverse environmental impact. The most common and time-tested example is the use of *Rhizobium*-based nano-biofertilizers in accentuated legume crop yields by enhancing nitrogen fixation (Saheewala et al., 2023).

Nanopesticides originating from microbial agents such as *Bacillus thuringiensis* (Bt) offer a superior and sustainable alternative to chemical pesticides (Kumar et al., 2021). These nanopesticides provide a controlled and extended supply of biocidal agents that target and kill the pests in a tailor-made manner, while not harming the non-target organisms. Moreover, the use of nanocarriers is also empowering under various environmental conditions in terms of improving the stability and bioavailability of microbial pesticides (Vurro et al., 2019). Besides improving crop productivity, such an approach also ensures less consumption of harmful chemicals, promoting sustainable agriculture practices with reasonably enhanced outputs.

1.2.2 Combating Post-harvest Losses

As much as one-third of food produced globally is being wasted from post-harvest losses, and is among the threats to food security. Microbial nanotechnology offers ways to mitigate this wastage through better preservation strategies and/or innovative packaging. Antimicrobial nanoparticles derived from certain beneficial microbes hold the potential to extend the shelf life of perishable foods by blocking the growth of pathogens and spoilage organisms (Sivaprakash et al., 2022). A study revealed that integration of microbially produced silver nanoparticles into food packaging films supported a remarkable hike in its shelf life. This was attributed to its antimicrobial action, which not only extended the shelf life of the products by powerfully reducing spoilage but also extended the freshness of packaged foods.

Another area of scientific concern, nanotechnology-enabled microbial enzymes or whole cell-based biosensors have been developed which can reliably detect spoilage indicators and contaminants in food products (Tovar-Lopez, 2023). These biosensor tools are capable of sensitive, rapid, and timely detection of microbial contamination, based on which the required interventions to avoid reduction of waste as well as prevention of foodborne illnesses may be implemented. Such advanced technologies provide much hope and promise towards the worthy goal of reducing post-harvest losses, thereby improving the abundance and availability of food.

1.2.3 Improving Nutritional Quality

Microbial nanotechnology may also be of immense use in enhancing the nourishing value of food. Probiotics and other beneficial microbes were demonstrated to become much more robust via microencapsulation in nanosized carriers, which was found to protect them from harsh conditions prevailing in the gastrointestinal tract. This is indeed critical if the goal is their safe delivery in functional condition in the gut. These microbes can discharge their beneficial effect if and only if they are made to retain their vitality and functionality (Xie et al., 2023). All probiotics, nutraceuticals, and functional foods are solely dependent on this prerequisite for contributing positively towards better overall gut health. Microencapsulating *Lactobacillus* strains in alginate nanoparticles have been demonstrated to improve their viability and therapeutic benefits in the gut (Li et al., 2023).

Additionally, microbial nanotechnology can be employed to fortify foods with essential micronutrients. Nanoparticles carrying vitamins, minerals, and bioactive compounds are being used to develop fortified foods with enhanced bioavailability and stability. Iron-fortified nanoparticles synthesized by microbes, for instance, have demonstrated improved absorption and effectiveness in combating iron deficiency anemia (Nadeem et al., 2021). By addressing micronutrient deficiencies, these innovations contribute to better nutrition and health outcomes, aligning with the goals of SDG 2.

1.2.4 Sustainable Agricultural Practices

Microbial nanotechnology supports sustainable agricultural practices by diminishing environmental impact of undesirable factors. General health of the soil in terms of moisture content, nutrient levels, presence of beneficial microbes and their numbers, combinations etc., and the prevailing net microbial activities occurring at any given time may be effectively monitored by nano-biosensors (Dar et al., 2020; Pudake et al., 2021) This data-driven modus operandi helps to optimize agricultural practices and saves the wastage of important resources including fertilizers, pesticides, and water. This adds up cumulatively to the farmers' advantage. Additionally, extra presence of chemical fertilizers in the soil is not able to accumulate there, maintaining the soil fertile and productive. A common example of the same is biosensors utilizing *Pseudomonas fluorescens,* which can reliably monitor soil phosphate, thereby indicating the levels of phosphorus fertilizers needed now (Yadav et al., 2023).

Furthermore, microbial nanotechnology facilitates bioremediation of contaminated soils, enhancing their fertility and productivity. Microbes engineered at the nanoscale can degrade pollutants and toxins in soil, restoring its health and suitability for agriculture. For example, nano-bioremediation using *Pseudomonas putida* has been employed to clean up soils contaminated with heavy metals and pesticides, enabling safe cultivation of crops (Aliyari Rad et al., 2023). These practices contribute to the sustainability and resilience of agricultural systems, ensuring long-term food security.

1.3 Microbial Nanotechnology in Health and Well-Being (SDG 3)

Microbial nanotechnology has shown immense potential in addressing various aspects of health and well-being, which are encompassed in SDG 3. The latter aims to ensure healthy lives and promote well-being for all at all ages. The integration of

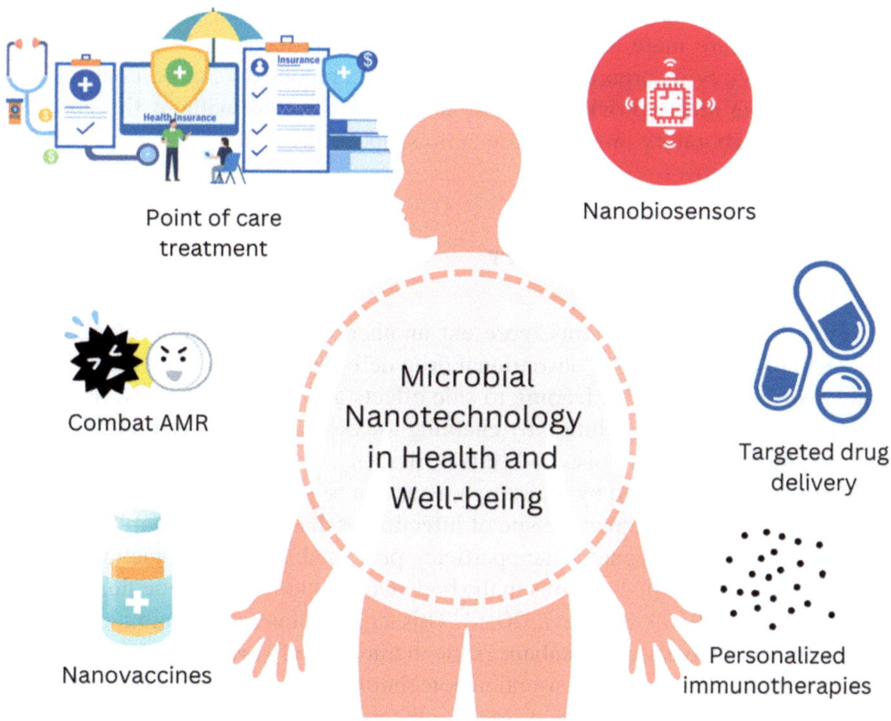

Fig. 1.3 Schematic depiction of how microbial nanotechnology finds unique applications in SDG 3 (Health and well-being)

microbial nanotechnology in healthcare can revolutionize diagnostics, treatment, and prevention of diseases, thereby significantly contributing to this global objective (Fig. 1.3).

1.3.1 Advanced Diagnostics

The most significant contribution of microbial nanotechnology to health and well-being is in the realm of diagnostics. Traditional diagnostic methods often face limitations in terms of sensitivity, specificity, and speed. On the other hand, microbial nanotechnology offers multiple innovative solutions to these challenges.

For instance, the development of biosensors using microbial nanocomponents can prove useful for rapid and highly sensitive detection of pathogens (Alafeef et al., 2020). These biosensors utilize nanoparticles synthesized by microorganisms, which can be functionalized with specific biomolecules to recognize and bind to target pathogens. This technology enables the detection of diseases at an early stage, which is crucial for effective treatment and management.

Moreover, microbial nanotechnology can improve point-of-care diagnostics, making healthcare more accessible, especially in low-resource settings. Portable diagnostic devices incorporating microbial nanocomponents can be conveniently used in remote areas, providing quick and reliable results without the need for sophisticated laboratory infrastructure at those locations.

1.3.2 Targeted Drug Delivery

Targeted drug delivery systems represent another vital contribution of microbial nanotechnology to SDG 3. Conventional drug delivery methods often result in systemic distribution of drugs, leading to side effects and reduced efficacy. Microbial nanotechnology offers a solution by enabling the design of nanoscale drug carriers that can specifically target diseased cells or tissues.

Nanoparticles produced by microorganisms can be engineered to carry therapeutic agents and release them at the site of infection or disease. For example, bacterial magnetosomes, *i.e.*, magnetic nanoparticles produced by magnetotactic bacteria, can be guided to specific locations in the body using external magnetic fields (Jacob & Suthindhiran, 2016; Mathuriya, 2016). This targeted approach minimizes the side effects on healthy tissues and enhances the therapeutic efficacy of drugs.

Furthermore, the use of microbial nanotechnology in drug delivery can improve the solubility and stability of drugs, making them more effective. This is particularly beneficial for drugs that have poor bioavailability or are rapidly degraded in the body.

1.3.3 Antimicrobial Resistance

Antimicrobial resistance (AMR) is an ever-growing global health threat, undermining the effectiveness of antibiotics and other antimicrobial agents. Microbial nanotechnology offers novel strategies to combat AMR. Nanoparticles synthesized by microorganisms possess unique physicochemical properties that can be exploited to develop new antimicrobial agents (Kungwani et al., 2024).

For example, silver nanoparticles (AgNPs) produced by bacteria have demonstrated potent antimicrobial activity against a wide range of pathogens, including multidrug-resistant bacteria (Tariq et al., 2020). These nanoparticles can disrupt microbial cell membranes, generate reactive oxygen species (ROS), and interfere with microbial metabolic processes, thereby effectively killing the pathogens.

Moreover, microbial nanotechnology can be used to enhance the efficacy of existing antibiotics. Nanoparticles can be conjugated with antibiotics to create nanocomposites that exhibit synergistic effects, overcoming resistance mechanisms in pathogens (León-Buitimea et al., 2022). This approach not only repurposes the use of older antibiotics but also extends their lifespan.

1.3.4 Vaccines and Immunotherapies

Vaccination is a critical component of global health initiatives aimed at preventing infectious diseases. Microbial nanotechnology is playing a pivotal role in the development of next-generation vaccines and immunotherapies. Nanoparticles produced by microorganisms can serve as adjuvants or carriers for antigens, enhancing the immune response.

For instance, biodegradable nanoparticles synthesized by bacteria can be used to deliver vaccine antigens in a controlled manner, improving the immunogenicity and longevity of the vaccine response (Akagi et al., 2011). These nanovaccines can elicit strong and durable immune responses, providing better protection against infectious diseases (Priyanka et al., 2023).

Additionally, microbial nanotechnology can facilitate the development of personalized immunotherapies for cancer treatment. Nanoparticles can be engineered to carry tumor-specific antigens and deliver them to the immune system, stimulating a targeted immune response against cancer cells (Gao et al., 2019).

1.3.5 Environmental and Public Health

Beyond direct medical applications, microbial nanotechnology also contributes to health and well-being through its impact on environmental health. Microorganisms can be used to synthesize nanoparticles that aid in the detection and removal of environmental pollutants, such as heavy metals and organic contaminants (Kapoor et al., 2021).

For example, biosynthesized iron oxide nanoparticles can be used in water treatment processes to remove arsenic, a toxic element that poses significant health risks (Majumder et al., 2019). By improving water quality, microbial nanotechnology helps prevent waterborne diseases and promotes overall public health.

Moreover, microbial nanotechnology can contribute to the development of sustainable agricultural practices, which indirectly impact human health favorably. Nanoparticles produced by soil microorganisms can enhance nutrient uptake in plants, reducing the need for chemical fertilizers and pesticides. This leads to the production of safer and more nutritious food, contributing to better health outcomes.

1.4 Microbial Nanotechnology in Clean Water and Sanitation (SDG 6)

Clean water and sanitation are fundamental human rights and essential components of sustainable development. SDG 6 aims to ensure the availability and sustainable management of water and sanitation for one and all by 2030. Microbial nanotechnology offers innovative solutions to address water contamination, improve

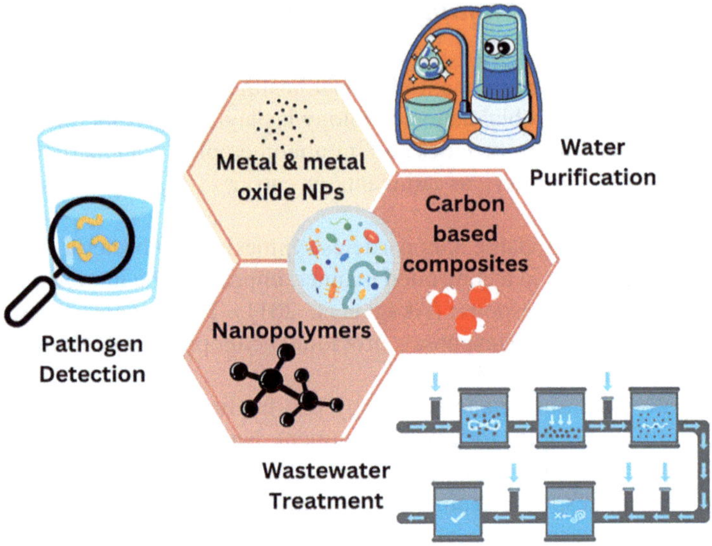

Fig. 1.4 Contributions of microbial nanotechnology towards achieving clean water and sanitation (SDG 6)

sanitation, and promote efficient water management (Fig. 1.4). We discuss the contributions of microbial nanotechnology towards achieving clean water and sanitation in below section.

1.4.1 Water Purification and Treatment

Microbial nanotechnology has emerged as a powerful tool in water purification and treatment processes. Nanoparticles and nanocomposites synthesized by microbes possess unique properties that enhance the removal of contaminants from water. For instance, iron oxide nanoparticles produced by bacteria such as *Shewanella oneidensis* have been employed for the removal of heavy metals like arsenic and lead from contaminated water sources (Agarwal & Darbar, 2023). These nanoparticles exhibit high surface area and reactivity, enabling efficient adsorption and sequestration of toxic metals.

Moreover, microbial nanotechnology enables the development of bio-inspired filtration systems. Biofilms of microbial communities can be engineered with nanomaterials to create advanced filtration membranes that efficiently remove pollutants, pathogens, and organic matter from water. For example, graphene oxide-based nanocomposites integrated with bacterial biofilms have shown enhanced filtration efficiency, providing a sustainable and cost-effective solution for water purification in rural and urban areas (Rana et al., 2024).

1.4.2 Wastewater Treatment

Effective wastewater treatment plants are crucial for maintaining clean water resources, potability, and preventing environmental pollution. Microbial nanotechnology offers innovative approaches to enhance the efficiency of wastewater treatment processes. Nanoparticles synthesized by microbes can be used to degrade and remove organic pollutants, pharmaceuticals, and other contaminants from wastewater. For instance, silver nanoparticles produced by *Pseudomonas stutzeri* have demonstrated effective degradation of antibiotics and dyes in wastewater, reducing their environmental impact (Wu et al., 2020).

Additionally, microbial nanotechnology enables the development of nano bioreactors for wastewater treatment. These bioreactors incorporate nanoscale microbial consortia that degrade pollutants through synergistic interactions. By optimizing the microbial community structure and function at the nanoscale, these systems achieve higher treatment efficiencies and shorter processing times. For example, nano bioreactors utilizing mixed cultures of *Pseudomonas* and *Bacillus* species have been shown to enhance the biodegradation of industrial effluents, providing a sustainable solution for wastewater management (Malik et al., 2022).

1.4.3 Pathogen Detection and Control

Microbial nanotechnology plays a critical role in detecting and controlling waterborne pathogens, ensuring the safety and quality of drinking water. Nanosensors incorporating microbial enzymes or whole cells can rapidly detect pathogenic bacteria, viruses, and protozoa in water samples (Ajayi et al., 2022). These biosensors offer high sensitivity and specificity, enabling real-time monitoring of water quality. For instance, biosensors utilizing enzymes from *Escherichia coli* have been developed to detect coliform bacteria, a common indicator of fecal contamination in water (Offenbaume et al., 2020).

Furthermore, microbial nanotechnology enables the development of antimicrobial nanomaterials for pathogen control in water systems. Silver nanoparticles synthesized by microbes exhibit strong antimicrobial activity against a wide range of pathogens, including bacteria, viruses, and fungi. These nanoparticles can be integrated into water treatment systems, coatings, and filters to prevent microbial contamination and ensure safe drinking water. For example, silver nanoparticle-coated filters have been used to inactivate *Escherichia coli* and other pathogens in drinking water, providing a reliable and cost-effective disinfection method (Praveena & Aris, 2015).

1.4.4 Sustainable Water Management

Microbial nanotechnology contributes to sustainable water management by promoting efficient use and recycling of water resources. Nanobiosensors can monitor water quality parameters such as pH, dissolved oxygen, and nutrient levels in real-time, enabling precise control of water treatment processes (Mahmoud & Fawzy, 2021). This data-driven approach ensures optimal use of water resources, reducing wastage and enhancing the efficiency of water treatment systems. For example, biosensors utilizing *Pseudomonas fluorescens* have been developed to monitor nitrate levels in water, guiding the application of nutrient management practices in agriculture (Yadav et al., 2023).

Additionally, microbial nanotechnology supports the development of innovative water recycling and reuse systems. Microbial fuel cells (MFCs) incorporating nanomaterials can treat wastewater while simultaneously generating electricity (Attia et al., 2024). These systems utilize the metabolic activities of electroactive bacteria to degrade organic pollutants and produce bioelectricity, providing a sustainable solution for water and energy recovery.

1.5 Microbial Nanotechnology in Energy Production (SDG 7)

Achieving SDG 7 needs a multifaceted approach that depends on increasing renewable energy capacity, improving energy efficiency, and management of energy resources. By developing innovative technologies, related policies, and investments, a significant development in the direction of a more sustainable and resilient energy future can be made. Therefore, microbial-assisted nanotechnology is a promising area in sustainable energy production, which integrates the efficiency of microorganisms with the unique properties of nanoparticles to improve the energy conversion processes (Fig. 1.5). The use of nanoparticles in the energy sector is focused on solar energy, hydrogen conversion, and thermoelectric devices.

1.5.1 Enhancing Energy Storage

Nanoparticles have shown applications in the development of solar cells, batteries, and supercapacitors to improve energy storage capacity, cycling stability, and charge/discharge rates. Studies suggested that nano-based photovoltaic solar cells have better efficiency with low manufacturing and electricity production costs.

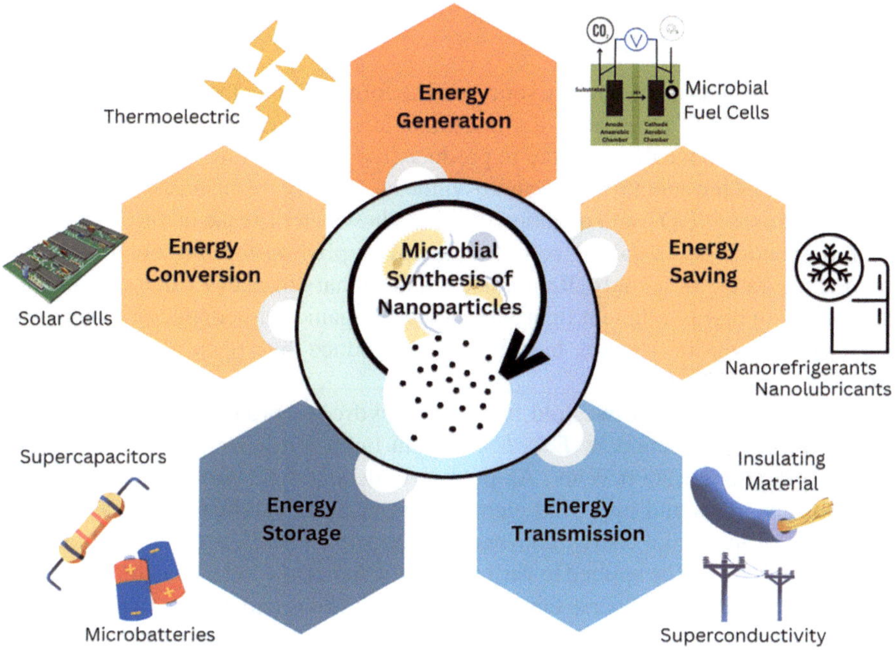

Fig. 1.5 Microbial nanotechnology for energy production (SDG 7)

1.5.2 Improving Microbial Fuel Cells (MFCs)

In recent developments, microbial fuel cell, MFC (which uses microorganisms which directly convert the chemical energy stored in organic matter to electricity) can be made more efficient by integrating with nanoparticles. With the incorporation of a variety of nanoparticles such as graphene, carbon nanotubes, and metal nanoparticles into MFCs, a significant improvement in the conductivity and surface area of electrodes can be achieved, thereby enhancing electron transfer efficiency and overall power output. As per the report, a new strategy to boost the charge-extraction efficiency in *Shewanella* MFCs has been developed (Cao et al., 2021).

The study highlighted a substantial improvement in the charge-extraction efficiency of MFC by introducing transmembrane and outer-membrane silver nanoparticles. The developed *Shewanella*-silver MFCs showed a maximum current density of 3.85 milliamperes per square centimeter, power density of 0.66 milliwatts per square centimeter, and single-cell turnover frequency of 8.6×10^5 per sec. The study highlighted that this hybrid MFC is best reported to date (Cao et al., 2021).

1.5.3 Biohydrogen Production

Further, a new alternative energy source in the form of biohydrogen has been developed. It is a clean, renewable, and sustainable source of energy. Some types of microorganisms have the potential to produce hydrogen gas through biological processes such as fermentation and photosynthesis. The use of nanoparticles, such as titanium dioxide (TiO_2) and quantum dots has shown an increase in the efficiency of light absorption and catalytic activity, thereby improving the microbial hydrogen production rates. In a study, Ramprakash and Incharoensakdi (2020) developed a system using highly efficient light-harvesting inorganic semiconductor (TiO_2) with *Escherichia coli* to form a biocatalyst to produce hydrogen (Ramprakash & Incharoensakdi, 2020).

The study revealed a two-fold increase in hydrogen production in this system. The hydrogen production by *E. coli* was light intensity-dependent showing maximum production at 2000 W/m^2. As per our understanding, nanomaterials provide high surface areas and improved electrical properties, significantly enhancing the efficiency of microbial energy conversion processes. Therefore, microbial-assisted nanotechnology can be applied to various organic materials and renewable resources, making it adaptable to diverse environmental conditions and energy requirements.

1.6 Microbial Nanotechnology in Environmental Sustainability (SDG 13)

A completely sustainable environment is achieved when the land, water, and air are free from toxic components. These compounds have certainly been contributing towards not only the degradation of the environment but also have caused great harm to the environmental biodiversity. Microbial nanotechnology has shown some innovative and smart ways to reclaim the environment as green, clean, serene, and safe (Fig. 1.6).

1.6.1 Environmental Cleansing

It has been observed that microbial-assisted nanomaterials possess the ability to perform oxidation, reduction, or sorption of pollutants. Because of the high surface area to volume ratio, nanomaterials can effectively bind with contaminants such as heavy metals, pathogens, dyes, and other such pollutions. Metallic nanoparticles, metal oxides, and nanotubes can be used as reactive materials for the removal of organic and inorganic pollutants. Iron oxide nanoparticles from *A. tubingenesis* were reported to remove as much as 90% of heavy metals [Pb (II), Ni (II), Cu (II), and Zn (II)] from wastewater (Mahanty et al., 2020). Further, copper nanoparticles

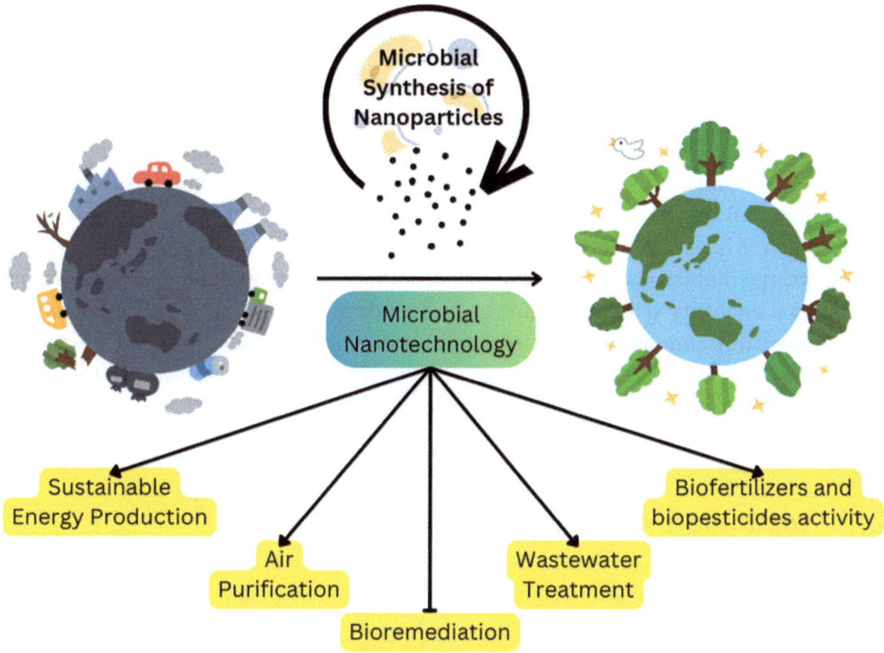

Fig. 1.6 Various ways in which microbial nanotechnology empowers environmental sustainability which is crucial for human lives

synthesized from *Escherichia* sp., have degraded azo dye and textile effluent at a lower concentration of 25 mg/L. These nanoparticles provide a sustainable, efficient, and economical way for the remediation of wastewater from harmful dyes (Noman et al., 2020). Similarly in soil also some specific nanomaterials (iron oxide and titanium oxide) can bind, degrade, and immobilize the organic pollutants of soil (Garg et al., 2020). As per the study (Romeh, 2018) on Fipronil, a pesticide known for neurotoxicity, carcinogenicity, and endocrine disruption. Silver nanoparticles synthesized from *Brassica* sp. showed up to 95.45% removal of Fipronil from water in 48 h. Further, in soil contaminated with Fibronil, the use of silver nanoparticles showed 68.8% reduction. To reduce air pollutants, a variety of nanosensors, nanocatalysts, and nanofilters have been developed. These nano filters with smaller pore sizes (1–10 nm) allow gases to pass but trap small particles such as bacteria, viruses, and organic pollutants (Taran et al., 2020). Recently a more advanced approach has been to use pollutant-degrading species of microorganisms and nanoparticles simultaneously for enhanced remediation. Nanoparticles can enhance microbial activity by increasing the surface area for microbial colonization and interaction with contaminants. As both nanoparticles and microorganisms work differently during interaction with pollutants, therefore a better understanding of the mechanism of bioremediation is very important in designing these systems (Vázquez-Núñez et al., 2020).

1.6.2 Enhancing Biofertilizers and Biopesticides Activity

Studies indicated that nanoparticles can increase the effectiveness of biofertilizers by enhancing nutrient delivery to plants and promoting better growth. Moreover, microbial-assisted nanotechnology can be used to develop biopesticides that can target specific pests with reduced environmental toxicity. These methods reduce the use of chemical fertilizers and pesticides and also the risks associated with it (Garg et al., 2020). Microbial nanotechnology provides comprehensive solutions to various environmental issues. To achieve the sustainable development goal for environmental sustainability it is important to further develop cost-effective and scalable technologies utilizing nanotechnology and microbial processes. With such developments, it is also important to perform a comprehensive risk assessment for maximum utilization of microbial nanotechnology soon.

1.7 Advantages of Microbial Nanotechnology

Utilizing the natural innate capacities of microorganisms such as fungi, algae, and bacteria, the application of severe chemicals and procedures that consume energy can be reduced (Koul et al., 2021). Further, as ambient conditions are frequently used for microbial synthesis, it reduces energy consumption and greenhouse gas emissions even further (Verstraete et al., 2022).

As this ecologically beneficial feature is in perfect harmony with the ideas of sustainable development, microbial nanotechnology is an appealing option for green industries and consumers. The efficacy and affordability of microbial nanotechnology are two of its main benefits (Tarkeshwar et al., 2023). Microorganisms are naturally able to manufacture complex structures with great selectivity and accuracy and to catalyze processes. The synthesis of nanoparticles with microbial assistance provides a simplified and resource-efficient manufacturing procedure. Moreover, microbial nanotechnology lowers capital investment and operating expenses by reducing the facilities and equipment that are generally involved with conventional nanoparticle synthesis techniques (Mandeep & Shukla, 2020). Microbial nanotechnology is the best alternative for companies looking to maximize production efficiency at a lower cost.

Innovation in the manufacturing of nanoparticles and their wide range of applications in many industries is stimulated by microbial nanotechnology. Technological innovation is made possible by the capacity to design microbes for specialized purposes, such as environmental cleanup or the manufacturing of nanoparticles (Koul et al., 2021). Furthermore, new applications in industries including healthcare, agriculture, energy, and environmental protection are made possible by specially designed nanomaterials using the potential of microorganisms. Microbial nanotechnology stimulates creativity and the creation of novel responses to difficult global problems by consistently pushing the envelope of what is feasible.

1.8 Ethical Considerations and Responsible Innovation

Microbial nanotechnology, like other nanotechnology fields, is developed and applied with ethical issues in mind. Strict ethical guidelines must be followed by researchers at every stage of the investigation from designing experiments and gathering data to publishing and disseminating findings. This includes getting informed consent before using human subjects in research, making sure lab animals are treated humanely, and upholding honesty and openness in the reporting of results. By following ethical research procedures, scientists may protect the welfare of people and communities impacted by their study and support the values of beneficence, fairness, and respect for autonomy.

It is necessary for responsible nanotechnology innovation to match technical breakthroughs with more expansive global goals, including the UN-SDGs. Through the incorporation of ethical, social, and environmental factors into the creation and use of nanotechnology solutions, scientists may guarantee that their research makes a beneficial impact on sustainable development (Guerra et al., 2022). This means giving initiatives that meet urgent societal needs, advance social fairness, and reduce negative environmental effects as a priority. Researchers may optimize the beneficial outcomes of their work and help create a fairer, and sustainable society by coordinating the development of nanotechnology with the SDGs.

1.9 Challenges and Barriers in Microbial Nanotechnology

Though microbial nanotechnology holds great promise, in order to reach its full potential, it must overcome a number of scientific and technical obstacles. Increasing control over the size and form of the nanoparticles, enhancing the production processes' scalability and repeatability, and maximizing microbial strains for higher synthesis efficiency are a few of these. Further study is necessary to fully realize the promise of microbial-assisted and green synthesis, as our present understanding of these approaches is still restricted. The frameworks for regulations and policies controlling the use of nanotechnology (microbial nanotechnology in particular) are still in the developing stage. To effectively address safety, environmental impact, and ethical issues, legislators, regulators, and industry stakeholders must take a proactive approach. Further, open communication, public outreach, and educational programs to broaden understanding of the potential benefits and risks of microbial nanotechnology are essential to establish public acceptance and confidence.

Many obstacles need to be cleared before microbial-mediated biosynthesis of nanomaterials being widely accepted as superior to "traditional" synthetic methods. These obstacles include reduced polydispersity of nanoparticles, improved biocapping layer agent characterization, purification of nanomaterials, standardization of microbial cultivation for successful nanoparticle synthesis, and production economics. For this reason, it is crucial to create trustworthy and environmentally acceptable methods for synthesizing nanomaterials to fully utilize the potential benefits of nanostructured applications in future (Dragone et al., 2017) and the responsible use of nanomaterials in diagnostic and therapeutic applications, such as drug administration, *in vivo* imaging, and creation of antibacterial and antitumoral medications (Kiessling et al., 2014).

1.10 Conclusions and Future Prospects

Microbial nanotechnology is the buzzword for the upcoming era to make revolutionary advances in almost every field toward achieving compliance with UN-SDGs more and more effectively. In response to the major issues of clean water and sanitation, affordable and clean energy, sustainable agriculture, food security, and outstanding health and well-being, it offers strongly transformative solutions steering us towards a more resilient, equitable, and sustainable future. In context with SDG-2, microbial nanotechnology is all geared up to be the game changer in addressing key challenges related to food security via enhancing crop production, reducing post-harvest losses, improving nutritional quality, and promoting sustainable agricultural practices. Likewise, this technology holds tremendous potential to advance health and well-being, directly contributing to the achievement of SDG 3. Through improved diagnostics, targeted drug delivery, novel antimicrobial strategies, enhanced vaccines, and contributions to environmental health, microbial nanotechnology offers innovative solutions to some of the most pressing global health challenges. SDG-6 goal of clean water and sanitation freely accessible to one and all is achievable by microbial nanotechnology via betterment in water purification, improving wastewater treatment, enabling pathogen detection and control, and promoting sustainable water management, Likewise, microbial-assisted nanotechnology enhances renewable energy production by integrating the microbial competence with unique properties of nanoparticles for improving energy conversion processes. This technology, adaptable to various organic materials and renewable resources, offers promising advancements in solar energy, hydrogen production, and microbial fuel cells, contributing significantly to achieving SDG 7. Microbial nanotechnology offers sustainable and efficient solutions for removing environmental pollutants, replacing chemical fertilizers with biofertilizers, and developing novel nanoparticle-based biopesticides, following SDG-13.

Microbial nanotechnology is a potent instrument for promoting sustainable development, but at the same time, comes with a great responsibility of tackling global issues ethically and cautiously without contributing any negative impact on the environment and health in the short or long term. Through the utilization of nanotechnology principles and the special skills of microbes, scientists can enhance human well-being, and protect the environment. However, to fully utilize microbial nanotechnology, it will be necessary to get beyond technological, legal, and financial obstacles as well as to coordinate research efforts with more focus on general social objectives and moral issues. Supportive policies and investments in this area are as important as the research and innovation in this field. Together, these can unlock new opportunities to fulfill the basic needs of the ever-growing global population with dignity.

To overcome the obstacles that microbial nanotechnology is now facing, concentrated efforts are also required to construct better R&D infrastructure, good investments, and the creation of efficient human resources. It has been observed that technological developments in other fields, *i.e.*, biotechnology, synthetic biology, physical science, medical science, and genetic engineering are also empowering in the expansion and use of microbial-assisted nanotechnology. Therefore, interdisciplinary partnerships between researchers of different fields will also provide momentum, stimulation, and will create synergies for a better resolve of global challenges. Further, the next generation of researchers will be developed through multidisciplinary training programs spanning the fields of biology, chemistry, engineering, and materials science. These researchers will be prepared to take on challenging scientific and technical problems in a more strategic and better way.

Collaborations at the international level are also very important to share the acquired knowledge, technologies, experience and best practices to maximize the beneficiary outcomes. Microbial nanotechnology has the vision and power to contribute towards the fulfillment of UN-SDGs, hence bringing about a future that is way more progressive, egalitarian, and wealthy for future generations, if researchers, academicians, industries, policymakers, and all stakeholders join hands to translate microbial nanotechnology innovations into practical addressing of all UN-SDGs. Continued research, innovation, and collaboration in this field are essential to harnessing the full potential of microbial nanotechnology for sustainable development. A lot of future research would have to be focused on more refinement and optimization of the synthesis and functionalization of microbial nanoparticles, understanding their interactions with biological systems, and conducting comprehensive preclinical and clinical studies.

Acknowledgements Authors are grateful to the International Centre of Genetic Engineering and Biotechnology (ICGEB), New Delhi, India; Swami Shraddhanand College and Sri Aurobindo College, University of Delhi, New Delhi, India, as well as Colipi GmbH, Hamburg, Federal Republic of Germany for support and encouragement.

References

Agarwal, S., & Darbar, S. (2023). Microorganism assisted synthesized metal and metal oxide nanoparticles for removal of heavy metal ions from the wastewater effluents. In *Environmental Applications of Microbial Nanotechnology* (pp. 127–148). Elsevier. https://doi.org/10.1016/B978-0-323-91744-5.00017-5

Ajayi, R. F., Barry, S., Nkuna, M., Ndou, N., Rakgotho, T., Nqunqa, S., Ngema, N., Thipe, V., & Muluadzi, T. (2022). Nanoparticles in biosensor development for the detection of pathogenic bacteria in water. In *Emerging freshwater pollutants* (pp. 331–358). Elsevier. https://doi.org/10.1016/B978-0-12-822850-0.00004-1

Akagi, T., Baba, M., & Akashi, M. (2011). *Biodegradable nanoparticles as vaccine adjuvants and delivery systems: Regulation of immune responses by nanoparticle-based vaccine* (pp. 31–64). https://doi.org/10.1007/12_2011_150

Alafeef, M., Moitra, P., & Pan, D. (2020). Nano-enabled sensing approaches for pathogenic bacterial detection. *Biosensors and Bioelectronics, 165*, 112276. https://doi.org/10.1016/j.bios.2020.112276

Aliyari Rad, S., Nobaharan, K., Pashapoor, N., Pandey, J., Dehghanian, Z., Senapathi, V., Minkina, T., Ren, W., Rajput, V. D., & Asgari Lajayer, B. (2023). Nano-microbial remediation of polluted soil: A brief insight. *Sustainability, 15*(1), 876. https://doi.org/10.3390/su15010876

Attia, Y. A., Samer, M., Mohamed, M. S., Salah, M., Moustafa, E., Hameed, R. M. A., Elsayed, H., & Abdelsalam, E. M. (2024). Enhancing bioelectricity generation from wastewater in microbial fuel cells using carbon nanomaterials. *Journal of Chemical Technology & Biotechnology, 99*(5), 1172–1180. https://doi.org/10.1002/jctb.7620

Cao, B., Zhao, Z., Peng, L., Shiu, H.-Y., Ding, M., Song, F., Guan, X., Lee, C. K., Huang, J., Zhu, D., Fu, X., Wong, G. C. L., Liu, C., Nealson, K., Weiss, P. S., Duan, X., & Huang, Y. (2021). Silver nanoparticles boost charge-extraction efficiency in *Shewanella* microbial fuel cells. *Science, 373*(6561), 1336–1340. https://doi.org/10.1126/science.abf3427

Costanza, R., Fioramonti, L., & Kubiszewski, I. (2016). The UN Sustainable Development Goals and the dynamics of well-being. *Frontiers in Ecology and the Environment, 14*(2), 59–59. https://doi.org/10.1002/fee.1231

Dar, F. A., Qazi, G., & Pirzadah, T. B. (2020). *Nano-biosensors: NextGen diagnostic tools in agriculture* (pp. 129–144). https://doi.org/10.1007/978-3-030-39978-8_7

Doolam, B., Mishra, B., Surabhi, D., Mandal, S. K., Sada, S., Reddy, N. R., Panda, J., Rustagi, S., Mishra, A. K., & Mohanta, Y. K. (2024). A systematic review of potential bioactive compounds from Saccharomyces cerevisiae: Exploring their applications in health promotion and food development. *Environment, Development and Sustainability*. https://doi.org/10.1007/s10668-024-04969-9

Dragone, R., Grasso, G., Muccini, M., & Toffanin, S. (2017). Portable bio/chemosensoristic devices: Innovative systems for environmental health and food safety diagnostics. *Frontiers in Public Health, 5*. https://doi.org/10.3389/fpubh.2017.00080

Gao, S., Yang, D., Fang, Y., Lin, X., Jin, X., Wang, Q., Wang, X., Ke, L., & Shi, K. (2019). Engineering nanoparticles for targeted remodeling of the tumor microenvironment to improve cancer immunotherapy. *Theranostics, 9*(1), 126–151. https://doi.org/10.7150/thno.29431

Garg, D., Sarkar, A., Chand, P., Bansal, P., Gola, D., Sharma, S., Khantwal, S., Surabhi, Mehrotra, R., Chauhan, N., & Bharti, R. K. (2020). Synthesis of silver nanoparticles utilizing various biological systems: Mechanisms and applications—A review. *Progress in Biomaterials, 9*(3), 81–95. https://doi.org/10.1007/s40204-020-00135-2

Gobinath, R., Bandeppa, Manasa, V., Rajendiran, S., Kumar, K., Paul, R., & Basavaraj, K. (2021). *Nanoparticle-mediated adsorption of pollutants: A way forward to mitigation of environmental pollution* (pp. 317–348). https://doi.org/10.1007/978-981-15-7455-9_13

Guerra, J. B. S. O. A., Hoffmann, M., Bianchet, R. T., Medeiros, P., Provin, A. P., & Iunskovski, R. (2022). Sustainable development goals and ethics: Building "the future we want." *Environment, Development and Sustainability, 24*(7), 9407–9428. https://doi.org/10.1007/s10668-021-01831-0

Jacob, J. J., & Suthindhiran, K. (2016). Magnetotactic bacteria and magnetosomes—Scope and challenges. *Materials Science and Engineering: C, 68*, 919–928. https://doi.org/10.1016/j.msec.2016.07.049

Kapoor, R. T., Salvadori, M. R., Rafatullah, M., Siddiqui, M. R., Khan, M. A., & Alshareef, S. A. (2021). Exploration of microbial factories for synthesis of nanoparticles—A sustainable approach for bioremediation of environmental contaminants. *Frontiers in Microbiology, 12*. https://doi.org/10.3389/fmicb.2021.658294

Kiessling, F., Mertens, M. E., Grimm, J., & Lammers, T. (2014). Nanoparticles for imaging: Top or flop? *Radiology, 273*(1), 10–28. https://doi.org/10.1148/radiol.14131520

Koul, B., Poonia, A. K., Yadav, D., & Jin, J.-O. (2021). Microbe-mediated biosynthesis of nanoparticles: Applications and future prospects. *Biomolecules, 11*(6), 886. https://doi.org/10.3390/biom11060886

Kumar, P., Pandhi, S., Mahato, D. K., Kamle, M., & Mishra, A. (2021). Bacillus-based nano-bioformulations for phytopathogens and insect–pest management. *Egyptian Journal Of Biological Pest Control, 31*(1), 128. https://doi.org/10.1186/s41938-021-00475-6

Kungwani, N. A., Panda, J., Mishra, A. K., Chavda, N., Shukla, S., Vikhe, K., Sharma, G., Mohanta, Y. K., & Sharifi-Rad, M. (2024). Combating bacterial biofilms and related drug resistance: Role of phyto-derived adjuvant and nanomaterials. *Microbial Pathogenesis*, 106874. https://doi.org/10.1016/j.micpath.2024.106874

León-Buitimea, A., Garza-Cárdenas, C. R., Román-García, M. F., Ramírez-Díaz, C. A., Ulloa-Ramírez, M., & Morones-Ramírez, J. R. (2022). Nanomaterials-based combinatorial therapy as a strategy to combat antibiotic resistance. *Antibiotics, 11*(6), 794. https://doi.org/10.3390/antibiotics11060794

Li, Q., Lin, H., Li, J., Liu, L., Huang, J., Cao, Y., Zhao, T., McClements, D. J., Chen, J., Liu, C., Liu, J., Shen, P., & Zhou, M. (2023). Improving probiotic (Lactobacillus casei) viability by encapsulation in alginate-based microgels: Impact of polymeric and colloidal fillers. *Food Hydrocolloids, 134*, 108028. https://doi.org/10.1016/j.foodhyd.2022.108028

Mahanty, S., Chatterjee, S., Ghosh, S., Tudu, P., Gaine, T., Bakshi, M., Das, S., Das, P., Bhattacharyya, S., Bandyopadhyay, S., & Chaudhuri, P. (2020). Synergistic approach towards the sustainable management of heavy metals in wastewater using mycosynthesized iron oxide nanoparticles: Biofabrication, adsorptive dynamics and chemometric modeling study. *Journal of Water Process Engineering, 37*, 101426. https://doi.org/10.1016/j.jwpe.2020.101426

Mahmoud, A. E. D., & Fawzy, M. (2021). *Nanosensors and nanobiosensors for monitoring the environmental pollutants* (pp. 229–246). https://doi.org/10.1007/978-3-030-68031-2_9

Majumder, A., Ramrakhiani, L., Mukherjee, D., Mishra, U., Halder, A., Mandal, A. K., & Ghosh, S. (2019). Green synthesis of iron oxide nanoparticles for arsenic remediation in water and sludge utilization. *Clean Technologies and Environmental Policy, 21*(4), 795–813. https://doi.org/10.1007/s10098-019-01669-1

Malik, S., Dhasmana, A., Preetam, S., Mishra, Y. K., Chaudhary, V., Bera, S. P., Ranjan, A., Bora, J., Kaushik, A., Minkina, T., Jatav, H. S., Singh, R. K., & Rajput, V. D. (2022). Exploring microbial-based green nanobiotechnology for wastewater remediation: A sustainable strategy. *Nanomaterials, 12*(23), 4187. https://doi.org/10.3390/nano12234187

Mandeep, & Shukla, P. (2020). Microbial nanotechnology for bioremediation of industrial wastewater. *Frontiers in Microbiology, 11*. https://doi.org/10.3389/fmicb.2020.590631

Mathuriya, A. S. (2016). Magnetotactic bacteria: Nanodrivers of the future. *Critical Reviews in Biotechnology, 36*(5), 788–802. https://doi.org/10.3109/07388551.2015.1046810

Mishra, B., Mishra, A. K., Mohanta, Y. K., Yadavalli, R., Agrawal, D. C., Reddy, H. P., Gorrepati, R., Reddy, C. N., Mandal, S. K., Shamim, M. Z., & Panda, J. (2024). Postbiotics: The new horizons of microbial functional bioactive compounds in food preservation and security. *Food Production, Processing and Nutrition, 6*(1), 28. https://doi.org/10.1186/s43014-023-00200-w

Nadeem, M., Khan, R., Shah, N., Bangash, I. R., Abbasi, B. H., Hano, C., Liu, C., Ullah, S., Hashmi, S. S., Nadhman, A., & Celli, J. (2021). A review of microbial mediated iron nanoparticles (IONPs) and its biomedical applications. *Nanomaterials, 12*(1), 130. https://doi.org/10.3390/nano12010130

Noman, M., Shahid, M., Ahmed, T., Niazi, M. B. K., Hussain, S., Song, F., & Manzoor, I. (2020). Use of biogenic copper nanoparticles synthesized from a native Escherichia sp. as photocatalysts for azo dye degradation and treatment of textile effluents. *Environmental Pollution, 257*, 113514. https://doi.org/10.1016/j.envpol.2019.113514

Offenbaume, K. L., Bertone, E., & Stewart, R. A. (2020). Monitoring approaches for faecal indicator bacteria in water: Visioning a remote real-time sensor for E. coli and Enterococci. *Water, 12*(9), 2591. https://doi.org/10.3390/w12092591

Praveena, S. M., & Aris, A. Z. (2015). Application of low-cost materials coated with silver nanoparticle as water filter in Escherichia coli removal. *Water Quality, Exposure and Health, 7*(4), 617–625. https://doi.org/10.1007/s12403-015-0167-5

Priyanka, Abusalah, M. A. H., Chopra, H., Sharma, A., Mustafa, S. A., Choudhary, O. P., Sharma, M., Dhawan, M., Khosla, R., Loshali, A., Sundriyal, A., & Saini, J. (2023). Nanovaccines: A game changing approach in the fight against infectious diseases. *Biomedicine & Pharmacotherapy, 167*, 115597. https://doi.org/10.1016/j.biopha.2023.115597

Pudake, R. N., Jain, U., & Kole, C. (Eds.). (2021). *Biosensors in agriculture: Recent trends and future perspectives*. Springer.

Ramprakash, B., & Incharoensakdi, A. (2020). Light-driven biological hydrogen production by Escherichia coli mediated by TiO_2 nanoparticles. *International Journal of Hydrogen Energy, 45*(11), 6254–6261. https://doi.org/10.1016/j.ijhydene.2020.01.011

Rana, V., Sheokand, M., Jain, K., Dhaka, S., Godara, S. K., Urmaliya, D. N., Madhav, S., Singh, K. P., & Dhaka, R. K. (2024). Effective and affordable water purification technologies for rural development. In *Water resources management for rural development* (pp. 91–106). Elsevier. https://doi.org/10.1016/B978-0-443-18778-0.00016-7

Romeh, A. A. A. (2018). Green silver nanoparticles for enhancing the phytoremediation of soil and water contaminated by fipronil and degradation products. *Water, Air, & Soil Pollution, 229*(5), 147. https://doi.org/10.1007/s11270-018-3792-3

Saheewala, H., Mahawar, S., Sanadhaya, S., Chauhan, S., Kollah, B., Mohanty, S. R., Chauhan, P. K., Upadhyay, S. K., & Jain, D. (2023). *Advances in biofertilizer production technologies: Paradigm for improving agriculture through microbial resources* (pp. 323–345). https://doi.org/10.1007/978-3-031-35147-1_16

Sivaprakash, G., Raja, R. K., Mohanrasu, K., Dinesh, G. H., & Arun, A. (2022). Microbial nanotechnology in food industry: Antimicrobial packaging. In *Handbook of microbial nanotechnology* (pp. 311–329). Elsevier. https://doi.org/10.1016/B978-0-12-823426-6.00002-4

Taran, M., Safaei, M., Karimi, N., & Almasi, A. (2020). Benefits and application of nanotechnology in environmental science: An overview. *Biointerface Research in Applied Chemistry, 11*(1), 7860–7870. https://doi.org/10.33263/BRIAC111.78607870

Tariq, F., Ahmed, N., Afzal, M., Khan, M. A. U., & Zeshan, B. (2020). Synthesis, characterization and antimicrobial activity of Bacillus subtilis-Derived silver nanoparticles against multidrug-resistant bacteria. *Jundishapur Journal of Microbiology, 13*(5). https://doi.org/10.5812/jjm.91934

Tarkeshwar, Pandit, M. A., Kapinder, Bhardwaj, K., & Kaur, J. (2023). Microbial nanotechnology: A potential tool for a sustainable environment. In *Environmental applications of microbial nanotechnology* (pp. 217–230). Elsevier. https://doi.org/10.1016/B978-0-323-91744-5.00010-2

Tovar-Lopez, F. J. (2023). Recent progress in micro- and nanotechnology-enabled sensors for biomedical and environmental challenges. *Sensors, 23*(12), 5406. https://doi.org/10.3390/s23125406

Vázquez-Núñez, E., Molina-Guerrero, C. E., Peña-Castro, J. M., Fernández-Luqueño, F., de la Rosa-Álvarez, M., & G. (2020). Use of nanotechnology for the bioremediation of contaminants: A review. *Processes, 8*(7), 826. https://doi.org/10.3390/pr8070826

Verstraete, W., Yanuka-Golub, K., Driesen, N., & De Vrieze, J. (2022). Engineering microbial technologies for environmental sustainability: Choices to make. *Microbial Biotechnology, 15*(1), 215–227. https://doi.org/10.1111/1751-7915.13986

Vurro, M., Miguel-Rojas, C., & Pérez-de-Luque, A. (2019). Safe nanotechnologies for increasing the effectiveness of environmentally friendly natural agrochemicals. *Pest Management Science, 75*(9), 2403–2412. https://doi.org/10.1002/ps.5348

Wu, L., Zhu, G., Zhang, X., & Si, Y. (2020). Silver nanoparticles inhibit denitrification by altering the viability and metabolic activity of Pseudomonas stutzeri. *Science of the Total Environment, 706*, 135711. https://doi.org/10.1016/j.scitotenv.2019.135711

Xie, A., Zhao, S., Liu, Z., Yue, X., Shao, J., Li, M., & Li, Z. (2023). Polysaccharides, proteins, and their complex as microencapsulation carriers for delivery of probiotics: A review on carrier types and encapsulation techniques. *International Journal of Biological Macromolecules, 242*, 124784. https://doi.org/10.1016/j.ijbiomac.2023.124784

Yadav, A., Yadav, K., Ahmad, R., & Abd-Elsalam, K. A. (2023). Emerging frontiers in nanotechnology for precision agriculture: Advancements, hurdles and prospects. *Agrochemicals, 2*(2), 220–256. https://doi.org/10.3390/agrochemicals2020016

Chapter 2
Microbial Nanotechnology: A Potentially Useful Tool for Environmental Sustainability

Niraj Singh, Pankaj Losan Sarma, Anuj Kumar Borah, Jibanjyoti Panda, and Jay Kumar

Abstract The biosynthesis of nanoparticles using microorganisms has recently attracted the great interest of the scientific community; it is considered an excellent approach for the synthesis of nanoparticles as compared to the traditional chemical and physical methods. Microorganisms-mediated production of nanomaterial provides an even greater boost to green biotechnology. Biogenic nanomaterials have shown enhanced properties in terms of mechanical, physical, and chemical due to their unique nanoscale behavior. Nanoparticles synthesized with the help of microorganisms have primarily been used as a carrier for economically important biomolecules for example enzymes, proteins, therapeutic drugs, etc. Over and above, in recent years microorganism-assisted nanomaterial has been recruited for the bioremediation of various industrial effluents and toxic pollutants. Nowadays, soil and water contamination heavily increases with organic/inorganic composites, and it is a worldwide issue for food safety and the environment. In this chapter, we are giving an overview of microbial nanotechnology and its applications, due to the small size, and catalytic activities nanoparticles have been used effectively to remediate the contaminated soil and water. Microbial nanotechnology may act as a sustainable, cost-effective, and eco-friendly solution to most of the environmental cleanup issues, agriculture and food.

N. Singh (✉) · P. L. Sarma
Department of Microbiology, The Assam Royal Global University, Guwahati, Assam, India

A. K. Borah
Department of Biochemistry, The Assam Royal Global University, Guwahati, Assam, India

J. Panda
Nano-Biotechnology and Translational Knowledge Laboratory, Department of Applied Biology, University of Science and Technology Meghalaya (USTM), Ri-Bhoi, Meghalaya, India

J. Kumar
Department of Ophthalmology, Visual and Anatomical Sciences, Wayne State University, Detroit, MI, USA

Y. K. Mohanta et al. (eds.), *Nano-microbiology for Sustainable Development*,
https://doi.org/10.1007/978-3-031-78845-1_2

Keywords Microbial · Nanotechnology · Sustainable · Bioremediation · Nanoparticles

2.1 Introduction

Nanotechnology refers to the manipulation and utilization of materials at the nanoscale, typically ranging from 1 to 100 nanometers. This burgeoning field has seen significant growth in both scientific exploration and commercial applications in recent years. Nanomaterials, a key component of nanotechnology, are increasingly integrated into various products and technologies across multiple industrial sectors. Nanotechnology holds promise for addressing pressing environmental and health challenges (Parveen et al., 2016; Huang et al., 2024). It is widely regarded as a promising avenue for sustainable technological advancements, aiming to address emerging environmental and health-related challenges in various sectors such as biomedical, cosmetics, healthcare, food, environment remediation, chemical industries, electronics, agriculture, sports equipment, sensors, and energy-storage devices have become indispensable commodities (Balasooriya et al., 2017). Despite their significance, the environmental sustainability of nanomanufacturing processes remains a major concern. In recent years, there has been a significant surge in research dedicated to the synthesis of nanomaterials to develop new products (Hu et al., 2016) Typically, nanomaterial synthesis relies on chemical and physical methods, which often result in the generation of toxic by-products, posing environmental and health hazards (Christopher et al., 2019; Kiran et al., 2011) Consequently, there has been a heightened focus within the scientific community on mitigating the potential risks associated with both nanomaterial synthesis and its application. As a response to these concerns, there is growing interest in utilizing biological approaches for nanomaterial synthesis. This emerging trend offers a promising "green" alternative by substituting hazardous chemicals with sustainable and environmentally friendly reagents (Ying et al., 2022).

In the realm of nanoparticle (NP) synthesis, conventional methods relying on physical and chemical processes are prevalent, yet they pose drawbacks such as costliness, time intensiveness, and potential hazards. In response to these challenges, the concept of green synthesis has emerged as a pivotal advancement in nanotechnology (Srivastava & Bhargava, 2022). Green synthesis entails the creation of nanomaterials by utilizing biological entities like microbes and plants (Mohanta et al., 2023). Particularly, the microbial approach to NP synthesis has garnered significant interest due to its eco-friendly and sustainable characteristics (Balasooriya et al., 2017). Microbes serve as natural factories for NP synthesis, leveraging their biological processes (Jadoun et al., 2022). Recent studies have delved into the utilization of various microbial species including bacteria, actinobacteria, fungi, yeast, microalgae, and viruses for NP synthesis, yielding promising outcomes. The versatility in microbial selection offers distinct advantages, as microbes exhibit resilience

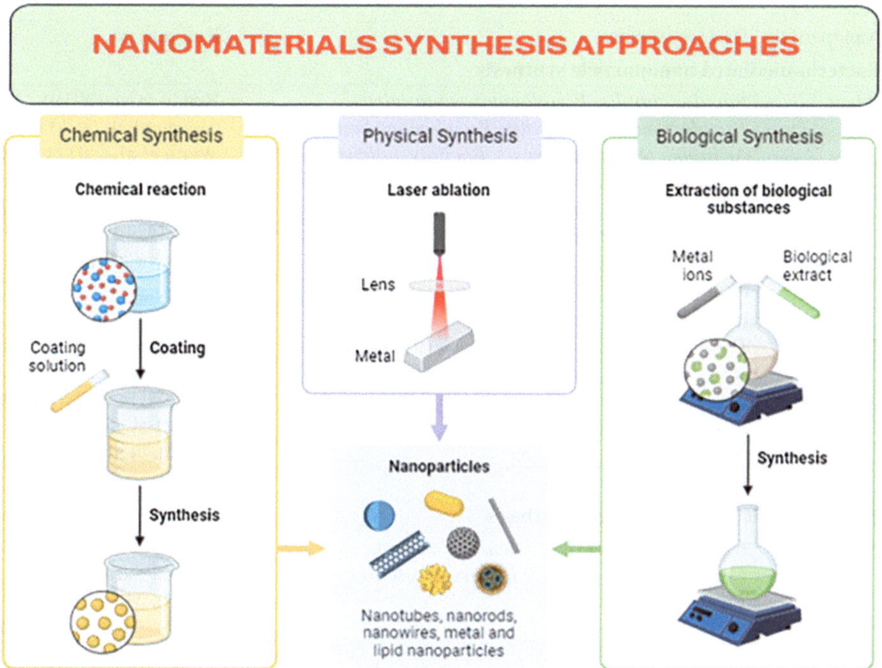

Fig. 2.1 Nanomaterials synthesis approaches. (Image templet created in Biorender: https://www.biorender.com/)

to extreme conditions, thus capturing the attention of the scientific community and researchers (Vetchinkina et al., 2019)

The green synthesis method employs natural and environmentally friendly substances as both capping and reducing agents, reducing energy consumption and eliminating the need for harmful chemicals (Ying et al., 2022). Microbial-mediated biosynthesis of nanomaterials has emerged as a promising 'green' alternative to traditional physical and chemical methods for nanoparticle synthesis (Fig. 2.1). Utilizing bacteria, fungus, yeast, molds, and microalgae, microbial-mediated biosynthesis has been extensively documented for producing metallic, non-metallic, and metal oxide nanoparticles (Table 2.1)

Among microbial candidates, bacteria stand out as a preferred choice for NP synthesis due to their high yield, cost-effective purification, ease of handling and maintenance. Additionally, fungi are favored for their ease of scale-up, downstream processing, economic viability, and increased surface area facilitated by mycelia presence (Dikshit et al., 2021). Microbial-mediated green and eco-friendly approaches have demonstrated the successful synthesis of cadmium, gold, iron, palladium, platinum, silver, and zirconium nanoparticles, employing both prokaryotic and eukaryotic microbes (Hulkoti & Taranath, 2014). These microbe-mediated nanoparticles exhibit high catalytic activity, stability, potential biocompatibility,

Table 2.1 List of a few important microbes recruited for nanoparticle synthesis

Nanoparticle	Bacteria species	References
Bacteria-mediated nanoparticle synthesis		
Silver	*Bacillus subtilis, Pseudomonas antarctica, Pseudomonas proteolytica, Pseudomonas meridiana, Arthrobacter gangotriensis, Bacillus indicus, Bacillus cecembensis, Streptomyces kasugaensis, Streptomyces celluloflavus, Streptomyces kasugaensis, Streptomyces celluloflavus, Escherichia coli, Bacillus megaterium, Bacillus siamensis, Pseudomonas aeruginosa, Staphylococcus aureus, Acinetobacter baumannii, Bacillus cereus*	Reddy et al. (2010), Shivaji et al. (2011), Wypij et al. (2017), Zaki et al. (2011), Ibrahim et al. (2019), Peiris et al. (2018), and Sunkar and Nachiyar (2012)
Gold	*Paracoccus haeundaensis, Shewanella algae, Brevibacillus Formosus, Pseudomonas aeruginosa, Bacillus subtili, Geobacillus stearothermophilus, Streptomyces clavuligerus*	Mohammed Fayaz et al. (2011), Kumar et al. (2015), He et al. (2008), Patil et al. (2019), and Srinath et al. (2017)
	Fungal species	
Fungus-mediated nanoparticle synthesis		
Silver	*Nigrospora sphaerica, Phenerochaete chrysosporium, Candida guilliermondii, Pleurotus ostreatusm, Fusarium acuminatum*	Muhsin and Hachim (2014), Saravanan et al. (2018), Mishra et al. (2011), Al-Bahrani et al. (2017), and Ingle et al. (2008)
Gold	*Fusarium solani, Penicillium brevicompactum, Hormoconis resinae, Aspergillus sydowii, Aspergillus niger, Candida albicans, Candida guilliermondii*	Clarance et al. (2020), Mishra et al. (2011), Mishra et al. (2010), Vala (2015), Soni and Prakash (2012), and Chauhan et al. (2011)

and minimal toxicity concerns. Furthermore, nanoparticle synthesis within microbial cells may occur intracellularly or extracellularly (Hulkoti & Taranath, 2014; Mann, 2001). The extracellular synthesis of nanoparticles has garnered significant recognition and interest, primarily due to its ability to streamline downstream processing steps. Furthermore, metal resistance genes, proteins, and reducing co-factors play crucial roles in green synthesis. They serve as natural reducing agents, aiding in nanoparticle synthesis by acting as inherent capping agents, preventing aggregation, and imparting stability to the nanoparticles (Messaoudi & Bendahou, 2020; Purohit et al., 2019; Rana et al., 2020). Additionally, certain microorganisms demonstrate the capability to synthesize distinctive nanostructured materials, including biomineralized nanostructures like silicified frustules, calcified coccoliths, magnetosomes, and organic nanomaterials such as bacterial nanocellulose exopolysaccharide nanoparticles and bacterial nanowires. Moreover, specific bacteria and microalgae exhibit significant potential in fabricating unique nanomaterials like nanoplates and nanowires. Researchers are actively exploring genetic engineering approaches to enhance microorganisms' ability to synthesize nanoparticles, indicating a promising future for microbe-mediated synthesis. This chapter offers an overview of nanoparticle green synthesis utilizing microbial cells and discusses their applications across various sectors including agriculture and the environment.

2.2 Microbial-Mediated Green Synthesis of Nanoparticles

Microbial production of NP can be done via two routes: intracellular and extracellular synthesis (Koul et al., 2021). In the extracellular process, microbe-secreted extra-cellular enzymes (reductase) in culture broth are allowed to interact with metal ions and subsequent bio-reduction of the metal ions resulting in the production of nano-sized particles (Marooufpour et al., 2019). In the intracellular approach, on the other hand, a solution containing microbial biomass is allowed to interact with metal ions at desired incubation conditions till specific coloration of the biomass-metal ion solution takes place indicating the formation of nano-particles e.g., yellow coloration indicates the formation of Zinc and Manganese NP (Waghmare et al., 2015). In such a process, metal ions (cations) trapped within the cell wall of microbial cells are reduced by intracellular enzymes leading to the formation of NP which then diffuses out to the culture solution (Marooufpour et al., 2019). The short life cycle and inherent capacity of microorganisms to grow in atmospheric conditions of temperature, pH, and pressure are other benefits of green synthesis of NP by microbes. While a wide group of microorganisms including bacteria, yeast, fungi, actinomycetes, and even viruses have been explored as bio-nano-factories (Salem & Fouda, 2021), current state of art delves into demonstrating a concise update on green synthesis of nano-particles using the merit of bacteria and fungi.

2.2.1 Bacteria as Bio-nano-factories

The primary players in external and intracellular biocatalytic production (including potential excretion) of NPs are oxidoreductase enzymes (such as cysteine desulfhydrase, NADPH-dependent sulfite-reductase flavoprotein subunit, and NADH-dependent nitrate reductase) and cellular transporters. There have also been descriptions of physicochemical processes mediated by biomolecules (such as proteins and carbohydrates), such as biosorption, complexation, nucleation, growth, and stability (Grasso et al., 2019). Extant pieces of evidence reveal that bacterial cell systems have been employed in the production of metal as well as metal oxide nanoparticles of a wide range (Koul et al., 2021).

Since the first report of fabrication of silver nanoparticles (AgNPs) (size less than 200 nm) using the bacterial strain *Pseudomonas stutzeri* AG259, a lot of progress has been made in discovering numerous other factories that can create nano-sized particles of silver metals (Klaus et al., 1999). Biomass as well as cell-free supernatant of *Escherichia coli* and *Klebsiella pneumoniae* are reported to be utilized to reduce the aqueous solution of silver (Shahverdi et al., 2007b). Culture supernatant of *Enterobacter cloacae* (Shahverdi et al., 2007a), *Staphylococcus aureus* (Nanda & Saravanan, 2009), *Lactobacillus casei ssp. casei* CCM 7088 (Singh et al., 2015), *Lactobacillus plantarum* (Garmasheva et al., 2016), and *Brevibacterium casei* (Kalishwaralal et al., 2010) are reported to possess the ability of a speedy generation of AgNP. Studies made with *Morganella* sp. RP-42 revealed that genes encoding periplasmic silver-binding protein (silE, silP, and silS) are pivotal in AgNP production by bacteria apart from the oxidoreductase enzymes (Parikh et al., 2008). Further, CusCFBA silver/copper efflux system present in *E. coli* strain (116AR) was also linked with enhanced AgNP production (Ramanathan et al., 2013). Extremophilic bacteria and genetically modified bacteria are also under scanner as bionanofactories for AgNP production. Among the diverse group of bacterial species linked with the extracellular and intracellular synthesis of gold nanoparticles (AuNPs), *Bacillus subtilis* 168 (Beveridge & Murray, 1980), *Escherichia coli* K12 (Srivastava et al., 2013), *Lactobacillus casei* JCM1134 (Kikuchi et al., 2016), *Lactobacillus kimchicus* DCY51T (Markus et al., 2016), *Shewanella algae* (Konishi et al., 2004), *Rhodopseudomonas capsulate* (He et al., 2008) are notable. It is now known that both enzymatic and non-enzymatic processes are involved in AuNP production in these factories. Genetically modified or recombinant strains of bacteria such as *P. savastanoi*, *R. solanacearum*, *P. syringae*, and *V. fischeri* are also now known that produce AuNP of various sizes utilizing the merit of NADH and NADH-dependent reductase (Attaran et al., 2016). Selenium nanoparticles (SeNPs) have been produced *Bacillus subtilis* (Wang et al., 2010), *Streptomyces minutisclero-ticus* M10A62 (Ramya et al., 2015), *Pantoea agglomerans* UC-32 (Torres et al., 2012), *Streptomyces bikiniensis* Ess_amA-1 (Ahmad et al., 2015), *Lactobacillus acidophilus* (Eszenyi et al., 2011), *Lactobacillus casei* (Eszenyi et al., 2011), and *Bifidobacterium sp.* (Eszenyi et al., 2011) for wide application including anti-biofilm, anti-oxidant, anti-proliferative as well as nano-fertilizer to be used in agriculture. Non-probiotic bacteria *Streptomyces griseus* (Ponmurugan et al., 2016) and probiotic *Lactobacillus casei* (Kouhkan et al., 2020) are linked to the production of Copper Nanoparticles (CuNPs) for a wide range of application including agricultural (micronutrient). Similarly, Cadmium nanoparticles are created using *Lactobacillus acidophilus* DSMZ 20079T through an extracellular mechanism that has been used as a nano-biosensor (Riaz et al., 2020).

Currently, the production of metal oxide nanoparticles has gained immense scientific attention owing to a wide range of applications in the biomedicine, biosensors, and food industry (Koul et al., 2021). Ag_2O nanoparticles (size 2–20 nm) with strong anti-microbial and wound-healing properties have been synthesized in the biofactories such as *Lactobacillus mindensis* (Ag-resistant) (Dhoondia & Chakraborty, 2012) and *Bacillus thuringiensis* SSV1 (Vithiya et al., 2014). Zinc oxide nanoparticles (ZnONPs), with vast application in biomedicine, the food industry, agriculture, and cosmetics, are synthesized through cell-based and cell-free approaches in bacteria such as *Lactobacillus plantarum* TA4 (Mohd Yusof et al., 2020) and *Paenibacillus polymyxa* strain Sx3 (Ogunyemi et al., 2020). The bacterium *Paenibacillus polymyxa* strain Sx3 is also linked with the production of MgO NP (Ogunyemi et al., 2020). Magnetic nanoparticles such as Iron oxide nanoparticles (IONP) (Fe_3O_4) are now synthesized using the bio-factory of *Proteus vulgaris* ATCC-29905 (Majeed et al., 2021) and *Bacillus subtilis* (Sundaram et al., 2012). Evidence reveal that mamAB, mamGFDC, mamXY, and mms6 are some of the genes involved in magnetic NP synthesis in the bacterium *Magnetospirillum gryphiswaldense* (Kolinko et al., 2014). Genetically modified organisms such as *Magnetospirillum magneticum* AMB-1 is also now available for the production of IONP (Boucher et al., 2017). Bacteria are also considered as factories for the production of organic nanoparticles in recent times. For instance, *Gluconacetobacter* is linked to the production of nano-structured cellulose (Golmohammadi et al., 2017). It is now known that growth conditions including temperature, pH, and media compositions are pivotal in modulating the ability of bacteria's ability to design and synthesize nanoparticles of various dimensions (Koul et al., 2021).

2.2.2 Fungi as Bio-nano-factories

Fungi, as bio-nanofatories, are an appealing area of nanotechnology and nanotechnology-based application owing to the inherent bioreduction and bioaccumulation potential of fungi. Cell-based as well as cell-free systems of metal and metal-oxide nanoparticle synthesis have been recorded for fungi-mediated green NP production (Koul et al., 2021). *Trichoderma viride and Hypocrea lixii were reported to produce g*old nanoparticles (AuNPs) of size ranging from 20 and 30 nm in which pH and temperature played important roles in their ease and speed of NP production (Mishra et al., 2014). Another fungus, *Cladosporium oxysporum,* operating through intracellular mechanism, was recorded for AuNP production ability. The yeast *Magnusiomyces ingens* LH-F1 was also associated with the production of AuNP. The fungi *Schizophyllum radiatum* could produce AgNPs (<50 nm) via mechanisms such as extracellular biomineralization (Metuku et al., 2014). While modulating the pH of the growth medium *Fusarium oxysporum* could be employed in producing nanoparticles of various shapes such as triangular, spherical, rod, and other irregular shapes (Rajput et al., 2016). AgNPs synthesiszed in *Penicillium polonicum* could display strong antibacterial and anti-biofilm formation (Neethu et al., 2018). Extant evidences revealed that AgNP production in the marine yeast *Yarrowia lipolytica* is linked to the presence of melanin in yeast cells (Apte et al., 2013). AgNPs of size as small as 2–10 nm have been synthesized using *Candida lusitaniae* found in termite gut (Eugenio et al., 2016). Similarly, *Pichia pastoris* (Elahian et al., 2017) and *Saccharomyces cerevisiae* (Sriramulu & Sumathi, 2018) are utilized for the production of NP of silver and palladium respectively. Extracellular secretion of fungi such as *Trichoderma harzianum* (SKCGW008) is further employed in the production of Chitosan nanoparticles for anti-oxidant and bactericidal activity (Saravanakumar et al., 2018).

Similar to bacteria, fungi are also seen as a factory for the production of metal oxide nanoparticles for various applications. Aluminum oxide NPs (Al_2O_3 NPs) biosynthesized using *Colletotrichum* sp. and more activated by plant-derived oils demonstrated antimicrobial activity against food-spoiling pathogens (Lakshmayya et al., 2023; Suryavanshi et al., 2017). *Aspergillus terreus* AF1, *Aspergillus niger* strain (G3-1) and *F. keratoplasticum* strain (A1-3) were employed in creating ZnO nanoparticles of various shapes (Fouda et al., 2018; Mohamed et al., 2019). *Penicillium chrysogenum* (Mohamed et al., 2021), *Penicillium corylophilum* As-1 (Fouda et al., 2020), *Aspergillus niger* G3-1 (Mohamed et al., 2019) were utilized in producing CuO NP. Iron oxide nanoparticles (IONPs) with application in biomedicine, biosensor, agriculture and bioremediation have now been synthesized using *Aspergillus japonicus* (AJP01), *Fusarium incarnatum, Trichoderma asperellum,* and *Phialemoniopsis ocularis* (Bhargava et al., 2014; Mahanty et al., 2019a, b). Interestingly, Cobalt oxide NPs (CoO NPs) biosynthesized using *Aspergillus nidulans* provided a strategy of energy storage (Vijayanandan & Balakrishnan, 2018).

2.2.3 Fungi or Bacteria as Bio-nano-factories?

As evident from extant literature both fungi and bacteria are envisaged as production houses for nanomaterials (Bahrulolum et al., 2021; Grasso et al., 2019). However, fungi display a clear edge over bacteria as bionanofactories (Nayak et al., 2024). Because fungi can withstand higher metal concentrations and have a high bioaccumulation efficiency, they can produce more NPs than bacterial cells. This also provides a more inexpensive and efficient approach in comparison to bacteria. The fungus-based biosynthesis of NPs simplifies the handling of biomass and downstream processing of NPs. Furthermore, the fungi system of NP production does not require the use of detergents, any physical factors (such as ultrasound), and the use of biomolecules. Fungi have been thoroughly researched for their potential to synthesize various NPs and in recent times, extracellular production of NPs with fungi has become a trend (Bahrulolum et al., 2021; Koul et al., 2021).

2.3 Sustainability and Eco-Friendly Nature of Microbial Technology

The traditional method of nanoparticle synthesis through routes such as physical and chemical has a bit of disadvantages in comparison to the biological production of nanomaterials. Nanoparticle (NP) synthesis through biological routes is a fast-growing area owing to ease of production, ecofriendly nature, and easy manipulation of toxicity and size-related characteristics (Koul et al., 2021). It has been an open secret that bio-prepared nanoparticles harbor a wide range of biological applications including cell imaging, diagnostics, biofilm formation, biolabeling, and bio-sensor development. Further, nano-particles NP are with enhanced antioxidants, antimicrobials, anticancer, anticoagulants, antimigration agents, and antiproliferative effects. Among the biological methods of nanoparticle synthesis, microbial synthesis of nanoparticles is of fundamental advantage due to ease of cultivation as well as manipulation of microbial biomass through simple changes in media pH, growth temperature, and time of incubation (Bahrulolum et al., 2021; Koul et al., 2021). Along with other conventional applications, microbe as bionanofactories offer sustainable and eco-friendly solutions for various environmental applications, particularly in wastewater treatment (Shukla, 2020). The following are some attributes that make microbial nano-factorys is the first choice sustainable and eco-friendly approach to tackle environmental issues.

2.3.1 *Green Synthesis Process*

Because microbial-based green nanotechnology does away with the need for dangerous chemicals and significant energy consumption associated with conventional physical and chemical methods of nanoparticle manufacturing, this process is naturally environmentally benign.

2.3.2 *Cost Effectiveness*

Naturally occurring nano-factories, microbes can produce nanoparticles without the need for costly machinery or labor-intensive post-processing procedures.

2.3.3 *Environmental Safety*

Since they are usually non-toxic and biocompatible, green nanoparticles produced from microbial sources are acceptable for use in environmental applications as well as human health.

2.3.4 *Reusability and Regeneration*

Green nanomaterials aided by microbes frequently possess regeneration and reusability qualities, allowing for repeated cycles of treatment with minimal efficiency loss.

2.3.5 *Reduced Energy Consumption*

Traditional approaches for NP synthesis and subsequent application are energy-intense processes which is not so in the case of microbial NP.

2.3.6 *Inherent Capabilities of Microorganisms*

Microorganisms are blessed with bioaccumulation, bioreduction, biosorption, and other attributes that enhance metal nanoparticle formation and while doing so can also be utilized as a means of metal remediation from contaminated environmental samples.

2.4 Application of Microbial Nanotechnology in Environmental Issues

Offering exciting answers to a range of environmental problems, microbial nanotechnology is a unique combination of biology and nanoscience. Biological synthesis techniques can circumvent the drawbacks of both chemical and physical approaches to nanoparticle creation. Furthermore, the application of microorganisms has provided a dependable, environmentally benign, secure, and long-lasting method of nano-synthesis. Microbial nanotechnology has several potential applications in domains including environmental monitoring and management, agriculture and food sector, industrial applications, rapid diagnostics, imaging, biopharmaceuticals, drug delivery systems, antimicrobials, biomaterials for tissue regeneration, biosensors, etc.(Ghosh et al., 2021) Through the utilization of the distinct characteristics of microorganisms and nanomaterials, scientists have created inventive uses in several fields, to reduce environmental problems and advance sustainability. It is known that the cells of bacteria, algae, fungi, and yeasts can take metals from their surroundings and transform them into forms of elemental nanoparticles that can be released or accumulated. Furthermore, viruses have also been used to create strong nanocarriers (Arul et al., 2023). During biosynthesis, capping agents are frequently released to shield the nanoparticles from aggregation and to encourage stability. In this article, we'll go over the numerous applications of microbial nanotechnology in the food sector, sustainable agriculture, and the removal of pollutants from the environment (Ray et al., 2023) (Fig. 2.2). The concept of environmental nanobiotechnology and sustainable environment will be discussed concerning microbial-mediated nanoparticles. Due to the unique properties of nano-sized materials, applications of nanotechnology for environmental applications have many advantages. The environment is currently facing serious pollution because of various human activities, which prompts real and unconventional efforts to maintain the sustainability of the

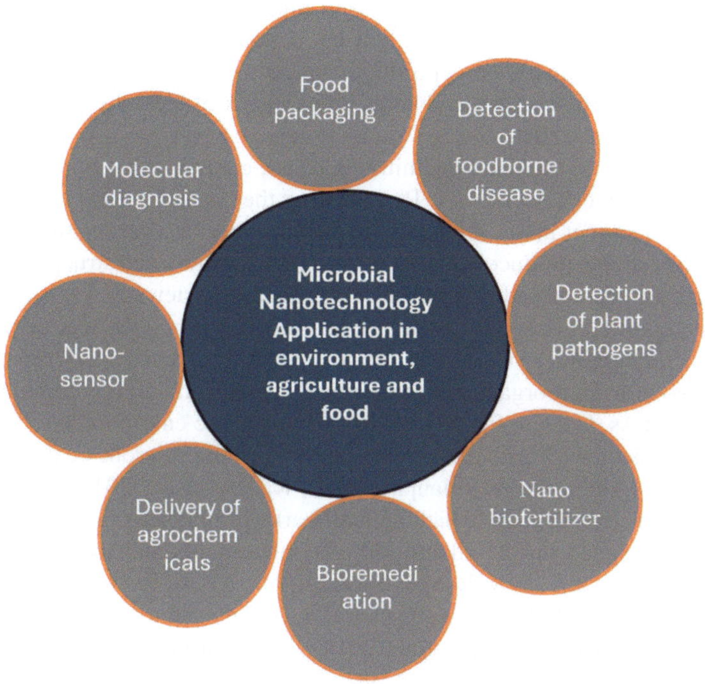

Fig. 2.2 Applications of microbial nanotechnology in environment, agriculture, and food

environment (Hourdequin, 2024). Numerous NP-assisted procedures linked to environmental sustainability have been examined and evaluated. Since biosynthesized NPs are more sustainable and favorable to the environment than chemically synthesized ones, as was previously noted, many researchers have concentrated on using them in various environmental studies (Darwesh et al., 2019; Hussain et al., 2019). The most prevalent applications of NPs in environmental sustainability are bioremediation, biosensing, microbial nano-pesticides, and microbial nano-fertilizers.

2.4.1 Microbial Nanoparticles in Bioremediation

Since the survival of life on Earth depends on water, it is equally important to remove pollution from it. Water use has been under tremendous strain from industrialization because of its use in industry. An enormous volume of industrial effluents is produced because of increased output. For industries and the environment to develop sustainably, these industrial effluents must be treated strictly and economically. Diverse electrochemical methods, sophisticated oxidation procedures, and valorization approaches have been implemented to lessen wastewater effluent toxicity and enable sustainable wastewater use (Gupta & Shukla, 2020). The creation of green nanomaterials through the synthesis of microbes and other organism extracts has cleared the way for environmentally friendly pollution cleanup. Because of their non-toxic nature, magnetic susceptibility, and redox potential when reacting with water, iron nanoparticles, also known as green nanoparticles, are utilized in remediation (Bolade et al., 2020).

Famously, arsenic (As) and cadmium (Cd) are contaminants of water and play major roles in numerous illnesses. Remediating the As and Cd from contaminated water is a project that many scientists are working on. In a recent study, Antarctic pseudomonas strains produced quantum dots (QDs), nano-sized structures that were used to treat arsenic and cadmium from contaminated wastewater. The generated Cd and As nanoparticles are significant because of their possible spectroscopic and semiconductor qualities (Glatstein et al., 2018). It has been reported that NP-synthesizing microorganisms themselves can absorb and accumulate heavy metals such as As, Co, Cu, Fe, Ni, and Mg from aquatic environments. This is particularly the case for a particular species of NP-forming microbes called magnetotactic bacteria (Ambashta & Sillanpää, 2010; Kabary et al., 2017). Large surface areas and magnetic qualities, which are hallmarks of magnetosomes and magnetotactic bacteria, speed up the adsorption of contaminants and their removal from wastewater (Ali et al., 2018; Ranjan Sarker et al., 2019). Even while an external magnetic field can just affect these bacteria, it can be used to remove heavy metals like these from wastewater (Qu et al., 2014). It was discovered that the magnetosome could effectively extract up to 38 g kg^{-1} of fluoride water and could also adsorb Cu^{2+} from wastewater by creating a magnetic polymer with polyvinyl acetate (Wang et al., 2020).

Apart from heavy metals, dye effluents released into the environment pose a significant threat to environmental contamination and the sustainability of natural resources, as noted by Darwesh et al. (2014). Dye degradation could be accomplished using NPs in a safe manner. In this process, biogenic nanoparticles produced by microorganisms can be a powerful tool (Srinivasan & Sadasivam, 2021). Sharma and Shirkot (2019) verified the use of biogenic iron NP for the bioremediation of azo dye from the environment.

Petroleum is regarded as one of the most hazardous environmental contaminants because of its extreme toxicity and widespread prevalence in the biosphere. Petroleum and its byproducts rank second in terms of harmful effects, right behind radioactivity. Ecosystems are severely harmed by petroleum spills resulting from mining and processing mishaps. Soil is primarily impacted in these situations because of its vast adsorbing surface area, which allows it to absorb massive amounts of contaminants. The chemical composition, structure, and qualities of soil are significantly altered by petroleum pollution, which also lowers soil fertility and arable value. These effects have a negative impact on soil biocenosis. Petroleum spills have the potential to transform soils into typical technogenic deserts devoid of most biological life. Microbes can selectively oxidize petroleum components, therefore for a more thorough cleanup, a variety of microbes—including fungi and various bacteria—are required (Al-Hawash et al., 2018). Microbial cultures are frequently isolated from soils contaminated with petroleum. For example, two bacterial strains,

Bacillus thuringiensis B3, *B. cereus* B6, and two fungal strains, *Geomyces pannorum* HR and *Geomyces* sp. strain HV, isolated from a northwestern Ecuador oil mine, may be used effectively for the large-scale bioremediation of crude oil-polluted land. Cyanobacteria like *Spirulina*, *Spirogyra*, *Oscillatoria*, and *Synechocystis*, as well as unicellular algae like *Chlorella*, *Scenedesmus*, *Chlorococcum*, *Nannochloropsis*, and *Selenastrum*, have been used to remove petroleum and its byproducts (Stepanova et al., 2022). The most significant algae for petroleum removal are *Chlorella* and *Scenedesmus*, which have been shown to efficiently break down crude oil and polycyclic aromatics like naphthalene, phenanthrene, and pyrene (Lei et al., 2007; Kalhor et al., 2017; Kuttiyathil et al., 2021). Numerous writers highlight the potentially beneficial pairings of microalgae and bacteria. First, algae will provide oxygen and encourage bacteria to oxidize hydrocarbons. The bioavailability of hydrocarbons and bacterial proliferation may be enhanced by microalgal exudates (Stepanova et al., 2022). Microalgae have generally been employed in numerous research to remove petroleum and its byproducts from aquatic environments. Nonetheless, a few studies have attempted to address the use of microalgae for removing petroleum and its byproducts from aquatic habitats and cleaning up soil.

2.4.2 *Microbial Nanoparticles in Biosensing*

To address, preserve, and protect an environment for the present and future generations, it is crucial to identify and monitor contaminants in it. This is because recognizing and evaluating the issue is a critical first step toward finding a solution. Conventional chromatography and spectroscopy techniques are typically used in the analysis of pollution detection and quantification procedures. These traditional methods are thought to be costly, time-consuming, and require expensive equipment and highly skilled personnel. Additionally, the environmental samples ought to be brought into the laboratory for examination. Consequently, to address the shortcomings of the conventional approaches, the researchers concentrated on creating novel analysis tools, the most significant of which are biosensors and nano-biosensors. As optical, electrical, or mechanical signals generated by a particular "receptor" and transferred through a "transducer" into a "detector" or read-out unit, a "sensor" is any device that is used to detect or measure an element or parameter (Arora, 2018; Sinha et al., 2020). "Bio-receptors" found in biosensors may include enzymes, antibodies, nucleic acid, or even entire microbial, plant, or animal cells—living or dead.

As "biomimetic receptors," synthetic molecules like aptamers, molecularly imprinted polymers, and synthetic antibodies may also be employed as bio-receptor elements (Ejeian et al., 2018; Zare & Shekari, 2019; Sinha et al., 2020). Many environmental pollutants, including pesticides, herbicides, and heavy metals, can be quickly and accurately detected, and monitored in soil, water, and the air using biosensors and nano-biosensors. Several intriguing nanomaterials with huge surface areas and strong luminescence, electrical, magnetic, and plasmon resonance have been reported to have emerged successfully on probes conjugated to biomolecules for biological tests. Magnetic nanoparticles, gold nanoparticles, nano-sized porous alumina, hybrid gold-platinum nanocomposites, multiwall carbon nanotubes, carbon nanoparticles, and graphene are examples of nanomaterials that are frequently utilized to create nano-biosensors (Chen & Park, 2016). The usage of nanoparticles green-synthesized by microbes to create these biosensors is thought to be more sustainable, given that the use of nano-biosensors contributes to environmental sustainability by detecting environmentally dangerous compounds. Numerous attempts have been made to use microorganisms' actions to create the nanomaterial that makes up nano-biosensors. However there isn't any available evidence to date that the manufacturing process of nanoparticles, whether chemical, physical, or biological, can have an impact on the effectiveness of nano-sensors. Figure 2.3 reflects an

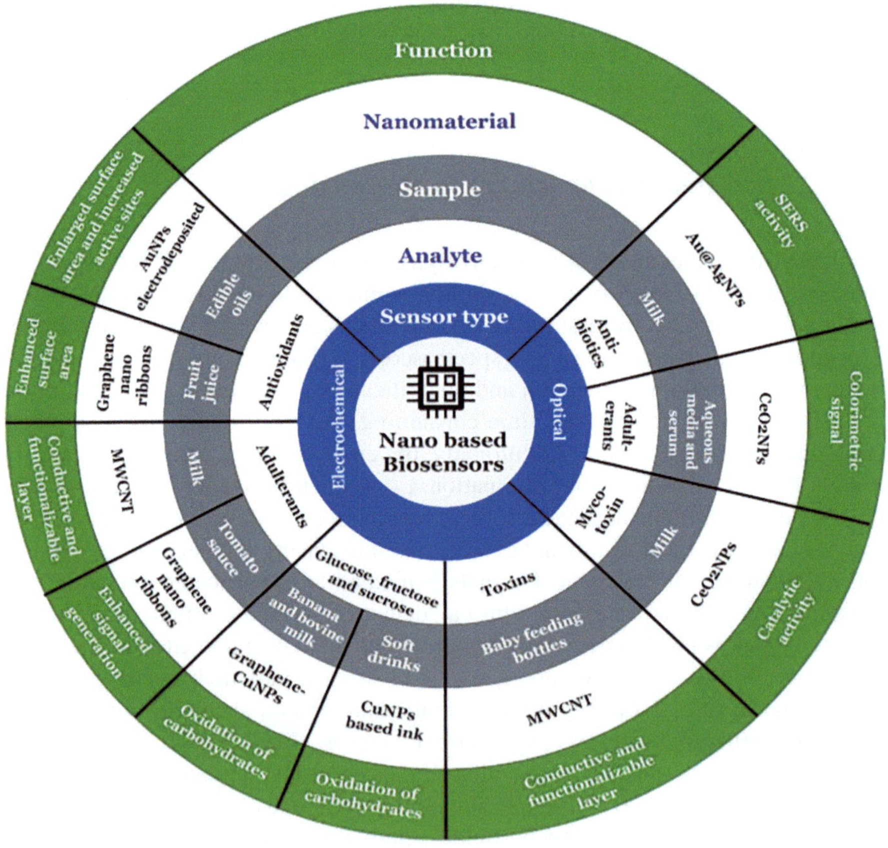

Fig. 2.3 Several kinds of nanosensors and their useful applications in food product preservation. (Figure adapted from Kumar et al. (2021) and it is an open article (copyright © 2021 by authors) distributed under the terms and conditions of the Creative Commons Attribution (CC BY) License)

overall idea about different types of nano-based biosensors and their application associated with the food industry.

In addition to being able to be produced chemically, magnetic nanoparticles (NPs) are also widely used in a variety of applications, including nano-sensors. They are also efficiently biosynthesized by a variety of microbes. It is utilized in the production of antibiotics like kanamycin (Chen et al., 2013) and various nano-sensors that may identify harmful bacteria (such as magnetic nanoparticle-urease nano-sensors) (Sun et al., 2018). Fe_3O_4 magnetic nanoparticles have been integrated with alkaline phosphatase into a "sol-gel/chitosan biosensor membrane" to detect and quantify herbicides, such as 2,4-dichlorophenoxyacetic acid (2,4-D) (Loh et al., 2008).

2.5 Role of Microbial Nanotechnology in Agriculture and Food

Nanotechnology, a rapidly evolving branch of science and technology, has the potential to be a very useful tool for improving agricultural regulating procedures. Agroecosystem sustainability may be enhanced by nanotechnology through the production of safer and better food, a reduction in agricultural inputs, and the use of nanoscale nutrients. The use of nanotechnology in pest management and plant nutrition may have a significant impact on maintaining the agricultural environment (Prasad et al., 2017). The efficient use of agrochemicals (pesticides and fertilizers) in agriculture is increased by the advancement of nanotechnology applications, which creates intelligent properties in agri-inputs. Applying microbial nanotechnology to improve agriculture processes such as targeted delivery, controlled release, improving solubility, and long shelf life not only improves their efficacy but also reduces the risk of environmental contamination (Chhipa, 2019). The significance of using microbial nanoparticles and nanotechnology in the identification of food-borne pathogens and the diagnosis of food-borne illnesses should be emphasized, much like in the field of agriculture.

2.5.1 Microbial Nanoparticles in Agriculture

Nano-fertilizers are one of the most recent developments in the use of nanotechnology in agriculture. Numerous studies found that nano-fertilizers were more effective than conventional fertilizers, which could be attributed to their nano size and high surface area-to-volume ratio. Applications of silver, iron, zinc, calcium, titanium, carbon nanotubes, molybdenum, and silica as nano-fertilizers for various crop systems have resulted in a variety of trials. Figure 2.4 exhibited different types of possible applications of the nanomaterials in the agricultural sector. Several

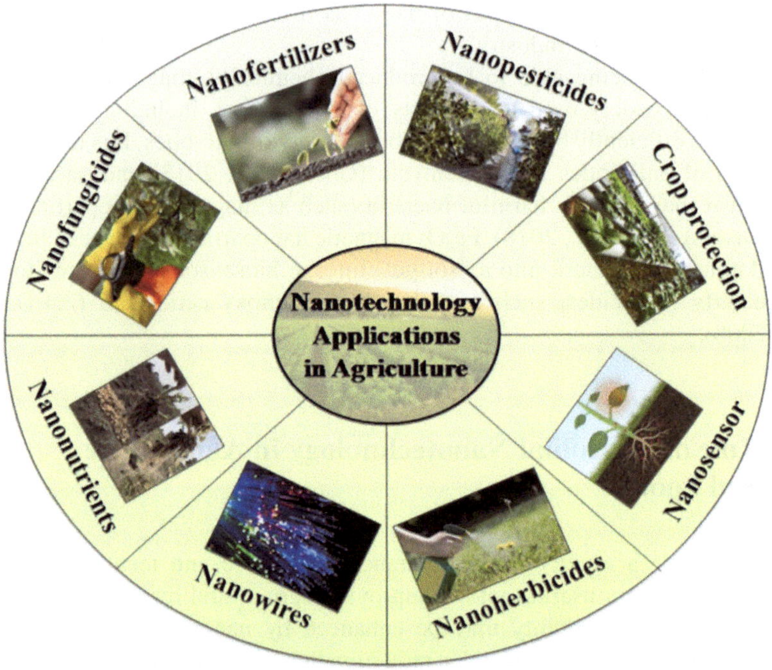

Fig. 2.4 Versatile application of microbial mediated nanoparticles in the agricultural sector. (Figure adapted from Singh et al. (2023) and it is an open article (copyright © 2023 by authors) distributed under the terms and conditions of the Creative Commons Attribution (CC BY) License)

investigations revealed more than 50% more plant growth (for example, in tobacco and maize) because of applying carbon nanotubes and copper oxide, respectively, as nano-fertilizers (Adhikari et al., 2016; Khodakovskaya et al., 2012). Furthermore, according to Shebl et al. (2019), the significance of nanotechnology in the production of nano-fertilizers for agriculture depends not only on raising the fertilization efficacy and bioavailability of such fertilizers but also on lowering their contaminating effect on the surrounding environment.

The production of a few nano-micronutrients, including zinc, iron, magnesium, and cobalt, mediated by several microorganisms have broad applications in the agricultural sector (plant and/or animal production) (Sohair et al., 2018; Mohd Yusof et al., 2019). After 14 days of cultivation, TiO_2 NPs mediated by *Aspergillus flavus* TFR 7 were sprayed on leaves. Raliya et al. (2015) reported a significant increase in plant vegetative parameters of mung beans (shoot length (17.02%), root length (49.6%), root area (43%), root nodule (67.5%), chlorophyll content (46.4%), as well as total protein content (94%).

Pathogenic microorganisms have an impact on plants and crops, which tragically reduces crop yield and results in the loss of plant products (Mohanta et al., 2024). In addition to a few other techniques, people employ chemical pesticides to control plant diseases. Chemical pesticide use has a highly harmful impact on the environment, many biological species, and humans. Nanoparticles have recently been used in the pest control industry as a more environmentally friendly substitute for chemical pesticides (Masum et al., 2019; Elshahawy et al., 2018). Several studies have already demonstrated the effectiveness of various microbially generated nanoparticles against a range of phytopathogens. According to Ali et al. (2020), biologically mediated metallic nanoparticles (NPs) such as Ag, Cu, Au, and Zn demonstrated a preventive effect against pathogenic fungi like *A. niger*, *F. oxysporum*, and *A. fumigatus*, as well as other disease-causing microbes and both Gram (+) and (−) bacteria like *B. subtilis*, *E. coli*, and *Staphylococcus aureus*. The interest in the green manufacture of silver nanoparticles using plants, bacteria, fungi, or yeasts has increased (Rafique et al., 2017). To lessen the harmful effects on the environment, nanoparticles could be used as carriers to regulate the release of chemical pesticides from their formulation in addition to acting as pesticides (Ali et al., 2020; Khan & Rizvi, 2014).

2.5.2 *Microbial Nanotechnology in Food*

Foodborne illnesses, which are usually accompanied by gastrointestinal symptoms are caused by infections, toxins, or pollutants found in food products. Most foodborne illness outbreaks that have been recorded in recent years have been linked to bacteria such *Salmonella* (Chin et al., 2022; Thomas et al., 2020) *Campylobacter* (Thomas et al., 2020), *Norovirus* (Chin et al., 2022), *Listeria monocytogenes* (Dos Santos et al., 2021), and *E. coli* that produces Shiga toxin (EFSA Biohaz Panel et al., 2020). Pathogen-detection techniques are essential for guaranteeing the safety of food and water, but real-time monitoring is not possible with them because they require lengthy procedures that often take days to get an accurate result (Kaya et al., 2021; Kumar et al., 2020). Even though molecular biology techniques are used to ensure the security of food and water, if we consider the advancements in pathogen-related technology in the health sector detection, this industry has already created cutting-edge products like biosensors that use nanomaterials and informatics technology. To create more appropriate and sensitive in situ detection systems for diseases, currently scientists are using nanomaterials such as nanoparticles, nanorods, nano-conjugates (Au–Fe_3O_4), silicon nanonet FET, and nanospheres (RNs@Au) in a biosensor device, along with molecular detection techniques (ELISA, qPCR), and informatic technologies. This technology has been used to enhance the ability of the healthcare system to respond to emergencies involving the presence of pathogens, such as the influenza A virus and HIV (as discussed by Jain et al., 2021). These technologies now used in the health sector should be expanded to other areas like food and water security considering the advancements made in health security and the demands for increased food and water safety. Figure 2.5 showing versatile application of nanotechnology in food sector.

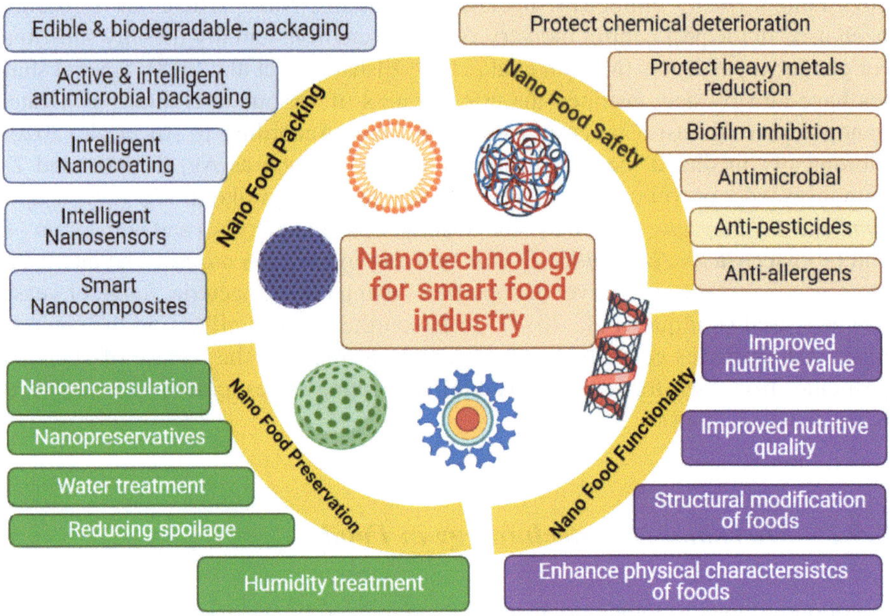

Fig. 2.5 Role of nanotechnology in food. (Figure adopted from Ansari (2023) and it is open article (copyright © 2023 by authors) distributed under the terms and condition of the Creative Commons Attribution (CC BY) License)

2.6 Conclusion and Prospect

Microbial nanotechnology exhibits vast potential concerning pollutant sensing, degradation, and removal of heavy metal ions from contaminated sites. Moreover, it finds extensive applications in the food sector, including food processing, packaging, safety, detection of foodborne pathogens, and extending the shelf life of food products. While recent studies highlight the significant promise of microbes in synthesizing novel nanoparticles (NPs) and their utilization, the process of nanomaterial biosynthesis using microbes requires refinement for commercial production, necessitating scale-up compared to conventional methods. Furthermore, a thorough investigation is imperative to assess the toxicity, biocompatibility, and potential impact of NPs on human health and other life domains. The utilization of genetically engineered microorganisms for crafting NPs of specific shapes and sizes introduces novelty in this domain. Despite several advantages over conventional nanoparticle biosynthesis methods, such as enhanced biosynthesis rates, energy efficiency, environmental safety, cost-effectiveness, and feasibility of NP production, the use of genetically engineered microbes remains relatively limited. Challenges like reproducibility and the survival of recombinant strains require further exploration to enhance NP synthesis utilizing genetically engineered organisms.

Additionally, concerns regarding public acceptance, biosafety, biosecurity, and transgene escape pose challenges for researchers employing genetically engineered organisms in NP fabrication. Nevertheless, it is anticipated that shortly, biofabrication of NPs using genetically engineered organisms will garner considerable attention and emerge as a preferred method for sustainable nanomaterial production. Thus, microbial nanotechnology and microbe-mediated NP synthesis hold immense potential across various applications. Nonetheless, addressing issues like toxicity and related concerns is imperative before NPs can be commercialized.

References

Adhikari, T., Sarkar, D., Mashayekhi, H., & Xing, B. J. (2016). Growth and enzymatic activity of maize (Zea mays L.) plant: Solution culture test for copper dioxide nano particles. *Journal of Plant Nutrition, 39*(1), 99–115.

Ahmad, M. S., Yasser, M. M., Sholkamy, E. N., Ali, A. M., & Mehanni, M. M. (2015). Anticancer activity of biostabilized selenium nanorods synthesized by Streptomyces bikiniensis strain Ess_amA-1. *International Journal of Nanomedicine, 10*, 3389–3401.

Al-Bahrani, R., Raman, J., Lakshmanan, H., Hassan, A. A., & Sabaratnam, V. J. M. L. (2017). Green synthesis of silver nanoparticles using tree oyster mushroom Pleurotus ostreatus and its inhibitory activity against pathogenic bacteria. *Materials Letters, 186*, 21–25.

Al-Hawash, A. B., Dragh, M. A., Li, S., Alhujaily, A., Abbood, H. A., Zhang, X., & Ma, F. J. (2018). Principles of microbial degradation of petroleum hydrocarbons in the environment. *The Egyptian Journal of Aquatic Research, 44*(2), 71–76.

Ali, I., Peng, C., Khan, Z. M., Naz, I., & Sultan, M. J. (2018). An overview of heavy metal removal from wastewater using magnetotactic bacteria. *Journal of Checmical Technology and Biotechnology, 93*(10), 2817–2832.

Ali, M. A., Ahmed, T., Wu, W., Hossain, A., Hafeez, R., Islam Masum, M. M., et al. (2020). Advancements in plant and microbe-based synthesis of metallic nanoparticles and their antimicrobial activity against plant pathogens. *Nanomaterials, 10*(6), 1146.

Ambashta, R. D., & Sillanpää, M. J. (2010). Water purification using magnetic assistance: A review. *Journal of Hazardous Materials, 180*(1–3), 38–49.

Ansari, M. A. (2023). Nanotechnology in food and plant science: Challenges and future prospects. *Plants, 12*(13), 2565. https://www.mdpi.com/2223-7747/12/13/2565

Apte, M., Sambre, D., Gaikawad, S., Joshi, S., Bankar, A., Kumar, A. R., & Zinjarde, S. (2013). Psychrotrophic yeast Yarrowia lipolytica NCYC 789 mediates the synthesis of antimicrobial silver nanoparticles via cell-associated melanin. *AMB Express, 3*, 1–8.

Arora, K. J. N. (2018). Advances in nano based biosensors for food and agriculture. In *Nanotechnology, food security and water treatment* (pp. 1–52).

Arul, S. S., Balakrishnan, B., Handanahal, S. S., & Venkataraman, S. J. B. (2023). Viral nanoparticles: Current advances in design and development. *Biochimie, 219*, 33–50.

Attaran, N., Eshghi, H., Rahimizadeh, M., Mashreghi, M., & Bakavoli, M. (2016). Genetically modified luminescent bacteria Ralostonia solanacerum, pseudomonas syringae, pseudomonas savastanoi, and wild type bacterium Vibrio fischeri in biosynthesis of gold nanoparticles from gold chloride trihydrate. *Artificial Cells, Nanomedicine, and Biotechnology, 44*(1), 263–269.

Bahrulolum, H., Nooraei, S., Javanshir, N., Tarrahimofrad, H., Mirbagheri, V. S., Easton, A. J., & Ahmadian, G. (2021). Green synthesis of metal nanoparticles using microorganisms and their application in the agrifood sector. *Journal of Nanobiotechnology, 19*, 1–26.

Balasooriya, E. R., Jayasinghe, C. D., Jayawardena, U. A., Ruwanthika, R. W. D., Mendis de Silva, R., & Udagama, P. V. J. (2017). Honey mediated green synthesis of nanoparticles: New era of safe nanotechnology. *Journal of Nanomaterials, 2017*(1), 5919836.

Beveridge, T., & Murray, R. (1980). Sites of metal deposition in the cell wall of Bacillus subtilis. *Journal of Bacteriology, 141*(2), 876–887.

Bhargava, A., Jain, N., Barathi, L. M., Akhtar, M. S., Yun, Y.-S., & Panwar, J. (2014). Synthesis, characterization and mechanistic insights of mycogenic iron oxide nanoparticles. In *Nanotechnology for sustainable development* (pp. 337–348). Springer.

Bolade, O. P., Williams, A. B., & Benson, N. U. J. E. N. (2020). Green synthesis of iron-based nanomaterials for environmental remediation: A review. *Monitoring & Management, 13*, 100279.

Boucher, M., Geffroy, F., Prévéral, S., Bellanger, L., Selingue, E., Adryanczyk-Perrier, G., et al. (2017). Genetically tailored magnetosomes used as MRI probe for molecular imaging of brain tumor. *Biomaterials, 121*, 167–178.

Chauhan, A., Zubair, S., Tufail, S., Sherwani, A., Sajid, M., Raman, S. C., et al. (2011). Fungus-mediated biological synthesis of gold nanoparticles: Potential in detection of liver cancer. *International Journal of Nanomedicine, 6*, 2305–2319.

Chen, J., & Park, B. J. J. (2016). Recent advancements in nanobioassays and nanobiosensors for foodborne pathogenic bacteria detection. *Journal of Food Protection, 79*(6), 1055–1069.

Chen, Y. P., Qiang Zou, M., Qi, C., Xie, M.-X., Wang, D.-N., & Wang, Y.-F. (2013). Immunosensor based on magnetic relaxation switch and biotin–streptavidin system for the detection of kanamycin in milk. *Biosensors and Bioelectronics, 39*(1), 112–117.

Chhipa, H. (2019). Applications of nanotechnology in agriculture. In *Methods in microbiology* (pp. 115–142). Elsevier.

Chin, N. A., Salihah, N. T., Shivanand, P., & Ahmed, M. U. (2022). Recent trends and developments of PCR-based methods for the detection of food-borne salmonella bacteria and norovirus. *Journal of Food Science Technology*, 1–13.

Christopher, F. C., Ponnusamy, S. K., Ganesan, J. J., & Ramamurthy, R. (2019). Investigating the prospects of bacterial biosurfactants for metal nanoparticle synthesis–a comprehensive review. *IET Nanobiotechnology, 13*(3), 243–249.

Clarance, P., Luvankar, B., Sales, J., Khusro, A., Agastian, P., Tack, J.-C., et al. (2020). Green synthesis and characterization of gold nanoparticles using endophytic fungi *Fusarium solani* and its in-vitro anticancer and biomedical applications. *Saudi Journal of Biological Sciences, 27*(2), 706–712.

Darwesh, O., Moawad, H., El-Rahim, W., Barakat, O., & Sedik, M. (2014). Bioremediation of textile reactive blue (RB) azo dye residues in wastewater using experimental prototype bioreactor. *Research Journal of Pharmaceutical Biological and Chemical Sciences, 5*(4), 1203–1219.

Darwesh, O. M., Matter, I. A., Eida, M. F., Moawad, H., & Oh, Y.-K. (2019). Influence of nitrogen source and growth phase on extracellular biosynthesis of silver nanoparticles using cultural filtrates of Scenedesmus obliquus. *Applied Sciences, 9*(7), 1465.

Dhoondia, Z. H., & Chakraborty, H. (2012). Lactobacillus mediated synthesis of silver oxide nanoparticles. *Nanomaterials and Nanotechnology, 2*, 15.

Dikshit, P. K., Kumar, J., Das, A. K., Sadhu, S., Sharma, S., Singh, S., et al. (2021). Green synthesis of metallic nanoparticles: Applications and limitations. *Catalysts, 11*(8), 902.

Dos Santos, J. S., Biduski, B., & Dos Santos, L. R. (2021). Listeria monocytogenes: Health risk and a challenge for food processing establishments. *Archives of Microbiology*, 1–13.

Ejeian, F., Etedali, P., Mansouri-Tehrani, H.-A., Soozanipour, A., Low, Z.-X., Asadnia, M., et al. (2018). Biosensors for wastewater monitoring: A review. *Biosensors & Bioelectronics, 118*, 66–79.

Elahian, F., Reiisi, S., Shahidi, A., & Mirzaei, S. A. (2017). High-throughput bioaccumulation, biotransformation, and production of silver and selenium nanoparticles using genetically engineered Pichia pastoris. *Nanomedicine: Nanotechnology, Biology and Medicine, 13*(3), 853–861.

Elshahawy, I., Abouelnasr, H. M., Lashin, S. M., & Darwesh, O. M. J. (2018). First report of Pythium aphanidermatum infecting tomato in Egypt and its control using biogenic silver nanoparticles. *Journal of Plant Protection Research, 58*(2), 137–151.

Eszenyi, P., Sztrik, A., Babka, B., & Prokisch, J. (2011). Elemental, nano-sized (100-500 nm) selenium production by probiotic lactic acid bacteria. *International Journal of Bioscience, Biochemistry and Bioinformatics, 1*(2), 148.

Eugenio, M., Müller, N., Frasés, S., Almeida-Paes, R., Lima, L. M. T., Lemgruber, L., et al. (2016). Yeast-derived biosynthesis of silver/silver chloride nanoparticles and their antiproliferative activity against bacteria. *RSC Advances, 6*(12), 9893–9904.

Fayaz, A. M., Girilal, M., Rahman, M., Venkatesan, R., & Kalaichelvan, P. J. (2011). Biosynthesis of silver and gold nanoparticles using thermophilic bacterium Geobacillus stearothermophilus. *Process Biochemistry, 46*(10), 1958–1962.

Fouda, A., Saad, E., Salem, S. S., & Shaheen, T. I. (2018). In-vitro cytotoxicity, antibacterial, and UV protection properties of the biosynthesized zinc oxide nanoparticles for medical textile applications. *Microbial Pathogenesis, 125*, 252–261.

Fouda, A., Salem, S. S., Wassel, A. R., Hamza, M. F., & Shaheen, T. I. (2020). Optimization of green biosynthesized visible light active CuO/ZnO nano-photocatalysts for the degradation of organic methylene blue dye. *Heliyon, 6*(9).

Garmasheva, I., Kovalenko, N., Voychuk, S., Ostapchuk, A., Livins'ka, O., & Oleschenko, L. (2016). Lactobacillus species mediated synthesis of silver nanoparticles and their antibacterial activity against opportunistic pathogens in vitro. *BioImpacts: BI, 6*(4), 219.

Ghosh, S., Ahmad, R., Zeyaullah, M., & Khare, S. K. (2021). Microbial nano-factories: Synthesis and biomedical applications. *Frontiers in Chemistry, 9*, 626834.

Glatstein, D. A., Bruna, N., Gallardo-Benavente, C., Bravo, D., Carro Pérez, M. E., Francisca, F. M., & Pérez-Donoso, J. M. (2018). Arsenic and cadmium bioremediation by antarctic bacteria capable of biosynthesizing CdS fluorescent nanoparticles. *Journal of Environmental Engineering, 144*(3), 04017107.

Golmohammadi, H., Morales-Narváez, E., Naghdi, T., & Merkoçi, A. (2017). Nanocellulose in sensing and biosensing. *Chemistry of Materials, 29*(13), 5426–5446.

Grasso, G., Zane, D., & Dragone, R. (2019). Microbial nanotechnology: Challenges and prospects for green biocatalytic synthesis of nanoscale materials for sensoristic and biomedical applications. *Nanomaterials, 10*(1), 11.

Gupta, G. K., & Shukla, P. J. (2020). Insights into the resources generation from pulp and paper industry wastes: Challenges, perspectives and innovations. *Bioresource Technology, 297*, 122496.

He, S., Zhang, Y., Guo, Z., & Gu, N. (2008). Biological synthesis of gold nanowires using extract of *Rhodopseudomonas capsulata. Biotechnology Progress, 24*(2), 476–480.

Hourdequin, M. (2024). *Environmental ethics: From theory to practice*. Bloomsbury Publishing.

Hu, X., Li, D., Gao, Y., Mu, L., & Zhou, Q. (2016). Knowledge gaps between nanotoxicological research and nanomaterial safety. *Environment International, 94*, 8–23.

Huang, X., Auffan, M., Eckelman, M.J., Elimelech, M., Kim, J.H., Rose, J., Zuo, K., Li, Q. & Alvarez, P.J. (2024). Trends, risks and opportunities in environmental nanotechnology. *Nature Reviews Earth & Environment, 5*(8), 572–587.

Hulkoti, N. I., & Taranath, T. J. C. (2014). Biosynthesis of nanoparticles using microbes—A review. *Colloids and Surfaces B Biointerfaces, 121*, 474–483.

Hussain, M., Raja, N. I., Iqbal, M., & Aslam, S. J. (2019). Applications of plant flavonoids in the green synthesis of colloidal silver nanoparticles and impacts on human health. *Iranian Journal of Science and Technology. Transaction A, Science, 43*, 1381–1392.

Ibrahim, E., Fouad, H., Zhang, M., Zhang, Y., Qiu, W., Yan, C., . . . Chen, J. (2019). Biosynthesis of silver nanoparticles using endophytic bacteria and their role in inhibition of rice pathogenic bacteria and plant growth promotion. RSC Advances 9(50), 29293–29299.

Ingle, A., Gade, A., Pierrat, S., Sonnichsen, C., & Rai, M. J. (2008). Mycosynthesis of silver nanoparticles using the fungus Fusarium acuminatum and its activity against some human pathogenic bacteria. *Current Nanoscience, 4*(2), 141–144.

Jadoun, S., Chauhan, N. P. S., Zarrintaj, P., Barani, M., Varma, R. S., Chinnam, S., & Rahdar, A. J. (2022). Synthesis of nanoparticles using microorganisms and their applications: A review. *Environmental Chemistry Letters, 20*(5), 3153–3197.

Jain, S., Nehra, M., Kumar, R., Dilbaghi, N., Hu, T., & Kumar, S. (2021). Internet of medical things (IoMT)-integrated biosensors for point-of-care testing of infectious diseases. *Biosensors and Bioelectronics, 179*, 113074.

Kabary, H., Eida, M. F., Attia, M., Awad, N., & Easa, S. (2017). Magnetotactic characterization and environmental application P. Aeruginosa kb1 isolate. *Annual Research & Review in Biology, 20*(1), 1–10.

Kalhor, A. X., Movafeghi, A., Mohammadi-Nassab, A. D., Abedi, E., & Bahrami, A. (2017). Potential of the green alga Chlorella vulgaris for biodegradation of crude oil hydrocarbons. *Marine Pollution Bulletin, 123*(1–2), 286–290.

Kalishwaralal, K., Deepak, V., Pandian, S. R. K., Kottaisamy, M., BarathManiKanth, S., Kartikeyan, B., & Gurunathan, S. (2010). Biosynthesis of silver and gold nanoparticles using Brevibacterium casei. *Colloids and Surfaces B: Biointerfaces, 77*(2), 257–262.

Kaya, H. O., Cetin, A. E., Azimzadeh, M., & Topkaya, S. N. (2021). Pathogen detection with electrochemical biosensors: Advantages, challenges and future perspectives. *Journal of Electroanalytical Chemistry, 882*, 114989.

Khan, M. R., & Rizvi, T. F. (2014). Nanotechnology: Scope and application in plant disease management. *Plant Pathology Journal, 13*(3), 214–231.

Khodakovskaya, M. V., De Silva, K., Biris, A. S., Dervishi, E., & Villagarcia, H. (2012). Carbon nanotubes induce growth enhancement of tobacco cells. *ACS Nano, 6*(3), 2128–2135.

Kikuchi, F., Kato, Y., Furihata, K., Kogure, T., Imura, Y., Yoshimura, E., & Suzuki, M. (2016). Formation of gold nanoparticles by glycolipids of lactobacillus casei. *Scientific Reports, 6*(1), 34626.

Kiran, G. S., Selvin, J., Manilal, A., & Sujith, S. (2011). Biosurfactants as green stabilizers for the biological synthesis of nanoparticles. *Critical Reviews in Biotechnology, 31*(4), 354–364.

Klaus, T., Joerger, R., Olsson, E., & Granqvist, C.-G. (1999). Silver-based crystalline nanoparticles, microbially fabricated. *Proceedings of the National Academy of Sciences, 96*(24), 13611–13614.

Kolinko, I., Lohße, A., Borg, S., Raschdorf, O., Jogler, C., Tu, Q., et al. (2014). Biosynthesis of magnetic nanostructures in a foreign organism by transfer of bacterial magnetosome gene clusters. *Nature Nanotechnology, 9*(3), 193–197.

Konishi, Y., Nomura, T., Tsukiyama, T., & Saitoh, N. (2004). Microbial preparation of gold nanoparticles by anaerobic bacterium. *Transactions-Materials Research Society of Japan, 29*, 2341–2344.

Kouhkan, M., Ahangar, P., Babaganjeh, L. A., & Allahyari-Devin, M. (2020). Biosynthesis of copper oxide nanoparticles using lactobacillus casei subsp. casei and its anticancer and antibacterial activities. *Current Nanoscience, 16*(1), 101–111.

Koul, B., Poonia, A. K., Yadav, D., & Jin, J.-O. (2021). Microbe-mediated biosynthesis of nanoparticles: Applications and future prospects. *Biomolecules, 11*(6), 886.

Kumar, A., Sharma, S., Mishra, S., & Dames, J. (2015). Arbuscular mycorrhizal inoculation improves growth and antioxidative response of Jatropha curcas (L.) under Na2SO4 salt stress. *Plant Biosystems—An International Journal Dealing with all Aspects of Plant Biology, 149*(2), 260–269.

Kumar, H., Kuča, K., Bhatia, S. K., Saini, K., Kaushal, A., Verma, R., et al. (2020). Applications of nanotechnology in sensor-based detection of foodborne pathogens. *Sensors, 20*(7), 1966.

Kumar, A., Choudhary, A., Kaur, H., Mehta, S., & Husen, A. (2021). Metal-based nanoparticles, sensors, and their multifaceted application in food packaging. *Journal of Nanobiotechnology, 19*(1), 256. https://doi.org/10.1186/s12951-021-00996-0

Kuttiyathil, M. S., Mohamed, M. M., & Al-Zuhair, S. J. (2021). Using microalgae for remediation of crude petroleum oil–water emulsions. *Biotechnology Progress, 37*(2), e3098.

Lakshmayya, N. S. V., Mishra, A. K., Mohanta, Y. K., Panda, J., Naik, B., Mishra, B., & Varma, R. S. (2023). Essential oils-based nano-emulsion system for food safety and preservation: Current status and future prospects. *Biocatalysis and Agricultural Biotechnology, 53*, 102897. https://doi.org/10.1016/j.bcab.2023.102897

Lei, A.-P., Hu, Z.-L., Wong, Y.-S., & Tam, N. F.-Y. (2007). Removal of fluoranthene and pyrene by different microalgal species. *Bioresource Technology, 98*(2), 273–280.

Loh, K.-S., Lee, Y. H., Musa, A., Salmah, A. A., & Zamri, I. J. (2008). Use of Fe3O4 nanoparticles for enhancement of biosensor response to the herbicide 2, 4-dichlorophenoxyacetic acid. *Sensors, 8*(9), 5775–5791.

Mahanty, S., Bakshi, M., Ghosh, S., Chatterjee, S., Bhattacharyya, S., Das, P., et al. (2019a). Green synthesis of iron oxide nanoparticles mediated by filamentous fungi isolated from Sundarban mangrove ecosystem, India. *BioNanoScience, 9*, 637–651.

Mahanty, S., Bakshi, M., Ghosh, S., Gaine, T., Chatterjee, S., Bhattacharyya, S., et al. (2019b). Mycosynthesis of iron oxide nanoparticles using manglicolous fungi isolated from Indian sundarbans and its application for the treatment of chromium containing solution: Synthesis, adsorption isotherm, kinetics and thermodynamics study. *Environmental Nanotechnology, Monitoring & Management, 12*, 100276.

Majeed, S., Danish, M., Mohamad Ibrahim, M. N., Sekeri, S. H., Ansari, M. T., Nanda, A., & Ahmad, G. (2021). Bacteria mediated synthesis of iron oxide nanoparticles and their antibacterial, antioxidant, cytocompatibility properties. *Journal of Cluster Science, 32*(4), 1083–1094. https://doi.org/10.1007/s10876-020-01876-7

Mann, S. (2001). *Biomineralization: Principles and concepts in bioinorganic materials chemistry.* Oxford University Press.

Markus, J., Mathiyalagan, R., Kim, Y.-J., Abbai, R., Singh, P., Ahn, S., et al. (2016). Intracellular synthesis of gold nanoparticles with antioxidant activity by probiotic lactobacillus kimchicus DCY51T isolated from Korean kimchi. *Enzyme and Microbial Technology, 95*, 85–93.

Marooufpour, N., Alizadeh, M., Hatami, M., & Asgari Lajayer, B. (2019). Biological synthesis eof nanoparticles by different groups of bacteria. In *Microbial Nanobionics: Volume 1, state-of-the-art* (pp. 63–85).

Masum, M. M. I., Siddiqa, M. M., Ali, K. A., Zhang, Y., Abdallah, Y., Ibrahim, E., et al. (2019). Biogenic synthesis of silver nanoparticles using Phyllanthus emblica fruit extract and its inhibitory action against the pathogen Acidovorax oryzae strain RS-2 of rice bacterial brown stripe. *Frontiers in Microbiology, 10*, 820.

Messaoudi, O., & Bendahou, M. (2020). Biological synthesis of nanoparticles using endophytic microorganisms: Current development. In *Nanotechnology and the environment* (pp. 1–19). Intechopen.

Metuku, R. P., Pabba, S., Burra, S., Hima Bindu, N., S., Gudikandula, K., & Singara Charya, M. (2014). Biosynthesis of silver nanoparticles from *Schizophyllum radiatum* HE 863742.1: Their characterization and antimicrobial activity. *3 Biotech, 4*, 227–234.

Mishra, A. N., Bhadauria, S., Gaur, M. S., & Pasricha, R. (2010). Extracellular microbial synthesis of gold nanoparticles using fungus Hormoconis resinae. *Journal of the Minerals, Metals & Materials Society, 62*, 45–48.

Mishra, A., Tripathy, S. K., Wahab, R., Jeong, S.-H., Hwang, I., & Yang, Y.-B. (2011). Microbial synthesis of gold nanoparticles using the fungus Penicillium brevicompactum and their cytotoxic effects against mouse mayo blast cancer C 2 C 12 cells. *Biotechnology, 92*, 617–630.

Mishra, A., Kumari, M., Pandey, S., Chaudhry, V., Gupta, K., & Nautiyal, C. (2014). Biocatalytic and antimicrobial activities of gold nanoparticles synthesized by Trichoderma sp. *Bioresource Technology, 166*, 235–242.

Mohamed, A. A., Fouda, A., Abdel-Rahman, M. A., Hassan, S. E.-D., El-Gamal, M. S., Salem, S. S., & Shaheen, T. I. (2019). Fungal strain impacts the shape, bioactivity and multifunctional properties of green synthesized zinc oxide nanoparticles. *Biocatalysis and Agricultural Biotechnology, 19*, 101103.

Mohamed, A. A., Abu-Elghait, M., Ahmed, N. E., & Salem, S. S. (2021). Eco-friendly mycogenic synthesis of ZnO and CuO nanoparticles for in vitro antibacterial, antibiofilm, and antifungal applications. *Biological Trace Element Research, 199*(7), 2788–2799.

Mohanta, Y. K., Mishra, A. K., Panda, J., Chakrabartty, I., Sarma, B., Panda, S. K., et al. (2023). Promising applications of phyto-fabricated silver nanoparticles: Recent trends in biomedicine. *Biochemical and Biophysical Research Communications, 688*, 149126. https://doi.org/10.1016/j.bbrc.2023.149126

Mohanta, Y. K., Nongbet, A., Panda, J., Chakrabartty, I., Mohanta, T. K., & Mishra, B. (2024). Chapter 10—Nanobiotechnology in genetic engineering for abiotic stress resistance in crops. In R. N. Pudake, R. M. Tripathi, & S. S. Gill (Eds.), *Nanotechnology for abiotic stress tolerance and Management in Crop Plants* (pp. 153–164). Academic.

Mohd Yusof, H., Mohamad, R., Zaidan, U. H., & Abdul Rahman, N. A. (2019). Microbial synthesis of zinc oxide nanoparticles and their potential application as an antimicrobial agent and a feed supplement in animal industry: A review. *Journal of Animal Science and Biotechnology, 10*, 1–22.

Mohd Yusof, H., Abdul Rahman, N. A., Mohamad, R., Zaidan, U. H., & Samsudin, A. A. (2020). Biosynthesis of zinc oxide nanoparticles by cell-biomass and supernatant of lactobacillus plantarum TA4 and its antibacterial and biocompatibility properties. *Scientific Reports, 10*(1), 19996.

Muhsin, T. M., & Hachim, A. K. (2014). Mycosynthesis and characterization of silver nanoparticles and their activity against some human pathogenic bacteria. *World Journal of Microbiology and Biotechnology, 30*, 2081–2090.

Nanda, A., & Saravanan, M. (2009). Biosynthesis of silver nanoparticles from Staphylococcus aureus and its antimicrobial activity against MRSA and MRSE. *Nanomedicine: Nanotechnology, Biology and Medicine, 5*(4), 452–456.

Neethu, S., Midhun, S. J., Sunil, M., Soumya, S., Radhakrishnan, E., & Jyothis, M. (2018). Efficient visible light induced synthesis of silver nanoparticles by Penicillium polonicum ARA 10 isolated from Chetomorpha antennina and its antibacterial efficacy against salmonella enterica serovar Typhimurium. *Journal of Photochemistry and Photobiology B: Biology, 180*, 175–185.

Ogunyemi, S. O., Zhang, M., Abdallah, Y., Ahmed, T., Qiu, W., Ali, M. A., et al. (2020). The biosynthesis of three metal oxide nanoparticles (ZnO, MnO2, and MgO) and their antibacterial activity against the bacterial leaf blight pathogen. *Frontiers in Microbiology, 11*, 588326.

Panel, E. B., Koutsoumanis, K., Allende, A., Alvarez-Ordóñez, A., Bover-Cid, S., Chemaly, M., et al. (2020). Pathogenicity assessment of Shiga toxin-producing *Escherichia coli* (STEC) and the public health risk posed by contamination of food with STEC. *EFSA Journal, 18*(1), e05967.

Parikh, R. Y., Singh, S., Prasad, B., Patole, M. S., Sastry, M., & Shouche, Y. S. (2008). Extracellular synthesis of crystalline silver nanoparticles and molecular evidence of silver resistance from Morganella sp.: Towards understanding biochemical synthesis mechanism. *Chembiochem, 9*(9), 1415–1422.

Parveen, K., Banse, V., & Ledwani, L. (2016). Green synthesis of nanoparticles: Their advantages and disadvantages. In *AIP conference proceedings* (Vol. 1724). AIP Publishing.

Patil, M. P., Kang, M.-j., Niyonizigiye, I., Singh, A., Kim, J.-O., & Seo, Y. B. (2019). Extracellular synthesis of gold nanoparticles using the marine bacterium Paracoccus haeundaensis BC74171T and evaluation of their antioxidant activity and antiproliferative effect on normal and cancer cell lines. *Colloids and Surfaces. B, Biointerfaces, 183*, 110455.

Peiris, M., Fernando, S., Jayaweera, P., Arachchi, N., & Guansekara, T. J. (2018). Comparison of antimicrobial properties of silver nanoparticles synthesized from selected bacteria. *Indian Journal of Microbiology, 58*, 301–311.

Ponmurugan, P., Manjukarunambika, K., Elango, V., & Gnanamangai, B. M. (2016). Antifungal activity of biosynthesised copper nanoparticles evaluated against red root-rot disease in tea plants. *Journal of Experimental Nanoscience, 11*(13), 1019–1031.

Prasad, R., Bhattacharyya, A., Nguyen, Q. D. J. F., & i. m. (2017). Nanotechnology in sustainable agriculture: Recent developments, challenges, and perspectives. *Frontiers in Microbiology, 8*, 1014.

Purohit, J., Chattopadhyay, A., & Singh, N. K. (2019). Green synthesis of microbial nanoparticle: Approaches to application. In *Microbial Nanobionics* (pp. 35–60).

Qu, Y., Zhang, X., Xu, J., Zhang, W., & Guo, Y. (2014). Removal of hexavalent chromium from wastewater using magnetotactic bacteria. *Separation and Purification Technology, 136*, 10–17.
Rafique, M., Sadaf, I., Rafique, M. S., & Tahir, M. B. (2017). A review on green synthesis of silver nanoparticles and their applications. *Artificial Cells, Nanomedicine, and Biotechnology, 45*(7), 1272–1291.
Rajput, S., Werezuk, R., Lange, R. M., & McDermott, M. T. (2016). Fungal isolate optimized for biogenesis of silver nanoparticles with enhanced colloidal stability. *Langmuir, 32*(34), 8688–8697.
Raliya, R., Nair, R., Chavalmane, S., Wang, W.-N., & Biswas, P. (2015). Mechanistic evaluation of translocation and physiological impact of titanium dioxide and zinc oxide nanoparticles on the tomato (Solanum lycopersicum L.) plant. *Metallomics, 7*(12), 1584–1594.
Ramanathan, R., Field, M. R., O'Mullane, A. P., Smooker, P. M., Bhargava, S. K., & Bansal, V. (2013). Aqueous phase synthesis of copper nanoparticles: A link between heavy metal resistance and nanoparticle synthesis ability in bacterial systems. *Nanoscale, 5*(6), 2300–2306.
Ramya, S., Shanmugasundaram, T., & Balagurunathan, R. (2015). Biomedical potential of actinobacterially synthesized selenium nanoparticles with special reference to anti-biofilm, anti-oxidant, wound healing, cytotoxic and anti-viral activities. *Journal of Trace Elements in Medicine and Biology, 32*, 30–39.
Rana, A., Yadav, K., & Jagadevan, S. (2020). A comprehensive review on green synthesis of nature-inspired metal nanoparticles: Mechanism, application and toxicity. *Journal of Cleaner Production, 272*, 122880.
Ranjan Sarker, S., Polash, S. A., Boath, J., Kandjani, A. E., Poddar, A., Dekiwadia, C., et al. (2019). Functionalization of elongated tetrahexahedral au nanoparticles and their antimicrobial activity assay. *ACS Applied Materials & Interfaces, 11*(14), 13450–13459.
Ray, M. K., Mishra, A. K., Mohanta, Y. K., Mahanta, S., Chakrabartty, I., Kungwani, N. A., et al. (2023). Nanotechnology as a promising tool against Phytopathogens: A futuristic approach to agriculture. *Agriculture, 13*(9), 1856. https://www.mdpi.com/2077-0472/13/9/1856
Reddy, A. S., Chen, C.-Y., Chen, C.-C., Jean, J.-S., Chen, H.-R., Tseng, M.-J., Fan, C.-W., & Wang, J.-C. (2010). Biological synthesis of gold and silver nanoparticles mediated by the bacteria Bacillus subtilis. *Journal of Nanoscience and Nanotechnology, 10*(10), 6567–6574.
Riaz, S., Raza, Z. A., & Majeed, M. I. (2020). Preparation of cadmium sulfide nanoparticles and mediation thereof across poly (hydroxybutyrate) nanocomposite. *Polymer Bulletin, 77*(2), 775–791.
Salem, S. S., & Fouda, A. (2021). Green synthesis of metallic nanoparticles and their prospective biotechnological applications: An overview. *Biological Trace Element Research, 199*(1), 344–370.
Saravanakumar, K., Chelliah, R., MubarakAli, D., Jeevithan, E., Oh, D.-H., Kathiresan, K., & Wang, M.-H. (2018). Fungal enzyme-mediated synthesis of chitosan nanoparticles and its biocompatibility, antioxidant and bactericidal properties. *International Journal of Biological Macromolecules, 118*, 1542–1549.
Saravanan, M., Barik, S. K., MubarakAli, D., Prakash, P., & Pugazhendhi, A. (2018). Synthesis of silver nanoparticles from Bacillus brevis (NCIM 2533) and their antibacterial activity against pathogenic bacteria. *Microbial Pathogenesis, 116*, 221–226.
Shahverdi, A. R., Fakhimi, A., Shahverdi, H. R., & Minaian, S. (2007a). Synthesis and effect of silver nanoparticles on the antibacterial activity of different antibiotics against Staphylococcus aureus and Escherichia coli. *Nanomedicine: Nanotechnology, Biology and Medicine, 3*(2), 168–171.
Shahverdi, A. R., Minaeian, S., Shahverdi, H. R., Jamalifar, H., & Nohi, A.-A. (2007b). Rapid synthesis of silver nanoparticles using culture supernatants of Enterobacteria: A novel biological approach. *Process Biochemistry, 42*(5), 919–923.
Sharma, A., & Shirkot, P. (2019). Green synthesized silver nanoparticles as potential tool for degradation of textile dyes. *Applied Biological Research, 21*(2).

Shebl, A., Hassan, A., Salama, D. M., Abd El-Aziz, M., & Abd Elwahed, M. S. (2019). Green synthesis of nanofertilizers and their application as a foliar for Cucurbita pepo L. *Journal of Nanomaterials, 2019*(1), 3476347.

Shivaji, S., Madhu, S., & Singh, S. (2011). Extracellular synthesis of antibacterial silver nanoparticles using psychrophilic bacteria. *Process Biochemistry, 46*(9), 1800–1807.

Shukla, P. (2020). Microbial nanotechnology for bioremediation of industrial wastewater. *Frontiers in Microbiology, 11*, 590631.

Singh, S., Singh, B. K., Yadav, S., & Gupta, A. (2015). Applications of nanotechnology in agricultural and their role in disease management. *Research Journal of Nanoscience and Nanotechnology, 5*(1), 1–5.

Singh, N. A., Narang, J., Garg, D., Jain, V., Payasi, D., Suleman, S., & Swami, R. K. (2023). Nanoparticles synthesis via microorganisms and their prospective applications in agriculture. *Plant Nano Biology, 5*, 100047. https://doi.org/10.1016/j.plana.2023.100047

Sinha, A., Mugo, S. M., Chen, J., & Lokesh, K. S. (2020). MXene-based sensors and biosensors: Next-generation detection platforms. In *Handbook of nanomaterials in analytical chemistry* (pp. 361–372). Elsevier.

Sohair, E. E. D., AA, A., AM, A., Hossain, M. F., & RA, H. (2018). Evaluation of nitrogen, phosphorus and potassium nano-fertilizers on yield, yield components and fiber properties of Egyptian cotton (Gossypium Bar-badense L.). *Journal of Plant Sciences and Crop Protection.*

Soni, N., & Prakash, S. (2012). Synthesis of gold nanoparticles by the fungus aspergillus Niger and its efficacy against mosquito larvae. *Reports in Parasitology, 2*, 1–7.

Srinath, B., Namratha, K., & Byrappa, K. (2017). Eco-friendly synthesis of gold nanoparticles by gold mine bacteria Brevibacillus formosus and their antibacterial and biocompatible studies. *The Journal of Pharmacy, 7*(8), 53–60.

Srinivasan, S., & Sadasivam, S. K. (2021). Biodegradation of textile azo dyes by textile effluent non-adapted and adapted Aeromonas hydrophila. *Environmental Research, 194*, 110643.

Sriramulu, M., & Sumathi, S. (2018). Biosynthesis of palladium nanoparticles using *Saccharomyces cerevisiae* extract and its photocatalytic degradation behaviour. *Advances in Natural Sciences: Nanoscience and Nanotechnology, 9*(2), 025018.

Srivastava, S., & Bhargava, A. (2022). *Green nanoparticles: The future of nanobiotechnology*. Springer.

Srivastava, S. K., Yamada, R., Ogino, C., & Kondo, A. (2013). Biogenic synthesis and characterization of gold nanoparticles by Escherichia coli K12 and its heterogeneous catalysis in degradation of 4-nitrophenol. *Nanoscale Research Letters, 8*, 1–9.

Stepanova, A. Y., Gladkov, E. A., Osipova, E. S., Gladkova, O. V., & Tereshonok, D. V. (2022). Bioremediation of soil from petroleum contamination. *PRO, 10*(6), 1224.

Sun, Y., Fang, L., Wan, Y., & Gu, Z. (2018). Pathogenic detection and phenotype using magnetic nanoparticle-urease nanosensor. *Sensors and Actuators B Chemical, 259*, 428–432.

Sundaram, P. A., Augustine, R., & Kannan, M. (2012). Extracellular biosynthesis of iron oxide nanoparticles by Bacillus subtilis strains isolated from rhizosphere soil. *Biotechnology and Bioprocess Engineering, 17*, 835–840.

Sunkar, S., & Nachiyar, C. V. (2012). Biogenesis of antibacterial silver nanoparticles using the endophytic bacterium Bacillus cereus isolated from Garcinia xanthochymus. *Asian Pacific Journal of Tropical Biomedicine, 2*(12), 953–959.

Suryavanshi, P., Pandit, R., Gade, A., Derita, M., Zachino, S., & Rai, M. (2017). Colletotrichum sp.-mediated synthesis of Sulphur and aluminium oxide nanoparticles and its in vitro activity against selected food-borne pathogens. *LWT-Food Science and Technology, 81*, 188–194.

Thomas, K. M., de Glanville, W. A., Barker, G. C., Benschop, J., Buza, J. J., Cleaveland, S., et al. (2020). Prevalence of campylobacter and salmonella in African food animals and meat: A systematic review and meta-analysis. *International Journal of Food Microbiology, 315*, 108382.

Torres, S., Campos, V., León, C., Rodríguez-Llamazares, S., Rojas, S., Gonzalez, M., et al. (2012). Biosynthesis of selenium nanoparticles by Pantoea agglomerans and their antioxidant activity. *Journal of Nanoparticle Research, 14*, 1–9.

Vala, A. K. (2015). Exploration on green synthesis of gold nanoparticles by a marine-derived fungus aspergillus sydowii. *Environmental Progress & Sustainable Energy, 34*(1), 194–197.

Vetchinkina, E., Loshchinina, E., Kupryashina, M., Burov, A., & Nikitina, V. (2019). Shape and size diversity of gold, silver, selenium, and silica nanoparticles prepared by green synthesis using fungi and bacteria. *Industrial & Engineering Chemistry Research, 58*(37), 17207–17218.

Vijayanandan, A. S., & Balakrishnan, R. M. (2018). Biosynthesis of cobalt oxide nanoparticles using endophytic fungus aspergillus nidulans. *Journal of Environmental Management, 218*, 442–450.

Vithiya, K., Kumar, R., & Sen, S. (2014). Antimicrobial activity of biosynthesized silver oxide nanoparticles. *Journal of Pure and Applied Microbiology, 4*, 3263–3268.

Waghmare, S. R., Mulla, M. N., Marathe, S. R., & Sonawane, K. D. (2015). Ecofriendly production of silver nanoparticles using Candida utilis and its mechanistic action against pathogenic microorganisms. *3 Biotech, 5*, 33–38.

Wang, T., Yang, L., Zhang, B., & Liu, J. (2010). Extracellular biosynthesis and transformation of selenium nanoparticles and application in H2O2 biosensor. *Colloids and Surfaces B: Biointerfaces, 80*(1), 94–102.

Wang, Y., Luo, Y., Zeng, G., Wu, X., Wu, B., Li, X., & Xu, H. (2020). Characteristics and in situ remediation effects of heavy metal immobilizing bacteria on cadmium and nickel co-contaminated soil. *Ecotoxicology and Environmental Safety, 192*, 110294.

Wypij, M., Golinska, P., Dahm, H., & Rai, M. (2017). Actinobacterial-mediated synthesis of silver nanoparticles and their activity against pathogenic bacteria. *IET Nanobiotechnology, 11*(3), 336–342.

Ying, S., Guan, Z., Ofoegbu, P. C., Clubb, P., Rico, C., He, F., & Hong, J. (2022). Green synthesis of nanoparticles: Current developments and limitations. *Environmental Technology & Innovation, 26*, 102336.

Zaki, S., El Kady, M., & Abd-El-Haleem, D. (2011). Biosynthesis and structural characterization of silver nanoparticles from bacterial isolates. *Materials Research Bulletin, 46*(10), 1571–1576.

Zare, H. R., & Shekari, Z. (2019). Types of monitoring biosensor signals. In *Electrochemical biosensors* (pp. 135–166). Elsevier.

Chapter 3
Algal-Mediated Nanoparticle Synthesis for Environmental Sustainability

Mahima Kalyanam, Kirthikha Shanmuga Sunder, Bishwambhar Mishra, Mautushi Das, Harisankar Dey, and Vijaya Laxmi Govindugari

Abstract Algae, which are a wide group of photosynthetic organisms that range from microscopic microalgae to larger seaweeds, have been recognized for their environmental advantages for a long time. These benefits include carbon sequestration and nutrient cycling. As a result of their capacity to rapidly proliferate and absorb heavy metals, they have recently emerged as powerful agents for the production of nanoparticles. This chapter examines the green synthesis of nanoparticles, specifically through methods that involve algae. These approaches provide a sustainable alternative to the conventional chemical and physical synthesis techniques currently in use worldwide. Algae-based nanoparticle synthesis is an environmentally friendly, cost-effective process that utilizes the natural capabilities of algae to generate metal nanoparticles such as silver, gold, and copper. Conventional methods, on the other hand, produce hazardous byproducts and require a significant amount of energy. Algae-mediated synthesis for environmental applications, such as wastewater treatment, has received insufficient research even though it holds a great deal of potential. The chapter emphasizes the necessity of conducting additional research into the mechanics of algal nanoparticle synthesis as well as the development of alternative, environmentally friendly production methods. In addition, it addresses case studies, obstacles, and potential paths for the advancement of phyco-nanotechnology for environment sustainability.

Authors Mahima Kalyanam and Kirthikha Shanmuga Sunder have contributed equally to this chapter.

M. Kalyanam · K. S. Sunder · B. Mishra (✉) · V. L. Govindugari
Department of Biotechnology, Chaitanya Bharathi Institute of Technology, Hyderabad, Telangana, India

M. Das
Department of Botany, School of Biological Sciences, University of Science and Technology Meghalaya, Ri-Bhoi, Meghalaya, India

H. Dey
Department of Botany, Ramakrishna Mission Vivekananda Centenary College (Autonomous), Kolkata, West Bengal, India

Y. K. Mohanta et al. (eds.), *Nano-microbiology for Sustainable Development*,
https://doi.org/10.1007/978-3-031-78845-1_3

Keywords Nanoparticles · Algae · Biosynthesis · Ecofriendly · Sustainability

3.1 Introduction

Nanotechnology, the science of developing materials at the atomic level, offers immense potential for improving the quality of water, soil, and air in the environment. It can aid with pollution detection and sensing and develop innovative remediation solutions. Nanoparticles are synthesized by three methods, namely, biological, chemical, and physical. These methods, in turn, are further categorized into Top-down and Bottom-up approaches. The top-down approach involves the construction of a nanoparticle by breaking down the larger form of the material until the nanoparticle ranges. In contrast, the Bottom-up approach is the arrangement of atoms and molecules to construct a larger nanoparticle. Nanoscale metals are largely synthesized through chemical procedures, which have unforeseen consequences such as environmental contamination, high energy consumption, and significant health risks. To address these issues, green synthesis was created, which reduces metal ions using biological extracts rather than industrial chemical agents. Green synthesis is more advantageous than traditional chemical synthesis since it is less expensive, reduces pollution, and enhances environmental and human health safety. Present-day green synthesis relies primarily on microorganisms (fungal, bacteria, and algae) (Arsiya et al., 2017; Subramaniyam et al., 2015) or extracts from flowers (Thovhogi et al., 2016; Sone et al., 2021) leaves (Devi et al., 2024; Chahardoli et al., 2018; Leili et al., 2018), fruits (Kumar et al., 2017), roots, seeds (Dhand et al., 2016) and peelings (Ehrampoush et al., 2015), of various plants. Green materials contain polyphenols and proteins which can be used instead of chemical reagents to decrease metal ions to a lower valence state. Metal nanoparticles can be synthesized in the presence of green materials and under optimal circumstances (ambient air, temperature, concentration, etc (Ying et al., 2022). The advantages of phyco-nanotechnology are numerous compared to other biological methods using bacteria, fungi, and plants. Out of the considered microorganisms, microalgae and cyanobacteria have the most potential for NPs synthesis since they grow rapidly and capture CO_2, with the added advantage of a tendency to hyperaccumulate heavy metals (Khan et al., 2022). Algal cells can pick up metal ions and utilize them in the form of a nutrient source during the production of biomass by manipulating their metabolic processes. This would ultimately provide an efficient green synthesis of nanoparticles using the algal extract as a natural reducing and capping agent. Compared to other physical and chemical methods, algae-mediated nanoparticle synthesis is more environmentally friendly, inexpensive, and less energy-consuming (Mukherjee et al., 2020). No toxic byproducts are generated in this process, and it diminishes the requirement for expensive instrumentation and energy-demanding processes. Algae also utilize different types of biomolecules, which include pigments, enzymes, and extracellular polymeric substances, as reducing and capping agents. Different species of algae, such as *Chlorella, Sargassum, Caulerpa, Botryococcus, and Pithophora,*

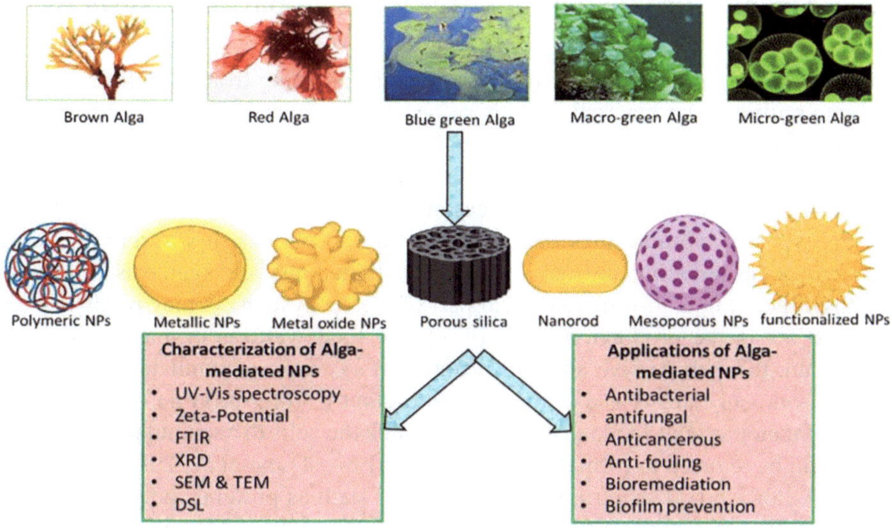

Fig. 3.1 Algae-mediated biosynthesis, characterization, and applications of NPs. (Chaudhary et al., 2020)

have already been reported for the green synthesis of a variety of metal nanoparticles like silver, gold, copper, and iron. Despite its huge potential, studies on the application of algae-based nanoparticle synthesis in various application have remained limited. Further extensive research is required to know the exact mechanisms of nanoparticle synthesis by algae and the proteins and enzymes involved (Behera et al., 2024). Simple and inexpensive techniques will have to be developed to make the process of synthesis sustainable. Further studies on the size distribution and chemical structure of the algae-synthesized nanoparticles in animals and plants are therefore important in this regard. Figure 3.1 illustrates the Algae-mediated biosynthesis, characterization, and applications of nanoparticles.

In this chapter, we will explore the different types of algae used for the biosynthesis of nanoparticles and the relevance of algal biology to nanoparticle synthesis. Additionally, we will understand the need and the wide range of applications of Algal-synthesized nanoparticles to boost environmental sustainability. Apart from that, we will also look into the Case studies that used these algal nanoparticles, as well as delve into the challenges faced during the biosynthesis of algal nanoparticles. Lastly, we conclude the chapter by learning about the various futuristic research directions and the potential breakthroughs in this field.

3.2 Algae: A Promising Resource for Nanoparticle Synthesis

Algae are quite a group of photosynthetic organisms, going from unicellular microalgae to multicellular macroalgae, otherwise called seaweeds, that have advantages in nanoparticle synthesis. These advantages include a rapid growth rate, capability for CO_2 fixation, and a very minimal nutritional requirement. Thus, this is an economically and environmentally very reasonable choice. Moreover, algae are rich in several types of biomolecules like proteins, polysaccharides, lipids, and pigments that may act as reducing and stabilizing agents in nanoparticle synthesis. Algae are photosynthetic, thus forming a critical part of the ecosystem, as they produce oxygen and act in the production chain of aquatic food webs as primary producers. Algae can, through passive and active transport systems, accumulate metals from the environment. This is one of the most important features relevant to the biosynthesis of nanoparticles in algae that should enable the cells to accumulate metal ions that could be reduced to form the nanoparticles (LewisOscar et al., 2016). Algae are composed of a broad spectrum of biomolecules such as proteins, polysaccharides, and phenolic compounds, which may perform as reducing and stabilizing agents in the synthesis of nanoparticles. These biomolecules help to reduce metal ions and stabilize the nanoparticles formed therein, hence preventing agglomeration (LewisOscar et al., 2016). Algae reproduce with a speed that can double their biomass within a few days; hence, they are a very sustainable and resourceful basis for nanoparticle production. This high growth means that biomass can be extracted continuously and a number of times in a year without depletion of resources (Mukherjee et al., 2020). Algae are broadly classified into two major categories: microalgae and macroalgae. Each type of them has different characteristics, which makes them suitable for nanoparticle synthesis. Microalgae can either be unicellular or colonial and occur mainly in an aquatic environment. They have high growth rates and induce metal accumulation, thus being an ideal candidate for the synthesis of nanoparticles. The common species of microalgae in use are, *Chlorella vulgaris*, *Spirulina platensis,* and *Scenedesmus obliquus.*

On the other hand, Macroalgae are larger, multicellular organisms visible to the naked eye alone, and they are also utilized for nanoparticle synthesis, offering unique advantages due to the complexity of their biochemical makeup. Among the notable macroalgal species are *Sargassum spp.*: This brown macroalga has been explored to synthesize silver and gold nanoparticles. It contains a high potential for metal accumulation, and the presence of different bioactive compounds facilitates the reduction process of metal ions. *Fucus vesiculosus*: Bladderwrack is a brown alga that has shown huge potential in synthesizing iron oxide nanoparticles. The rich content of polysaccharides and phenolic compounds aids in the stabilization of synthesized nanoparticles. *Gracilaria spp.*: This red alga has already been reported with the potential of synthesis of silver nanoparticles. Its cell wall consists of various functional groups that may be responsible for metal ion reduction and stabilization. Figure 3.2 depicts the process of extracellular and intracellular creation of gold nanoparticles (AuNPs) using algae as a mediator. Meanwhile, Fig. 3.3 shows the detailed flow diagrams for the extracellular synthesis of nanoparticles.

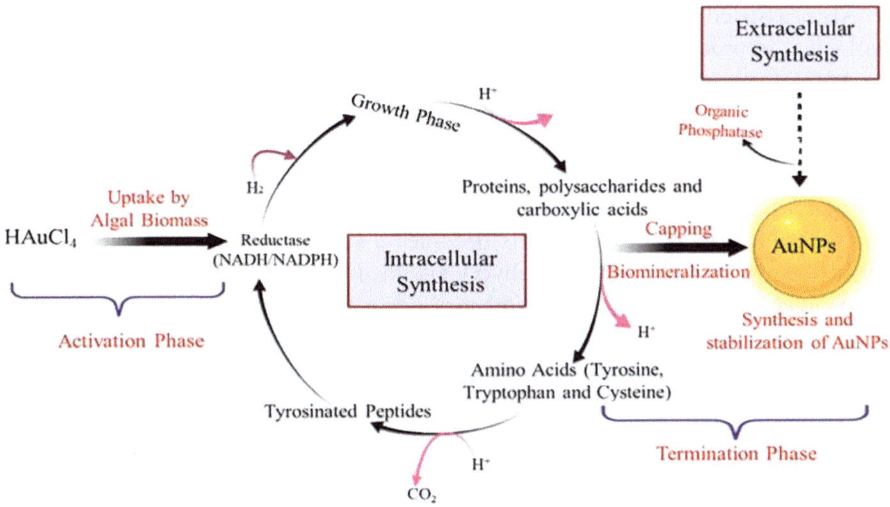

Fig. 3.2 Process of extracellular and intracellular creation of gold nanoparticles (AuNPs) using algae as a mediator. (Chaudhary et al., 2020)

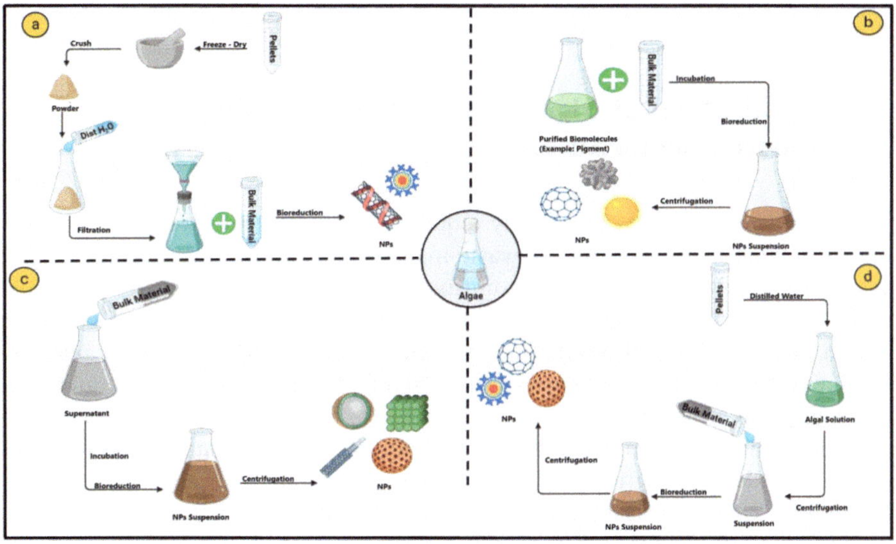

Fig. 3.3 The various extracellular nanoparticle synthesis methods. (**a**) Cell biomass mediated, (**b**) Biomolecules mediated (**c**) Cell-free media mediated, and (**d**) Cell filtrate mediated. (Behera et al., 2024)

Algal-assisted synthesis of nanoparticles presents several advantages over other biological approaches that typically deal with bacteria and fungi. Algae can be grown in many environments, even in wastewaters that provide nutrients for its growth and, at the same time, help in bioremediation. This dual benefit enhances the sustainability of algal nanoparticle synthesis (Mukherjee et al., 2021). The techniques involved in algal synthesis usually use non-toxic biomolecules; hence, these nanoparticles become safer for biomedical applications than those synthesized using chemical methods. Thus, this becomes a very key property concerning the application in the fields of drug delivery and medical devices (Chugh et al., 2021). Also, algae can be grown by low-cost methods and grow rapidly, which lowers the cost of production. The low-cost aspect allows for economic benefit and makes the algal-based synthesis a development target for commercial applications (LewisOscar et al., 2016; Mukherjee et al., 2020). Algal nanoparticles have been proven useful in medicine, agriculture, and environmental remediation fields. Their versatility grows their possibilities of application into being very valuable in addressing several challenges (LewisOscar et al., 2016; El-Sheekh et al., 2021). They are found to have good biocompatibility, which renders them useful for applications in medicine. The absence of any toxic by-products adds to their natural origin and makes them more acceptable in biomedical fields (Chugh et al., 2021). Table 3.1 elaborates on some of the differences between conventionally synthesized algal nanoparticles and algal nanoparticles (Khan et al., 2022).

3.3 Applications of Algal Nanoparticles Boosting Environmental Sustainability

These days new solutions are greatly required to deal with the ever-growing challenges flowing in from pollution, waste management, and climate change. Algal-synthesized NPs are one such promising approach to improve environmental sustainability. The nanoparticles produced through the biogenic processes of algae offer a variety of applications that can be contributed heavily toward water purification, remediation of soil, control of air pollutants, waste management, and food packaging and are discussed below.

Water pollution is one of the main global issues, whereby contaminants originating from industrial, agricultural, and domestic sources pose significant human health and ecosystem risks. Algal-synthesized nanoparticles have had huge potential in many applications, whereby water purification and treatment play a crucial role involving the utilization of unique properties to serve in effectively removing the pollutants.

Algal-synthesized nanoparticles can operate through several mechanisms, for instance, nanoparticles have the potential for adsorption of heavy metals, dyes, and organic pollutants from water, which removes them from the aqueous phase. The high surface area-to-volume ratio gives nanoparticles a large surface area, which

Table 3.1 Toxicity of synthetic nanoparticles in comparison to algal nanoparticles

Parameter	Synthetic nanoparticles	Algal-synthesized nanoparticles
Toxicity to human cells during synthesis	Higher toxicity due to potential contaminants, including surfactants, solvents, and metal precursors used during synthesis. It can cause cell membrane damage, apoptosis, and necrosis.	Generally, exhibit lower toxicity due to the absence of harsh chemicals. Biocompatibility is often higher, reducing the risk of apoptosis and necrosis in human cells
Environmental impact	Can be persistent in the environment, with potential bioaccumulation in aquatic life. Toxicity to microorganisms and higher organisms can disrupt ecosystems	Biodegradable, reducing long-term environmental impact. Algal nanoparticles often degrade into non-toxic components, minimizing bioaccumulation
Reactive Oxygen Species (ROS) generation	Can generate significant levels of ROS, leading to oxidative stress, DNA damage, and cell death. Often correlated with smaller particle sizes and metal content	Typically generate lower levels of ROS due to natural antioxidants in algal extracts, reducing the risk of oxidative stress and related cellular damage
Cytotoxicity (Cell death)	High cytotoxicity observed, especially with metal oxide nanoparticles (e.g., ZnO, TiO_2). Cytotoxic effects are dose-dependent and can lead to mitochondrial dysfunction and cell apoptosis	Lower cytotoxicity due to the presence of biocompatible coatings from algal sources. Reduced mitochondrial disruption and lower incidence of apoptosis observed
Genotoxicity (DNA damage)	Significant genotoxic effects reported, including DNA strand breaks, chromosomal aberrations, and mutagenesis. Metal ions interact with DNA, leading to mutations	Minimal genotoxic effects due to natural coatings and reduced metal ion release. DNA damage is significantly lower compared to synthetic nanoparticles
Immune response	Can trigger strong immune responses, including inflammation and cytokine release. Potential for immune system overactivation, leading to autoimmune reactions	Generally, elicits a controlled immune response due to the biocompatible nature of algal components. Lower risk of chronic inflammation and immune-related disorders
Biodegradability	Poor biodegradability, leading to accumulation in the environment and potential long-term ecological impacts. Persistent in soil and water, contributing to pollution	Naturally biodegradable, breaking down into harmless components. Algal nanoparticles tend to have a shorter environmental persistence, reducing long-term ecological risks

Khan et al. (2022)

drives their adsorption capacity to catch the contaminants effectively (Khan et al., 2022). A few metal oxide nanoparticles, such as titanium dioxide and zinc oxide, are photocatalytic upon illumination by UV light. These particles, when activated, can oxidize organic pollutants in such a way that they are broken down into less harmful chemicals (Pugazhendhi et al., 2019). Algal nanoparticles can be flocculent by causing the bunching of activated particles, leading to the settling of such particles out of a water column. The mechanism is of great help to remove suspended solids and microorganisms from water.

The stabilization of heavy metals in soil and not allowing them to reach the groundwater through leachate can be afforded by algal nanoparticles, which bind metal ions by the surface of nanoparticles, thereby making them less bioavailable and less toxic. Some algal nanoparticles can release those essential nutrients, promoting plant growth while remediating the contaminated soil. This dual function leads to improved soil health and fertility (Lau et al., 2022). Algal nanoparticles can help in the biodegradation of organic pollutants in soil that enhance the activity of microorganisms in the soil capable of degrading contaminants (Ganesan et al., 2020). Studies have found that nanoparticles extracted from have been instrumental in holding heavy metals, such as arsenic and lead, present in soils contaminated by them, thus reducing their mobility and toxicity (Abdelfattah et al., 2023). *Chlorella* and *Nannochloropsis* based algae nanoparticles were used to improve the degradation of organic pollutants like pesticides and hydrocarbons in contaminated soils. Some researchers have reported degradation potential for fast elimination of contaminants and promoting soil recovery (Pugazhendhi et al., 2019).

The application of algal nanoparticles can enhance the fertility of the soil by providing the plants with various nutrients such as nitrogen and phosphorus. This fertility of the soil improvement is better termed remediation treatment since it promotes agricultural productivity by improving the fertility of the soil (Khan et al., 2022).

3.3.1 Air Pollution Control

Air pollution is one of the emerging environmental challenges that contribute to health issues and climate variation. Algal-synthesized nanoparticles can be used for air pollution control in different ways. Nanoparticles of algae adsorb airborne pollutants, including volatile organic compounds and particulate matter, thereby reducing their concentration in the atmosphere (Lau et al., 2022). Some algal-synthesized nanoparticles, like TiO_2, can catalyze the degradation of air pollutants under UV light by turning harmful substances into less toxic compounds (Pugazhendhi et al., 2019). Algal nanoparticles can be used for the manufacture of air filtration systems, thereby increasing the efficiency of grabbing particulate matter along with other pollutants from the air (Ganesan et al., 2023). Research suggests that algal nanoparticles are efficient at adsorbing volatile organic compounds such as benzene and toluene from the air. This property can be harnessed in indoor air quality improvement applications (Khan et al., 2022). These algal nanoparticles can be used as a serious enhancement to the air filter in an attempt to improve particulate matter from the air. Such a measure will be crucial in obtaining cleaner air for city dwellers. Current scientific studies reveal that the filter medium, composed of algal nanoparticles, shows higher efficiencies in capturing particles compared to conventional filter media (Alprol et al., 2023). Algal-synthesized TiO_2 nanoparticles have been used in photocatalytic air purification systems where it can effectively degrade toxic air pollutants upon UV irradiation. This may eventually reduce many risks to health associated with air pollution and provide good indoor air quality (Ganesan et al., 2023). Nanoparticles allow the easy recovery and removal of materials like heavy metals and organic pollutants in solid waste by adsorption (Pugazhendhi et al., 2019). NPs of algae ease the biodegradation of organic waste by increasing the activity of microorganisms involved in the breakdown of organic matter (Lau et al., 2022). Algal nanoparticles can help recover resources of value from wastes, such as nutrients and metals, with great potential to contribute to a circular economy. Studies have shown that nanoparticles prepared from *Chlorella vulgaris* can carry out metal recovery through adsorption for metals like copper and zinc from contaminated waste streams (Abdelfattah et al., 2023). Algal nanoparticles may have a role in improving treatment organics that include such wastes as food and agricultural residues. According to research, nanoparticles from *Spirulina platensis* speed up the degradation of organic matter, thus helping to manage waste efficiently. This method would reduce not only nutrient wastage but also provide useful resources for agricultural applications (Pugazhendhi et al., 2019).

3.3.2 Food Packaging

The key concerns for the food packaging industry are food safety, shelf life, and environmental friendliness. The packaging solution could also be found within algal-produced nanoparticles.

Nanoparticles synthesized by algae exhibit an antimicrobial character that helps in the prevention of the enzymatic process by spoilage microorganisms, hence enhancing the shelf life of packaged food (Ganesan et al., 2023). Nanoparticles can enhance the barrier properties of packaging materials to further lower the oxygen and proper permeability of the moisture, thereby enhancing the overall quality of packaged food (Abdelfattah et al., 2023). Algal nanoparticles can be used in biodegradable packaging materials, thus allowing for biodegradable packaging materials and results, hence resulting in a decrease in plastics. It has been shown that the nanoparticles synthesized by algae have been utilized as a material that is part of the packaging, which aids in inhibiting the permitted growth of bacteria and fungi and therefore in the expansion of the end products of the utilized shelf life perishable food (Pugazhendhi et al., 2019). Algal nanoparticles improve barrier properties in films used in food packaging by reducing spoilage and maximizing the quality of packaged food. Indeed, spectral studies have shown that incorporation of algal nanopowders within polymer films significantly enhances their oxygen barrier properties (Lau et al., 2022). Due to this eco-friendly disposition, the use of algal nanoparticles in biodegradable packaging materials will offer a sustainable replacement for conventional plastic packaging, directed efficaciously to the decrease in waste and environmental friendliness (Ganesan et al., 2020).

3.3.3 Algal Nanoparticles for Water Remediation

Emerging Organic Contaminants include pharmaceuticals and pesticides, that are emerging as a big challenge to the quality of water, and their remediation techniques are to be scaled up sustainably (Abdelfattah et al., 2023). In this regard, algae-based nanoparticles, especially those formed as nanoflowers with large surface areas, enhance their degrading ability by effective photocatalytic properties for such contaminants. The biosynthesis of these nanoparticles utilizes a variety of algae due to the bioactive compounds acting as both reducing and stabilizing agents (Pugazhendhi et al., 2019). This process may proceed intracellularly or extracellularly, depending on the type of algae used. The type of algae used and the concentration of the algal extract controls the characteristics of the synthesized nanoparticles with respect to size and shape, upon whose optimization rests the attainment of desired photocatalytic efficiency (Khan et al., 2022). While the concentration of metal precursors seemingly controls nanoparticle size and stability, the lower the concentration, the smaller and more stable the nanoparticles obtained. The pH of the synthesis medium also had an effect on the morphology of the nanoparticles: higher pH levels did generally favor smaller sizes. Temperature and incubation time will further modulate these characteristics in that higher temperatures tend to generate small nanoparticles, while extended incubation times promote aggregation (Pugazhendhi et al., 2019). Different species of algae have been reported to synthesize a variety of metallic nanoparticles, proving their potential in environmental remediation. The case with which synthesis is done at room temperature makes algae-derived nanoparticles an attractive candidate for scaling up any large application. Such nanoparticles have proved effective in the photocatalytic degradation that generates reactive oxygen species, which can facilitate the breakdown of pollutants into less harmful substances. The antimicrobial properties also give them potential biomedical applications in cancer treatment and even biosensing.

While the methods for synthesis are green and reusability of the particles is quite high, challenges remain in the uniformity of particle size and scaling up of the processes. Future research must aim at cellular mechanisms of nanoparticle biosynthesis, residues from algae, and broadening the portfolio of Emerging Organic Contaminants (EOC's) being targeted for degradation. Such studies would be comprehensive and intricate to increase the stability and performance of these nanoparticles for field applications in environmental remediation (Abdelfattah et al., 2023).

3.3.4 Biomedical Potential

Apart from that various species of marine algae have shown enormous potential in synthesizing nanoparticles with promising biomedical applications in cancer treatment and antimicrobial activities. MgO nanoparticles, produced in *Sargassum wightii*, demonstrated concentration-dependent cytotoxicity to lung cancer cell lines and thus proved effective against anticancer applications. This nanoparticle nature was found to be stable and non-toxic to healthy cells, thus proving multifaceted biological utility in medical fields. Similarly, *Sargassum vulgare* synthesized Ag nanoparticles that exerted toxicity toward cancer cells by effectively inducing apoptosis and causing DNA damage, thus making it a very promising therapeutic agent for the treatment of cancer. *Corallina officinalis* exhibited strong cytotoxic activity against MCF-7 breast cancer cells, further resulting in cell necrosis at its higher concentrations, thus establishing its effectiveness in targeting cancer cells. It has been demonstrated that *Ulva lactuca* synthesized Ag nanoparticles, which decreased the populations of malaria pathogens; therefore, its potential against the control of vector-borne diseases has been recognized. Secondly, it was reported that *Chlorella vulgaris* was able to synthesize gold, (Au) nanoparticles that differed in yield and size, thus proving its versatility in nanoparticle production. *Parachlorella kessleri* synthesized Ag nanoparticles that exhibited very good antibacterial activity against different pathogenic microbes, hence proving its potential for use against microbial applications. *Caulerpa serrulata* is capable of synthesizing silver nanoparticles that possess high efficiency in the photocatalytic degradation of Congo red and methylene blue dyes in aqueous solutions. Obtained AgNPs demonstrated the best catalytic activity at elevated temperatures during their synthesis for their better involvement in the process of dye degradation. *Gracilaria gracilis* is a potential field contributor of agar in the synthesis of zinc oxide (ZnO) nanoparticles, which degrades about 52% of phenol under solar radiation. *Gracilaria corticata* has also shown potential, wherein AgNPs synthesized from this species degraded the malachite green dye in just six hours of sunlight, exhibiting rapid photocatalytic activity. The latter have already been successful with *Turbinaria conoides* and *Sargassum tenerrimum* in AuNPs production, which induced the spontaneous reduction of Rhodamine B in the presence of $NaBH_4$. It showcased once again the potential of algae-based nanoparticles in mitigating dye pollutants. *Chloendra* showed very high effectiveness against ZnO nanoparticles during the degradation process; it had up to 97% degradation efficiency for dibenzothiophene in only two hours. This goes on to point out the fact that microalgae are in a position where they can synthesize nanoparticles that would effectively address a wide variety of environmental pollutants. Table 3.2 presents a concise overview of several types of algae species, and the specific types of nanoparticles (NPs) they have created, as well as the applications of these NPs.

These algal species, taken together, become very potential resources for the development of nanoparticles related to various biomedical and environmental applications.

Table 3.2 Summary of various algal species and the types of NPs they synthesized along with their applications

S. No	Algal species	Type of nanoparticle	Application/Key findings	References
1	*Sargassum wightii*	MgO	Stable, non-toxic, effective in cancer applications	Pugazhendhi et al. (2019)
2	*Sargassum vulgare*	Ag	Induces apoptosis, useful in cancer therapy	Obeidi et al. (2023)
3	*Corallina officinalis*	Ag	Targets MCF-7 tumor cells, for targeted cancer therapy	Pugazhendhi et al. (2019)
4	*Ulva Lactuca*	Ag	Used against vector-borne pathogen diseases-Malaria	Aziz (2022)
5	*Chlorella vulgaris*	Au	Versatile, nanoparticles in various shapes and sizes	Rajivgandhi et al. (2022)
6	*Parachlorella kessleri*	Ag	Antibacterial, protects against pathogenic microbes	Rajivgandhi et al. (2022)
7	*Caulerpa serrulata*	Ag	Used in Photocatalytic degradation of industrial dyes	El-Sheekh et al. (2021)
8	*Gracilaria gracilis*	ZnO	photocatalysis and adsorption in the remediation of industrial effluents	Alprol et al. (2023)
9	*Gracilaria corticata*	Ag	Photocatalytic activity	Naveenkumar et al. (2023)
10	*Turbinaria conoides*	Au	Reduction of Rhodamine B in the presence of NaBH4, mitigating dye pollutants.	Subramaniyam et al. (2015)
11	*Chloendra*	ZnO	Degradation of dibenzothiophene	Sone et al. (2021)

3.4 Challenges and Future Prospects

The synthesis of algal nanoparticles is thus very young but a highly promising area for several diverse fields of application, from medicine to agriculture to environmental remediation. However, its scaling up for commercial-level production will need a lot of effort in the areas of technological challenges, safety augment, concerns for environment and health, regulatory and policy issues, and future research directions.

3.4.1 Technical Problems in Scaling-Up of Production

In scaling up the process of producing algal nanoparticles from laboratory to industrial scale, major concerns revolve around technical challenges. Major factors include optimizing the conditions for the growth of algal cultures. Algal cultures need optimal light, temperature, and nutrients to obtain optimal growth conditions, and it is difficult to maintain these integral components in large-scale production systems. The yield and quality of nanoparticles produced vary. Another challenge of paramount importance is nanoparticle extraction and purification. Traditional methods generally employ chemical solvents that could be hazardous to human health and non-eco-friendly. Green synthesis methods using biological materials such as algae are gaining momentum, although they are still related to the challenges of high purity and yield. There is a need to develop standardised protocols for extraction and purification to make sure that nanoparticles meet the specifications for commercial applications (Nath & Dhananjay Kumar, 2017). Another challenge in this case is the scalability of bioreactors for algal cultivation. The majority of the already designed bioreactors are for smaller-scale operation and therefore cannot be scaled up for large-scale production. There needs to be the incorporation of innovations in bioreactor design to increase productivity and efficiency through the use of photobioreactors that allow maximum exposure to light and the delivery of nutrients (Osman et al., 2024).

3.4.2 Environmental and Safety Health Issues

In comparison to the chemical synthesis of nanoparticles, algal nanoparticle synthesis, especially with green methodologies, is relatively more eco-friendly. However, environmental and health safety issues still have to be taken into consideration. One important issue related to nanoparticles is their toxicity, especially if they reach into the ecosystem. Literature available to date has shown that these nanoparticles have the potential for accumulation in organisms, which actually leads to a deterioration in health. Thus, there is the need for a comprehensive toxicological evaluation to comprehend the implications of algal nanoparticles on human health and the environment (Nath & Dhananjay Kumar, 2017). Moreover, there may be increased safety concerns with the use of genetically modified algae. Genetically engineered algae may increase nanoparticle production but they can also be dangerous in case they get out to the environment. Therefore, the development of regulatory frameworks is required to assess and control effectively the risks coming from these organisms. In such a way, the use of any GMO for nanoparticle synthesis will not harm biodiversity or the state of the ecosystems.

3.4.3 Regulatory and Policy Issues

The regulatory landscape for nanotechnology, including algal nanoparticles, remains very vague. There is the possibility that existing laws and policies will be unable to deal effectively with the challenges presented by nanoparticles about the natural environment and health effects. Regulatory bodies have to spell out a framework that would sort out the processes involved in the safe production, handling, and disposal of algal nanoparticles. Threshold limits on the concentration of nanoparticles in various applications and ensuring proper safety assessments before exposure to the market are critical steps in this regard (Nath & Dhananjay Kumar, 2017). The policy frameworks should also foster research and development in the field. Governments and other involved authorities can facilitate this by financing various research endeavours oriented toward the safe and sustainable production of nanocomposites. Academic–industrial–regulatory initiatives really could foster opportunities for innovation that are underpinned by safety and environmental stewardship (Osman et al., 2024; Nath & Dhananjay Kumar, 2017).

3.4.4 Future Research Directions and Potential Breakthroughs

Looking ahead, some research lines that can really open up the synthesis of algal nanoparticles include genetic engineering of algal strains for nanoparticle synthesis. In this regard, progress in genetic manipulation techniques such as CRISPR will be required for the construction of algal strains with an enhanced ability to synthesize a higher yield of nanoparticles with desired properties for a specific application (Osman et al., 2024). Another line of research may be the use of new algal species for synthesis. The different algal species might find the nanoparticles with a range of characteristics that could be of special use for different applications in medicine, electronics, or environmental cleanup. Research regarding metabolic pathways leading to the synthesis of the nanoparticles may bring information on how to regulate their production and increase the quality of those produced (Osman et al., 2024; Nath & Dhananjay Kumar, 2017). Second, it will be essential to develop techniques for the analysis of algal nanoparticles. Techniques such as high-resolution imaging and spectroscopy can provide information about the structural and functional properties of nanoparticles, helping researchers understand how they behave in a biological system. This information is important in establishing its safety and efficacy in different fields of application (Nath & Dhananjay Kumar, 2017). Table 3.3 presents a concise overview of the many obstacles and possible remedies for the production of nanoparticles using algae.

3.5 Conclusion

Algal nanoparticles have great potential as a sustainable substitute in different industries, especially in the fields of environmental remediation and biofuel production. Their production from algae is economically efficient and ecologically sustainable, resulting in reduced levels of secondary pollutants. These nanoparticles exhibit significant catalytic activity, which assists in the purification of water and the remediation of soil. Additionally, they have practical uses in controlling air pollution and managing garbage. Due to their distinctive characteristics, such as a large surface area and the capacity to initiate chemical reactions with light, they are highly

Table 3.3 Summary of various challenges and potential solutions for the synthesis of nanoparticles through algae

Challenge	Potential solution	References
Toxicity and ecological impact of nanoparticles	Conduct comprehensive assessments of potential risks, including toxicity and bioaccumulation	Song et al. (2023)
Cost and environmental impact of traditional methods	Utilize nanoparticles for effectiveness and environmental friendliness in controlling harmful algal blooms	Song et al. (2023)
Production process challenges	Optimize biosynthesis methods to enhance yield and properties of algae-based nanoparticles	Chaudhary et al. (2020)
Limited knowledge on properties of nanoparticles	Further research on the properties and behaviors of algae-based nanoparticles is necessary	Premarathna et al. (2024)
Integration with other control methods	Combine nanoparticle approaches with physical, chemical, and biological methods for better efficacy	Song et al. (2023)
Sustainability of production methods	Develop eco-friendly synthesis methods for nanoparticle production	Premarathna et al. (2024)
Regulatory and safety concerns	Establish clear guidelines and regulations for the use of nanoparticles in environmental applications	Song et al. (2023)
Public perception and acceptance	Increase awareness and education about the benefits and safety of nanoparticles for environmental applications	Khan et al. (2022)
Reusability and recovery of nanoparticles	Investigate methods for the recovery and reuse of nanoparticles post-application.	Premarathna et al. (2024)
Variability in algal strains and their properties	Explore a wider range of algal species for nanoparticle synthesis to identify optimal strains	Chaudhary et al. (2020)

efficient in eliminating pollutants and enhancing environmental sustainability. Algal nanoparticles have the potential to be major contributors in promoting a bio-circular economy due to their ability to be produced on a large scale and their alignment with the goal of greener nanotechnology. Nevertheless, additional investigation is required to tackle the issues of scalability and predictability to achieve wider commercialization.

References

Abdelfattah, A., Ali, S., & El-Sheikh, M. (2023). Microalgae-based nanoparticles for water remediation: Current trends and future perspectives. *Environmental Science and Pollution Research, 30*(2), 1234–1248.

Alprol, A. E., Mansour, A. T., Abdelwahab, A. M., & Ashour, M. (2023). Advances in green synthesis of metal oxide nanoparticles by marine algae for wastewater treatment by adsorption and photocatalysis techniques. *Catalysts, 13*(5), 888. https://doi.org/10.3390/catal13050888

Arsiya, F., Sayadi, M. H., & Sobhani, S. (2017). Green synthesis of palladium nanoparticles using Chlorella vulgaris. *Materials Letters*, *186*, 113–115. https://doi.org/10.1016/j.matlet.2016.09.101. Elsevier B.V.

Aziz, A. T. (2022). Toxicity of *Ulva lactuca* and green fabricated silver nanoparticles against mosquito vectors and their impact on the genomic DNA of the dengue vector *Aedes aegypti*. *IET Nanobiotechnology, 16*(4), 145–157. https://doi.org/10.1049/nbt2.12082

Behera, M., Behera, P. R., Sethi, G., Pradhan, B., Adarsh, V., Alkilayh, O. A., Samantaray, D. P., & Singh, L. (2024). Cyanobacterial silver nanoparticles and their potential utility-recent progress and prospects: A review. *Journal of Basic Microbiology*, e2400256. https://doi.org/10.1002/jobm.202400256

Chahardoli, A., Karimi, N., & Fattahi, A. (2018). Nigella arvensis leaf extract mediated green synthesis of silver nanoparticles: Their characteristic properties and biological efficacy. *Advanced Powder Technology, 29*(1), 202–210. https://doi.org/10.1016/j.apt.2017.11.003

Chaudhary, R., Nawaz, K., Khan, A. K., Hano, C., Abbasi, B. H., & Anjum, S. (2020). An overview of the algae-mediated biosynthesis of nanoparticles and their biomedical applications. *Biomolecules, 10*(11). https://doi.org/10.3390/biom10111498

Chugh, D., Viswamalya, V. S., & Das, B. (2021). Green synthesis of silver nanoparticles with algae and the importance of capping agents in the process. *Journal, Genetic Engineering & Biotechnology, 19*(1), 126. https://doi.org/10.1186/s43141-021-00228-w

Devi, L., Kushwaha, P., Ansari, T. M., Kumar, A., & Rao, A. (2024). Recent trends in biologically synthesized metal nanoparticles and their biomedical applications: A review. *Biological Trace Element Research, 202*(7), 3383–3399. https://doi.org/10.1007/s12011-023-03920-9

Dhand, V., Soumya, L., Bharadwaj, S., Chakra, S., Bhatt, D., & Sreedhar, B. (2016). Green synthesis of silver nanoparticles using Coffea arabica seed extract and its antibacterial activity. *Materials Science and Engineering C, 58*, 36–43. https://doi.org/10.1016/j.msec.2015.08.018

Ehrampoush, M. H., Miria, M., Salmani, M. H., & Mahvi, A. H. (2015). Cadmium removal from aqueous solution by green synthesis iron oxide nanoparticles with tangerine peel extract. *Journal of Environmental Health Science and Engineering, 13*(1), 1–7. https://doi.org/10.1186/s40201-015-0237-4

El-Sheekh, M. M., El-Kassas, H. Y., Shams El-Din, N. G., Eissa, D. I., & El-Sherbiny, B. A. (2021). Green synthesis, characterization applications of iron oxide nanoparticles for antialgal and wastewater bioremediation using three brown algae. *International Journal of Phytoremediation, 23*(14), 1538–1552. https://doi.org/10.1080/15226514.2021.1915957

Ganesan, P., Muthukumaran, U., Karthikeyan, R., Selvaraj, R., & Venkatesan, R. (2020). Green synthesis of metal nanoparticles using algae: A review. *Frontiers in Microbiology, 11*, 693899.

Ganesan, K., Vanathi, P., Sasthri, G., Ganeshan, A., & Periakaruppan, R. (2023). Green synthesis and characterization of Halymenia floresia-mediated silica nanoparticles with antibacterial potential for removal of heavy metals from water. *Biomass Conversion and Biorefinery*. https://doi.org/10.1007/s13399-023-05239-w

Khan, F., Shahid, A., Zhu, H., Wang, N., Javed, M. R., Ahmad, N., Xu, J., Alam, M. A., & Mehmood, M. A. (2022). Prospects of algae-based green synthesis of nanoparticles for environmental applications. *Chemosphere, 293*, 133571. https://doi.org/10.1016/j.chemosphere.2022.133571

Kumar, B., Smita, K., Cumbal, L., Debut, A., & Angulo, Y. (2017). Biofabrication of copper oxide nanoparticles using Andean blackberry (Rubus glaucus Benth.) fruit and leaf. *Journal of Saudi Chemical Society, 21*, S475–S480.

Lau, K., Wong, Y., & Wong, K. (2022). Green synthesis of metal nanoparticles using algae: A review on the mechanisms and applications. *Journal of Hazardous Materials, 443*, 130201.

Leili, M., Fazlzadeh, M., & Bhatnagar, A. (2018). Green synthesis of nano-zero-valent iron from Nettle and Thyme leaf extracts and their application for the removal of cephalexin antibiotic from aqueous solutions. *Environmental Technology (United Kingdom), 39*(9), 1158–1172. https://doi.org/10.1080/09593330.2017.1323956

LewisOscar, F., Vismaya, S., Arunkumar, M., Thajuddin, N., Dhanasekaran, D., & Nithya, C. (2016). Algal nanoparticles: Synthesis and biotechnological potentials. *Algae-Organisms for Imminent Biotechnology, 7*, 157–182. https://doi.org/10.5772/62909

Mukherjee, S., Dutta, S., Verma, A. S., & Saha, M. K. (2020). Nanoparticle synthesis harnessing benign green routes. In *Animal biotechnology* (pp. 689–708). Elsevier. https://doi.org/10.1016/B978-0-12-811710-1.00038-0

Mukherjee, A., Sarkar, D., & Sasmal, S. (2021). A review of green synthesis of metal nanoparticles using algae. *Frontiers in Microbiology, 12*, 693899. https://doi.org/10.3389/fmicb.2021.693899. Frontiers Media S.A.

Nath, B., & Dhananjay Kumar, T. (2017). *Prospects and challenges in algal biotechnology* (Vol. 10, pp. 1–326). Springer.

Naveenkumar, S., Kamaraj, C., Ragavendran, C., et al. (2023). Gracilaria corticata red seaweed mediate biosynthesis of silver nanoparticles: Larvicidal, neurotoxicity, molecular docking analysis, and ecofriendly approach. *Biomass Conversion and Biorefinery*. https://doi.org/10.1007/s13399-023-04026-x

Obeidi, N., Babolei, M. D., Khamisipour, G., Mohammadi, S. A., Farhangdost, F., & Valipor, S. (2023). The effect of Sargassum Angustifolium-derived AgNPs on apoptosis-associated genes in acute lymphoblastic leukemia cells. *Asian Pacific Journal of Cancer Prevention: APJCP, 24*(8), 2691–2696. https://doi.org/10.31557/APJCP.2023.24.8.2691

Osman, A. I., Zhang, Y., Farghali, M., Rashwan, A. K., Eltaweil, A. S., Abd El-Monaem, E. M., Mohamed, I. M. A., Badr, M. M., Ihara, I., Rooney, D. W., & Yap, P. S. (2024). Synthesis of green nanoparticles for energy, biomedical, environmental, agricultural, and food applications: A review. *Environmental Chemistry Letters, 22*(Issue 2), 841–887. https://doi.org/10.1007/s10311-023-01682-3. Springer Science and Business Media Deutschland GmbH.

Premarathna, K. S. D., Lau, S. Y., Chiong, T., Show, P.-L., Vithanage, M., & Lam, M. K. (2024). Greening up the fight against emerging contaminants: Algae-based nanoparticles for water remediation. *Clean Technologies and Environmental Policy*. https://doi.org/10.1007/s10098-024-02862-7

Pugazhendhi, A., Prabhu, R., Muruganantham, K., Shanmuganathan, R., & Natarajan, S. (2019). Anticancer, antimicrobial and photocatalytic activities of green synthesized magnesium oxide nanoparticles (MgONPs) using aqueous extract of Sargassum wightii. *Journal of Photochemistry and Photobiology B: Biology, 190*, 86–97. https://doi.org/10.1016/j.jphotobiol.2018.11.014

Rajivgandhi, G. N., Ramachandran, G., Kannan, M. R., Velanganni, A. A. J., Siddiqi, M. Z., Alharbi, N. S., Kadaikunnan, S., & Li, W.-J. (2022). Photocatalytic degradation and anti-cancer activity of biologically synthesized Ag NPs for inhibit the MCF-7 breast cancer cells. *Journal of King Saud University, Science, 34*(1), 101725. https://doi.org/10.1016/j.jksus.2021.101725

Sone, B. T., Makamu, E., Mohamed, H. E. A., Oputu, O., & Fester, V. (2021). Green-synthesized ZnO via Hyphaene thebaica fruit extracts: Structure & catalytic effect on the ozonation of Coralene Rubine-S2G azo disperse dye. *Environmental Nanotechnology, Monitoring & Management, 16*. https://doi.org/10.1016/j.enmm.2021.100515

Song, J., Xu, Z., Chen, Y., & Guo, J. (2023). Nanoparticles, an emerging control method for harmful algal blooms: Current technologies, challenges, and perspectives. *Nanomaterials, 13*(16), 2384. https://doi.org/10.3390/nano13162384

Subramaniyam, V., Subashchandrabose, S. R., Thavamani, P., Megharaj, M., Chen, Z., & Naidu, R. (2015). Chlorococcum sp. MM11—A novel phyco-nanofactory for the synthesis of iron nanoparticles. *Journal of Applied Phycology, 27*(5), 1861–1869. https://doi.org/10.1007/s10811-014-0492-2

Thovhogi, N., Park, E., Manikandan, E., Maaza, M., & Gurib-Fakim, A. (2016). Physical properties of CdO nanoparticles synthesized by green chemistry via Hibiscus Sabdariffa flower extract. *Journal of Alloys and Compounds, 655*, 314–320. https://doi.org/10.1016/j.jallcom.2015.09.063

Ying, S., Guan, Z., Ofoegbu, P. C., Clubb, P., Rico, C., He, F., & Hong, J. (2022). Green synthesis of nanoparticles: Current developments and limitations. In *Environmental technology and innovation* (Vol. 26). Elsevier B.V. https://doi.org/10.1016/j.eti.2022.102336

Chapter 4
Sustainable Production of Nanoparticles from Fungi and Their Agricultural Applications

C. R. Meera, Lyudmila Zaynitdinova, and Udensi Great

Abstract Adverse impacts on the ecosystem and human health due to the generation of toxic byproducts are major limitations of traditional methods such as physical and chemical nanoparticle (NP) synthesis. Hence, biosynthesis of nanoparticles has become a hot spot of research in the past decade as it offers an effective alternative to conventional synthesis methods. Green synthesis of nanoparticles from biogenic sources such as plants, bacteria, fungi, algae, and actinomycetes is more attractive due to their simplicity, cost-effectiveness, and ease of large-scale production. "Myconanotechnology" is an emerging field that utilizes fungal biomass for the production of nanoparticles. Diverse fungal strains are widely utilized to produce NPs as they form eco-friendly bio-factories capable of converting metal ions into their zerovalent or nanoforms. A variety of enzymes and proteins present in fungi act as reducing agents in this process. Fungi outperform bacteria and other sources in nanoparticle synthesis due to enzyme secretion, extensive mycelia surface area, and unique cell wall properties aiding metal ion absorption and reduction. Various fungal species including *Penicillium, Aspergillus, Alternaria, Neurospora, Fusarium, Chrysosporium, Cladosporium, Trichoderma, Candida, Cryptococcus, Pleurotus, Agaricus, Coriolus, Ganoderma, Verticillium* etc. are widely employed in nanoparticle synthesis. Myco-nanoparticles aid plant disease control with eco-friendly nano-insecticides, nano-fungicides, nano-pesticides, and nano-fertilizers, boosting agricultural productivity. This chapter evaluates the fungal nanoparticle synthesis, optimal production conditions, and their agricultural use for enhanced yields and sustainable farming.

C. R. Meera (✉)
Department of Microbiology, St Mary's College (Autonomous), Thrissur, Kerala, India

L. Zaynitdinova
Laboratory of Biodiversity of Microorganisms, Institute of Microbiology, Tashkent, Uzbekistan

U. Great
Molecular Biosciences (Micro Track), University of Kansas (KU), Lawrence, KS, USA

Y. K. Mohanta et al. (eds.), *Nano-microbiology for Sustainable Development*,
https://doi.org/10.1007/978-3-031-78845-1_4

Keywords Nanoparticles · Myconanotechnology · Nano-insecticides · Nano-fungicides · Nano-pesticides · Nano-fertilizers

4.1 Introduction

Fungi are eukaryotic heterotrophic organisms encompassing yeast, molds, and macrofungi, including mushrooms. Fungi comprise one of the most diverse groups of organisms on Earth, and the exact count of fungal species remains a topic of ongoing debate (Panda et al., 2024; Shamim et al., 2023). Existing literature suggests a range between 2.2 and 3.8 million species, yet only approximately 150,000 fungal species have been formally described to date. Moreover, about 5.1 million species of fungi have been identified through advanced sequencing methods. Regardless of the exact species count, the role of fungi in nutrient biogeochemical cycles within ecosystems is crucial, primarily as saprophytes. They break down complex organic molecules into simpler ones through the production of versatile proteins and extracellular enzymes. Many such enzymes can be utilized for rapid and sustainable nanoparticle (NP) production.

Nanomaterials find applications across multiple fields, including healthcare and medical fields, agriculture, science, electronics, engineering, and more. Metals or alloys forming nanocrystalline grains with particle sizes between 5 and 100 nm are referred to as nanometallic materials. The major advantage of nanometals is their greater surface area compared to their bulk or non-nano counterparts. Their unique physical and chemical properties, arising from the phenomenon of the small size, surface characteristics, interface properties, and quantum effects, contribute to their distinct characteristics absent in non-nano metals (Al-Musawi & Al-Saadi, 2021). Several techniques are utilized for nanomaterial synthesis, which include physical, chemical, and green synthesis. Physical techniques involve methods such as aerosol processing, ultraviolet radiation exposure, and thermal decomposition, typically conducted under high temperatures and pressures (Al-Harbi & Abd-Elrahman, 2024). Conversely, in chemical methods expensive and toxic reagents like sodium borohydride, along with organic solvents and dispersant stabilizers are used (Hussain et al., 2016). Hence, green synthesis approaches have gradually supplanted conventional physical and chemical methods, which often entail the consumption of significant energy, reliance on complex types of equipment, emission of harmful chemicals, and stringent synthesis conditions. Green synthesis harnesses natural sources including microorganisms like bacteria, algae, fungi, and extracts from various plants for synthesizing nanoparticles. These green materials serve as reducing agents or capping agents and dispersants. Green synthesis offers numerous advantages over physical and chemical methods. It is non-hazardous, cost-effective, pollution-free, eco-friendly, and inherently sustainable (Osman et al., 2024). The quality and properties of metal nanoparticles synthesized from green materials under appropriate temperatures, metal concentrations, and other optimal conditions can even surpass those synthesized by conventional methods. Consequently, much research is now focused on exploring the synthesis and beneficial properties of green nanoparticles.

4.2 Supremacy of Fungi as Nanoparticle Bio-factories

The biogenic synthesis of nanoparticles offers a straightforward, cost-effective, environmentally friendly, and sustainable approach, enhancing their biocompatibility for various applications. Various microorganisms including bacteria, microalgae, fungi, and yeast, have been harnessed for the green synthesis of nanoparticles. Among the microorganisms utilized in nanoparticle synthesis, fungi stand out due to their numerous advantages over other microbes (Fig. 4.1). Microorganisms considered for nanomaterial synthesis are expected to be generally recognized as safe (GRAS) category. Most of the fungi used in nanotechnology belong to this GRAS category. Fungi emerge as potent tools in nanotechnology owing to their ease of isolation and cultivation in laboratory settings. They can be cultured in large amounts in laboratories. Additionally, they offer a larger surface area for metal ion accumulation and nanoparticle synthesis compared to bacteria, attributed to their elongated and branched mycelial network (Adebayo et al., 2021). Low cost and high productivity are added advantages of fungi in the synthesis of nanoparticles. With their high metal ion tolerance as well as biological metal ion accumulation potential, fungi gained much importance in nanometal synthesis. The strong affinity for wall binding and intracellular metal absorption capability of fungi makes it

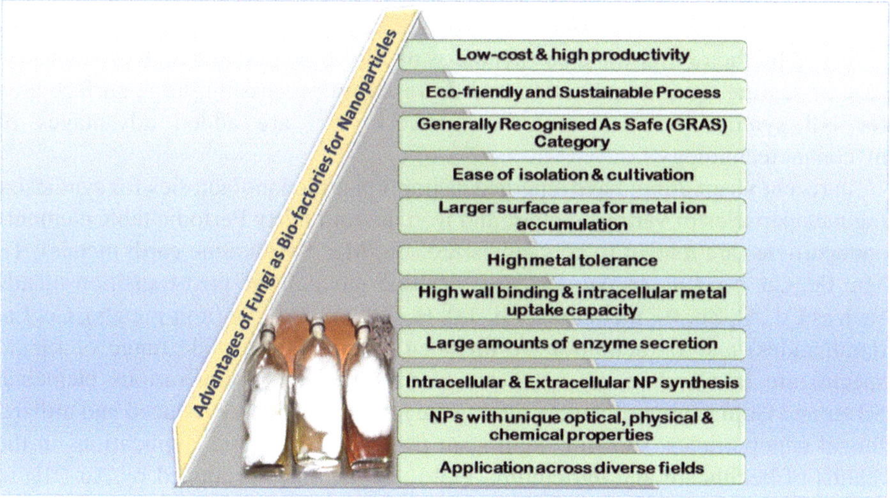

Fig. 4.1 Enhancing nano-production: advantages of fungi as bio-factories compared to other microorganisms

suitable for nanobiotechnological functions (Bourzama et al., 2021). Metal ion accumulation by fungi is through a range of physical, chemical, and biological processes, including extracellular polymers/metabolites binding, specific polypeptide binding, and metabolism-driven accumulation. Compared to bacteria and other microorganisms, fungi secrete huge amounts of enzymes that can hydrolyze metals, thus increasing the yield of nanoparticles. Fungi can produce nanoparticles as well as nanostructures composed of metals through intracellular and extracellular reducing enzymes, along with a biomimetic mineralization process (Mughal et al., 2021).

Nanoparticles derived from fungal biomasses showcase unique optical, physical, and chemical properties. These include high quantum yield, exceptional biocompatibility, superior photostability, and sufficient near-infrared (NIR) light-absorbing capacity. Improved biological properties, shelf life, and stability are added attributes of fungal NPs despite their eco-friendly nature. These features collectively contribute to the superior performance of fungi in the synthesis of nanoparticles, positioning them as favorable and making them highly versatile for applications across diverse fields including medicine, pharmaceuticals, therapeutics, agriculture, DNA sequencing, biosensors, molecular detection and diagnostics, catalysis, and food packaging.

4.3 Green Synthesis of Metal Nanoparticles from Fungi

Myconanotechnology, harnessing fungi for nanoparticle synthesis is progressing as a promising technology with wide-ranging applications beneficial to mankind. The process of designing and synthesizing nanoparticles using fungi is termed **mycofabrication**. Fungi secrete large amounts of proteins, mainly enzymes and other chemical compounds compared to other microorganisms which makes them an attractive biofactory for the green synthesis of NPs. Low cost and high productivity, ease of scaling up and cultivation and downstream processing, large surface area for NP synthesis due to mycelial network, etc. are added advantages of myconanotechnology.

In recent years, fungi have emerged as important bionanofactories for synthesizing nanoparticles of various metals, and from around thirty Periodic table elements nanoparticles are known to be synthesized. Ba, Mg, Sr (alkaline earth metals); Ti, Mn, Fe, Co, Ni, Cu, Zr, Ag, Pt, Au (transition metals), the post-transition metals such as Cd, Al, Zn, Bi, metalloids (Si, Te), H, C, N, O, P, S, Se (non-metals), Ce, La, (lanthanides) and U (actinide) are included in this list. A wide range of fungal species are involved in the synthesis of nanomaterials from various elements. Silver and Gold nanoparticles are among the most extensively produced and utilized fungal nanoparticles, showing significant potential for various applications in the realms of healthcare and agriculture. Gold nanoparticles produced by Au (III) to Au(0) reduction, nanoparticles of silver by Ag(I) to Ag(0) reduction, nanomaterials of palladium by Pd(IV, II) to Pd(0) reduction, selenium and tellurium metalloids through Se(IV, VI)/Te(IV, VI) to Se(0)/Te(0) reduction etc. are the most significant fungal nanoparticles (Li et al., 2022).

Top-down or bottom-up methods are common processes that are used in nanoparticle synthesis (Roy et al., 2021). The top-down techniques start with bulk materials that are subjected to compression bit by bit using different physical methods including milling, grinding, lithography, etching, thermal ablation, etc) and using some chemical methods for the generation of fine NPs. In bottom-up synthesis, also known as self-assembly, NPs assemble from minute components like atoms, molecules, and smaller particles. Green or Biosynthesis of NPs through biological agents including bacteria, viruses, fungi, yeast, algae, and plants, falls under the bottom-up process as the metals in their atomic form cluster with each other to form NPs with the aid of biological agents (Sharma et al., 2020). Fungi accumulate metal ions from their surroundings and convert them into their elemental form by using their enzymes. Based on the site of formation of these nanoparticles, the process could be classified as intracellular and extracellular NP synthesis. In intracellular synthesis, metal ions transported inside the microbial cell are converted to NPs by the intracellular enzymes. In extracellular NP synthesis, metal ions are entrapped on the cell surfaces and undergo reduction by enzymes to form the NPs (Saleh & Hassan, 2021).

4.3.1 Intracellular Method

In intracellular myconanoparticle synthesis, metal precursors are introduced into the fungal culture and the fungal biomass internalizes them subsequently. Since the NPs are inside the fungi during intracellular synthesis, extraction procedures viz., centrifugation, filtration, etc and other methods become essential to release the synthesized NPs from the fungal biomass (Badwaik, 2022).

Mukherjee et al. (2001) reported a novel fungi-induced green synthesis method for nanoparticles of Silver (Ag) by the fungus *Verticillium.* On exposure to Ag_+ ions, *Verticillium* biomass produced AgNPs of dimensions 25 ± 12 nm intracellularly. The study states that the electron microscopic analysis of fungal cells revealed the formation of AgNPs below the cell wall surface, possibly by the reduction of metal ions by the enzymes in the cell wall membrane. Sastry et al. (2003) isolated an acidophilic species of *Verticillium* from Texus plant and analyzed its ability to produce silver (Ag) and gold (Au) nanoparticles. The authors assert that AuNPs and AgNPs were synthesized within the fungal biomass, manifesting striking colors that were readily identifiable through visual observation. A pink to blue color was formed in fungal biomass by AuNPs and light yellow to brown by AgNPs which is claimed to be through the excitation of surface plasmon vibrations within the particles. Verma et al. (2010) studied the NP synthesis ability of *Aspergillus clavatus*, an endophytic fungus found in *Azadirachta indica.* Incubation of fungal mass with an aqueous solution of chloroaurate ions resulted in the formation of intracellular AuNPs. Intracellular nanoparticle synthesis, mainly using Ag and Au by many other fungi like *Aspergillus flavus* (Vigneshwaran et al., 2007), *Penicillium brevicompactum*

(Shaligram et al., 2009), *Neurospora crassa* (Castro et al., 2011), *Aspergillus niger, Ganoderma lucidium G. applanatum* (Aygün et al., 2020) have been reported by various researchers in the last few decades.

4.3.2 Extracellular Method

In the extracellular synthesis method, fungal culture filtrate or aqueous filtrate having macromolecules of fungi is treated with metal precursors for the production of NPs. The fungal biomolecules including the extracellular enzymes in the aqueous filtrate facilitate the conversion of metal ions to NPs. The extracellular approach is more widely used than the intracellular method as it is cost-effective, and the complicated downstream processing involved in the intracellular method is not required for the separation of NPs. Extracellular methods are widely used in extensive silver nanoparticle (AgNPs) production. Extracellular synthesis of AgNPs by the fungus *Cladosporium cladosporioides* produced NPs of 10–100 nm (Balaji et al., 2009). Various steps in the extracellular production of AgNPs include the production of fungal biomass in a suitable liquid culture media, filtration to remove the fungal biomass, and silver nitrate solution ($AgNO_3$) addition to the filtrate. The filtrate color is transformated to a dark brown color as the fungal biomolecules cause the reduction of silver ions (Ag+) to elemental silver (Ag°) which is directly visible and can be measured using UV spectroscopy (Fig. 4.2). Extensive research was carried out on the extracellular NPs production by fungi and few examples include AgNP formation by *Aspergillus terreus* (Mohmed et al., 2017), *Cladosporium cladosporioides* (Hulikere & Joshi, 2019), *Ganoderma lucidium* and *G. applanatum,* gold NPs by *Cladosporium cladosporioides* (Joshi et al., 2017), copper Nps by *Aspergillus niger* (Noor et al., 2020), etc.

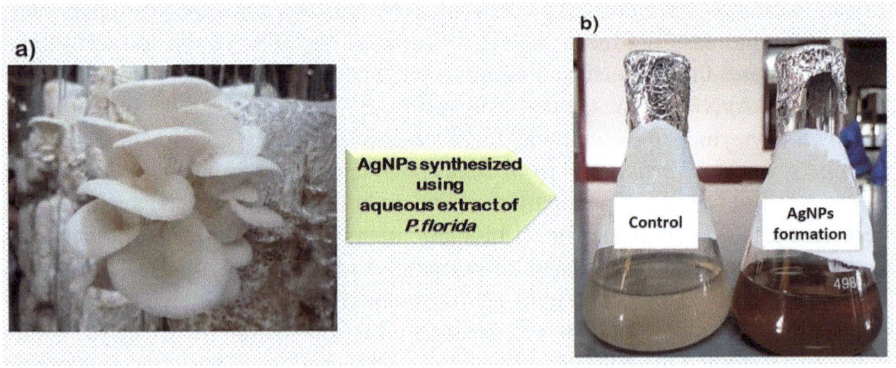

Fig. 4.2 Extracellular synthesis of silver nanoparticles using the aqueous extract of the mushroom *Pleurotus florida* (**a**) Fruiting body of *P. florida* (**b**) Dark brown color formation indicating the development of AgNPs due to the reduction of silver ions. (Reproduced from Radhakrishnan & Peter, 2020)

4.3.3 The Mechanism Behind Fungal Nanoparticle Formation

The exceedingly branched fungal mycelial network possesses a significant surface area relative to its mass, the gelatinous hydrated sheath surrounding the hyphae, a considerable quantity of metal-binding functional groups presents in the cell wall, along with extracellular polymeric substances (EPS). All these make the fungal network a suitable template for nucleation and nanoparticle formation.

Fungi show high metal tolerance developed as part of the evolutionary adaptation process; hence, their cell walls have high metal ion binding capacity. Fungi have an overall negative charge due to the plenty of sorption sites with functional groups –OH, –COOH, –O–CH_3, –SH, =NH, and –NH_2 groups on their cell surface. Hence, fungi attract the metal ions with a positive charge toward their surface. The cell wall enzymes and enzymes in the cytoplasm of fungi as well as the extracellular enzymes secreted by them facilitate the conversion of toxic metal ions into their NPs (Chatterjee et al., 2020). To defend against the chemical toxins in the surrounding environment, fungi employ many chemical reactions to convert the metal ions to their NPs by reduction. These reactions include immobilization, precipitation and coprecipitation, complexation, bio-coupling, biosorption, ion-form alteration etc. (Patel et al., 2021). In this reduction process, fungi use their intracellular and extracellular enzymes, proteinaceous molecules, as well as molecules bound to their cell membrane as electron donors to quickly transform the poisonous ions into their NPs. This process takes place either intracellularly or extracellularly.

In the intracellular conversion of metal ions into NPs, fungi commonly make use of reductase enzymes like cellular ATPases and hydrogenases that are present in the fungal cell wall. *Fusarium oxysporum* produced intracellular cytoplasmic vacuoles of Au-nanoparticles, using glyceraldehyde-3-phosphate dehydrogenase, plasma membrane-ATPase, and 3-glucan binding enzyme. Cytoplasmic hydrogen generated by hydrogenases are known to be involved in NP precipitation and metal ion detoxification (Riddin et al., 2009). Other intracellular proteins with significant redox and nucleophilic properties can also initiate the biological reduction of metal ions. Intracellular glutathione as well as proteins with metal-binding potential (metallothionein and phytochelatin) are utilized by the yeast cells in the reduction process.

Fungal cells also make use of their antioxidative defense system to protect themselves from the oxidative stress caused by the metal ions. Exposure of fungal cells to heavy metals leads to intracellular oxygen depletion which causes the generation of Reactive Oxygen Species (ROS). This process in turn leads to the generation of highly reactive and deleterious hydrogen free radicals (H·) within the cells. However, fungi employ their antioxidative enzymatic system to prevent such harmful reactions (Morano et al., 2012; Yaakoub et al., 2022). The free radical defense mechanisms present in fungi are catalase (CAT), glutathione (GSH), and glutathione peroxidases (GP_x), superoxide dismutase (SOD), transferases, methionine sulfoxide reductase (MSR), peroxiredoxins (Prxs), thioredoxins (Trx), etc.

Fungi also possess various membrane channels and transporter proteins to facilitate the transport of essential nutrients, micronutrients, and ions across the cell membrane (Ujam et al., 2021). Toxic metals present in the surroundings may also make use of these channels to enter the cytoplasm, and the fungal cells try to prevent the entry of these toxic metals by obstructing or even removing the membrane transport systems. However, some metal ions can reach the cytoplasm through various other channels, and under such a situation, the enzyme system of fungi is activated to precipitate the entered metal ions into their NPs, thus preventing the generation of harmful free radicals.

In the extracellular metal ion reduction as well as precipitation into NPs, the fungal cell membrane bound compounds like proteins, peptides, polysaccharides, quinones, oxidoreductases, etc., play a significant role. Nitrate and nitrite reductase, NAD(P)H-dependent reductase, sulfate, and sulfite reductases are the enzymes mainly participating in the metal ion reduction process. NADH-dependent reductase is the main protein responsible for the conversion of $Ag+$ to Ag^0. The extracellular secretion of sulfate reductase by *F. oxysporum* initiated the synthesis of ZnS, CdS, PbS, and MoS_2 nanomaterials in an aqueous solution. NADPH-dependent nitrate reductase with 214 kDa molecular weight isolated from the *F. oxysporum* culture supernatant produced AgNPs of ~50 nm size. Moreover, hydrogenases, nitrate reductases, and FAD-dependent glutathione reductases have emerged as key contributors to the extracellular synthesis of metal nanoparticles by fungi. Anthraquinones as well as quinine derivatives of naphthoquinones also act as redox centers in metal ion reduction leading to NP synthesis (Zhao et al., 2018).

Capping of proteins onto the NP surface is the last stage of NP synthesis in many fungi. Protein capping on inorganic mineral surfaces relies on various factors including the amino acid sequence, secondary structure, presence of metal cofactors, and the specific positioning of certain amino acids. This process can stop the nanocrystal growth further by binding to the NP surface. Hence, protein capping is considered as an important feature in the generation of NPs with controlled size and morphology during biomineralization.

In their 2003 study, Ahmad et al. investigated the silver nanoparticle (Ag^0) synthesis by *F. oxysporum*. Their findings revealed a diverse range of morphologies for the nanoparticles, which could exhibit either spherical or triangular shapes. Transmission Electron Microscopy (TEM) studies confirmed that even within aggregates, the NPs did not directly contact each other and were exceptionally stable due to the presence of proteins surrounding the nanoparticles, secreted by the fungus. Metal toxicity-induced production of certain extracellular compounds is also reported to be involved in fungi which facilitates nanoparticle synthesis. Cd (II) and Ni (II) toxicity are found to initiate the extracellular production of pullulan, a polysaccharide in *Aureobasidium pullulans*. Extracellular mucilage was secreted by *Curvularia lunata* on exposure to Cu (II), Zn (II) and Pb (II).

Along with fungi and mushrooms, significant attention from scientists is paid to yeasts for the biosynthesis of NPs (Mohanta et al., 2016). The ability of yeast cells to absorb and accumulate harmful metal ions from the environment, if present in higher concentrations, is a potential factor in nanoparticle formation (Bachheti et al., 2021). Literary data on the use of yeast in metal nanoparticle formation shows that the process is associated with various mechanisms in yeast cells. Detoxification mechanisms in yeast cells, including chelation, bioprecipitation, and intracellular sequestration help them to cope with the metal toxicity. Many researchers use these properties of yeast cells.

4.3.4 Optimal Conditions for Nanoparticle Synthesis and Methods of Characterization

Nanoparticles are synthesized using physical, chemical as well as biological methods (Green synthesis) (Fig. 4.3). Optimizing specific physiological parameters is crucial for the synthesis of nanoparticles with diverse and desirable physicochemical properties. These parameters encompass temperature, pH, pressure, light exposure, agitation, concentration of metal precursor, incubation duration, biomass quantity, and microbial source, among others. Fine-tuning these parameters leads to the creation of nanoparticles with precise size, morphology, and chemical composition. The speed of formation, size, and stability of NPs is affected by the temperature employed for the mycosynthesis. In a study, *Colletotrichum sp.,* an endophytic fungus showed a considerable increase in NP formation at elevated temperatures and the synthesis was finished in less than 20 min (Azmath et al., 2016). AbdelRahim et al. (2017) conducted a study employing Rhizopus *stolonifer* filtrate, which showed that no synthesis of NPs occurred at 80 or 10 °C, which they attributed to the enzyme and other components' inactivation or denaturation at higher and lower temperatures. Most of the studies have demonstrated that the higher temperatures faster the rate of synthesis. In addition to its effect on the kinetics of synthesis, temperature can also control the properties like stability and size of nanomaterials formed. In intracellular NP synthesis, the temperature selected is for the optimal growth conditions of the mycelia, whereas in the extracellular method, the temperature is adjusted for the greatest yield of nanoparticles with extracellular metabolites (Sebesta et al., 2022).

pH also plays an important role in NP synthesis as it can control certain features like intensity of dispersion and stability. Each type of nanoparticle has its optimal temperature. In general, most of the research has reported an increase in the rate of synthesis and stability of NP with an increase in pH as the metal ions and protons compete with negatively charged regions for bond formation. Azmath et al. (2016) reported the completion of the synthesis of AgNP by *Colleotrichum sp.* in 20 min when the temperature was higher and pH was alkaline. An investigation by Balakumaran et al. (2015) with the fungus *Guignardia mangiferae* showed the successful synthesis of AgNPs at neutral and alkaline pH. Color change due to the formation of AgNPs was not observed at a pH between 1 and 4, whereas, change in color started at a pH of 5 and 6, and NPs with stronger monodispersion and stability were developed as the pH was increased to 7.

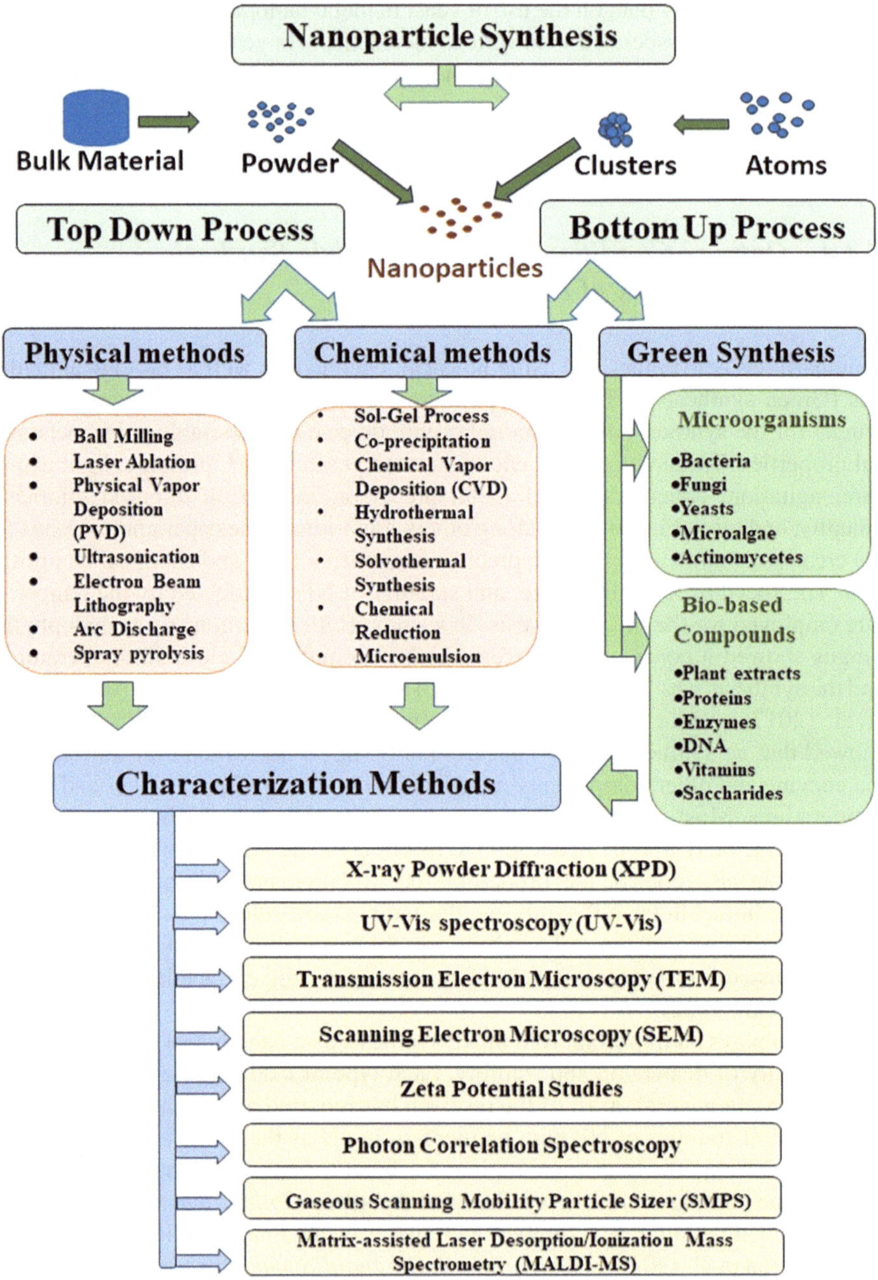

Fig. 4.3 Different approaches in the synthesis of nanoparticles and characterization

Similarly, the amount of fungal biomass and metal precursors used in the NP synthesis also influence the formation and nanoparticle properties. Research by Elamawi et al. (2018), with the fungi *Trichoderma longibrachiatum* revealed that lower biomass concentration leads to increased nanoparticle formation, whereas a higher biomass amount promotes the rate of synthesis of nanomaterials. In many investigations on the extracellular synthesis of NPs by fungi, a metal precursor concentration of 1 mM is suggested (Diko et al., 2020). Smaller nanoparticles with better dispersion were formed under a low metal precursor concentration, whereas a high metal precursor concentration produced very large nanoparticles with uneven shapes. AgNPs produced under high metal precursor levels using *Aspergillus fumigatus BTCB10* were large and with uneven shapes which was attributed to the competition among the metal ions and functional groups of the fungal filtrate, along with the high metal toxicity (Shahzad et al., 2019). Selection of the right metal precursor with the appropriate state and coordination is also essential as the precursor transformation may take place when reacting with media components used for nanoparticle synthesis.

The study of size, chemical composition, and other physical properties of the biogenic molecules is essential to determine their application in various fields. Various analytical methods like UV-Vis spectroscopy (UV-Vis), Transmission Electron Microscopy (TEM), Scanning Electron Microscopy (SEM), zeta potential analysis, X-ray Powder Diffraction (XPD) etc. are usually employed to study the properties of NPs. Electron microscopic techniques are key to determining the particle and cluster size, shape, and surface morphology of NPs. In addition to TEM and SEM, gaseous Scanning Mobility Particle Sizer (SMPS) as well as photon correlation spectroscopy are also popular and provide faster and more accurate measures than other methods (Michael et al., 2022) (Fig. 4.3).

The chemical composition of NPs is another important factor that determines the efficacy of biogenic molecules. Secondary reactions of NPs lead to the accumulation of unwanted components which reduce the efficiency of NPs and also hamper the overall manufacturing process. The extensively used methods for determining the composition include X-ray photoelectron spectroscopy (XPS), ion chromatography, atomic emission spectroscopy, etc. It is also important to determine the surface charge of the nanoparticle as it decides the interaction of NP with the precursor. To evaluate the surface charge of NPs when in solution and to forecast their stability Zeta potentiometer is usually employed (Sebesta et al., 2022).

4.4 Metal-Based Nanoparticles from Fungi

Fungi are the prime organisms on the earth that synthesize nanoparticles in a safe and eco-friendly manner. The major mechanisms of NP synthesis by fungi is summarized in Fig. 4.4. Fungal synthesis of NPs is frequently favored, because of the numerous characteristics seen during the process of soil formation also entail the biogenic creation of fungal nanoparticles. Comparing fungi-mediated nanoparticle

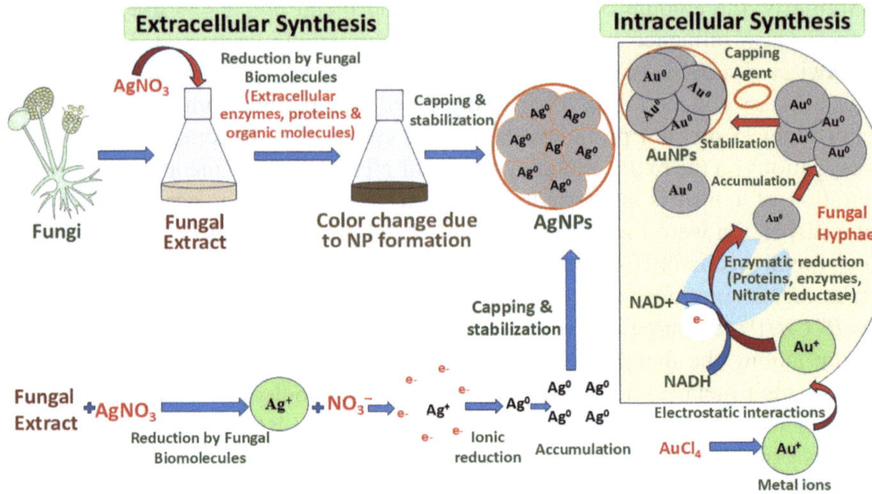

Fig. 4.4 Extracellular and intracellular nanoparticle synthesis by fungi. In the extracellular method metal precursor reduction is caused by biomolecules in the fungal extract whereas in the intracellular synthesis, fungal enzymes within the cell are involved

production to other microorganisms such as bacteria and actinomycetes reveals significant benefits. Among the many fantastic benefits provided by fungus are its ease of handling and downstream processing, economic viability, and ecological friendliness when it comes to covering a wide area (Adebayo et al., 2021).

Several fungal species have been exploited for the production of nanoparticles from metal ions, *Aspergillus, Penicillium,* and *Neurospora* being the most widely used among them. Filamentous fungi are considered more suitable for NP synthesis because of reasons like the large surface area of mycelia for interaction with metal ions, secretion of more amounts of enzymes and secondary metabolites, and ease of handling and cultivation in laboratory conditions. Despite being less exploited, endophytic fungi have the potential to serve as an excellent source of biogenic nanoparticles. Endophytic fungi, which form mutualistic relationships with plants, are known to inhabit various parts of plants. Endophytic fungi can be readily isolated from different plant parts and harnessed as nano-factories due to their capabilities in metal uptake, tolerance, and accumulation. Fungal endophytes have been documented to produce nanoparticles of silver, copper, zinc, magnesium, gold, etc. Some examples of endophytic fungi used in NP synthesis include *Guignardia mangiferae, Epicoccum nigrum, Aspergillus clavatus, Colletotrichum sp.,* and *Periconium sp.* (Verma et al., 2010; Balakumaran et al., 2015; Azmath et al., 2016; Ganesan et al., 2020)

Macrofungi including edible and medicinal mushrooms also act as eco-friendly biofactories of NPs as they secrete enzymes to reduce metal ions to their zero-valent form (Mohanta et al., 2018). Reports suggest that reductases and flavins in them play a major role in NP synthesis even though the exact mechanism is not fully known. Other phytomolecules of mushrooms like phenolic acids, tannins, alkaloids, terpenoids etc., and biomolecules like proteins, essential amino acids, and polysaccharides are also involved in NP formation. Mushroom biomolecules including proteins and polysaccharides secreted into the growth medium also act as reducing and capping agents that stabilize the NPs (Nayak et al., 2024). Thus, intracellular and extracellular production of NP synthesis is possible with mushrooms. Water extract from the edible mushroom *Volvariella volvacea* which contains reducing and capping agents has been reported to synthesize Au and AgNPs (Philip, 2009). Both metal (Au, Ag, Fe, Pt etc.) and non-metal (Se, CdS etc.) nanoparticles have been synthesized from a variety of mushrooms like *Volvoriella volvaceae, Ganoderma lucidum, Agaricus bisporus, P. florida, Pleurotus sajor caju, Lentinula edodes, Agrocybe aegerita, Calocybe sp.* and others (Kalia & Kaur, 2018). We go over a few examples of metal-based nanoparticles made by fungus in the sections that follow.

4.4.1 Silver Nanoparticles

Silver nanoparticles (AgNPs) have been identified for use in various applications over the years due to advancements in nanomaterials research. These include the creation of biosensors, water decontamination, and food packaging designed to ward against microorganisms (Munir et al., 2021). The oxidation reaction that leads to the reduction of Ag^+ by contact with a reducing agent and the subsequent transformation into neutral atoms (Ag^0) is the fundamental process that forms AgNPs. This was mediated using a variety of techniques, including biological techniques, and several reducing agents (Crisan et al., 2021). Silver nanoparticle production frequently uses microorganisms, such as filamentous fungus. Depending on the species involved in the synthesis process, the synthesis mediated by these microorganisms might take place in the extracellular or intracellular medium (Santos et al., 2021). Because the microorganisms must come into close touch with the silver ions that will be absorbed and digested within the cells, intracellular synthesis involves the production of nanoparticles inside the cells. Because the resultant nanoparticles stick to the fungus' surface, separation techniques—which are like those used in intracellular synthesis—must be used to extract the nanoparticles. This can result in additional expenses throughout the material processing process (Santos et al., 2021).

The process by which bacteria and fungi produce silver nanoparticles differs according to the type of microbe or strain that is employed. Some bacteria and fungi, including *Fusarium oxysporum*, facilitate the extracellular formation of AgNPs by the creation and release of NADH-dependent reductases; not all fungi employ this enzyme. *Schizophyllum radiatum*, a white rot pathogen, was found to be capable of synthesizing stable silver nanoparticles measuring 10–40 nm. These nanoparticles demonstrated significant antibacterial activity against various pathogenic bacterial cultures (Metuku et al., 2014). AgNPs with a size of 2–10 nm, produced by the yeast *Candida lusitaniae* isolated from the termite intestine proved to be a hopeful substitute for antibiotics as it prevented the growth of microorganisms *Klebsiella pneumoniae* and *Staphylococcus aureus.* A marine strain of *Yarrowia lipolytica* was used to synthesize AgNPs in a cell-associated manner and the AgNPs were active against *Salmonella paratyphi*. The study states that the property of NP synthesis could be attributed to the presence of melanin in these yeast cells (Apte et al., 2013). Elahian et al. using genetically modified yeast *Pichia pastoris* with an overexpressed gene responsible for metal resistance, cytochrome B5 reductase enzyme from *Mucor racemosus*, achieved successful metal ion transformation into NPs. The particular enzyme induced the production of stable AgNPs with particle sizes of 70–180 nm and with well-dispersion properties (Elahian et al., 2017). Extracellular synthesis of AgNPs having an average size of 45 nm was produced by the non-edible medicinal mushroom *Ganoderma lucidum* (Karwa et al., 2011). AgNps produced using aqueous mycelial extract from *G.lucidum* by the reduction of $AgNO_3$ were able to inhibit pathogens including *Staphylococcus aureus*, *Escherichia coli*, *Pseudomonas aeruginosa*, *Candida albicans*, and *Candida parapsilosis* (Constantin et al., 2023). *Pleurotus florida*, an edible oyster mushroom also showed the potential for extracellular synthesis of AgNPs by photoirradiation (Bhat et al., 2011).

The primary benefit of using biological synthesis to create silver nanoparticles is that it is environmentally friendly. Because this synthesis technique uses harmless or biodegradable materials, there is less chance of environmental harm (Dhaka et al., 2023). The potential for differences in the amounts of the biomolecules found in the biological source (plants or microbes) because of alterations in the environment or the availability of nutrients is a drawback of biological synthesis.

4.4.2 Gold Nanoparticles

A large range of fungi have been used recently to synthesize gold nanoparticles. Filamentous fungi have a strong ability for bioaccumulation and metal tolerance. Therefore, they have a distinct edge over other microorganisms in the creation of gold nanoparticles (Dinesh et al., 2023). The fungal microorganisms' bio-active chemicals can regulate physical properties like size and shape, also the biochemical makeup NPs. In fungi, the bioactive substances responsible for metabolic energy are proteins such as glyceraldehyde-3-phosphate dehydrogenase, ATPase, glucan-binding proteins and reducing sugars (Saravanan et al., 2021). Several authors have described the active involvement of fungi in the synthesis and production of AuNPs. According to Rabeea et al. (2020), incubation of hot water extract of the mushroom *Flammulina velutipes* with chloroauric acid initiated the nanoparticle synthesis in 30 min which was verified by the color change to deep violet. Also, *Fusarium oxysporum*, a phytopathogen fungus, is responsible for producing gold nanoparticles with an ideal size of 20 nm and a spherical or hexagonal shape. It has been also reported that combining Tetrachloroauric acid trihydrate ($HAuCl_4.3H_2O$) and cell-free extract prepared using a mesophilic fungus in an acetate buffer environment aided in the creation of AuNPs that remained stable for many days (Saravanan et al., 2021). Bhargava et al. conducted a detailed study on various parameters like pH, salt concentration, reaction time, etc. that control properties like size, yield, etc of AuNPs using *Cladosporium oxysporum* AJP03 (Bhargava et al., 2016). Mishra and colleagues studied the role of parameters like incubation time, temperature etc, in controlling the extracellular development of AuNPs using cultures of *Trichoderma viride* and *Hypocrea lixii*. It was found that AuNP production by *Trichoderma viride* culture occurred within 10 min at 30 °C, and they had the properties of a biocatalyst and a strong antimicrobial agent (Mishra et al., 2014). A recent research reported the production of AuNPs by the fungi *Candida rugopelliculosa* isolated from aerobic granular sludge, opening a sustainable solution for treatment of residual sludge (Zhao et al., 2024)

4.4.3 Other Nanoparticles

Fungal nanomaterials using various metal ions have been widely reported with beneficial applications in medical, agricultural, and industrial zones. Zinc nanoparticle synthesis is known to be mediated by an endophytic fungus including *Periconium* sp., which was isolated from the leaves of *Balanites aegyptiaca*. After incubating the fungus in Potato Dextrose Broth (PDA) for 21 days, the mycelial mass was separated and dried for 12 h at 60 °C. Dried fungal biomass was used to create the aqueous extract. The zinc nitrate combination was then added to the extract, and the mixture was allowed to evaporate to produce a sol. Water molecule slowly evaporated from the sol to produce the gel. Zinc oxide (ZnO) nanoparticles were formed under aerobic conditions by calcining dried gel at 700 °C for 4 h (Ganesan et al., 2020). A similar procedure was used when synthesizing copper oxide nanoparticles from indigenous fungus. Copper nitrate solution was added to *Trichoderma asperellum* aqueous mycelial free water extract to facilitate the formation of copper oxide nanoparticles. The idea at play is that copper hydroxide is created when a reaction occurs between copper and the hydroxyl anion of water. Copper hydroxide is then transformed into copper oxide nanoparticles (CuONPs) by the addition of other substances, enzymes, and proteins (Saravanakumar et al., 2019). An aqueous extract of *Saccharomyces cerevisiae* was used to synthesize 32 nm palladium nanoparticles. Nanoparticles contributed to the photocatalytic destruction of textile dye in ultraviolet up to 98% in 60 min (Sriramulu & Sumathi, 2018). Cerium oxide (CeO_2) nanoparticles were synthesized using *Aspergillus niger* culture filtrate by a very simple procedure. The functional groups present on the molecules in the culture filtrate induced the formation of NPs (Gopinath et al., 2015). Bioprecipitation of metal ions including Ni^{2+}, Co^{2+}, Cu^{2+}, Fe^{2+}, Zn^{2+}, and Mn^{2+} were achieved by the fungus *Neurospora crassa* in a medium containing urea. The results indicated that *N. crassa* could cause precipitation of these metals either as nano carbonates or oxides (Li et al., 2020; Liu et al., 2021). *Trichoderma citrinoviride* extract has been reported to act as a reducing agent in the formation of Titanium dioxide (TiO_2) nanoparticles. The process resulted in the generation of stable, polymorphic, and crystalline NPs ranging in dimension from 10 to 400 nm (Arya et al., 2021). A study using *Colletotrichum sp.* reported the formation of Al_2O_3 nanoparticles which were subsequently functionalized by essential oils derived from *Eucalyptus globulus* and *Citrus medica* which enhanced its antibacterial and antifungal properties (Suryavanshi et al., 2017). The marine yeast *Rhodosporidium diobovatum* showed added advantage over terrestrial yeasts in the production of NPs from the metal ions of Lead (Pb). Usually, the addition of metal ions is preferred in the mid-log phase growth of yeast culture for NP synthesis. *R. diobovatum* could produce PbSNPs even during the simultaneous addition of Pb^{2+} to the culture media which eliminated the necessity of growth phase monitoring (Seshadri et al., 2011).

4.5 Agricultural Applications of Fungal Nanoparticles

Use of a "green" approach makes it possible to obtain particles with more precise sizes and shapes than using physicochemical methods. Nanotechnology is rapidly advancing and making significant strides in the agricultural industry. One of its primary uses lies in plant protection, where nanoparticles act as potent antimicrobial agents. Additionally, they serve as effective tools in combating fungi, pests, and insects, thus functioning as nanofungicides, nanopesticides, and nanoinsecticides. Moreover, nanotechnology contributes to enhancing soil fertility and crop nutrition through the development of nanofertilizers. Furthermore, nanomaterials play a crucial role in the creation of nano biosensors, enabling the detection of various substances such as fertilizers, herbicides, pesticides, pathogens, moisture levels, and soil pH. The "green" method of producing metal nanoparticles is not without certain disadvantages, however, its environmental friendliness and energy efficiency, as well as the biocompatibility of the resulting nanoparticles, prevail in most cases (Madani et al., 2022). Some examples of myconanoparticles and their application in agriculture are summarized in Table 4.1.

4.5.1 Fungal Nanoparticles in Crop Improvement & Plant Disease Management

The great potential of nanoparticles can lead to various effects on living systems, including plants. Numerous studies have established that silver NPs are biologically active substances that affect the growth and development of plants and enhance seed germination. AgNPs increase growth (seedlings and length of roots and leaves) and biochemical parameters (chlorophyll, carbohydrates and proteins, content of antioxidant enzymes) in beans and corn. AgNPs can have both positive and negative effects on root elongation depending on the plant species. For example, root length increases in barley seedlings, but is suppressed in lettuce (Gruyer et al., 2013). Silver nanoparticles are regarded as a suitable candidate for increasing crop yields

Table 4.1 Some examples of myconanoparticles and their application in agriculture

Fungi	NP synthesized	Application in agriculture	References
Alternaria alternata	Silver nanoparticles (AgNPs)	Active against the fungal plant pathogens (*Phoma glomerata, Phoma herbarum, Fusarium semitectum)* in combination with Fluconazole	Gajbhiye et al. (2009)
Aspergillus niger	Gold nanoparticles (AuNPs)	Gold microwires as diagnostic and monitoring tools for chemicals and pathogens in agriculture Biosensors	Rehman et al. (2011) and Sadaf et al. (2024)
Aspergillus versicolor	Silver nanoparticles (AgNPs)	Active against White rot and Gray mould, commercially damaging diseases of strawberry plants	Elgorban et al. (2016)
Epicoccum nigrum	Silver nanoparticles (AgNPs)	Active against the plant pathogens like *Fusarium solani, Aspergillus flavus*	Qian et al. (2013)
Fusarium solani	Silver nanoparticles (AgNPs)	Use in textile fabrics and as an antifungal agent Active against different fungal contaminants of wheat, maize, and barley seeds	Ingle et al. (2009) and Abd El-Aziz et al. (2015)
Fusarium oxysporum	Silver nanoparticles (AgNPs)	Useful in the development of biosensors	Rajput et al. (2016)
Fusarium solani IOR 825	Silver nanoparticles (AgNPs) Zinc oxide nanoparticles (ZnONPs)	Active against plant pathogens like *Alternaria alternata, Fusarium culmorum, Fusarium oxysporum, Phoma lingam,* and *Sclerotinia sclerotiorum* Active against plant pathogens and stimulatory effect on seedlings' growth	Trzcińska-Wencel et al. (2023a)
Fusarium culmorum strain JTW1	Silver nanoparticles (AgNPs)	Inhibited fungal growth and spore germination of phytopathogens like *Botrytis cinerea, Phoma lingam,* and *Sclerotinia sclerotiorum*	Trzcińska-Wencel et al. (2023b)
Guignardia mangiferae	Silver nanoparticles (AgNPs)	Active towards the fungal phytopathogens like *Colletotrichum sp., Rhizoctonia solani*, and *Curvularia lunata* and phytopathogenic fungi causing grain contamination	Balakumaran et al. (2015)

(continued)

Table 4.1 (continued)

Fungi	NP synthesized	Application in agriculture	References
Grifola umbellata	Selenium nanocomposites (SeNCs)	Strong positive effect on the number of leaves, plant height, and germ length in potato tubers	Tsivileva and Perfileva (2022)
Ganoderma lucidum SIE1303	Selenium nanocomposites (SeNCs)	Increase in the potato vegetative biomass, active against potato ring rot causative agent *Clavibacter sepedonicus*	Tsivileva and Perfileva (2022)
Pleurotus ostreatus	Selenium nanocomposites (SeNCs)	Increase in the potato root biomass	Tsivileva and Perfileva (2022)
Phanerochaete chrysosporium	Cadmium sulfide (CdS) quantum dots	Detection of even traces of chemical herbicides and pesticides affecting soil quality and productivity	Chen et al. (2014)
Rhizopus oryzae	Gold nanoparticles (AuNPs)	Nanopesticides	Das et al. (2009)
Trichoderma viride	Silver nanoparticles (AgNPs)	Antibacterial activity, Useful in vegetable and fruit preservation Biosensor and bio-imaging	Fayaz et al. (2009, 2010)
Trichoderma harzianum	Silver, Copper oxide and Zinc oxide nanoparticles (AgNPs, CuONPs and ZnONPs)	Antifungal potential against phytopathogens	Consolo et al. (2020)

as they stimulate the growth of plants to a significant level and the germination of their seeds. AgNP concentrations influence plant growth positively or negatively. In this regard, it is believed that it is necessary to experimentally select the concentrations of nanoparticles to provide the desired effect (Zaynitdinova et al., 2022). There is data available regarding the impact of silver nanoparticles, sized at 20 nm, on fenugreek seeds (*Trigonella foenum-graecum*) (Sadak, 2019). Various concentrations of silver nanoparticles ranging between 10 and 40 μg mL^{-1} were employed, revealing that the highest seed germination & germination rate, root length, and fresh and dry root mass were observed at a concentration of 10 μg/ml.

Exposing plants to specific concentrations of silver nanoparticles may enhance plant growth in comparison to untreated plants, whereas both higher and lower concentrations could potentially hinder plant growth (Matras et al., 2023). The growth response was observed when using different concentrations of nanoparticles (0, 25, 50, 100, 200 and 400 ppm) on *Brassica juncea*, showing that the 50-ppm treatment optimal for plant weight, shoot and root lengths (Amerasan et al., 2016). Additionally, silver nanoparticles at a concentration of 50 ppm indirectly stimulated plant growth and increased root nodule formation in cowpeas (Amooaghaie et al., 2015). Results suggest the possibility of applying AgNPs in agriculture to boost germination potential, germination index and vigor index of seeds, fresh weight of seedlings, and reduced time of seed germination.

In agriculture, the most promising areas of application of mycogenic nanoparticles include the development of nano-preparations to increase crop yields and early detection of plant diseases, the creation of intelligent farming systems, and the use of biomimetic models to improve the efficiency of agricultural production. According to research, mycogenic nanoparticles of silver, gold, selenium, copper, titanium oxide, zinc oxide, and graphene may be the most promising for use in medicine, industry, and agriculture (Bhardwaj et al., 2020). Quantum dots (QD), the NPs with attractive physiochemical properties are gaining much importance nowadays in agriculture due to their possibilities in precise and ecofriendly farming. The photoluminescence property of QDs enables the detection of even traces of chemical herbicides and pesticides which remain undegraded for a long time affecting soil quality and productivity. The white rot fungus *Phanerochaete chrysosporium* showed the potential of extracellular biosynthesis of Cadmium sulfide (CdS) quantum dots having an average size of 2.6 nm (Chen et al., 2014).

Production of chitosan from fungi has gained much importance in recent years due to the unique biological properties of chitosan nanoparticles (ChNPs) in the field of agriculture. Chitosan, a β-1,4-linked glucosamine polymer, is a naturally occurring in crustacean shell as well as in fungal cell wall. Chitosan formed by the enzymatic deacetylation of chitin can be seen in the mycelia, cell walls, spores and stalks of some fungi coming under Ascomycetes, Basidiomycetes, and Zygomycetes. Zygomycetous fungi like Rhizopus oryzae, Absidia coerulea, Mucor racemosus, Benjaminiella poitrasii, Mucor rouxii, Cunninghamella elegans and Gongrenella butleri have reported chitosan production. Some edible mushrooms (basidiomycetous fungi) like *Pleurotus sajor-caju, Lentinula edodes* and *Agaricus bisporus* have been reported to produce chitosan. *Penicillium chrysogenum, Aspergillus niger, Saccharomyces cerevisiae* and other wine yeasts are also explored for the production of chitosan (Ghormade et al., 2017). ChNPs are widely used in agriculture for plant growth promotion, induction of immune response in plants, controlled release of fertilizers and pesticides, and precision farming. ChNPs are reported to be effectively inhibiting the growth of *Fusarium solani, F. oxysporum,* and *Alternaria terrens* (Sahab et al., 2015). Xing et al. (2017) reported the ability of ChNPs to inhibit the growth and seed germination of *Verticilium dahlia* which causes wilt disease in many plant species. They are known to promote plant growth and plant systemic resistance as in the case of infections like stem black rot and wilting of branches.

The nano biosensors offer a versatile and efficient means of monitoring agricultural conditions and optimizing resource management for improved crop yield and quality. A study using different strains of *Fusarium oxysporum* for the synthesis of silver nanoparticles demonstrated the effect of isolate choice, temperature and pH on the morphology of nanoparticles (Rajput et al., 2016). The study revealed the usefulness of various strains of *F. oxysporum* in the development of biosensors

A recent study revealed the ability of *A.niger* to act as a living template for Au-NP accumulation that could be used for biosensor development (Sadaf et al., 2024). The early detection of plant pathogens using nanostructured materials like fluorescent probes, nanowires, etc. synthesized by fungi is another revolution in agriculture that aids in crop protection and precision farming. Filamentous fungi including *Aspergillus nidulans*, *Aspergillus niger*, and *Neurospora crassa* were used to synthesize Au nanostructures (1–2 μm) by self-assemblance of colloidal gold on living fungal hyphae (Sugunan et al., 2007). Synthesis of porous gold microwires using fungal hyphae of *A. niger* was reported by Rehman et al., 2011. These nanowires could be harnessed for the development of diagnostic and monitoring tools for chemicals and pathogens in the agriculture field.

4.5.2 Fungal Nanoparticles as Insecticides, Fungicides & Pesticides

Research into controlling plant diseases in fruit plants, fruits, food crops, etc. holds significant economic importance and is imperative. Evenly dispersed and stable colloidal AgNPs have greater adhesion to the surface of bacterial and fungal cells; therefore, they act as the best bactericides and fungicides. To date, there have already been great successes in the fight against plant diseases using silver nanoparticles, including those synthesized by microscopic fungi. Therefore, DNA was immersed in a solution containing silver nanoparticles developed on graphene oxide, and the antibacterial effectiveness against *Xanthomonas perforans*, the pathogen responsible for bacterial spot disease in tomatoes, was investigated (Gupta et al., 2023).

Silver nanoparticles inhibited the growth of fungi *P. Concavum* and *B. Cinerea* with an inhibition potential of 28% and 65.4% respectively at a concentration of 0.01%. The mode of action of NPs against fungi remains incompletely understood, with some suggesting that silver nanoparticles create pores on the fungal cell membrane, leading to cell death and inhibition of the budding process. *B. cinerea* spore germination was inhibited completely by 0.01% AgNP solution. Copper nanoparticles were ineffective against fungi—they had virtually no effect on the growth of spores and fungi (Bayat et al., 2021). The antifungal potential of AgNPs, produced through the action of *Trichoderma longibrachiatum,* has been demonstrated against nine fungal strains, including *Aspergillus.* AgNPs synthesized by the fungus *Aspergillus versicolor* was observed to be active against the pathogenic fungi *Sclerotinia sclerotiorum* and *Botrytis cinerea* causing white rot and gray mould, commercially damaging diseases of strawberry plants (Elgorban et al., 2016). Qian et al. (2013) reported the antifungal activity of AgNPs synthesized by the endophytic fungus E*picoccum nigrum* against the plant pathogens like *Fusarium solani, Aspergillus flavus.* AgNPs developed using another endophytic fungus *Guignardia mangiferae* also demonstrated significant inhibitory activity towards the fungal phytopathogens like fungi *Colletotrichum sp., Rhizoctonia solani*, and *Curvularia lunata* (Balakumaran et al., 2015). AgNPs also could destroy the phytopathogenic fungi causing grain contamination. AgNPs produced by the pathogenic fungus *Fusarium solani,* isolated from wheat showed profound activity against different fungi contaminating wheat, maize, and barley seeds (Abd El-Aziz et al., 2015).

Combining biogenic nanoparticles with conventional biocides is known to augment their biological activities. NPs from the fungus *Alternaria alternata* when used in combination with fluconazole, a common antifungal agent in use, showed more activity against the fungal plant pathogens *Phoma glomerata, Phoma herbarum, Fusarium semitectum* (Gajbhiye et al., 2009).

Promising results were obtained for antifungal activity of silver nanoparticles against pathogenic fungi viz. *Fusarium oxysporum*, *Alternaria alternata* and *Aspergillus flavus*. The antifungal effectiveness of colloidal nanosilver (average diameter 1.5 nm) against rose powdery mildew caused by *Sphaerotheca pannosa var rosae* was studied. This disease affects both cultivated roses in greenhouses and wild roses in natural settings. Deformed and curled leaves, early defoliation, less flower production, etc. are the major indications of the disease. It is also noted that, in addition, silver exhibits excellent properties as a plant growth stimulant. Bacterial diseases are another cause of significant crop loss worldwide. It has been proven that silver nanoparticles are active against phytopathogenic bacteria. They were found to have higher antibacterial activity against *Erwinia cartovora* than antibiotics (Ingle et al., 2023).

4.5.3 Fungal Nanoparticles as Fertilizers

Currently, there is practically no area of agriculture left where the use of nanoparticles has not been proposed. Nanoparticles are offered for use as a means of preventing and treating plant diseases, a means of protecting and stimulating the development of plants, fertilizers, and stimulants of biochemical processes in the soil. The application of nanotechnology in the field of agriculture offers a promising alternative to the conventional use of toxic chemical fertilizers, insecticides, pesticides, etc. In agriculture, NPs offer several advantages over traditional methods, including a high ratio of surface area to volume, enhanced mass transfer capabilities, and the ability to deliver nutrients or pesticides in lower concentrations in a slow, controlled, and targeted manner to boost crop productivity, while proper use (Hussain et al., 2023). Even though the application of NPs in agriculture is a revolutionary event, the implication of nano-based addictives on microorganisms associated with crops and soil remains unclear. Similar to chemical fertilizers or pesticides, agricultural additives based on nanomaterials can affect microorganisms associated with plants. The effects of nanoparticles on plant microbes could be either beneficial or deleterious, depending on a variety of factors (Mgadi et al., 2023).

Scientists pay special attention to the use of metal nanoparticles in stimulating plant growth, development, and productivity. A team of Indian scientists published the results of the use of zinc as a nanofertilizer in the cultivation of American pinnasetum (*Pennisetum americanum*) (Tarafdar et al., 2014). As a result of the application of nanofertilizer, 6-week-old plants showed a noticeable increase in shoot length (15%), root length (4%), root surface area (24%), chlorophyll content (24%), total leaf protein (38%), dry matter plant weight (12%) compared to the control. Overall, the use of zinc nanoparticles increased grain yield by 37%. Singh et al. studied the effects of ZnONPs on the growth and seed germination of the tomato plant (*Solanum lycopersicum*) and found that compared with the control and the use of a $ZnSO_4$ solution, "green" nanoparticles at low concentrations had a positive effect on seedling vigor, pigment, protein, and sugar (Singh et al., 2016).

A study conducted by Ukrainian scientists on the effect of a mixture of nanoparticles (manganese, copper, iron, zinc and silver) on the growth and development of soybeans allowed them to recommend a colloidal solution of a mixture of nanoparticles for pre-sowing seed treatment and foliar feeding during the budding phase when cultivating early ripening soybean varieties on the chernozems of forest-steppe Ukraine (Batsmanova et al., 2020). When cultivating certain soybean varieties, the use of a mixture of nanoparticles made it possible to obtain an increased yield even with a reduced background of mineral fertilizers $N_{60}P_{60}K_{60}$.

Afzal and colleagues conducted studies on the effect of iron oxide nanoparticles on the germination of rice seeds and concluded that green nanoparticles at low concentrations can act as an effective treatment for rice seeds, increasing their germination and overall seedling development (Afzal et al., 2021). Another study evaluated the role of iron oxide nanomaterials on growth and development of rice and corn (Prerna et al., 2021). Laboratory studies have found that pre-sowing treatment of rice and corn seeds with iron oxide nanoparticles leads to an increase in the germination rate and improved growth of crop seedlings. Field studies, in turn, have shown that foliar application of low concentrations of nanoparticles stimulates the growth and yield of rice and corn crops.

Most scientific publications give priority to the participation of the extracellular component in the production of metal nanoparticles. The main advantage of extracellular synthesis of nanoparticles is the absence of impurities which eliminates the need for detergent treatment or ultrasound. In addition, knowledge of the mechanistic aspects of nanoparticle production is necessary to develop effective methods. To address this knowledge gap, Rajput et al. (2016) investigated various strains of *Fusarium oxysporum* for their ability to synthesize silver nanoparticles and also studied the effect of isolate choice, temperature, and pH on the morphology of nanoparticles. The study found that understanding the interaction between organic and interfacial layers is useful for identifying new areas of application of developments, particularly in the development of biosensors. To more fully study the nanoparticle formation by the "green" method, Kitching and colleagues extracted proteins from the cell surface of *Rhizopus oryzae* to *in vitro* produce gold nanoparticles which may have beneficial biocatalytic and biomedical applications (Kitching et al., 2016). They found that proteins extracted with a non-ionic detergent (Triton X-100) and a reducing agent (DTT) formed spherical gold nanoparticles with an average diameter of less than 20 nm, while protein extracted with an ionic detergent (SDS) formed nanoparticles in flower shape having size range between 19 and 43 nm. Thus produced AuNPs showed good colloidal stability in various physiological buffers with high ionic strength.

Utilization of nanoparticles in agriculture as nano-fertilizers has the potential to significantly enhance nutrient uptake by plants (Rabalao et al., 2022). Moreover, using nanoparticles as nano pesticides can prevent bacterial and fungal infections in crops (Yadav et al., 2023). However, as previously discussed, the effects of prolonged nanoparticle use on plant-associated microorganisms remain uncertain.

4.5.4 Bionanotechnology-Based Commercial Agriculture Products

In recent years, nanotechnology and nanoproducts have found applications in many areas of agriculture. Most of the developed drugs are used in crop production, where nano preparations are used as micro fertilizers, which can increase the yield and resistance to adverse weather conditions of many food and grain crops. It is believed that magnesium nanoparticles can have a positive effect on increasing photosynthetic productivity in plants. There are also great prospects for using nanoparticles for post-harvest processing of some crops, which will increase their shelf life. Scientists predict that the application of nanotechnology in the food production industry, including processing and packaging, as well as in agriculture will give rise to a completely new group of food products known as “nanoproducts”. These nanoproducts are expected to eventually replace genetically modified food products in the market.

Currently, drugs based on nanotechnology such as “Nano-Gro” have been developed and are being successfully used. “Nano-Gro” represents an organic plant growth regulator founded on a revolutionary method aimed at enhancing plant productivity, elevating crop quality, and bolstering crop immunity. The manufacturer and creator of this product which was invented in 2005 in U.S. is Agro Nanotechnology, Corp and was patented by the Organic Materials Review Institute (OMRI). Nano-Grow is commercially available as oligosaccharide pellets. It consists of oligosaccharide granules saturated with sulfates of elements like Fe, Co, Al, Mg, Mn, Ni, and Ag. Before the development of “Nano-Gro”, a decade of research was dedicated to investigating the impact of diverse bioactive nano compounds on plants and seeds. The findings enabled the determination of the influence of substances at nanomolecular concentrations on biological organisms. The technology underlying “Nano-Gro” was engineered as a natural, straightforward, and cost-effective approach to enhance agricultural production efficiency, devoid of synthetic chemical usage. Upon application, when a solution containing ingredients at nanomolecular concentrations comes into contact with a plant or seed, the plant interprets its presence as a specific stressor. This stress is purely perceived, yet the plant initiates preparations for an impending struggle for survival which is expressed in an increased production of growth hormones like auxins, gibberellins, and cytokinins. This leads to an increase in the mass and length of roots, an increase in absorbed atmospheric nitrogen, and, consequently, stimulation of protein synthesis by the plant, an increase in biomass, and an increase in yield. To apply "Nano-Gro" to grain crops such as rice, wheat, corn, oats, barley, and rye, the solution having a concentration of 24 granules/10 L water is recommended. Seeds are treated using a dressing machine, requiring 10 l of solution to process 1 ton of seeds. To streamline labor and conserve water, "Nano-Gro" can be seamlessly integrated with nearly all fertilizers and pesticides.

Another product is "Bioplant Flora", an organic liquid nano fertilizer formulation developed in Russia using the most modern technologies. Bioplant Flora is a combination of humic acids with various micro and macro elements beneficial for plants. The product was patented by SSI ARRI of Rice of the Russian Academy of Agricultural Sciences. Utilizing this nano fertilizer enables a significant enhancement in crop yields, with increases of at least 40–50%, while simultaneously reducing production costs by two-fold or more. The resultant molecular structures, when in a nanoscale state, exhibit superior absorption by using plant cells, leading to heightened biometric indicators across agricultural plants. "Bioplant Flora" enhances plants' immunity against a range of adverse external conditions such as rapid temperature shifts, drought, frost, suboptimal temperature conditions, waterlogging, etc. It possesses a potent growth-enhancing impact, catalyzing all the plant's physiological functions: enhancing seed germination energy and rates up to 100%, fostering root development, encouraging flowering and fruit formation, and triggering essential plant processes. Extensive trials of "Bioplant Flora" across diverse crops in various climatic regions of the Russian Federation have been validated by numerous scientific institutions. The State Scientific Institution TatNIISKH of the Russian Agricultural Academy in Kazan, highly praised the effectiveness of "Bioplant Flora" fertilizer for wheat, barley, and rapeseed cultivation. Similarly, GNUVNIPTIOU of the Russian Agricultural Academy located in Vladimir, endorsed its use for green crops, while GNUAFI of the Russian Agricultural Academy (St. Petersburg) recommended it for tomatoes. State Scientific Research Institute of Agriculture of the South-East, Saratov, appreciated its use of chickpeas and wheat.

The drug "Nanoplant", developed by Belarusian scientists, is also known. "Nanoplant" is a micro fertilizer-stimulator with the elements Co, Mn, Cu, Fe, Zn, Cr, Mo, and Se, a balanced composition of eight microelements for all types of plants, combined with biopolymer coating. The active substance of "Nanoplant" is nanoparticles of microelement compounds that have the property of superpermeability through the protective cell membranes of the plant. It is this property that makes it possible to provide the highest biological effect compared to traditional micro-fertilizers and growth activators. At the same time, the consumption of the nano preparation is hundreds of times less than traditional microfertilizers.

Recently, the list of drugs based on nanoparticles has been growing, among them: "Serebromedin"—a highly effective biological product, that contains an innovative complex of silver and copper nanoparticles along with arabinogalactan. It is used to protect all types of plants from a complex of fungal and bacterial diseases, as well as seed treatment. The most important feature of the drug is its absolute safety for humans, pollinators and soil. "Serebromedin" has the potential to form a protective layer on plants for up to 2 months that prevents a wide range of plant diseases. The fungicide "Leaf" is an innovative product developed based on silver and copper nanoparticles. These nanoparticles have strong antimicrobial properties and effectively fight fungal infections, preventing their development and healing plants. The "Nano-Ag Answer", a unique fertilizer and soil conditioner developed by Urth Agriculture, United States, is a rich consortium of microorganisms, sea kelp, and mineral electrolytes. It has the potential to replenish soil nutrients and also fulfills the nitrogen-phosphorus and potassium (NPK) requirements as a fertilizer. The "Nano-Ag Answer" is listed by OMRI for use in Certified Organic® production and processing. "Master Nano Chitosan Organic Fertilizer" containing water-soluble chitosan, phenolic compounds, organic acids, and salicylic acids was developed by Pannaraj Intertrade, Thailand. It stimulates the immunity of plants and thus resists fungal and bacterial diseases.

4.6 Challenges and Opportunities

Myconanotechnology has numerous applications in the field of agriculture. Nanoparticles (NPs) derived from fungi play a promising role as nano-fertilizers, nano-fungicides, nano-pesticides, and nano-insecticides. They are found to significantly facilitate plant growth, yield, nutrient uptake, immunity, disease resistance, phytochemical quantity, and resistance to abiotic and biotic stresses. Research on myconanoparticles as biosensors and nanowires is ongoing, which may contribute to the development of effective tools for detecting plant diseases, toxic chemicals, pesticides, and thus promote precise farming. Moreover, the synthesis of bionanoparticles is simple, cost-effective, and a sustainable, environmentally friendly approach. The use of fungal nanoparticles promotes both plant and soil health with minimal harm to the environment. Hence, the implementation of myconanotechnology in modern agriculture is highly recommended as it could boost the global economy. However, like all other techniques, myconanotechnology also has some limitations and drawbacks. Non-uniform particle size, low yield, and seasonal and regional variations in the availability of raw materials are matters of concern. The majority of research and studies in this field to date are confined to laboratory-scale setups. Therefore, a thorough analysis of the optimal conditions and requirements for the bulk production of bionanoparticles, along with possibilities for industrial collaborations, are urgently needed. Even though the synthesis of NPs is an eco-friendly process compared to conventional methods, the implications on humans, plants, and soil due to their accumulation and long-term persistence in the environment are not fully understood. NPs have high reactivity potential and can easily escape into the surroundings due to their nano-size. There are possibilities of reactions between these nanoparticles and various environmental components, which may pose threats to human health, food safety, and security. Hence, future research should focus on an in-depth analysis of the toxicity and effects of nanoparticle accumulation and their interaction with biotic and abiotic components of the environment to maximize the benefits of NPs in agriculture and other fields.

4.7 Conclusion

In conclusion, the sustainable production of nanoparticles from fungi presents a promising avenue for addressing agricultural challenges while minimizing environmental impact. This critical review has underscored the diverse applications of fungal-derived nanoparticles in agriculture, including their efficacy as antimicrobial agents, fungicides, pesticides, and fertilizers. Moreover, the development of nano biosensors using these nanoparticles opens up new possibilities for precision agriculture and environmental monitoring. By harnessing the unique properties of fungal-derived nanoparticles, we can achieve sustainable solutions for crop protection, soil fertility enhancement, and resource management in agriculture. However, further research is needed to optimize production methods, elucidate mechanisms of action, and assess long-term environmental implications. Overall, the integration of fungal-based nanoparticles into agricultural practices holds great potential for promoting both productivity and environmental stewardship in the face of global food security challenges.

Acknowledgments We would like to express our sincere gratitude to the American Society for Microbiology (ASM), USA, for providing a platform to interact with experts from various nations, which made the completion of this chapter possible.

Authors' Contributions All authors made significant contributions towards the completion of this review article. Abstract, Sects. 4.1, 4.2, 4.3, 4.3.1, 4.3.2, 4.3.3, 4.3.4, 4.5, Conclusion, and related references were prepared by CRM. Sections 4.5, 4.5.1, 4.5.2, 4.5.3, and 4.5.4 and the references coming under these sections were prepared by LZ. UG contributed the contents of Sects. 4.4, 4.4.1, 4.4.2, and 4.4.3, and their related references. Formatting, Figures, and Tables of this review article were done by CRM.

Conflicts of Interest The authors declare no conflicts of interest related to the publication of this review chapter.

References

Abd El-Aziz, A. R. M., Al-Othman, M. R., Mahmoud, M. A., & Metwaly, H. A. (2015). Biosynthesis of silver nanoparticles using *Fusarium solani* and its impact on grain-borne fungi. *Digest Journal of Nanomaterials and Biostructures, 10*, 655–662.

AbdelRahim, K., Mahmoud, S. Y., Ali, A. M., Almaary, K. S., Mustafa, A. E. Z. M., & Husseiny, S. M. (2017). Extracellular biosynthesis of silver nanoparticles using Rhizopus stolonifer. *Saudi Journal of Biological Sciences, 24*(1), 208–216.

Adebayo, E. A., Azeez, M. A., Alao, M. B., Oke, A. M., & Aina, D. A. (2021). Fungi as veritable tool in current advances in nanobiotechnology. *Heliyon, 7*(11).

Afzal, S., Sharma, D., & Singh, N. K. (2021). Eco-friendly synthesis of phytochemical-capped iron oxide nanoparticles as nano-priming agent for boosting seed germination in rice (Oryza sativa L.). *Environmental Science and Pollution Research, 28*, 40275–40287.

Ahmad, A., Mukherjee, P., Senapati, S., Mandal, D., Khan, M. I., Kumar, R., & Sastry, M. (2003). Extracellular biosynthesis of silver nanoparticles using the fungus Fusarium oxysporum. *Colloids and Surfaces B: Biointerfaces, 28*(4), 313–318.

Al-Harbi, N., & Abd-Elrahman, N. K. (2024). Physical methods for preparation of nanomaterials, their characterization, and applications: A review. *Journal of Umm Al-Qura University for Applied Sciences, 22*(3), 841–887. https://doi.org/10.1007/s43994-024-00165-7

Al-Musawi, Z. F., & Al-Saadi, N. H. (2021). Antitumor activities of biosynthesized silver nanoparticles using Dodonaea viscosa (L.) leaves extract. *Basrah Journal of Agricultural Sciences, 34*(2), 42–59.

Amerasan, D., Nataraj, T., Murugan, K., Panneerselvam, C., Madhiyazhagan, P., Nicoletti, M., & Benelli, G. (2016). Myco-synthesis of silver nanoparticles using Metarhizium anisopliae against the rural malaria vector Anopheles culicifacies Giles (Diptera: Culicidae). *Journal of Pest Science, 89*, 249–256.

Amooaghaie, R., Saeri, M. R., & Azizi, M. (2015). Synthesis, characterization and biocompatibility of silver nanoparticles synthesized from *Nigella sativa* leaf extract in comparison with chemical silver nanoparticles. *Ecotoxicology and Environmental Safety, 120*, 400–408.

Apte, M., Sambre, D., Gaikawad, S., Joshi, S., Bankar, A., Kumar, A. R., & Zinjarde, S. (2013). Psychrotrophic yeast Yarrowia lipolytica NCYC 789 mediates the synthesis of antimicrobial silver nanoparticles via cell-associated melanin. *Amb Express, 3*, 1–8.

Arya, S., Sonawane, H., Math, S., Tambade, P., Chaskar, M., & Shinde, D. (2021). Biogenic titanium nanoparticles (TiO_2NPs) from *Trichoderma citrinoviride* extract: Synthesis, characterization, and antibacterial activity against extremely drug-resistant *Pseudomonas aeruginosa. Nano Letters, 11*, 35–42.

Aygün, A., Özdemir, S., Gülcan, M., Cellat, K., & Şen, F. (2020). Synthesis and characterization of Reishi mushroom-mediated green synthesis of silver nanoparticles for the biochemical applications. *Journal of Pharmaceutical and Biomedical Analysis, 178*, 112970.

Azmath, P., Baker, S., Rakshith, D., & Satish, S. (2016). Mycosynthesis of silver nanoparticles bearing antibacterial activity. *Saudi Pharmaceutical Journal, 24*(2), 140–146.

Bachheti, R. K., Abate, L., Bachheti, A., Madhusudhan, A., & Husen, A. (2021). Algae-, fungi-, and yeast-mediated biological synthesis of nanoparticles and their various biomedical applications. In *Handbook of greener synthesis of nanomaterials and compounds* (pp. 701–734). Elsevier.

Badwaik, V. J. (2022). Bioinspired synthesis of metallic nanoparticles and applications. *Advances in Animal Science, I*, 97.

Balaji, D. S., Basavaraja, S., Deshpande, R., Mahesh, D. B., Prabhakar, B. K., & Venkataraman, A. (2009). Extracellular biosynthesis of functionalized silver nanoparticles by strains of Cladosporium cladosporioides fungus. *Colloids and Surfaces B: Biointerfaces, 68*(1), 88–92.

Balakumaran, M. D., Ramachandran, R., & Kalaicheilvan, P. T. (2015). Exploitation of endophytic fungus *Guignardia mangiferae* for extracellular synthesis of silver nanoparticles and their in vitro biological activities. *Microbiological Research, 178*, 9–17.

Batsmanova, L., Taran, N., Konotop, Y., Kalenska, S., & Novytska, N. (2020). Use of a colloidal solution of metal and metal oxide-containing nanoparticles as fertilizer for increasing soybean productivity. *Journal of Central European Agriculture, 21*(2), 311–319.

Bayat, M., Zargar, M., Chudinova, E., Astarkhanova, T., & Pakina, E. (2021). In vitro evaluation of antibacterial and antifungal activity of biogenic silver and copper nanoparticles: The first report of applying biogenic nanoparticles against *Pilidium concavum* and *Pestalotia sp.* fungi. *Molecules, 26*(17), 5402.

Bhardwaj, K., Sharma, A., Tejwan, N., Bhardwaj, S., Bhardwaj, P., Nepovimova, E., et al. (2020). Pleurotus macrofungi-assisted nanoparticle synthesis and its potential applications: A review. *Journal of Fungi, 6*(4), 351.

Bhargava, A., Jain, N., Khan, M. A., Pareek, V., Dilip, R. V., & Panwar, J. (2016). Utilizing metal tolerance potential of soil fungus for efficient synthesis of gold nanoparticles with superior catalytic activity for degradation of rhodamine B. *Journal of Environmental Management, 183*, 22–32.

Bhat, R., Deshpande, R., Ganachari, S. V., Huh, D. S., & Venkataraman, A. (2011). Photo-irradiated biosynthesis of silver nanoparticles using edible mushroom *Pleurotus florida* and their antibacterial activity studies. *Bioinorganic Chemistry and Applications*, Article ID 650979.

Bourzama, G., Ouled-Haddar, H., Marrouche, M., & Aliouat, A. (2021). Iron uptake by fungi isolated from arcelor mittal-annaba-in the northeast of Algeria. *Brazilian Journal of Poultry Science, 23*.

Castro-Longoria, E., Vilchis-Nestor, A. R., & Avalos-Borja, M. (2011). Biosynthesis of silver, gold and bimetallic nanoparticles using the filamentous fungus Neurospora crassa. *Colloids and Surfaces B: Biointerfaces, 83*(1), 42–48.

Chatterjee, S., Mahanty, S., Das, P., Chaudhuri, P., & Das, S. (2020). Biofabrication of iron oxide nanoparticles using manglicolous fungus Aspergillus niger BSC-1 and removal of Cr (VI) from aqueous solution. *Chemical Engineering Journal, 385*, 123790.

Chen, G., Yi, B., Zeng, G., Niu, Q., Yan, M., Chen, A., Du, J., Huang, J., & Zhang, Q. (2014). Facile green extracellular biosynthesis of CdS quantum dots by white rot fungus Phanerochaete chrysosporium. *Colloids and Surfaces. B, Biointerfaces, 117*, 199–205. https://doi.org/10.1016/j.colsurfb.2014.02.027

Consolo, V. F., Torres-Nicolini, A., & Alvarez, V. A. (2020). Mycosinthetized Ag, CuO and ZnO nanoparticles from a promising *Trichoderma harzianum* strain and their antifungal potential against important phytopathogens. *Scientific Reports, 10*, 1–9. https://doi.org/10.1038/s41598-020-77294-6

Constantin, M., Răut, I., Suica-Bunghez, R., Firinca, C., Radu, N., Gurban, A. M., Preda, S., Alexandrescu, E., Doni, M., & Jecu, L. (2023). *Ganoderma lucidum*-mediated green synthesis of silver nanoparticles with antimicrobial activity. *Materials (Basel, Switzerland), 16*(12), 4261. https://doi.org/10.3390/ma16124261

Crisan, C. M., Mocan, T., Manolea, M., Lasca, L. I., Tăbăran, F. A., & Mocan, L. (2021). Review on silver nanoparticles as a novel class of antibacterial solutions. *Applied Sciences, 11*(3), 1120.

Das, S. K., Das, A. R., & Guha, A. K. (2009). Gold nanoparticles: Microbial synthesis and application in water hygiene management. *Langmuir, 25*(14), 8192–8199.

Dhaka, A., Mali, S. C., Sharma, S., & Trivedi, R. (2023). A review on biological synthesis of silver nanoparticles and their potential applications. *Results in Chemistry, 6*, 101108. https://doi.org/10.1016/j.rechem.2023.101108

Diko, C. S., Qu, Y., Henglin, Z., Li, Z., Nahyoon, N. A., & Fan, S. (2020). Biosynthesis and characterization of lead selenide semiconductor nanoparticles (PbSe NPs) and its antioxidant and photocatalytic activity. *Arabian Journal of Chemistry, 13*(11), 8411–8423.

Dinesh, R. A., Raja, S., Kishanlal, N., Nachiyar, C. V., & Sunkar, S. (2023). Microbe-based synthesis of gold nanoparticles and its catalytic applications. *Biosciences Biotechnology Research Asia, 20*(2). https://bit.ly/3LcPfby

Elahian, F., Reiisi, S., Shahidi, A., & Mirzaei, S. A. (2017). High-throughput bioaccumulation, biotransformation, and production of silver and selenium nanoparticles using genetically engineered Pichia pastoris. *Nanomedicine: Nanotechnology, Biology and Medicine, 13*(3), 853–861.

Elamawi, R. M., Al-Harbi, R. E., & Hendi, A. A. (2018). Biosynthesis and characterization of silver nanoparticles using Trichoderma longibrachiatum and their effect on phytopathogenic fungi. *Egyptian Journal of Biological Pest Control, 28*(1), 1–11.

Elgorban, A. M., Aref, S. M., Seham, S. M., Elhindi, K. M., Bahkali, A. H., Sayed, S. R., et al. (2016). Extracellular synthesis of silver nanoparticles using Aspergillus versicolor and evaluation of their activity on plant pathogenic fungi. *Mycosphere, 7*(6), 844–852.

Fayaz, A. M., Balaji, K., Girilal, M., Kalaichelvan, P. T., & Venkatesan, R. (2009). Mycobased synthesis of silver nanoparticles and their incorporation into sodium alginate films for vegetable and fruit preservation. *Journal of Agricultural and Food Chemistry, 57*(14), 6246–6252.

Fayaz, M., Tiwary, C. S., Kalaichelvan, P. T., & Venkatesan, R. (2010). Blue orange light emission from biogenic synthesized silver nanoparticles using *Trichoderma viride*. *Colloids and Surfaces B: Biointerfaces, 75*(1), 175–178. https://doi.org/10.1016/j.colsurfb.2009.08.028

Gajbhiye, M., Kesharwani, J., Ingle, A., Gade, A., & Rai, M. (2009). Fungus-mediated synthesis of silver nanoparticles and their activity against pathogenic fungi in combination with fluconazole. *Nanomedicine: Nanotechnology, Biology and Medicine, 5*(4), 382–386.

Ganesan, V., Hariram, M., Vivekanandhan, S., & Muthuramkumar, S. (2020). Periconium sp.(endophytic fungi) extract mediated sol-gel synthesis of ZnO nanoparticles for antimicrobial and antioxidant applications. *Materials Science in Semiconductor Processing, 105*, 104739.

Ghormade, V., Pathan, E. K., & Deshpande, M. V. (2017). Can fungi compete with marine sources for chitosan production? *International Journal of Biological Macromolecules, 104*(Part B), 1415–1421. https://doi.org/10.1016/j.ijbiomac.2017.01.112

Gopinath, K., Karthika, V., Sundaravadivelan, C., & Ilangovan, R. (2015). Mycogenesis of cerium oxide nanoparticles using *Aspergillus niger* culture filtrate and their applications for antibacterial and larvicidal activities. *Journal of Nanostructure in Chemistry, 5*(3), 295–303. https://doi.org/10.1007/s40097-015-0161-2

Gruyer, N., Dorais, M., Bastien, C., Dassylva, N., & Triffault-Bouchet, G. (2013, October). Interaction between silver nanoparticles and plant growth. In *International symposium on new technologies for environment control, energy-saving and crop production in greenhouse and plant 1037* (pp. 795–800).

Gupta, A., Rayeen, F., Mishra, R., Tripathi, M., & Pathak, N. (2023). Nanotechnology applications in sustainable agriculture: An emerging ecofriendly approach. *Plant Nano Biology*, 100033.

Hulikere, M. M., & Joshi, C. G. (2019). Characterization, antioxidant and antimicrobial activity of silver nanoparticles synthesized using marine endophytic fungus-Cladosporium cladosporioides. *Process Biochemistry, 82*, 199–204.

Hussain, I., Singh, N. B., Singh, A., Singh, H., & Singh, S. C. (2016). Green synthesis of nanoparticles and its potential application. *Biotechnology Letters, 38*, 545–560.

Hussain, M., Zahra, N., Lang, T., Zain, M., Raza, M., Shakoor, N., et al. (2023). Integrating nanotechnology with plant microbiome for next-generation crop health. *Plant Physiology and Biochemistry, 196*, 703–711.

Ingle, A., Rai, M., Gade, A., & Bawaskar, M. (2009). *Fusarium solani:* A novel biological agent for the extracellular synthesis of silver nanoparticles. *Journal of Nanoparticle Research, 11*(8), 2079–2085. https://doi.org/10.1007/s11051-008-9573-y

Ingle, A. P., Moharil, M. P., Saxena, S., Patil, V., & Borse, S. (2023). Nanotechnology: the new perspective in agroecosystem and sustainable agriculture. In *Nanotechnology in agriculture and agroecosystems* (pp. 1–24). Elsevier.

Joshi, C. G., Danagoudar, A., Poyya, J., Kudva, A. K., & Dhananjaya, B. L. (2017). Biogenic synthesis of gold nanoparticles by marine endophytic fungus-Cladosporium cladosporioides isolated from seaweed and evaluation of their antioxidant and antimicrobial properties. *Process Biochemistry, 63*, 137–144.

Kalia, A., & Kaur, G. (2018). Biosynthesis of nanoparticles using mushrooms. In B. Singh, Lallawmsanga, & A. Passari (Eds.), *Biology of macrofungi* (pp. 387–404). Springer. https://doi.org/10.1007/978-3-030-02622-6_17

Karwa, A., Gaikwad, S., & Rai, M. K. (2011). Mycosynthesis of silver nanoparticles using Lingzhi or Reishi medicinal mushroom, *Ganoderma lucidum* (W. Curt.:Fr.) P. Karst. and their role as antimicrobials and antibiotic activity enhancers. *International Journal of Medicinal Mushrooms, 13*(5), 483–491. https://doi.org/10.1615/IntJMedMushr.v13.i5.80

Kitching, M., Choudhary, P., Inguva, S., Guo, Y., Ramani, M., Das, S. K., & Marsili, E. (2016). Fungal surface protein mediated one-pot synthesis of stable and hemocompatible gold nanoparticles. *Enzyme and Microbial Technology, 95*, 76–84.

Li, Q., Liu, D., Wang, T., Chen, C., & Gadd, G. M. (2020). Iron coral: Novel fungal biomineralization of nanoscale zerovalent iron composites for treatment of chlorinated pollutants. *Chemical Engineering Journal, 402*, 126263. https://doi.org/10.1016/j.cej.2020.126263

Li, Q., Liu, F., Li, M., Chen, C., & Gadd, G. M. (2022). Nanoparticle and nanomineral production by fungi. *Fungal Biology Reviews, 41*, 31–44.

Liu, F., Shah, D. S., & Gadd, G. M. (2021). Role of protein in fungal biomineralization of copper carbonate nanoparticles. *Current Biology, 31*, 358–368.

Madani, M., Hosny, S., Alshangiti, D. M., Nady, N., Alkhursani, S. A., Alkhaldi, H., Al-Gahtany, S. A., Ghobashy, M. M., & Gaber, G. A. (2022). Green synthesis of nanoparticles for varied applications: Green renewable resources and energy-efficient synthetic routes. *Nanotechnology Reviews, 11*(1), 731–759.

Matras, E., Gorczyca, A., Pociecha, E., Przemieniecki, S. W., Zeliszewska, P., & Ocwieja, M. (2023). Silver nanoparticles affect wheat (*Triticum aestivum* L.) germination, seedling blight and yield. *Functional Plant Biology: FPB, 50*(5), 390–406. https://doi.org/10.1071/FP22086

Metuku, R. P., Pabba, S., Burra, S., Hima Bindu, N., Gudikandula, K., & Singara Charya, M. A. (2014). Biosynthesis of silver nanoparticles from Schizophyllum radiatum HE 863742.1: Their characterization and antimicrobial activity. *3 Biotech, 4*, 227–234.

Mgadi, K., Ndaba, B., Roopnarain, A., Rama, H., & Adeleke, R. (2023). Nanoparticle applications in agriculture: Overview and response of plant-associated microorganisms. *Frontiers in Microbiology, 15*, 1354440.

Michael, A., Singh, A., Roy, A., & Islam, M. R. (2022). Fungal-and algal-derived synthesis of various nanoparticles and their applications. *Bioinorganic Chemistry and Applications*.

Mishra, A., Kumari, M., Pandey, S., Chaudhry, V., Gupta, K. C., & Nautiyal, C. S. (2014). Biocatalytic and antimicrobial activities of gold nanoparticles synthesized by Trichoderma sp. *Bioresource Technology, 166*, 235–242.

Mohanta, Y. K., Singdevsachan, S. K., Parida, U. K., Panda, S. K., Mohanta, T. K., & Bae, H. (2016). Green synthesis and antimicrobial activity of silver nanoparticles using wild medicinal mushroom Ganoderma applanatum (Pers.) Pat. from Similipal Biosphere Reserve, Odisha, India. *IET Nanobiotechnology, 10*(4), 184–189.

Mohanta, Y., Nayak, D., Biswas, K., Singdevsachan, S., Abd Allah, E., Hashem, A., Alqarawi, A., Yadav, D., & Mohanta, T. (2018). Silver nanoparticles synthesized using wild mushroom show potential antimicrobial activities against food borne pathogens. *Molecules, 23*(3), 655.

Mohmed, A., Hassan, S., Fouda, A., Elgamal, M., & Salem, S. (2017). Extracellular biosynthesis of silver nanoparticles using Aspergillus sp. and evaluation of their antibacterial and cytotoxicity. *International Journal of Applied Life Sciences, 11*(2), 1–12.

Morano, K. A., Grant, C. M., & Moye-Rowley, W. S. (2012). The response to heat shock and oxidative stress in Saccharomyces cerevisiae. *Genetics, 190*(4), 1157–1195.

Mughal, B., Zaidi, S. Z. J., Zhang, X., & Hassan, S. U. (2021). Biogenic nanoparticles: Synthesis, characterisation and applications. *Applied Sciences, 11*(6), 2598.

Mukherjee, P., Ahmad, A., Mandal, D., Senapati, S., Sainkar, S. R., Khan, M. I., et al. (2001). Fungus-mediated synthesis of silver nanoparticles and their immobilization in the mycelial matrix: A novel biological approach to nanoparticle synthesis. *Nano Letters, 1*(10), 515–519.

Munir, T., Mahmood, A., Imran, M., Sohail, A., Fakhar-e-Alam, M., Sharif, M., et al. (2021). Quantitative analysis of glucose by using (PVP and MA) capped silver nanoparticles for biosensing applications. *Physica B: Condensed Matter, 602*, 412564.

Nayak, D., Mishra, A. K., Malla, S., Swain, U., Panigrahi, K., Panda, J., Rustagi, S., Rauta, P. R., & Mohanta, Y. K. (2024). Eco-friendly mycogenic synthesis of iron oxide nanoparticles using Volvariella volvacea and unveiling its potential as antibiofilm and anticancer agents. *ECS Journal of Solid State Science and Technology, 13*(8), 087003.

Noor, S., Shah, Z., Javed, A., Ali, A., Hussain, S. B., Zafar, S., et al. (2020). A fungal based synthesis method for copper nanoparticles with the determination of anticancer, antidiabetic and antibacterial activities. *Journal of Microbiological Methods, 174*, 105966.

Osman, A. I., Zhang, Y., Farghali, M., et al. (2024). Synthesis of green nanoparticles for energy, biomedical, environmental, agricultural, and food applications: A review. *Environmental Chemistry Letters, 22*(3), 841–887. https://doi.org/10.1007/s10311-023-01682-3

Panda, J., Mishra, A. K., Nath, P. C., Mahanta, S., Sharma, M., Nayak, P. K., Mohanta, Y. K., & Sridhar, K. (2024). Wild edible mushrooms to achieve sustainable development goals: Novel sources for food security, health, and well-being. *Food Bioscience, 60*, 104277.

Patel, A., Enman, J., Gulkova, A., Guntoro, P. I., Dutkiewicz, A., Ghorbani, Y., Rova, U., Christakopoulos, P., & Matsakas, L. (2021). Integrating biometallurgical recovery of metals with biogenic synthesis of nanoparticles. *Chemosphere, 263*, 128306. https://doi.org/10.1016/j.chemosphere.2020.128306

Philip, D. (2009). Biosynthesis of Au, Ag and Au–Ag nanoparticles using edible mushroom extract. *Spectrochimica Acta Part A: Molecular and Biomolecular Spectroscopy, 73*(2), 374–381. https://doi.org/10.1016/j.saa.2009.02.037

Prerna, D. I., Govindaraju, K., Tamilselvan, S., Kannan, M., Vasantharaja, R., Chaturvedi, S., & Shkolnik, D. (2021). Influence of nanoscale micro-nutrient α-Fe_2O_3 on seed germination, seedling growth, translocation, physiological effects and yield of rice (Oryza sativa) and maize (Zea mays). *Plant Physiology and Biochemistry, 162*, 564–580.

Qian, Y., Yu, H., He, D., Yang, H., Wang, W., Wan, X., et al. (2013). Biosynthesis of silver nanoparticles by the endophytic fungus *Epicoccum nigrum* and their activity against pathogenic fungi. *Bioprocess and Biosystems Engineering, 36*(11), 1613–1619.

Rabalao, T. M., Ndaba, B., Roopnarain, A., & Vatsha, B. (2022). Towards a circular economy: The influence of extraction methods on phytosynthesis of metallic nanoparticles and their impact on crop growth and protection. *JSFA Reports, 2*(5), 208–221.

Rabeea, M. A., Owaid, M. N., Aziz, A. A., Jameel, M. S., & Dheyab, M. A. (2020). Mycosynthesis of gold nanoparticles using the extract of Flammulina velutipes, Physalacriaceae, and their efficacy for decolorization of methylene blue. *Journal of Environmental Chemical Engineering, 8*(3), 103841.

Radhakrishnan, M. C., & Peter, N. (2020). Biological properties of silver nanoparticles from β (1-3) glucan isolated from the edible mushroom Pleurotus florida. *Letters in Applied NanoBioScience, 10*, 2096–2106.

Rajput, S., Werezuk, R., Lange, R. M., & McDermott, M. T. (2016). Fungal isolate optimized for biogenesis of silver nanoparticles with enhanced colloidal stability. *Langmuir, 32*(34), 8688–8697.

Rehman, A., Majeed, M. I., Ihsan, A., Hussain, S. Z., Rehman, S., Ghauri, M. A., Khalid, Z. M., & Hussain, I. (2011). Living fungal hyphae-templated porous gold microwires using nanoparticles as building blocks. *Journal of Nanoparticle Research, 13*, 6747–6754.

Riddin, T. L., Govender, Y., Gericke, M., & Whiteley, C. G. (2009). Two different hydrogenase enzymes from sulphate-reducing bacteria are responsible for the bioreductive mechanism of platinum into nanoparticles. *Enzyme and Microbial Technology, 45*(4), 267–273.

Roy, A., Sharma, A., Yadav, S., Jule, L. T., & Krishnaraj, R. (2021). Nanomaterials for remediation of environmental pollutants. *Bioinorganic Chemistry and Applications*.

Sadaf, A., Afolayan, J., & Perry, C. (2024). Developing gold nanoparticle mycelial composites: Effect of nanoparticle surface functionality on *Aspergillus niger* viability and cell wall biochemistry. *Current Research in Biotechnology, 7*, 100185. https://doi.org/10.1016/j.crbiot.2024.100185

Sadak, M. S. (2019). Impact of silver nanoparticles on plant growth, some biochemical aspects, and yield of fenugreek plant (Trigonella foenum-graecum). *Bulletin of the National Research Centre, 43*(1), 1–6.

Sahab, A. F., Waly, A. I., Sabbour, M. M., & Nawar, L. S. (2015). Synthesis, antifungal and insecticidal potential of Chitosan (CS)-g-poly (acrylic acid) (PAA) nanoparticles against some seed borne fungi and insects of soybean. *International Journal of Chemistry Technology Research, 8*(12), 589–598. https://doi.org/10.13140/RG.2.1.1198.8325

Saleh, H. M., & Hassan, A. I. (2021). Nanomaterials fabrication: Bacteria-assisted synthesis. In *Handbook of greener synthesis of nanomaterials and compounds* (pp. 683–700). Elsevier.
Santos, T. S., Silva, T. M., Cardoso, J. C., Albuquerque-Júnior, R. L. D., Zielinska, A., Souto, E. B., et al. (2021). Biosynthesis of silver nanoparticles mediated by entomopathogenic fungi: Antimicrobial resistance, nanopesticides, and toxicity. *Antibiotics, 10*(7), 852.
Saravanakumar, K., Shanmugam, S., Varukattu, N. B., MubarakAli, D., Kathiresan, K., & Wang, M. H. (2019). Biosynthesis and characterization of copper oxide nanoparticles from indigenous fungi and its effect of photothermolysis on human lung carcinoma. *Journal of Photochemistry and Photobiology B: Biology, 190*, 103–109.
Saravanan, A., Kumar, P. S., Karishma, S., Vo, D. V. N., Jeevanantham, S., Yaashikaa, P. R., & George, C. S. (2021). A review on biosynthesis of metal nanoparticles and its environmental applications. *Chemosphere, 264*, 128580.
Sastry, M., Ahmad, A., Khan, M. I., & Kumar, R. (2003). Biosynthesis of metal nanoparticles using fungi and actinomycete. *Current Science*, 162–170.
Sebesta, M., Vojtková, H., Cyprichová, V., Ingle, A. P., Urík, M., & Kolenčík, M. (2022). Mycosynthesis of metal-containing nanoparticles—Fungal metal resistance and mechanisms of synthesis. *International Journal of Molecular Sciences, 23*(22), 14084.
Seshadri, S., Saranya, K., & Kowshik, M. (2011). Green synthesis of lead sulfide nanoparticles by the lead resistant marine yeast, *Rhodosporidium diobovatum. Biotechnology Progress, 27*, 1464–1469.
Shahzad, A., Saeed, H., Iqtedar, M., Hussain, S. Z., Kaleem, A., Abdullah, R., …, Chaudhary, A. (2019). Size-controlled production of silver nanoparticles by Aspergillus fumigatus BTCB10: Likely antibacterial and cytotoxic effects. *Journal of Nanomaterials.*
Shaligram, N. S., Bule, M., Bhambure, R., Singhal, R. S., Singh, S. K., Szakacs, G., & Pandey, A. (2009). Biosynthesis of silver nanoparticles using aqueous extract from the compactin producing fungal strain. *Process Biochemistry, 44*(8), 939–943.
Shamim, M. Z., Mishra, A. K., Kausar, T., Mahanta, S., Sarma, B., Kumar, V., Mishra, P. K., Panda, J., Baek, K.-H., & Mohanta, Y. K. (2023). Exploring edible mushrooms for diabetes: Unveiling their role in prevention and treatment. *Molecules (Basel, Switzerland), 28*(6), 2837.
Sharma, N., Sharma, M., Sajid Jamal, Q. M., Kamal, M. A., & Akhtar, S. (2020). Nanoinformatics and biomolecular nanomodeling: A novel move en route for effective cancer treatment. *Environmental Science and Pollution Research, 27*, 19127–19141.
Singh, P., Kim, Y. J., Zhang, D., & Yang, D. C. (2016). Biological synthesis of nanoparticles from plants and microorganisms. *Trends in Biotechnology, 34*(7), 588–599.
Sriramulu, M., & Sumathi, S. (2018). Biosynthesis of palladium nanoparticles using Saccharomyces cerevisiae extract and its photocatalytic degradation behaviour. *Advances in Natural Sciences: Nanoscience and Nanotechnology, 9*(2), 025018.
Sugunan, A., Melin, P., Schnürer, J., Hilborn, J. G., & Dutta, J. (2007). Nutrition-driven assembly of colloidal nanoparticles: Growing fungi assemble gold nanoparticles as microwires. *Advanced Materials, 19(1)*, 77–81.
Suryavanshi, P., Pandit, R., Gade, A., Derita, M., Zachino, S., & Rai, M. (2017). *Colletotrichum* sp.-mediated synthesis of sulphur and aluminium oxide nanoparticles and its in vitro activity against selected food-borne pathogens. *LWT—Food Science and Technology, 81*, 188–194.
Tarafdar, J. C., Raliya, R., Mahawar, H., & Rathore, I. (2014). Development of zinc nanofertilizer to enhance crop production in pearl millet (Pennisetum americanum). *Agricultural Research, 3*, 257–262.
Trzcińska-Wencel, J., Wypij, M., Rai, M., & Golińska, P. (2023a). Biogenic nanosilver bearing antimicrobial and antibiofilm activities and its potential for application in agriculture and industry. *Frontiers in Microbiology, 14*, 1125685. https://doi.org/10.3389/fmicb.2023.1125685
Trzcińska-Wencel, J., Wypij, M., Terzyk, A. P., Rai, M., & Golińska, P. (2023b). Biofabrication of novel silver and zinc oxide nanoparticles from *Fusarium solani* IOR 825 and their potential application in agriculture as biocontrol agents of phytopathogens, and seed germination and seedling growth promoters. *Frontiers in Chemistry, 11*, 1235437. https://doi.org/10.3389/fchem.2023.1235437

Tsivileva, O. M., & Perfileva, A. I. (2022). Mushroom-derived novel selenium nanocomposites' effects on potato plant growth and tuber germination. *Molecules (Basel, Switzerland), 27*(14), 4438. https://doi.org/10.3390/molecules27144438

Ujam, N. T., Ajaghaku, D. L., Okoye, F. B., & Esimone, C. O. (2021). Antioxidant and immunosuppressive activities of extracts of endophytic fungi isolated from Psidium guajava and Newbouldia laevis. *Phytomedicine Plus, 1*(2), 100028.

Verma, V. C., Kharwar, R. N., & Gange, A. C. (2010). Biosynthesis of antimicrobial silver nanoparticles by the endophytic fungus Aspergillus clavatus. *Nanomedicine, 5*(1), 33–40.

Vigneshwaran, N., Ashtaputre, N. M., Varadarajan, P. V., Nachane, R. P., Paralikar, K. M., & Balasubramanya, R. H. (2007). Biological synthesis of silver nanoparticles using the fungus Aspergillus flavus. *Materials Letters, 61*(6), 1413–1418.

Xing, K., Liu, Y., Shen, X., Zhu, X., Li, C., Miao, X., Feng, Z., Peng, X., & Qin, S. (2017). Effect of O-chitosan nanoparticles on the development and membrane permeability of Verticillium dahliae. *Carbohydrate Polymers, 165*, 334–343. https://doi.org/10.1016/j.carbpol.2017.02.063

Yaakoub, H., Mina, S., Calenda, A., Bouchara, J. P., & Papon, N. (2022). Oxidative stress response pathways in fungi. *Cellular and Molecular Life Sciences: CMLS, 79*(6), 333. https://doi.org/10.1007/s00018-022-04353-8

Yadav, S. A., Suvathika, G., Alghuthaymi, M. A., & Abd-Elsalam, K. A. (2023). Fungal-derived nanoparticles for the control of plant pathogens and pests. In *Fungal cell factories for sustainable nanomaterials productions and agricultural applications* (pp. 755–784).

Zaynitdinova, L. I., Juraeva, R. N., Tashpulatov, J. J., Kukanova, S. I., Lazutin, N. A., & Mavjudova, A. M. (2022). Influence of silver and copper nanoparticles on the enzymatic activity of soil-borne microorganisms. *Baghdad Science Journal, 19*(6 Suppl), 1487.

Zhao, X., Zhou, L., Riaz Rajoka, M. S., Yan, L., Jiang, C., Shao, D., et al. (2018). Fungal silver nanoparticles: Synthesis, application and challenges. *Critical Reviews in Biotechnology, 38*(6), 817–835.

Zhao, X., Hou, N., Wan, C., Zhang, L., & Liu, X. (2024). Gold nanoparticles synthesis mediated by fungus isolated from aerobic granular sludge: Process and mechanisms. *Heliyon, 10*(6), e28281. https://doi.org/10.1016/j.heliyon.2024.e28281

Chapter 5
Biorefining Agro-Industrial Waste into Green Nanomaterials for Sustainable Agriculture

Gajendra B. Singh, Ashmita Sethi, Manisha Kumari, Praneeta Panigrahi, Palak Chaudhary, and Anirudh Sharma

Abstract Agriculture, as a cornerstone of global food security, faces the intricate challenge of feeding a growing population while mitigating its environmental footprint. The expanding global population and food demands have led to an enormous increase in agro-industrial operations and products with a concomitant increase in non-consumable biomass—agro-industrial wastes. These wastes are largely left untreated resulting in environmental degradation and economic losses. Owing to their biochemical heterogeneity and presence of vital bioactive compounds like carbohydrates, proteins, and minerals, these can be used as feedstock for biorefinery for generation of a diversified variety of products including composts, organic fertilizers, and biochar, to name a few. With the advent of new techniques, the valorization of agro-industrial wastes as environment-friendly and sustainable tools for addressing the problems about conventional agricultural practices has grown tremendously. In recent decades, green nanotechnology has emerged as an acceptable approach for utilizing agro-industrial wastes to synthesize nanomaterials with excellent surface properties and biocompatibility which include, but are not limited to, nanoparticles and nanocomposites. This chapter focuses on examining and characterizing agro-industrial wastes as potential resources for generating cost-effective value-added nanomaterials, along with nano-management of agro-waste to improve the agricultural output. Furthermore, an assemblage of the approaches adopted for synthesis of various nanoparticles, along with the possible applications is put forth.

Keywords Agricultural waste · Industrial waste · Nanoparticles · Bio-refinery · Microorganism

---

G. B. Singh (✉) · A. Sethi · M. Kumari · P. Panigrahi · P. Chaudhary
University Institute of Biotechnology, Chandigarh University, Mohali, Punjab, India
e-mail: gajendra.singh@cumail.in

A. Sharma
Department of Biotechnology, Jaypee Institute of Information Technology, Noida, Uttar Pradesh, India

Y. K. Mohanta et al. (eds.), *Nano-microbiology for Sustainable Development*,
https://doi.org/10.1007/978-3-031-78845-1_5

5.1 Introduction

The global population is growing tremendously, crossing the bar of 8 billion recently and with an annual growth rate of 0.88 (UN, 2022). This global populace expansion has led to a substantial increase in the demand for food supply and consequentially, an inescapable expansion in the food and agricultural industrialization (Chakraborty et al., 2024). Unlike the traditional agricultural regimes, where the biowastes were put back to fulfil agricultural purposes thus, forming a circular economy, the agricultural industries currently practice intense production and distribution in a manner that generates large amounts of heterogeneous wastes, thereby leading to a linear-agricultural production system. Modern industrialization shows a few problems to address, and adequate waste disposal stands out amongst them. According to the reports of the Food and Agriculture Organization (FAO) 2013, 250 million tons of agro-industrial waste were generated globally per year (Heredia-Guerrero et al., 2017; Vinayak et al., 2022), which is expected to reach 3.40 billion metric tons by 2050 (Sodhi et al., 2022; Usmani et al., 2020). These wastes are largely left untreated and underutilized and are disposed of either by unplanned landfilling, open dumping, burning or incineration. This accumulation of waste has led to numerous environmental hazards including altered quality of surface and groundwater, intoxicated aquatic life, and deteriorated soil quality, among others. One of the most prominent consequences is the emission of greenhouse gases, which cause irreversible environmental deterioration and climate change by an elevated carbon footprint (Panda et al., 2024). These environmental hazards and economic losses have been the central subject of research for a long time now.

An intensive global investigation by researchers has started to consider these wastes as potential raw materials for various bio-based solutions, including bioenergy, bio-fuels, bio-products, and biorefinery (Nair et al., 2022; Leiva-Candia et al., 2014; Ortega et al., 2022; Yaashikaa et al., 2022; Mehndiratta et al., 2015). Given the innumerable benefits they offer, in addition to reduction in environmental hazards, these resources are considered environment-friendly, sustainable, low-cost, abundant availability, organic nature, carbon neutrality, and low sulfur content, to name a few.

Nanotechnology has gained popularity over time for conceivable reasons, and has revolutionized many industries, and therefore is seen as an excellent approach to valorize agro-industrial wastes by transformation into economically-valuable and inexpensive bio-composite-based nanoproducts (Abol-Fotouh et al., 2020; Vergara-Castañeda et al., 2020; El-Ramady et al., 2020). Green synthesis of nanomaterials offers a sustainable technique for valorization of organic biomass to generate high-performance green nanomaterials to remediate environmental and agricultural problems.

Agro-waste residues, being the most sustainable and abundantly available sources of cellulosic biomass, have been exploited for the generation of nanocellulose products (Ahmad Khorairi et al., 2021; Picot-Allain & Emmambux, 2023; Widiarto et al., 2017). Plant-based nanoparticle green synthesis is termed a gold standard when it comes to the biological synthesis of nanomaterials. Plant extracts include seeds, roots, stems, leaves, and fruits and are seen as potential sources to generate nanoparticles, with the ability to be defined in terms of shape, size, and composition (Yadav et al., 2022; Meral et al., 2022; Mohamed et al., 2021; Vinayak & Singh, 2022). Lignin, the second-largest plant polymer after cellulose, is exploited under green nanotechnology to generate lignin-based nanoparticles as an approach to generate high-value products that not only retain the properties of native lignin but also possess larger surface area (Duarah et al., 2020; Shalma et al., 2023). Numerous types of agro-industrial wastes are produced globally, and management of these wastes is one of the challenges for nations worldwide. With the idea of forming a circular bioeconomy, these wastes are needed to be handled efficiently, and nanotechnology may be seen as one of the promising tools for nano-management of waste through fabrications of numerous nanomaterials with applications in either environmental remediation or as value-added nanomaterials. Under the umbrella of the bioeconomy and industrial innovation, green nanotechnology has emerged as a steadfast approach for the utilization of agro-industrial wastes and the valorization of such wastes to synthesize nanomaterials with desirable characteristics and applications.

This chapter focuses on valorizing agro-industrial wastes in the view of green nanotechnology. The chapter then focuses on the challenges associated with the processing and disposal of agro-industrial wastes in view of a circular bioeconomy and the possible solutions using nano-management. Eventually, a detailed account on the applications of bio-based nanomaterials in sustainable agricultural practices is discussed, along with an assessment of challenges and risks associated with the utilization of biorefined green nanoparticles from agro-industrial wastes.

5.2 Types of Agro-Industrial Wastes

The large quantity of inedible biomass produced by various agro-industrial operations during processing of agricultural or livestock products is termed as agro-industrial waste. These wastes are organic substances that are rich in bioactive compounds like carbohydrates, proteins, vitamins, minerals, fiber. When seen as a feedstock, the compositional study focuses on the lignocellulosic content of the biomass, which constitutes the cellulose, hemicellulose, and lignin content. By virtue of such heterogeneous biochemical properties and vast nutritional composition, these 'wastes' are seen as alternatives to current raw materials for the production of various value-added products, which could be reused in the agricultural practices, thereby forming a sustainable circular bioeconomy.

The wastes from agriculture are divided into two sections namely, agricultural residues and industrial residues. Agriculture residues are further divided into field residues and process residues. Field residues are the waste materials that are left in the field after the crop is harvested. These mainly constitute seed pods, stems, stalks and leaves and are generated in large amounts. Process residues are residues that are

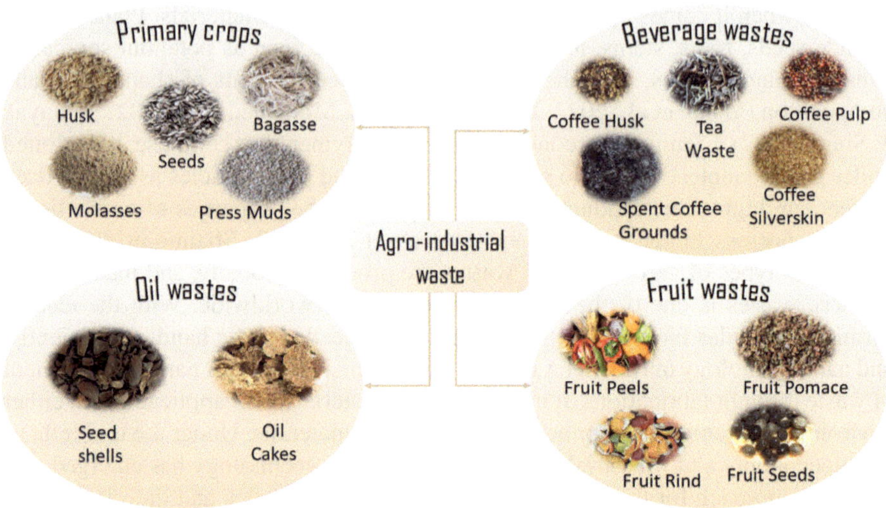

Fig. 5.1 Classification of different types of agro-industrial wastes

left over after the crop is processed into useful alternate resources and include husks, bagasse, molasses, seeds, and roots. Industrial residues include the wastes generated after industrial processing of an agricultural source and encompass a wide range of materials like fruit and vegetable wastes, cereal wastes, oil wastes, tea waste, coffee waste and many more. The agro-processing sector includes all the industrial-processing activities applied for the conversion of agricultural yield into final value-added products, with concomitant production of large quantities of waste. The term 'agro-industrial waste' in this chapter refers to the waste generated through and after agro-processing, and thus encompasses only the process residues and industrial residues (Fig. 5.1).

The agriculture sector accounts for a significant share of economic growth, nearly 4% of global Gross Domestic Product (GDP), and even 25% for developing nations (World Bank, 2023). The majority of the global food demand is met by three staple cereals, namely corn, followed by wheat and rice, accounting for 90% of cereal production with an estimated production of 1.2 billion tons of maize, and 0.8 billion tons each of wheat and rice, respectively (FAO, 2022). Such quantities of crops are processed industrially to yield the final consumable product, generating enormous amounts of waste. Milling of rice yields by-products such as rice husk (20%) and rice bran (8%) (Van Hoed et al., 2006). On a global scale, rice husk accounts for a significant amount of waste falling in the range of 150 million tons, and rice bran for more than 60 million tons. Wheat processing follows a complex series of steps—cleaning, gristing, separating, milling and blending—and produces waste such as wheat bran. Production of wheat bran as waste amounts to be nearly 150 million tons annually.

The aforementioned cereals, along with sugarcane account for nearly half the worldwide production of primary crops, wherein sugarcane accounts for 21% of the global crop production i.e., nearly 1.9 billion tons (FAO, 2022). Cane sugar production from sugarcane involves the generation of wastes including cane tops, molasses, press muds, and bagasse. Multiple nations are engaged in sugar extraction through industrial processing, and this generates around 513 million tons of sugarcane bagasse annually (Miranda et al., 2021) as well as press muds (processing of 1 ton of cane yields 0.03 ton of press mud) (Yadav & Solomon, 2006) in an approximated amount of more than 55 million tons annually.

Besides primary crops, beverages account for a major section of food crops produced globally. One such crop is coffee. Coffee grain processing results in over 10 million tons of coffee waste being produced worldwide (Echeverria & Nuti, 2017), which includes coffee husk, pulp, silverskin, and spent coffee grounds. Amongst these, the spent coffee grounds account for 0.6 tons per ton of coffee, coffee pulp for 0.5 tons per ton of coffee and husks account for 0.18 tons per ton of coffee (Anastopoulos et al., 2017; Janissen & Huynh, 2018). This waste is largely left underutilized and untreated, either by open dumping or burying for compost formation, thereby escalating environment crisis.

Another example of beverage consumed abundantly by humans is tea. Tea is produced globally by numerous countries and encompasses various kinds of tea. Global production of tea has reached 6.5 million tons annually (FAO, 2022). Nevertheless, tea processing of any kind generates waste that is untreated and underutilized and is discarded as residue. From global estimations, it is found that nearly 75% of the commercial tea produced globally is discarded as waste, which amounts to 4.8 million tons (Gao & Ogata, 2020).

Another industrially important crops with high economic value are the oil crops, for extraction of oil from either fruits or seeds. Oil crops form the second most important source of energy, coming just after cereals, for human usage. Extraction of oil generates a by-product commonly known as de-oiled seed cakes. These are organic residues left after processing and are high in proteins, carbohydrates, minerals, and lipids. Recent global trends show oil production to be 208 million tonnes globally (FAO, 2022), pointing towards the massive number of residues as waste.

Lastly, a huge section of agro-industrial waste is contributed by the processing of fruits and vegetables. These are processed for countless products for human consumption on a daily-basis at commercial levels, either in raw or processed forms. The wastes thus generated include peels, pomace, rind, and seeds. A major section of these wastes is covered by common fruits, accounting for 25–57 million tons of agro-industrial waste per year, (Leong & Chang, 2022) amongst which fruit peels contribute approximately 15–60% of waste including banana peel, orange peel, mango peel and seeds, pomegranate peel, lemon peel, apple peel and seeds, coconut husks, etc. Figure 5.2 depicts a graphical representation of agro-industrial wastes produced by various industries, with estimated global annual production in million tonnes (Kordi et al., 2023; Webber et al., 2014; Prueckler et al., 2014; Miranda et al., 2021; FAO, 2020, 2022; Echeverria & Nuti, 2017; Leong & Chang, 2022).

Over the decades, the intensification of agricultural and related agro-industrial activities has led to the production and accumulation of large quantities of waste matter. This adds to environmental problems like greenhouse effect, global

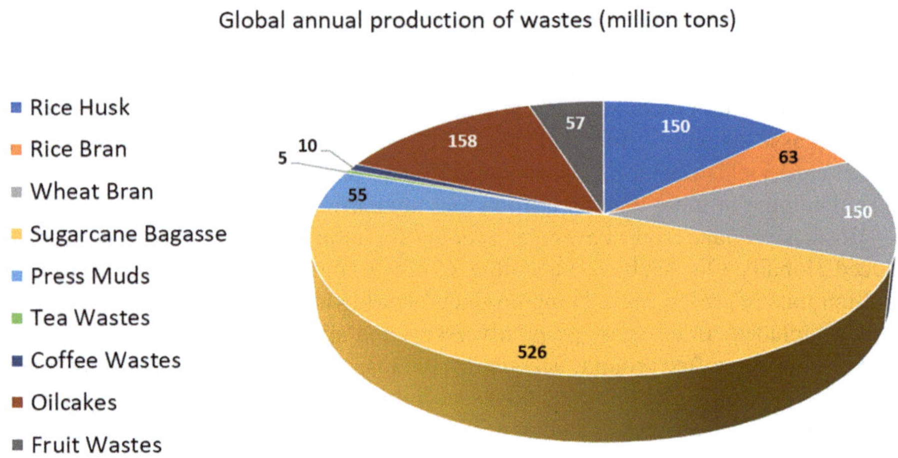

Fig. 5.2 Graphical representation of annual production of agro-industrial wastes (in million tons) globally through various industries

warming, deteriorating water and soil quality and eutrophication as well as economic problems. In light of adverse effects resulting from underutilized agro-industrial wastes, waste management is seen as a tool for a sustainable environment and agricultural economy *via* practices like clean technology and integrated waste management.

5.3 Ecological Consequences of Agro-Industrial Waste

Agro-industrial waste, often generated from agricultural and food processing activities, can have significant ecological consequences when not managed properly. The disposal and mismanagement of such waste can lead to a range of adverse impacts. The food production chain is one of the main concerns associated with agro-industrial waste. As the world's population has seen a substantial increase, growing from 3.7 billion in 1970 to 7.9 billion in 2021, projections indicate that it is expected to reach 9 billion by 2050 and 11 billion by 2100. Consequently, the challenge of ensuring food security is poised to become even more pressing in the years to come. Notably, China, India, and Africa have not only undergone rapid economic and population growth over the past century but have also witnessed a significant expansion in their agro-industrial waste production capacity (Pisello et al., 2021). As food processing and agriculture generate substantial amounts of waste, managing and getting rid of these by-products can be a resource-intensive process for producers. The disposal of agro-industrial waste has far-reaching environmental impacts. Traditional methods of waste disposal, such as dumping or burning, have raised concerns due to their adverse effects on the environment (Freitas et al., 2021).

Agro-industrial waste is composed of organic materials, including crop residues, food by-products, and another biomass. These organic components harbor stored energy in the form of carbon compounds. When burnt, this latent energy is liberated as heat, which can be harnessed for various purposes. Nevertheless, the release of heat through burning is usually unregulated and frequently goes to waste, dissipating into the environment. The loss of potential energy through combustion is a significant issue, especially in areas with limited access to economical and reliable energy sources. These waste, which includes pesticides, herbicides, and chemical fertilizers, often contains harmful chemicals that can leach into the soil and contaminate groundwater. Its disposal constitutes a global problem because much of it is currently burned or buried underground, causing air and water pollution that may have negative impacts on the environment. Improper disposal of crop residue produces greenhouse gases (GHG) such as carbon dioxide (CO_2), nitrous oxide (N_2O), methane (CH_4), and sulfur dioxide (SO_2) which pose a threat to humans and nature (Pisello et al., 2021). This untreated pollution can harm terrestrial and aquatic ecosystems, impacting biodiversity and making the land unsuitable for agriculture. When organic waste, such as crop residues and food by-products, is subjected to combustion, it releases carbon dioxide (CO_2) and other pollutants into the atmosphere. The emissions generated from burning waste can contain harmful pollutants, including particulate matter and toxic compounds, which degrade air quality (Karande et al., 2021). Moreover, the burning of agro-industrial waste exacerbates health crises in these areas. Due to their chemical stability and resistance to breakdown, these wastes have the potential to create severe toxic effects on humans (Arora, 2019). Sustained exposure to such pollutants can result in respiratory problems, cardiovascular diseases, and other health issues in local populations. Pollution stemming from waste incineration and mismanagement can contaminate soil and water sources, thereby impacting agricultural productivity and access to safe drinking water. This, in turn, affects food security, livelihoods, and overall well-being. The combination of environmental degradation, health risks, and economic challenges contributes to a reduced quality of life for those living in these areas. Rice husk, a by-product of rice milling, exhibits a notable high silica content, typically ranging between 8.7% and 12.1% (Karande et al., 2021). When rice husks are returned to the soil as agricultural mulch or used in composting, the silica gradually decomposes. During this process, silica aids in enhancing soil structure and fertility. It contributes to improved soil porosity, allowing better water retention and aeration, which subsequently promotes healthier root growth and nutrient uptake by plants. Moreover, the gradual release of silica aids in buffering soil pH, contributing to a more balanced soil environment.

Agro-industrial waste dumping often results in soil and water contamination. Soil pollution can hinder plant growth, disrupt nutrient cycles, and degrade the overall quality of soil, thus impacting agricultural productivity. Furthermore, the waste, when mishandled, can disrupt the growth and proliferation of microorganisms. These microorganisms play an indispensable role in the decomposition and recycling of organic matter within ecosystems, as they break down complex organic compounds into simpler forms, thereby releasing vital nutrients back into the environment.

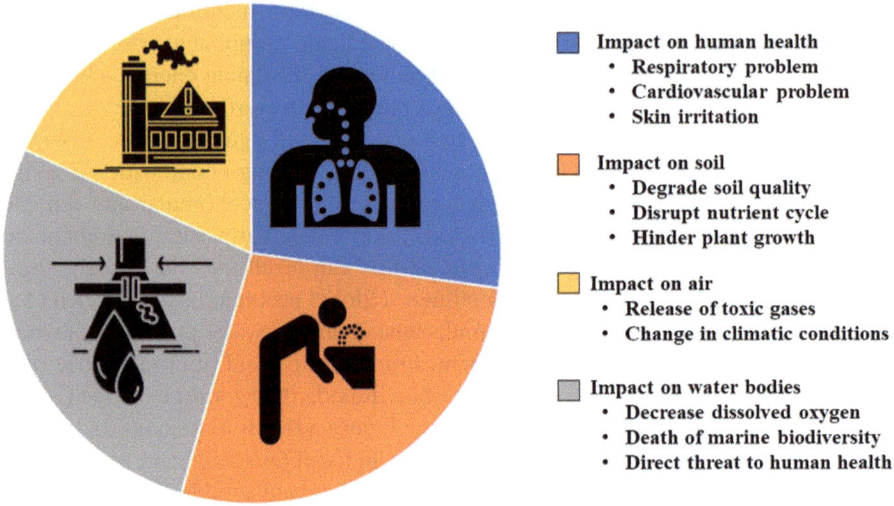

Fig. 5.3 Ecological consequences of agro-industrial waste on human health, air, soil, and water qualities

When agro-industrial waste is dumped into water bodies without proper treatment, it has a direct impact on aquatic biodiversity (flora and fauna). They promote pollution of the entire ecosystem, resulting in the death of aquatic life. They also impair photosynthetic activity and lead to a decrease in dissolved oxygen in water. The liquid form of these waste residues contributes significantly to wastewater contamination due to its high organic load, color, turbidity, and substantial oil and grease content (Arora, 2019). Water pollution poses a direct threat to human health because harmful chemicals and pathogens can leach into water sources used for drinking and irrigation. Figure 5.3 illustrates the ecological implications on human air, soil, and water due to agro industrial waste.

These agro-industrial waste when generated on a large scale and are managed improperly, it can give rise to significant ecological issues. Sustainable alternatives for the disposal of agro-industrial waste are available. Researchers have investigated numerous bio-based solutions, such as bioenergy and biorefinery processes, for the reuse of this waste. Bio-based solutions offer a transformative approach to agro-waste management, turning what was once considered waste into valuable resources. Compared with existing thermal and chemical pre-treatment processes, the bio-based methods are energy-efficient and eco-friendly. This not only solves the problem of waste disposal but also provides a renewable energy source, thereby reducing dependence on fossil fuel and minimizing greenhouse gas emissions. Researchers and industries have increasingly explored these innovative approaches as they align with sustainability goals. Furthermore, these approaches often lead to the production of biofuels, enzymes, organic acids, and other valuable compounds from agricultural residues, effectively converting waste into valuable resources.

5.4 Bio-Based Solution for Agro-Waste Management: From Waste to Wonder

In the field of agricultural waste management, there has been a notable paradigm shift towards bio-based solutions. Traditionally, for uses such as combustion, fodder, thatching, composting, mulching, and industrial applications such as match and paper production, crop residues, especially biomass lignocellulose, are being revaluated for their potential as sustainable sources of bioenergy and biofuels. The conversion of agro-waste into bioenergy, such as bioethanol and biogas are the prominent bio-based solution. These solutions employ biological processes, such as enzymatic treatments, microbial fermentation, and bioremediation, harnessing the power of natural systems to degrade and transform waste materials (Maraveas, 2020). Both biowastes and various by-product sources from industrial sectors such as the pulp, paper, and food industries can be utilized as sustainable assets for biodiesel and bioethanol production using fermentative bioconversion processes. Bioethanol is not only a biofuel and platform chemical, but also an alternative to traditional energy sources. It is created by fermentation of various carbon sources, including lignocellulosic materials, starchy plants, and sucrose-based feedstock (Pisello et al., 2021). Anaerobic digestion of organic waste, including crop residues, manure, and food processing by-products, can produce biogas primarily composed of methane and carbon dioxide. Biogas serves as a renewable energy source that can be utilized for electricity generation, heating, and cooking. (Manyi-Loh & Lues, 2023)

Biocomposting, also known as biological composting or bioconversion, is a natural and controlled process that utilizes microorganisms to decompose organic materials into a nutrient-rich soil amendment called compost. It involves the breakdown of organic matter, such as kitchen scraps, yard waste, agricultural residues, and other biodegradable materials, through the activity of bacteria, fungi, worms, and other decomposers. The characteristics of composites heavily rely on the interaction between the matrix (continuous phase) and filler (discontinuous phase), along with the composition, size, shape, and amount of reinforcement. These fillers can be classified into organic and inorganic fillers. Organic fillers play a pivotal role in the production of biocomposites, offering a sustainable and eco-friendly alternative to conventional materials. These fillers, derived from natural sources, are integrated into biopolymer matrices to create composite materials with enhanced properties. Several organic fillers are commonly utilized in biocomposite manufacturing. Natural fibers, such as jute, hemp, flax, and sisal, are prevalent choices due to their abundance, renewability, and favorable mechanical properties. These fibers are incorporated into biopolymers to reinforce the composite structure, improving strength, stiffness, and impact resistance. Moreover, lignocellulosic materials, including wood flour and cellulose fibers derived from wood pulp, are also employed as organic fillers. These materials offer good mechanical performance, dimensional stability, and low density, making them suitable for various applications in biocomposite manufacturing (Mishra et al., 2023; Pisello et al., 2021).

The extensive use of chemical fertilizers over an extended period has resulted in numerous environmental complications, adversely impacting both animals and humans. It has been revealed that approximately 10% of total greenhouse gas emissions can be attributed to these chemical fertilizers. Furthermore, they detrimentally affect soil quality and crop productivity. In the present scenario, biofertilizers emerge as a viable alternative to chemical fertilizers. They offer sustainable solutions and contribute to bioeconomic markets due to their cost-effectiveness and easy availability. Biofertilizers comprise living micro-organisms that efficiently supply essential nutrients to the soil for crop growth without causing detrimental effects on the environment (Mishra et al., 2023). Straw, a residue left after harvesting grains like wheat, rice, or barley, is rich in organic carbon. Returning straw to the field allows this organic matter to decompose naturally, enriching the soil with essential nutrients like carbon, nitrogen, and phosphorus (Bian et al., 2019). Soil health is also improved by the left-over straw as it can enhance soil structure hence decomposing straw adds organic matter to the soil, improving soil structure and texture. It increases soil porosity, aeration, and water retention capacity, thereby promoting better root growth and nutrient uptake by plants. The decomposition of straw by soil microorganisms increases microbial activity. This activity supports nutrient cycling, making nutrients more available to plants and supporting a healthy soil ecosystem (Arora, 2019).

Carbon-rich plant material is effective in producing soil amendments. Plant residues such as crop stalks, husks, straws, wood chips, and other biomass contain a significant amount of carbon. These materials are composed of organic compounds like cellulose, lignin, and other carbon-based molecules, making them valuable sources of carbon. Through a process called pyrolysis, these carbon-rich plant materials can be converted into biochar. Pyrolysis involves heating organic materials in the absence of oxygen, resulting in the breakdown of complex organic molecules into a stable form of carbon, i.e., biochar. Biochar acts as a soil conditioner by improving soil structure and stability. Its porous nature creates a network of small channels and pores within the soil, enhancing aeration and water retention. Its porous structure provides a conducive environment for beneficial soil microbes. This encourages microbial activity, which further aids in nutrient cycling and promotes healthier soil (Manyi-Loh & Lues, 2023).

Biopesticides can kill various pests and increase crop yields. Interestingly, compared to chemical pesticides, biological pesticides are very effective in killing a wide range of pests in very small quantities. Additionally, biopesticides have a long shelf life and require less space and maintenance, as long-term use at low concentrations ensures effectiveness. Furthermore, bioinsecticides are effective against some pests that can be applied by spraying. The production of biopesticides from organic waste offers an environmentally friendly and cost-effective approach to solid waste recycling, and the use of biopesticides in agriculture can bring about improvements in plant productivity and nutritional value (Mehariya et al., 2019; Ortega et al., 2022). Some agro-industrial waste, like certain plant extracts or by-products from food processing, can be used to create natural pesticides or insect repellents. These biopesticides offer an eco-friendly alternative to chemical pesticides, reducing environmental harm. To minimize agricultural environmental pollution and health risks caused by pesticides, a pesticide delivery device using Neem Leaf Powder (NLP), a biopesticide, has been developed (Singh et al., 2010). The utilization of agro-industrial waste in the fabrication of agricultural shade nets (anti-insect nets) and mulching films represents a significant advancement in sustainable agricultural practices. These nets and films play pivotal roles in modern farming techniques, offering various benefits derived from repurposed organic materials (Maraveas, 2020). Agro-industrial waste, such as discarded crop residues, husks, or fibers from agricultural produce, serves as the primary material for manufacturing these products. By employing these waste materials in the production of agricultural shade nets, the resulting nets offer dual functionality. They provide shade to crops, protecting them from excessive sunlight while also acting as a barrier against insects, minimizing infestation risks and reducing the need for chemical pesticides. Moreover, mulching films made from agro-industrial waste aid in weed suppression, moisture retention, and temperature regulation, fostering optimal growing conditions for crops. These films also contribute to soil health by preventing soil erosion and enhancing nutrient retention. The application of agro-industrial waste in the creation of these innovative agricultural aids not only mitigates environmental concerns associated with waste disposal but also promotes sustainable farming practices by offering cost-effective and eco-friendly solutions to farmers (Chiellini et al., 2004; Maraveas, 2020)

By exploring much more into the field of agricultural waste management, an intriguing and promising aspect emerges: the concept of biorefinery. Biorefineries represent a complex approach to extracting value from agricultural waste by integrating various advanced conversion processes. These facilities integrate various conversion processes to extract valuable compounds from agro-waste. Agro-waste can be refined into bio-based chemicals, bioplastics, and other high-value products. Biorefineries promote resource efficiency and minimize waste while promoting a circular economy (Chiellini et al., 2004; Mehariya et al., 2019). Agro-industrial waste possesses rich nutritional content and fosters the growth of microorganisms. Most of the agricultural waste comprises lignocellulosic materials abundant in carbohydrates. Consequently, these agricultural residues offer the potential to produce valuable products, including crucial industrial enzymes. Various agro-industrial wastes like sugar cane bagasse, corn cob, and rice bran have been extensively studied using diverse fermentation approaches to produce enzymes of significance. Enzymes extracted from agro-waste present opportunities for various industrial processes. For instance, enzymes derived from agricultural residues, such as cellulases and amylases, find applications in textile production, aiding in the breakdown of fibers and starches. These enzymes enable eco-friendly and efficient processing methods in industries (Ravindran et al., 2018).

The conversion of agro-industrial waste into green nanomaterials represents a sustainable approach in nanotechnology, offering both environmental benefits and resource efficiency. A high surface area to volume ratio gives nano-sized particles unique properties. Agro-waste, such as agricultural residues and by-products, can be utilized as precursors for the synthesis of nanomaterials, utilizing their inherent properties and chemical composition. Various nanoforms can be generated for their use in sustainable agriculture. These green nanomaterials find applications across various fields, including medicine, environmental remediation, and renewable energy, owing to their biocompatibility, low toxicity, and eco-friendly nature.

5.5 Agro-Industrial Waste Conversion into Green Nanomaterials

Recent developments and research in the field of nanotechnology hold substantial promise for providing an efficient alternative to traditional synthetic compounds that are widely used in the field of agriculture, thus lowering their adverse impact on the environment. Nanomaterials (NMs) exhibit various advantages when applied to crops at minimal concentrations, thereby lowering the frequency and quantities of treatment. Due to their large surface area, small size, and unique characteristic that lies in 1–100 nm range, the interest in nanomaterials arises, which enable NMs to penetrate and interact with different cells and tissues of plants in comparison to bulk material (Qazi & Dar, 2020). In agriculture, nanomaterials are widely known due to its multifaceted potential, as they can also be applied due to their antioxidant, antimicrobial, catalytic and larvicidal properties (Adelere & Lateef, 2016; Preetha et al., 2024). Furthermore, it has been proposed that both organic and inorganic nanomaterials can improve food quality, enhance the productivity of plants, and promote safety in the environment (McClements, 2020). The application of nanomaterials as nanofertilizers, nanobiosensors, nanopesticides, or plant growth stimulators (Fig. 5.4) enhances agricultural performance by strengthening plant defense against diseases and allowing effective nutrients or pesticide absorption (Rabalao et al., 2022; Vijayakumar et al., 2022; Pudake et al., 2024).

Conventionally, the synthesis of NMs involves expensive and time-consuming physical and chemical methods, which can also cause harmful effects on the environment. In contrast, environment-friendly alternatives based on biological methods are used for the synthesis of NMs, predominantly utilizing plant extracts and waste (Goswami et al., 2019; Saratale et al., 2018). The efficacy of the biogenic production of NMs, also known as green synthesis or sometimes photosynthesis, relies on the composition and quality of the biological agents employed for the production of NMs. For example, during the process of metal oxide nanomaterial synthesis a biological extract is generally reacted with a solution containing metal salt, where certain compounds act as capping agents that are present in the extract, then the processes of nanoparticle formation stabilize (Del Buono et al., 2021; Tripathi et al., 2018; Ying et al., 2022). Importantly, this approach is mainly characterized by minimal or no impact on the environment (Tolisano & Del Buono, 2023).

For the green synthesis of metallic as well as nonmetallic NMs, the growing interest in utilizing renewable biomass and resources has majorly focused the research and studies on using biobased materials. As a result, this green approach valorizes the employed biomass, thus offering an alternative method to its potential

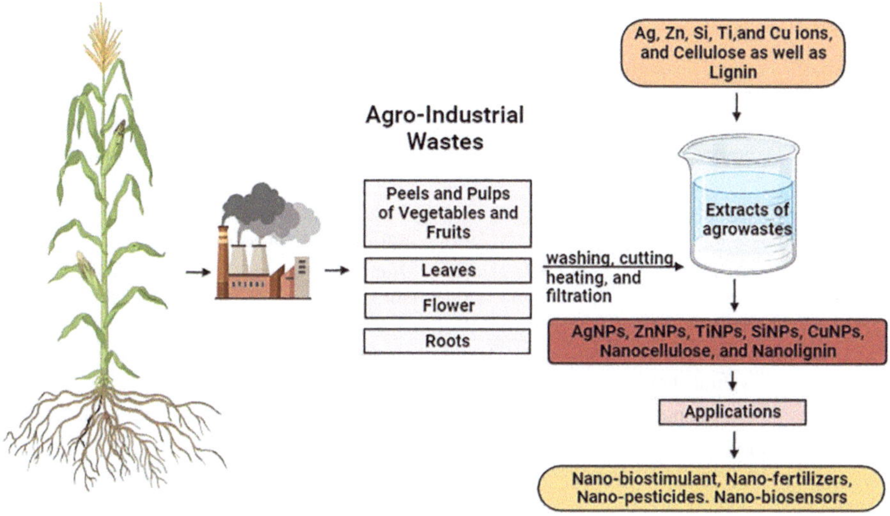

Fig. 5.4 Schematic representation of green synthesis of nanomaterials from agro-industrial wastes

disposal (Adelere & Lateef, 2016). The resultant NMs can contribute in enhancing agricultural uses, therefore aligning with the principle of circular economy. This increasing interest in producing novel NMs is due to the efficiency of nanostructured materials, particularly in the form of nanoparticles and nanofertilizers, which have shown the ability to increase crop productivity or to fight against the pests present in plants. Furthermore, nanomaterials for agricultural applications continue to be a focal point of intensive research, as indicated by the increasing number of applications to be patent (Ghosh et al., 2022). The prevailing perception is that nanotechnology could trigger a strategic revolution in the field of agriculture (Sabourin & Ayande, 2015; Pudake et al., 2019).

5.5.1 Green Nanomaterials Derived from Agricultural Biomass and Crop Residues

In the present context, the utilization of agricultural by-products such as agro-industrial waste (including peels of fruit, bagasse, fruit pulps, pulps and peels of vegetables, and cakes of oilseeds derived from soybeans, coconuts, and peanuts) and crop residues (e.g., seed pods, leaf litters, weeds, stems, etc.) rich in specific compounds has shown effectiveness in the synthesis of biogenic NMs (Koul et al., 2022). The research discussed and reported in the current paragraph and the subsequent one, focusing on the acquisition and testing of green nanomaterials derived from agro-industrial wastes and crop residues on plants, are summarized in Table 5.1 (Tolisano & Del Buono, 2023).

Table 5.1 List of various nanomaterials, their sources and applications

Agro-industrial waste	Source	Materials formulated	Applications	References
Peels and shells	Banana and orange peels	Nano-silica	Nano pesticide	Peerzada and Chidambaram (2020)
	Garlic peels	Ag/PNC	Nanobiosensors for agricultural purposes	Ji et al. (2021)
	Citrus aurantium Peels	ZnO NP	Nano-biostimulant for tomato	Punitha et al. (2023)
	Almond and walnut shells	Nano-adsorbent	wastewater treatment	Salem et al. (2020)
	Groundnut peels	Ag NP	Larvicidal activity	Nath et al. (2023)
	Citrus sinensis (orange) peel	MgO NP	Photolytic bioremediation properties	Nath et al. (2023)
	Pomelo peel	Carbon NP	Biosensor for pollutant detection	Abdelbasir et al. (2020)
Crop residues	Cotton residues	Nanocellulose	Nanoparticles in a biodegradable polymeric matrix of polyacid lactic	de Souza et al. (2020)
	Wheat and barley grass wastes	Nano-adsorbent	Bioremediation of agricultural water	Akhayere (2019)
	Wheat residues	Ag (coated with chitosan) nanostructure	Nano-biostimulant	Mondéjar-López et al. (2023)
	Pistachio crop residues	Ag NP	Nano biostimulant for *Solanum melongena*	Khan et al. (2021b)
Leaves	Eucalyptus Leaves	Mn NP	Nano-biostimulant for *Zea mays*	Gonçalves et al. (2022)
	Avicennia marina leaves	Cu NP	Nano-biostimulant for *Triticum aestivum*	Essa et al. (2021)
Husks and fibres	Coconut husk	Nano-silica	Nano-pesticide	Peerzada and Chidambaram (2020)
	Coconut husk fiber and rice husk	Nanocellulose	Environmental remediation	Nang An et al. (2020)
	Rice husk	Nanocomposite	Nano-adsorbent for bioremediation purpose	Dakroury et al. (2020)
	Rice husk	SiO_2 NPs	Nano-biostimulant for plant and the soil	Giordana et al., 2023
	Rice husk	Mesoporous Nanosilica	Good carrier of fertilizers and herbicides	Ali et al. (2021)

(continued)

Table 5.1 (continued)

Agro-industrial waste	Source	Materials formulated	Applications	References
Bagasse and fruit pomaces	Sugar cane bagasse and wheat straws	Nano-agrowaste	Better waste management	Fareed et al. (2020)
	Punica granatum pomace	Zn NP	Nano biostimulant	Bahmanzadegan et al., 2022
	Olive pomace	Nanocomposite	Bioremediation	Dakroury et al. (2020)
	Grape pomace	Different nanomaterials	Biofertilizer and other important applications	Caponio et al., 2023)
	Sugarcane straw	Lignocellulosic nanocrystals	Agrowaste management and potential future application	Bilatto et al. (2020)
Other plant wastes	*Nypa fruticans* trunk	Nanocellulose	Environmental remediation	Nang An et al. (2020)
	Tea stalk	Nanocellulose crystals	NP with different purposes	Guo et al. (2020)
	Grape (stalk, skin, and seed waste)	Au NP	Different agricultural purposes, antioxidant	Nath et al. (2023)
	Agrowaste (chitosan)	Alginate/chitosan nanocarriers	Nano- biostimulant	Magnabosco et al. (2023)
	Cuminum cyminum Seeds	TiO_2 NP	Nano-biostimulant for Mung beans	Mathew (2021)
	Different agrowaste	Nanostructure (Au-TiO_2 on chitosan)	Bioremediation of agrowaste water	Shoueir et al. (2019)

In an experiment carried out by Khan et al. (2021a, b), it was explored that the biogenic synthesis of AgNPs as a biological reducing agent can be done using a pistachio (*Pistacia vera*) seed coat which is an agro-industrial waste. The phenolic and different compounds present in this agro-waste facilitated the lowering of silver ions, preventing them from forming aggregation. The biosynthesized NMs exhibited distinct shapes (spherical, rectangular) having an approximate size of 20 nm (Khan et al., 2021a, b). Furthermore, the foliar application of these NMs at a concentration of 75 mg L^{-1} on eggplant (*Solanum melongena*) seedlings resulted in significant plant-growth stimulation, as evidenced by increased carotenoids and chlorophyll, as well as enhance growth of plants concerning, weight and length (Khan et al., 2021b).

During the extraction of raw materials from crops, industrial agricultural and food production activities generate huge quantities of waste throughout the processes of manufacturing (Tolisano & Del Buono, 2023). In a recent study by

Mondéjar-López et al. (2023), the synthesis of AgNPs was done using wheat residues. These nanoparticles were then coated with chitosan, resulting in the formation of a spherical-shaped complex (AgNPs-CH), with dimensions ranging from approximately 2–10 nm. The primary goal of this study was to examine the influence of nanopriming on the seeds of wheat (*Triticum vulgare*), describing the physiological responses induced by AgNPs-CH and determining any potential phytotoxicity during the stage of germination. Additionally, they also explored AgNPs-CH antifungal properties. Notably, the experiments also showed that it did not have any phytotoxic effects on the germination of seeds. Instead, the application of AgNPs-CH resulted in a significant rise in the content of total chlorophyll in seeds. This research highlights the potential of deriving nanomaterials from crop waste for use within the same crop, thereby aligning with the circular economy principles (Mondéjar-López et al., 2023).

Additional agricultural applications of nanomaterials are exemplified by nanofertilizers, the synthesis of which can also originate from waste materials such as silicon nanoparticle (SiNPs) (Giordana et al., 2023) and zinc nanoparticles (ZnNPs) (Tolisano & Del Buono, 2023). Bahmanzadegan et al. (2022) used this approach for synthesizing zinc nanostructures (Zn NS) from the waste of pomegranate pomace (*Punica granatum*) extracted after the processing of juice. By using the waste as a stabilizing and reducing agent, the researchers were able to produce Zn NS having dimensions around 50 nm. These nanostructures were then applied to the foliage of *Zataria multiflora*, a plant having traditional medicinal and culinary uses. The use of Zn NS resulted in the increased synthesis of essential oil and the main element of linalool. Moreover, high levels of flavonoid and phenolic compounds, total carotenoids, chlorophyll content, and antioxidant activity were observed. These findings highlight the ecological potential of these nanomaterials as sustainable nanofertilizers in agricultural systems (Bahmanzadegan et al., 2022).

5.5.2 Synthesis of Nanocellulose Using Agro-Waste Products

Lignocellulosic biomass (LCB) has undergone extensive research and studies, showing its potential as a primary substrate used for the synthesis of advanced materials such as biopolymers and biofuels (Mateo et al., 2021). Currently, the primary sources from where the lignocellulosic waste can be derived include agricultural, industrial, and forestry materials. Lignocellulosic wastes encompass cellulose as their primary component which is an unbranched natural polymer consisting of repetitive units of glucose. Residual biomass (such as cotton, argan press cake, apple pomace, olive tree pruning, or vine shoots) has undergone a variety of treatments to transform it into cellulose nanofibers and nanocrystals. Various industries such as biomedicine, food packaging, and mechanical reinforcement of membrane filtration and matrices have applied nanocellulose-based materials across diverse applications (Abitbol et al., 2016).

The term "nanocellulose" is referred as a collective term for micro-sized and nano-sized particles of cellulose, representing the breakdown product created by cellulosic fibers based on their length and diameter which depends on the structure (Mateo et al., 2021). This material has promoted extensive research across numerous domains and applications, encompassing nanocomposites, viscosity modifiers, aerogels, gels, films, fibers, barrier layers, filtering membranes, and energy applications (Heise et al., 2020). Due to the ordered structure of cellulose, it facilitates the isolation of nanocellulose from LCB through processes aiming to breakdown the fibers of cellulose into cellulose nanofibers (CNF) and cellulose nanocrystals (CNC), which mainly depends on the specific conditions maintained during processing.

Commonly employed methods for obtaining CNF and CNC involve mechanical treatments and acid hydrolysis, respectively. However, these techniques have a few disadvantages related to environmental and economic aspects such as the requirement of large concentration of water during the neutralization step and high energy demand (Ribeiro et al., 2019). In addition to this, enzymatic hydrolysis emerges as an eco-friendly and promising route for the production of nanocellulose, though the cost of the enzymes may pose challenges. To promote cellulose fibrillation and reduce energy consumption during the isolation of nanocellulose, some researchers have explored and combined enzymatic and mechanical treatment processes (Mateo et al., 2021).

Three distinct morphologies of nanocellulose are recognized: cellulose nanocrystal (CNC), which is a highly crystalline cellulose present in rod-shape primarily derived through hydrolysis with concentrated mineral acids (majorly H_2SO_4 or HCl); cellulose nanofibrils (CNF) typically obtained through mechanical and/or enzymatic disintegration processes; and bacterial nanocellulose (BNC), obtained mainly from the *Gluconoacetobacter xylinus* bacterial family. To standardize various terminologies that are used, the International Organization for Standardization (ISO), and the Technical Association of the Pulp and Paper Industry (TAPPI) created worldwide standards for nanocellulose, based on size distinctions (El Achaby et al., 2018).

5.5.3 Formulation of Green Nanomaterials from Plant Waste

Nanomaterials (NMs) can be efficiently biosynthesized using wastes from diverse plant species, offering a biobased approach that extends to plant valorization, which is not necessarily associated with food or, in some cases, thought to be invasive. This method aligns with the need to develop nanostructured materials through the valorization and utilization of renewable resources. Plant wastes have been shown to be beneficial for biogenic synthesis since they are rich in molecules with reducing and capping agent properties (Tolisano & Del Buono, 2023).

ZnO NPs gain particular interest because of the crucial role played by zinc as an essential element in various plant metabolic processes. The translocation and uptake mechanisms shown by ZnO NPs are mainly associated with the size of NMs and the releasing process of Zn^{2+} (Zinc ions). Following symplastic and apoplastic pathways, Zn, in this form, can enter into the leaves through cuticles, hydathodes, and stomata, and in the roots through lesions and lenticels (Singh et al., 2019).

A comparative study by Zaeem et al. (2020) was conducted to know the effectiveness of ZnO NPs in plants, which is bio-synthesized with the plant waste of flax (*Linum usitatissimum*) and then for the process of *in vitro* propagation that is done for both the stem-derived callus and seedlings of flax plants NMs were added in the growth media. At a concentration of 10 mg L^1, ZnO NPs exhibited stimulatory effects on the seedlings, enhancing the production of biomass such as dry and fresh weight, shoot, and root length. Additionally, ZnO NPs also elevated the enzymatic and non-enzymatic antioxidants, positively impacting stem-derived callus by increasing the accumulation of secondary metabolites and enhancing the activity of radical scavenging (Zaeem et al., 2020).

In addition to zinc oxide, different metallic nanoparticles, including silver nanoparticles (Ag NPs), manganese nanoparticles (Mn NPs), copper nanoparticles (Cu NPs), titanium dioxide nanoparticles (TiO_2 NPs), etc., can be biogenically synthesized from plant waste, offering potential benefits to crops (Table 5.1) (Tolisano & Del Buono, 2023). For example, in plants, Ag NPs have been reported to promote seed vigor, seed germination, dry and fresh weight, and enhance the metabolic and physiological aspects. However, their effects are observed to be dose-dependent by their ability of translocation and diffusion which mainly depends on their physical (surface charge and size) and chemical characteristics (Kale et al., 2021).

5.5.4 Agro-Waste Based Nanolignin Synthesis

Over the years, lignin has gained considerable attention as a promising source of bio-polymers when compared to synthetic polymers because of its eco-friendly characteristics, abundance in phenolic compounds, and cost competitiveness. A significant effort has been made to explore the use of lignin in nano form, exploiting its special qualities at the nanoscale range, to accelerate their development (Duarah et al., 2020). Nanotechnology has enabled the utilization, production, and development of nanomaterials, which has led to a great deal of research on their synthesis and applications. The synthesis of nanomaterials has been a subject of numerous studies, driven by the exceptional characteristics exhibited at the nanoscale, including high surface area, enhanced control of light spectrum, extraordinary chemical reactivity, excellent strength and lightweight properties when compared to their micro-scale counterparts (Lee, 2018; Muñoz-Bonilla & Fernández-García, 2015; Sinha & Purkait, 2015).

According to several research, lignin has been acknowledged as a great precursor for the creation of nano-scale materials that are biocompatible. Strategically synthesized lignin nanoparticles (LNPs) with controllable morphology and size which are obtained by altering the type of lignin precursor have shown various physical properties (hydrophobicity and surface charge) and morphologies (Richter et al., 2016). LNPs, due to their tremendous characteristics, hold significant promise for applications in different fields, including derma product manufacturing, adsorption, hydrogel formation, drug delivery, and other value-added formations. The synthesis process of LNPs derived from the biomass involves different strategies, with notable processes such as alkaline precipitation, ultrasonication, acid precipitation, and solvent exchange. The nature of the reaction broadly categorizes the process of nanoparticle synthesis into three types: chemical, biological, and mechanical (Duarah et al., 2020). However, there are numerous studies on various aspects of lignin, though there is a scarcity of recent reviews and studies on the progress in the application and synthesis of LNPs. Additionally, none of the existing literature provides an in-depth analysis of the commercial aspects of LNPs. Therefore, a detailed study of various synthesis routes for the efficient production of LNPs is important (Duarah et al., 2020).

5.5.5 Nano Management of Agro-Industrial Wastes

Compared to developed countries, agro-industrial waste management is observed as a major challenge, particularly in developing nations. This comprehensive management includes processes such as collection, recovery, production, transportation, and disposal of wastes, requiring close supervision both before and after the process of disposal (Elbasiouny et al., 2020; Khan & Faisal, 2020). The management process of agricultural industrial wastes has gained attention in recent years, which resulted in the development of new methods for dealing with organic wastes (such as biochar, compost inorganic fertilizer) and eliminating the pollutants present in the environment. However, certain agro-wastes, such as oils, might not be appropriate for composting due to decrease in landfills and economic concerns. Therefore, over the last century, significant evolution has undergone in agro-waste management system, aligning with the evolution of recent civilization (Elbasiouny et al., 2020).

Adhering to the concept of the circular economy, effective utilization of agricultural wastes should consider the concept of the 3Rs—reduce, reuse, and recycle. This approach mainly aims for low consumption, high efficiency, and low emissions (Shaaban & Nasr, 2020). The concept of reduction of agro-waste includes lowering the inputs of materials, energy, and water, minimizing unnecessary manufacturing of products, and reducing high demand without compromising the quality of life. The "Reuse" concept promotes the multipurpose use of materials, utilizing waste as a raw material, the growth of the remanufacturing industry, and favoring renewable resource use over non-renewable resources. To promote resource utilization and recycling, the concept of "Recycle" focuses on transforming waste materials into an initiative for creating a system of circular economy, constructing technology parks, and circulation of interior material that differs from conventional models (Elbasiouny et al., 2020; Khan & Faisal, 2020). Numerous research has evaluated the management and utilization of agro-industrial wastes, exploring a diverse range of applications such as nanocellulose production for pharmaceutical purposes (Kamel et al., 2020), biochar production (Tolisano & Del Buono, 2023), nanosilica production, and Si-nanoparticles (Giordana et al., 2023), and the manufacturing of supercapacitors with high performance based on porous graphene oxide (Tamilselvi et al., 2020) and nanocarbon (Bhat et al., 2020). Furthermore, Agro-waste management strategies have also been studied, including food processing and agriculture waste (Pandey & Dwivedi, 2020), applications in antibiotic or antimicrobial resistance (Zalewska & Popowska, 2020), environmental impact and management approaches for agro-wastes (Maji et al., 2020), and production of agro-composites for packaging (Suffo et al., 2020). Therefore, nanomaterials that are derived from agro-wastes have gained substantial attention, with numerous studies demonstrating their potential in various applications, including farmland wastewater treatment and as replacements for fossil fuels (Javad et al., 2020). Table 5.1 summarizes various nanomaterials, their sources, and applications.

5.6 Potential Applications of Agro-Industrial-Waste Derived Green Nanomaterials for Sustainable Agriculture Practices

Agro-industrial waste encompasses diverse categories of residual materials, including peels, bagasse, straws, pomaces, etc. These waste materials arise as byproducts of the production and processing of agricultural commodities, including fruits, vegetables, sugarcane, and grains. Their generation imposes substantial environmental challenges attributable to both their voluminous quantities and the potential for environmental pollution if inadequately managed. A noteworthy approach in this regard involves the valorization of these waste materials i.e., transforming them into functional nanomaterials with potential applications. As elucidated previously, various types of nanomaterials have been synthesized from the aforementioned agro-industrial waste streams. This innovative approach not only facilitates effective waste management but also enables the utilization of the derived nanomaterials across various applications. The facile formulation process, eco-friendly attributes, minimal toxicity, and non-pathogenic characteristics collectively position these nanomaterials as promising candidates for deployment in diverse fields. Subsequent sections illustrate some significant applications in which these nanomaterials demonstrate considerable potential.

5.6.1 Nano-biostimulators

Despite the presence of natural phytohormones and nutrients, achieving optimal plant growth and development often requires additional stimuli. This is where biostimulants play a crucial role. Biostimulants, characterized as organic compounds that enhance plant growth by improving physiological processes, nutrient absorption, and stress responses, have garnered attention in sustainable agricultural practices. Agro-waste derived nanomaterials exhibit notable potential as bio stimulants due to their capacity to augment nutrient uptake, promote plant growth, and enhance stress tolerance, all while maintaining ecological soundness and sustainability.

Numerous studies have substantiated the diverse beneficial effects of these nano-biostimulants. For instance, research employing nonencapsulated silicon (Si) demonstrated positive impacts on both plant growth and nutrient uptake. Silicon, functioning as a micronutrient, acts as a cofactor in various physiological reactions within plant cells. The study not only elucidated the advantageous effects of Si on soil but also proposed its role in enhancing phosphorus availability (Giordana et al., 2023; Magnabosco et al., 2023). In a distinct application, mesoporous silica nanoparticles served as a delivery system for the plant hormone abscisic acid (ABA) in *Arabidopsis thaliana*. This methodology exhibited significant enhancements in drought tolerance, underscoring the role of these nanoparticles in supporting plant development and survival (Nath et al., 2023).

Chitosan, another notable agrowaste component, has been extensively examined for its efficacy. In field trials, experiments employing alginate/chitosan nanocarriers stacked with gibberellic acid (GA) produced promising outcomes, demonstrating significant enhancements in plant growth and yield under stress conditions.

Investigations about Cuminum seed have indicated its potential as a source for synthesizing TiO_2 nanoparticles. The application of these nanoparticles has demonstrated advantageous nano-biostimulant effects for Mung beans, as reported by Mathew (2021). Recent studies have substantiated numerous potential applications, as outlined in the accompanying table. Consequently, agro-waste-derived nanomaterials are emerging as promising and sustainable contributors to plant growth stimulation in agriculture

5.6.2 Nano-fertilizers

Fertilizers play an important role in increasing agricultural output. While conventional industrial-grade fertilizers are useful, they frequently contain toxic chemicals that not only pollute the agricultural ecology but also produce biohazards such as biomagnification and biotoxification. As a result, the search for environmentally benign, less hazardous alternatives has led to the investigation of biobased fertilizers. The synthesis of these fertilizers entails the extraction of chemicals from agro-industrial waste as well as the incorporation of nanotechnology, resulting in a more sustainable agricultural paradigm.

Lignocellulosic biomass, found in agro-waste leftovers such as peels, leaves, and stems, is a significant resource. When lignin from this waste is combined with nanoparticles, it produces carriers for fertilizers capable of controlled and sustained nutrient release. Nanocellulose, another important component of several agro wastes, can be modified to allow for the progressive encapsulation and release of nutrients, improving soil water retention and increasing nutrient accessibility for plant uptake (Mahapatra et al., 2022). Recent research has shown that nanoparticles and waste-derived biomass are capable of being utilized to biofortify crops (Shang et al., 2019). In a recent investigation conducted by Caponio et al. (2023), an exploration into olive pomace revealed its capacity to create diverse nanoparticles, which can be tailored for utilization as a biofertilizer. Subsequent research endeavors have further illuminated the promising potential of these nanoparticles as nanofertilizers, with detailed findings documented in the accompanying table.

The environmentally conscious use of nanoparticles and waste-derived biomass in agriculture has tremendous promise for supporting sustainable farming practices. Beyond environmental benefits, these novel technologies present prospects to reduce reliance on synthetic fertilizers, thereby decreasing the environmental pollution associated with their excessive use.

5.6.3 Nano-pesticides

Weeds and insects pose significant challenges to agricultural yield, while bacteria and fungi further hinder proper plant growth. To mitigate these factors, farmers often resort to the use of various chemicals that either kill or impede their effects. However, the widespread use of these chemicals has adverse effects on immediate abiotic components like the soil and water. Bioaccumulation of these chemicals in plants leads to adverse effects when consumed by different life forms.

Researchers have been exploring innovative solutions, including the development of nano-pesticides. Nano-pesticides have been formulated from different biologically available sources, as well as derived from the waste produced from an agricultural field. Their adaptability to different environmental conditions, ability to manipulate as per the requirements, and targeted delivery reducing off-target effects along with being an eco-friendly option make them a reliable option to use (Yin et al., 2023).

Noteworthy studies involve the formulation of nano-pesticides containing chitosan, a component of agricultural waste, and zinc oxide [ZnO]. The application of these nano-pesticides in fields has proven successful in warding off major pests (Nakum & Bhattacharya, 2022). In a separate study, cellulose, a major agro-waste product, was integrated with nanotechnology to create a successful nano-herbicide. In a study by Peerzada and Chidambaram (2020), the efficacious impact of nano-silica, sourced from coconut husk, was demonstrated as a pesticide. Additionally, AgNP derived from groundnut peels exhibited noteworthy effectiveness as a larvicidal agent, as reported by Nath et al. in 2023.

The multifunctionality of nanoparticles enables the development of nano-pesticides with synergistic mechanisms and enhanced safety profiles. Controlled-release nano-pesticides, derived from agro-waste, can be designed to be more effective, requiring fewer applications. This approach not only lessens their overall environmental impact but also promotes waste recycling.

5.6.4 Nano-bioremediation

The widespread utilization of fertilizers and pesticides has escalated soil and water contamination, presenting imminent risks to both crop well-being and the broader ecosystem. A promising avenue of inquiry involves the transformation of agro-wastes into environmentally friendly nanomaterials, which are then applied for soil and water remediation. This approach offers a potential long-term solution for environmental protection. Chitosan, extracted from agricultural waste, emerges as a particularly effective remediation agent when combined with nanoparticles. Its outstanding absorbent properties, non-toxic nature, and lack of harm to living organisms make it a compelling option for addressing environmental pollution (Nakum & Bhattacharya, 2022; Bhatnagar et al., 2016; Singh et al., 2011).

Studies focusing on the phenomenon of photolysis, a method for breaking down pollutants into simpler compounds such as CO_2, H_2O, and N_2, have been conducted (Nakum & Bhattacharya, 2022). The nanoscale production of TiO_2 and ZnO particles, in conjunction with chitosan, has shown notable success in the photolysis of organochlorine pesticides—whose prevalent utilization has led to the emergence of notable contaminants in agricultural land (Yu et al., 2007). Another effective remediation strategy involves a nanostructure comprising gold, titanium oxide, and chitosan, demonstrating promise in degrading pollutants like carbofuran and dyes as well as reducing toxic minerals in wastewater (Shoueir et al., 2019). Additionally, a nanocomposite of carbon nanofiber and cellulose coated with $CoFe_2O_4$ acts as a catalyst in the deterioration of phthalates, a common pollutant found in agricultural water sources. This nanocomposite showed high efficiency and stability in the degradation process. A study by Dakroury et al. (2020) presented the formulation of nanoadsorbents derived from rice husks, demonstrating their efficacy in bioremediation. Ongoing research efforts are consistently revealing the potential valorization of agro-waste into nanomaterials, positioning them as agents for the remediation of soil and water. This highlights their viability as a sustainable agricultural practice.

5.6.5 Nano Biosensor

A fundamental element of sustainable agricultural practices involves meticulous monitoring of field conditions to detect variations. These fluctuations encompass changes in temperature, soil pH, atmospheric humidity, pollutants in irrigation water, and other factors affecting crop health and productivity. Smart farming, a cutting-edge approach that integrates nanotechnology with fundamental agricultural science, employs sensors to offer precise and optimized solutions for crop cultivation (Mittal et al., 2020). This innovative method extends to the conversion of agricultural waste into diverse biosensors, seamlessly integrating nanotechnology with compounds derived from agricultural waste. The use of nano-biosensors expedites detection processes compared to traditional methods.

For example, a study introduces microfluidic paper chips designed for the detection of 2,4-D, a chemical prevalent in pesticides and a significant organic pollutant. In this instance, fluorescent paper strips were crafted using a combination of cellulose and quantum nanodots to enhance sensitivity and streamline the detection process for the specified pesticide (Nasrollahzadeh et al., 2021). Another investigation showcases the fabrication of carbon nanoparticles from Pomelo peel, serving as detectors for pesticide pollutants (Abdelbasir et al., 2020). Additionally, nanobiosensors find application in the detection and mitigation of plant pathogens and diseases (Thakur et al., 2022).

Nano biosensors play a pivotal role in augmenting agricultural yield by integrating nanotechnology with the internet, neural networks, and artificial vision to develop intelligent delivery systems for nutrient distribution. This convergence of technologies holds substantial promise for advancing precision agriculture and optimizing resource utilization in crop management, marking a significant stride toward sustainable and efficient agricultural practices.

5.7 Challenges and Implication of Solutions Associated in the Application of Green Nanomaterials for Sustainable Agriculture

While the utilization of green nanomaterials derived from agro-industrial waste holds promise, it is not without associated risks and challenges. A notable concern is the potential for environmental pollution due to the presence of toxic compounds or heavy metals in these nanomaterials. This raises the risk of leaching into the soil and groundwater, posing threats to plant and animal life as well as the overall ecosystem (Fig. 5.5). Additionally, the application of the said nanomaterials introduces uncertainties regarding their long-term impacts on soil health and biodiversity. Ensuring the safe and effective utilization of these nanomaterials presents another set of challenges. Optimal dosages and application methods may vary across different crops and soil types, necessitating a tailored approach for the best outcomes. Furthermore, the scale-up of production for these nanomaterials poses potential difficulties.

Thus, comprehensive research is needed to better comprehend both the potential risks and benefits associated with incorporating agrowaste-derived nanomaterials into sustainable agricultural practices. In response to the challenges posed by the adverse effects of nanomaterials, concerted efforts are underway to decrease

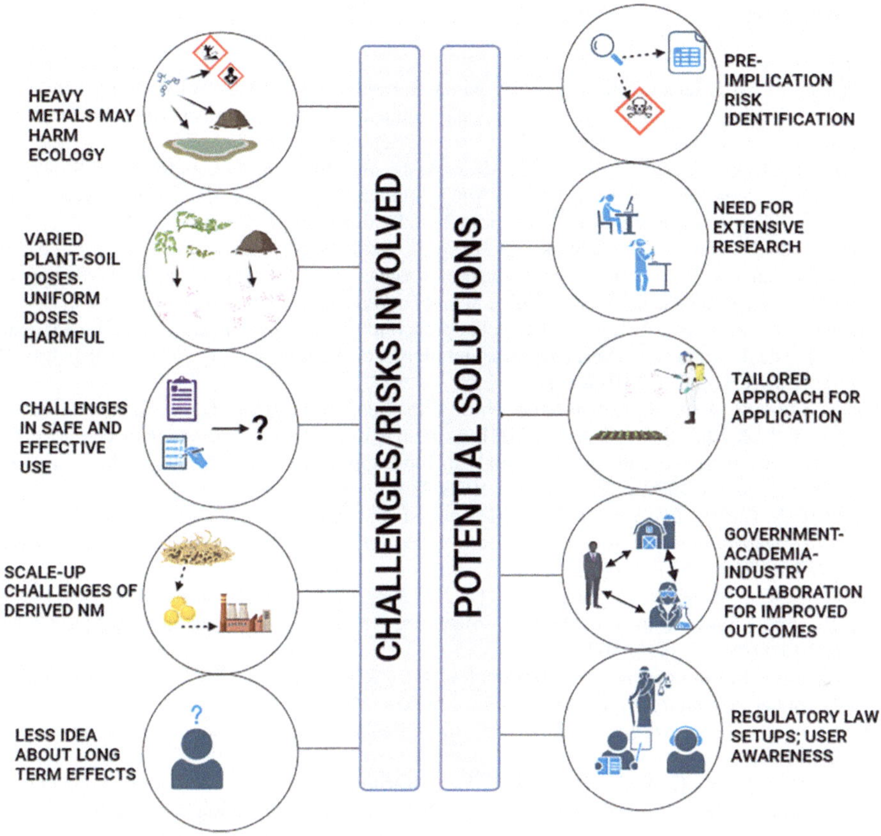

Fig. 5.5 Representation of different risks and potential solutions involved with the use of green nanomaterials for sustainable agricultural practices. (Image created by BioRender.com)

their impact and broaden their applications. A pivotal strategy involves a proactive reduction of their detrimental ecological effects, necessitating comprehensive risk assessments to identify potential hazards and the formulation of mitigation strategies. To ensure responsible deployment, the establishment stringent regulations and specific standards governing the utilization of these nanomaterials is imperative (Mahapatra et al., 2022).

Simultaneously, the advancement of research and innovation is pivotal to developing safer and more efficient application methods, fostering heightened awareness of their responsible use. This includes strategic investments in infrastructure and technology to bolster the scalable production of these materials. Moreover, collaborative initiatives involving researchers, industry experts, and government agencies are essential to address knowledge gaps and formulate best practices, thereby ensuring the secure and sustainable incorporation of agrowaste-derived nanomaterials in agricultural contexts.

References

Abdelbasir, S. M., McCourt, K. M., Lee, C. M., & Vanegas, D. C. (2020). Waste-derived nanoparticles: Synthesis Approaches, environmental applications, and sustainability considerations. *Frontiers in Chemistry, 8*. https://doi.org/10.3389/fchem.2020.00782

Abitbol, T., Rivkin, A., Cao, Y., Nevo, Y., Abraham, E., Ben-Shalom, T., Lapidot, S., & Shoseyov, O. (2016). Nanocellulose, a tiny fiber with huge applications. *Current Opinion in Biotechnology, 39*, 76–88. https://doi.org/10.1016/j.copbio.2016.01.002. Google Scholar.

Abol-Fotouh, D., Hassan, M. A., Shokry, H., Roig, A., Azab, M. S., & Kashyout, A. E. H. B. (2020). Bacterial nanocellulose from agro-industrial wastes: Low-cost and enhanced production by Komagataeibacter saccharivorans MD1. *Scientific Reports, 10*(1), 3491.

Adelere, I. A., & Lateef, A. (2016). A novel approach to the green synthesis of metallic nanoparticles: The use of agro-wastes, enzymes, and pigments. *Nanotechnology Reviews, 5*(6). https://doi.org/10.1515/ntrev-2016-0024

Ahmad Khorairi, A. N. S., Sofian-Seng, N.-S., Othaman, R., Rahman, H. A., Razali, N. S. M., Lim, S. J., & Mustapha, W. A. W. (2021). A review on agro-industrial waste as cellulose and nanocellulose source and their potentials in food applications. *Food Reviews International*. https://doi.org/10.1080/87559129.2021.1926478

Akhayere, E. (2019). *Effective and reusable nano-silica synthesized from barley and wheatgrass for the removal of nickel from agricultural wastewater*. Edidiong Augustine Essien. https://doi.org/10.1007/s11356-019-05759-x

Ali, S. H., Emran, M. Y., & Gomaa, H. (2021). *Rice husk-derived nanomaterials for potential applications*. Topics in Mining, Metallurgy and Materials Engineering. https://doi.org/10.1007/978-3-030-68031-2_19

Anastopoulos, I., Karamesouti, M., Mitropoulos, A. C., & Kyzas, G. Z. (2017). A review for coffee adsorbents. *Journal of Molecular Liquids, 229*, 555–565.

Arora, N. K. (2019). *Microorganisms for sustainability volume 14*. http://www.springer.com/series/14379

Bahmanzadegan, A., Tavallali, H., Tavallali, V., & Karimi, M. A. (2022). Variations in biochemical characteristics of Zataria multiflora in response to foliar application of zinc nano complex formed on pomace extract of Punica granatum. *Industrial Crops and Products, 187*, 115369. https://doi.org/10.1016/j.indcrop.2022.115369

Bhat, V. S., Kanagavalli, P., Sriram, G., Prabhu, B. R., John, N. S., Veerapandian, M., Kurkurie, M., & Hegde, G. (2020). Low cost, catalyst free, high performance supercapacitors based on porous nano carbon derived from agriculture waste. *Journal of Energy Storage, 32*, 101829. https://doi.org/10.1016/j.est.2020.101829

Bhatnagar, Y., Singh, G. B., Mathur, A., Srivastava, S., Gupta, S., & Gupta, N. (2016). Biodegradation of carbazole by Pseudomonas sp. GBS. 5 immobilized in polyvinyl alcohol beads. *Journal of Biochemical Technology, 6*(3), 1003–1007.

Bian, H., Gao, Y., Luo, J., Jiao, L., Wu, W., Fang, G., & Dai, H. (2019). Lignocellulosic nanofibrils produced using wheat straw and their pulping solid residue: From agricultural waste to cellulose nanomaterials. *Waste Management, 91*, 1–8. https://doi.org/10.1016/j.wasman.2019.04.052

Bilatto, S., Marconcini, J. M., Mattoso, L. H., & Farinas, C. S. (2020). Lignocellulose nanocrystals from sugarcane straw. *Industrial Crops and Products, 157*, 112938. https://doi.org/10.1016/j.indcrop.2020.112938

Caponio, G. R., Minervini, F., Tamma, G., Gambacorta, G., & De Angelis, M. (2023). Promising application of grape pomace and its agri-food valorization: Source of bioactive molecules with beneficial effects. *Sustainability*. https://doi.org/10.3390/su15119075

Chakraborty, A., Panda, J., Mishra, A. K., et al. (2024). Jackfruit waste for the production of sustainable biorefinery products: A step towards bio-economic advancement. *Biomass Conversion and Biorefinery*. https://doi.org/10.1007/s13399-024-05954-y

Chiellini, E., Cinelli, P., Chiellini, F., & Imam, S. H. (2004). Environmentally degradable bio-based polymeric blends and composites. *Macromolecular Bioscience, 4*(3), 218–231. https://doi.org/10.1002/mabi.200300126

Dakroury, G. A., Allan, K. F., Attallah, M. F., & El Afifi, E. M. (2020). Sorption and separation performance of certain natural radionuclides of environmental interest using silica/olive pomace nanocomposites. *Journal of Radioanalytical and Nuclear Chemistry, 325*(2), 625–639. https://doi.org/10.1007/s10967-020-07237-y

de Souza, A. G., Barbosa, R. F. S., & Rosa, D. S. (2020). Nanocellulose from industrial and agricultural waste for further use in PLA composites. *Journal of Polymers and the Environment, 28*(7), 1851–1868. https://doi.org/10.1007/s10924-020-01731-w

Del Buono, D., Di Michele, A., Costantino, F., Trevisan, M., & Lucini, L. (2021). Biogenic ZnO nanoparticles synthesized using a novel plant extract: Application to enhance physiological and biochemical traits in maize. *Nanomaterials, 11*(5), 1270. https://doi.org/10.3390/nano11051270

Duarah, P., Haldar, D., & Purkait, M. K. (2020). Technological advancement in the synthesis and applications of lignin-based nanoparticles derived from agro-industrial waste residues: A review. *International Journal of Biological Macromolecules, 163*, 1828–1843. https://doi.org/10.1016/j.ijbiomac.2020.09.076

Echeverria, M. C., & Nuti, M. (2017). Valorisation of the residues of coffee agro-industry: Perspectives and limitations. *The Open Waste Management Journal, 10*(1).

El Achaby, M. E., Kassab, Z., Barakat, A., & Aboulkas, A. (2018). Alfa fibers as viable sustainable source for cellulose nanocrystals extraction: Application for improving the tensile properties of biopolymer nanocomposite films. *Industrial Crops and Products, 112*, 499–510. https://doi.org/10.1016/j.indcrop.2017.12.049. Google Scholar.

Elbasiouny, H., Elbanna, B. A., Al-Najoli, E., Alsherief, A., Negm, S., Abou El-Nour, E., Nofal, A., & Sharabash, S. (2020). Agricultural waste management for climate change mitigation: Some implications to Egypt. In A. M. Negm, N. Shareef, et al. (Eds.), *Waste management in MENA regions, springer water* (pp. 149–169). https://doi.org/10.1007/978-3-030-18350-9_8

El-Ramady, H., El-Henawy, A., Amer, M., Omara, A. E. D., Elsakhawy, T., Elbasiouny, H., et al. (2020). Agricultural waste and its nano-management: Mini review. *Egyptian Journal of Soil Science, 60*(4), 349–364.

Essa, H. L., Abdelfattah, M. S., Marzouk, A. S., Shedeed, Z., Guirguis, H. A., & El-Sayed, M. M. H. (2021). Biogenic copper nanoparticles from Avicennia marina leaves: Impact on seed germination, detoxification enzymes, chlorophyll content and uptake by wheat seedlings. *PLoS ONE, 16*(4), e0249764. https://doi.org/10.1371/journal.pone.0249764

Fareed, A., Zaidi, S. B. A., Ahmad, N., Hafeez, I., Ali, A., & Ahmad, M. F. (2020). Use of agricultural waste ashes in asphalt binder and mixture: A sustainable solution to waste management. *Construction and Building Materials, 259*, 120575. https://doi.org/10.1016/j.conbuildmat.2020.120575

Food and Agricultural Organization (FAO). (2020). http://www.fao.org/economic/est/est-commodities/oilcrops/en/

Food and Agricultural Organization (FAO). (2022). *World food and agriculture—Statistical yearbook 2022.*

Freitas, L. C., Barbosa, J. R., da Costa, A. L. C., Bezerra, F. W. F., Pinto, R. H. H., & Carvalho Junior, R. N. d. (2021). From waste to sustainable industry: How can agro-industrial wastes help in the development of new products? In *Resources, Conservation and Recycling* (Vol. 169). Elsevier B.V.. https://doi.org/10.1016/j.resconrec.2021.105466

Gao, P., & Ogata, Y. (2020). CHAMU: An effective approach for improving the recycling of tea waste. In *IOP conference series: Materials science and engineering* (Vol. 711, No. 1, p. 012024). IOP Publishing.

Ghosh, S., Sarkar, B., Kumar, A., & Thongmee, S. (2022). Regulatory affairs, commercialization, and economic aspects of nanomaterials used for agriculture. In S. Ghosh, S. Thongmee, & A. Kumar (Eds.), *Agricultural nanobiotechnology, Woodhead publishing series in food science, technology and nutrition, 18* (pp. 479–502). Woodhead Publishing. https://doi.org/10.1016/B978-0-323-91908-1.00008-0

Giordana, A., Malandrino, M., Zambon, A., Lusvardi, G., Operti, L., & Cerrato, G. (2023). Biostimulants derived from organic urban wastes and biomasses: An innovative approach. *Frontiers in Chemistry, 11*, 969865. https://doi.org/10.3389/fchem.2023.969865

Gonçalves, J. P. Z., Seraglio, J., Macuvele, D. L. P., Padoin, N., Soares, C., & Riella, H. G. (2022). Green synthesis of manganese based nanoparticles mediated by Eucalyptus robusta and Corymbia citriodora for agricultural applications. *Colloids and Surfaces A: Physicochemical and Engineering Aspects, 636*, 128180. https://doi.org/10.1016/j.colsurfa.2021.128180

Goswami, P., Yadav, S., & Mathur, J. (2019). Positive and negative effects of nanoparticles on plants and their applications in agriculture. *Plant Science Today, 6*(2), 232–242. https://doi.org/10.14719/pst.2019.6.2.502

Guo, F., Bao, L., Wang, H., Larson, S. L., Ballard, J. H., Knotek-Smith, H. M., Zhang, Q., Su, Y., Wang, X., & Han, F. (2020). A simple method for the synthesis of biochar nanodots using hydrothermal reactor. *MethodsX, 7*, 101022. https://doi.org/10.1016/j.mex.2020.101022

Heise, K., Kontturi, E., Allahverdiyeva, Y., Tammelin, T., Linder, M. B., & Nonappa, I. (2020). Nanocellulose: Recent fundamental advances and emerging biological and biomimicking applications. *Advanced Materials, 33*, 2004349. Google Scholar.

Heredia-Guerrero, J. A., Heredia, A., Domínguez, E., Cingolani, R., Bayer, I. S., Athanassiou, A., & Benítez, J. J. (2017). Cutin from agro-waste as a raw material for the production of bioplastics. *Journal of Experimental Botany, 68*(19), 5401–5410. https://doi.org/10.1093/jxb/erx272

Janissen, B., & Huynh, T. (2018). Chemical composition and value-adding applications of coffee industry by-products: A review. *Resources, Conservation and Recycling, 128*, 110–117.

Javad, S., Akhtar, I., & Naz, S. (2020). Nanomaterials and agrowaste. In S. Javad (Ed.), *Nanoagronomy* (pp. 197–207). Springer Nature Switzerland AG. https://doi.org/10.1007/978-3-030-41275-3_11

Ji, T., Han, K., Teng, Z., Li, J., Wang, M., Zhang, J., Cao, Y., & Qi, J. (2021). Synthesis of activated carbon derived from garlic peel and its electrochemical properties. *International Journal of Electrochemical Science, 16*(1), 150653. https://doi.org/10.20964/2021.01.61

Kale, S. K., Parishwad, G. V., Husainy, A. S. N., & Patil, A. S. (2021). Emerging agriculture applications of silver nanoparticles. *ES Food and Agroforestry, 3*, 17–22. https://doi.org/10.30919/esfaf438

Kamel, R., El-Wakil, N. A., Dufresne, A., & Elkasabgy, N. A. (2020). Nanocellulose: From an agricultural waste to a valuable pharmaceutical ingredient. *International Journal of Biological Macromolecules, 163*, 1579–1590. https://doi.org/10.1016/j.ijbiomac.2020.07.242

Karande, S. D., Jadhav, S. A., Garud, H. B., Kalantre, V. A., Burungale, S. H., & Patil, P. S. (2021). Green and sustainable synthesis of silica nanoparticles. In *Nanotechnology for environmental engineering* (Vol. 6, Issue 2). Springer Science and Business Media Deutschland GmbH. https://doi.org/10.1007/s41204-021-00124-1

Khan, H. N., & Faisal, M. (2020). Planning and engineering strategies of agricultural wastes and their remediation strategies. In M. Naeem et al. (Eds.), *Contaminants in agriculture* (pp. 219–232). https://doi.org/10.1007/978-3-030-41552-5_10

Khan, I., Han, L., Khan, H., & Kim Oanh, L. T. (2021a). Analyzing renewable and nonrenewable energy sources for environmental quality: Dynamic investigation in developing countries. *Mathematical Problems in Engineering, 1*, 1–2. https://doi.org/10.1155/2021/3399049

Khan, M., Khan, A. U., Moon, I. S., Felimban, R., Alserihi, R., Alsanie, W. F., & Alam, M. (2021b). Synthesis of biogenic silver nanoparticles from the seed coat waste of pistachio (Pistacia vera) and their effect on the growth of eggplant. *Nanotechnology Reviews, 10*(1), 1789–1800. https://doi.org/10.1515/ntrev-2021-0107

Kordi, M., Farrokhi, N., Pech-Canul, M. I., & Ahmadikhah, A. (2023). Rice husk at a glance: From agro-industrial to modern applications. *Rice Science*.

Koul, B., Yakoob, M., & Shah, M. P. (2022). Agricultural waste management strategies for environmental sustainability. *Environmental Research, 206*, 112285. https://doi.org/10.1016/j.envres.2021.112285

Lee, E. (2018). Recent advances in protein-based nanoparticles. *Korean Journal of Chemical Engineering, 35*(9), 1765–1778. https://doi.org/10.1007/s11814-018-0102-0

Leiva-Candia, D. E., Pinzi, S., Redel-Macías, M. D., Koutinas, A., Webb, C., & Dorado, M. P. (2014). The potential for agro-industrial waste utilization using oleaginous yeast for the production of biodiesel. *Fuel, 123*, 33–42.

Leong, Y. K., & Chang, J. S. (2022). Valorization of fruit wastes for circular bioeconomy: Current advances, challenges, and opportunities. *Bioresource Technology, 359*, 127459.

Magnabosco, P., Masi, A., Shukla, R., Bansal, V., & Carletti, P. (2023). Advancing the impact of plant biostimulants to sustainable agriculture through nanotechnologies. *Chemical and Biological Technologies in Agriculture, 10*(1). https://doi.org/10.1186/s40538-023-00491-8

Mahapatra, D. M., Satapathy, K. C., & Panda, B. (2022). Biofertilizers and nanofertilizers for sustainable agriculture: Phycoprospects and challenges. *Science of the Total Environment, 803*, 149990. https://doi.org/10.1016/j.scitotenv.2021.149990

Maji, S., Dwivedi, D. H., Singh, N., Kishor, S., & Gond, M. (2020). Agricultural waste: Its impact on environment and management approaches. In R. N. Bharagava (Ed.), *Emerging eco-friendly green technologies for wastewater treatment, microorganisms for Sustainability 18* (pp. 329–351). https://doi.org/10.1007/978-981-15-1390-9_15

Manyi-Loh, C. E., & Lues, R. (2023). Anaerobic digestion of lignocellulosic biomass: Substrate characteristics (challenge) and innovation. *Fermentation, 9*(8). https://doi.org/10.3390/fermentation9080755. Multidisciplinary Digital Publishing Institute (MDPI).

Maraveas, C. (2020). Production of sustainable and biodegradable polymers from agricultural waste. *Polymers, 12*(5). https://doi.org/10.3390/POLYM12051127. MDPI AG.

Mateo, S., Peinado, S., Morillas-Gutiérrez, F., La Rubia, M. D., & Moya, A. J. (2021). Nanocellulose from agricultural wastes: Products and applications—A review. *Processes, 9*(9), 1594. https://doi.org/10.3390/pr9091594

Mathew, S. S. (2021). Green synthesis of anatase titanium dioxide nanoparticles using Cuminum cyminum seed extract; effect on Mung bean (Vigna radiata) seed germination. *Elsevier, 126*, 108485. https://doi.org/10.1016/j.inoche.2021.108485

McClements, D. J. (2020). Nanotechnology approaches for improving the healthiness and sustainability of the modern food supply. *ACS Omega, 5*(46), 29623–29630. https://doi.org/10.1021/acsomega.0c04050

Mehariya, S., Iovine, A., Casella, P., Musmarra, D., Chianese, S., Marino, T., Figoli, A., Sharma, N., & Molino, A. (2019). Bio-based and agriculture resources for production of bioproducts. In *Current trends and future developments on (bio-) membranes: Membranes in environmental applications* (pp. 263–282). Elsevier Inc. https://doi.org/10.1016/B978-0-12-816778-6.00012-6

Mehndiratta, P., Jain, A., Singh, G. B., Sharma, S., Srivastava, S., Gupta, S., & Gupta, N. (2015). Magnetite nanoparticle aided immobilization of Pseudomonas sp. GBS. 5 for carbazole degradation. *Journal of Biochemical Technology, 5*(4), 823–825.

Meral, R., Kose, Y. E., Ceylan, Z., & Cavidoglu, İ. (2022). The potential use of agro-industrial by-products as sources of bioactive compounds: A nanotechnological approach. *Studies in Natural Products Chemistry, 73*, 435–466.

Miranda, N. T., Motta, I. L., Maciel Filho, R., & Maciel, M. R. W. (2021). Sugarcane bagasse pyrolysis: A review of operating conditions and products properties. *Renewable and Sustainable Energy Reviews, 149*, 111394.

Mishra, K., Siwal, S. S., Nayaka, S. C., Guan, Z., & Thakur, V. K. (2023). Waste-to-chemicals: Green solutions for bioeconomy markets. *Science of the Total Environment, 887*. https://doi.org/10.1016/j.scitotenv.2023.164006

Mittal, D., Kaur, G., Singh, P., Yadav, K., & Ali, S. A. (2020). Nanoparticle-based sustainable agriculture and food science: Recent advances and future outlook. *Frontiers in Nanotechnology, 2*. https://doi.org/10.3389/fnano.2020.579954

Mohamed, H., Shah, A. M., & Song, Y. (2021). Conversion of agro-industrial wastes into value-added products. In *Conversion of renewable biomass into bioproducts* (pp. 197–217). American Chemical Society.

Mondéjar-López, M., López-Jimenez, A. J., Ahrazem, O., Gómez-Gómez, L., & Niza, E. (2023). Chitosan coated—Biogenic silver nanoparticles from wheat residues as green antifungal and nanoprimig in wheat seeds. *International Journal of Biological Macromolecules, 225*, 964–973. https://doi.org/10.1016/j.ijbiomac.2022.11.159

Muñoz-Bonilla, A., & Fernández-García, M. (2015). The roadmap of antimicrobial polymeric materials in macromolecular nanotechnology. *European Polymer Journal, 65*, 46–62. https://doi.org/10.1016/j.eurpolymj.2015.01.030

Nair, L. G., Agrawal, K., & Verma, P. (2022). An overview of sustainable approaches for bioenergy production from agro-industrial wastes. *Energy Nexus, 6*, 100086.

Nakum, J., & Bhattacharya, D. (2022). Various green nanomaterials used for wastewater and soil treatment: A mini-review. *Frontiers in Environmental Science; Frontiers Media*. https://doi.org/10.3389/fenvs.2021.724814

Nang An, V., Chi Nhan, H. T., Tap, T. D., Van, T. T. T., Van Viet, P., & Van Hieu, L. (2020). Extraction of high crystalline nanocellulose from biorenewable sources of Vietnamese agricultural wastes. *Journal of Polymers and the Environment, 28*, 1465–1474. https://doi.org/10.1007/s10924-020-01695-x

Nasrollahzadeh, M., Sajjadi, M., Iravani, S., & Varma, R. S. (2021). Starch, cellulose, pectin, gum, alginate, chitin and chitosan derived (nano)materials for sustainable water treatment: A review. *Carbohydrate Polymers, 251*, 116986. https://doi.org/10.1016/j.carbpol.2020.116986

Nath, P. C., Ojha, A., Debnath, S., Sharma, M., Sridhar, K., Nayak, P. K., & Inbaraj, B. S. (2023). Biogeneration of valuable nanomaterials from agro-wastes: A comprehensive review. *Agronomy, 13*(2), 561. https://doi.org/10.3390/agronomy13020561

Ortega, F., Versino, F., López, O. V., & García, M. A. (2022). Biobased composites from agro-industrial wastes and by-products. *Emergent Materials, 5*(3), 873–921. https://doi.org/10.1007/s42247-021-00319-x. Springer Nature.

Panda, J., Mishra, A. K., Mohanta, Y. K., et al. (2024). Exploring biopolymer for food and pharmaceuticals application in the circular bioeconomy: An agro-food waste-to-wealth approach. *Waste Biomass Valor*. https://doi.org/10.1007/s12649-024-02452-0

Pandey, S., & Dwivedi, N. (2020). Utilisation and management of agriculture and food processing waste. In P. Mishra et al. (Eds.), *Innovations in food technology* (pp. 269–288). Springer Nature. https://doi.org/10.1007/978-981-15-6121-4_19

Peerzada, J. G., & Chidambaram, R. (2020, August 18). A statistical approach for biogenic synthesis of nano-silica from different agro-wastes. *Silicon, 13*(7), 2089–2101. https://doi.org/10.1007/s12633-020-00629-5

Picot-Allain, M. C. N., & Emmambux, M. N. (2023). Isolation, characterization, and application of nanocellulose from agro-industrial by-products: A review. *Food Reviews International, 39*(2), 941–969.

Pisello, A. L., Hawkes, D., Bougdah, H., Rosso, F., Mohareb, N., Mesbah, S., Bozonnet, E., Pignatta, G., De Bonis, L., Kostopoulou, S., & Mannan, M. A. (2021). *IEREK interdisciplinary series for sustainable development editorial board*. http://www.springer.com/series/15883

Preetha, J. S. Y., Sriram, D., Premasudha, P., Pudake, R. N., Anand, A. V., & Arun, M. (2024). Recent advancements and applications of nano-agrochemicals against plant fungal diseases. In *Nanoparticles in plant biotic stress management* (pp. 403–418). Singapore, Springer Nature.

Prueckler, M., Siebenhandl-Ehn, S., Apprich, S., Hoeltinger, S., Haas, C., Schmid, E., & Kneifel, W. (2014). Wheat bran-based biorefinery 1: Composition of wheat bran and strategies of functionalization. *LWT- Food Science and Technology, 56*(2), 211–221.

Pudake, R. N., Chauhan, N., & Kole, C. (Eds.). (2019). *Nanoscience for sustainable agriculture* (Vol. 711). Springer.

Pudake, R. N., Mohanta, T. K., & Mahato, N. (2024). Opportunities and challenges for nanotechnology in sustainable agri-food production. *Frontiers in Nanotechnology, 6*, 1420192.

Punitha, V. N., Vijayakumar, S., Alsalhi, M. S., Devanesan, S., Nilavukkarasi, M., Vidhya, E., Kumar, S. P., & Kim, W. (2023). Biofabricated ZnO nanoparticles as vital components for agriculture revolutionization—A green approach. *Applied Nanoscience, 13*(9), 5821–5827. https://doi.org/10.1007/s13204-023-02765-x

Qazi, G., & Dar, F. A. (2020). Nano-agrochemicals: Economic potential and future trends. *Nanotechnology Life Sciences*. https://doi.org/10.1007/978-3-030-39978-8_11

Rabalao, T. M., Ndaba, B., Roopnarain, A., & Vatsha, B. (2022). Towards a circular economy: The influence of extraction methods on phytosynthesis of metallic nanoparticles and their impact on crop growth and protection. *JSFA Reports, 2*(5), 208–221. https://doi.org/10.1002/jsf2.44

Ravindran, R., Hassan, S. S., Williams, G. A., & Jaiswal, A. K. (2018). A review on bioconversion of agro-industrial wastes to industrially important enzymes. *Bioengineering, 5*(4). https://doi.org/10.3390/bioengineering5040093. MDPI AG.

Ribeiro, R. S. A., Pohlmann, B. C., Calado, V., Bojorge, N., & Pereira, N. (2019). Production of nanocellulose by enzymatic hydrolysis: Trends and challenges. *Engineering in Life Sciences, 19*(4), 279–291. https://doi.org/10.1002/elsc.201800158. [Green Version]. Google Scholar.

Richter, A. P., Bharti, B., Armstrong, H. B., Brown, J. S., Plemmons, D., Paunov, V. N., Stoyanov, S. D., & Velev, O. D. (2016). Synthesis and characterization of biodegradable lignin nanoparticles with tunable surface properties. *Langmuir, 32*(25), 6468–6477. https://doi.org/10.1021/acs.langmuir.6b01088

Sabourin, V., & Ayande, A. (2015). Commercial opportunities and market demand for nanotechnologies in agribusiness sector. *Journal of Technology Management & Innovation, 10*(1), 40–51. https://doi.org/10.4067/S0718-27242015000100004

Salem, S., Teimouri, Z., & Salem, A. (2020, October). Fabrication of magnetic activated carbon by carbothermal functionalization of agriculture waste via microwave-assisted technique for cationic dye adsorption. *Advanced Powder Technology, 31*(10), 4301–4309. https://doi.org/10.1016/j.apt.2020.09.007

Saratale, R. G., Saratale, G. D., Shin, H. S., Jacob, J. M., Pugazhendhi, A., Bhaisare, M., & Kumar, G. (2018). New insights on the green synthesis of metallic nanoparticles using plant and waste biomaterials: Current knowledge, their agricultural and environmental applications. *Environmental Science and Pollution Research International, 25*(11), 10164–10183. https://doi.org/10.1007/s11356-017-9912-6

Shaaban, S., & Nasr, M. (2020). Toward three R's agricultural waste in MENA: Reduce, reuse, and recycle. In A. M. Negm & N. Shareef (Eds.), *Waste management in MENA regions, springer water* (pp. 337–353). https://doi.org/10.1007/978-3-030-18350-9_17

Shalma, S., Shabbirahmed, A. M., Haldar, D., Patel, A. K., & Singhania, R. R. (2023). Influence of reaction conditions on synthesis and applications of lignin nanoparticles derived from agricultural wastes. *Environmental Technology & Innovation*, 103163.

Shang, Y., Hasan, M. K., Ahammed, G. J., Li, M., Yin, H., & Zhou, J. (2019). Applications of nanotechnology in plant growth and crop protection: A review. *Molecules, 24*(14), 2558. https://doi.org/10.3390/molecules24142558

Shoueir, K., Kandil, S., El-hosainy, H., & El-Kemary, M. (2019). Tailoring the surface reactivity of plasmonic Au@TiO_2 photocatalyst bio-based chitosan fiber towards cleaner of harmful water pollutants under visible-light irradiation. *Journal of Cleaner Production, 230*, 383–393. https://doi.org/10.1016/j.jclepro.2019.05.103

Singh, B., Sharma, D. K., Kumar, R., & Gupta, A. (2010). Development of a new controlled pesticide delivery system based on neem leaf powder. *Journal of Hazardous Materials, 177*(1–3), 290–299. https://doi.org/10.1016/j.jhazmat.2009.12.031

Singh, G. B., Srivastava, A., Saigal, A., Aggarwal, S., Bisht, S., Gupta, S., Srivastava, S., & Gupta, N. (2011). Biodegradation of carbazole and dibenzothiophene by bacteria isolated from petroleum-contaminated sites. *Bioremediation Journal, 15*(4), 189–195.

Singh, A. C., Asif, M. H., Bacher, G., Danielsson, B., Willander, M., & Bhand, S. (2019). Nanoimmunosensor based on ZnO nanorods for ultrasensitive detection of 17β-estradiol. *Biosensors and Bioelectronics, 126*, 15–22. https://doi.org/10.1016/j.bios.2018.10.004

Sinha, M. K., & Purkait, M. K. (2015). Preparation of fouling resistant PSF flat sheet UF membrane using amphiphilic polyurethane macromolecules. *Desalination, 355*, 155–168. https://doi.org/10.1016/j.desal.2014.10.017

Sodhi, A. S., Sharma, N., Bhatia, S., Verma, A., Soni, S., & Batra, N. (2022). Insights on sustainable approaches for production and applications of value added products. *Chemosphere, 286*, 131623.

Suffo, M., de la Mata, M., & Molina, S. I. (2020). A sugar- beet waste based thermoplastic agrocomposite as substitute for raw materials. *Journal of Cleaner Production, 257*, 120382. https://doi.org/10.1016/j.jclepro.2020.120382

Tamilselvi, R., Ramesh, M., Lekshmi, G. S., Bazaka, O., Levchenko, I., Bazaka, K., & Mandhakini, M. (2020). Graphene oxide—Based supercapacitors from agricultural wastes: A step to mass production of highly efficient electrodes for electrical transportation systems. *Renewable Energy, 151*, 731–739. https://doi.org/10.1016/j.renene.2019.11.072

Thakur, M., Wang, B., & Verma, M. L. (2022). Development and applications of nanobiosensors for sustainable agricultural and food industries: Recent developments, challenges and perspectives. *Environmental Technology & Innovation, 26*, 102371. https://doi.org/10.1016/j.eti.2022.102371

Tolisano, C., & Del Buono, D. (2023). Biobased: Biostimulants and biogenic nanoparticles enter the scene. *Science of the Total Environment, 885*, 163912. https://doi.org/10.1016/j.scitotenv.2023.163912

Tripathi, R. M., Pudake, R. N., Shrivastav, B. R., & Shrivastav, A. (2018). Antibacterial activity of poly (vinyl alcohol)—Biogenic silver nanocomposite film for food packaging material. *Advances in Natural Sciences: Nanoscience and Nanotechnology, 9*(2), 025020.

United Nations, Department of Economic and Social Affairs, Population Division. (2022). *World population prospects 2022: Data sources* (UN DESA/POP/2022/DC/NO. 9).

Usmani, Z., Sharma, M., Sudheer, S., Gupta, V. K., & Bhat, R. (2020). Engineered microbes for pigment production using waste biomass. *Current Genomics, 21*(2), 80–95.

Van Hoed, V., Depaemelaere, G., Ayala, J. V., Santiwattana, P., Verhé, R., & De Greyt, W. (2006). Influence of chemical refining on the major and minor components of rice brain oil. *Journal of the American Oil Chemists' Society, 83*, 315–321.

Vergara-Castañeda, H. A., Luna-Bárcenas, G., & Pool, H. (2020). Emerging and potential bio-applications of agro-industrial by-products through implementation of nanobiotechnology. In *Food wastes and by-products: Nutraceutical and health potential* (pp. 413–443).

Vijayakumar, M. D., Surendhar, G. J., Natrayan, L., Patil, P. P., Ram, P. M. B., & Paramasivam, P. (2022). Evolution and recent scenario of nanotechnology in agriculture and food industries. *Journal of Nanomaterials, 2022*(1), –17. https://doi.org/10.1155/2022/1280411

Vinayak, A., & Singh, G. B. (2022). Biodecolorization of reactive black 5 using magnetite nanoparticles coated Bacillus sp. RA5. *Materials Today Proceedings, 48*, 1523–1526.

Vinayak, A., Sharma, S., & Singh, G. B. (2022). Biopolymers from industrial waste. In *Biopolymers: Recent updates, challenges and opportunities* (pp. 129–149). Springer.

Webber, A., Hettiarachchy, N., Webber, D., Sivarooban, T., & Horax, R. (2014). Heat-Stabilized Defatted Rice Bran (HDRB) as an alternative growth medium for Saccharomyces cerevisiae. *Journal of Food and Nutrition, 1*, 1–6. https://doi.org/10.17303/jfn.2014.103

Widiarto, S., Yuwono, S. D., Rochliadi, A., & Arcana, I. M. (2017, February). Preparation and characterization of cellulose and nanocellulose from agro-industrial waste-cassava peel. In *IOP conference series: Materials science and engineering* (Vol. 176, No. 1, p. 012052). IOP Publishing.

World Bank, Agriculture and Food. (2023). https://www.worldbank.org/en/topic/agriculture/overview

Yaashikaa, P. R., Kumar, P. S., & Varjani, S. (2022). Valorization of agro-industrial wastes for biorefinery process and circular bioeconomy: A critical review. *Bioresource Technology, 343*, 126126.

Yadav, R. L., & Solomon, S. (2006). Potential of developing sugarcane by-product based industries in India. *Sugar Tech, 8*(2/3), 104–111.

Yadav, M., Dwibedi, V., Sharma, S., & George, N. (2022). Biogenic silica nanoparticles from agro-waste: Properties, mechanism of extraction and applications in environmental sustainability. *Journal of Environmental Chemical Engineering*, 108550.

Yin, J., Su, X., Yan, S., & Shen, J. (2023). Multifunctional nanoparticles and nanopesticides in agricultural application. *Nanomaterials, 13*(7), 1255. https://doi.org/10.3390/nano13071255

Ying, S., Guan, Z., Ofoegbu, P. C., Clubb, P., Rico, C., He, F., & Hong, J. (2022). Green synthesis of nanoparticles: Current developments and limitations. *Environmental Technology and Innovation, 26*, 102336. https://doi.org/10.1016/j.eti.2022.102336

Yu, B., Zeng, J., Gong, L., Zhang, M., Zhang, L., & Chen, X. (2007). Investigation of the photocatalytic degradation of organochlorine pesticides on a nano-TiO_2 coated film. *Talanta, 72*(5), 1667–1674. https://doi.org/10.1016/j.talanta.2007.03.013

Zaeem, A., Drouet, S., Anjum, S., Khurshid, R., Younas, M., Blondeau, J. P., Tungmunnithum, D., Giglioli-Guivarch, N., Hano, C., & Abbasi, B. H. (2020). Effects of biogenic zinc oxide nanoparticles on growth and oxidative stress response in flax seedlings vs. in vitro cultures: A comparative analysis. *Biomolecules, 10*(6), 918. https://doi.org/10.3390/biom10060918

Zalewska, M., & Popowska, M. (2020). Antimicrobial/antibiotic resistance genes due to manure and agricultural waste applications. In M. Z. Hashmi (Ed.), *Antibiotics and antimicrobial resistance genes, emerging contaminants and associated treatment technologies* (pp. 139–161). https://doi.org/10.1007/978-3-030-40422-2_6

Chapter 6
Nanobiotechnology for Sustainable Food Preservations and Shelf-Life Enhancement

Nishithendu Bikash Nandi, Jibanjyoti Panda, Puja Roy, Rusmita Buragohain, Khashayar Sarabandi, and Pinku Chandra Nath

Abstract In the context of growing populations and depleting natural resources, food waste is a serious problem that impacts public health, the surroundings, along with the economy. Microbial creatures and moisture content in food are the primary causes of food waste. There are traditional preservations techniques exist, but they have drawbacks such as high production costs, inadequate shelf lives, undesired residue, etc. Nanotechnology-related shelf-life extension strategies have the potential to make up for the drawbacks of conventional preservation techniques due to several special qualities. This chapter outlines the fundamentals and more recent, incredibly effective uses of shelf-life extension techniques related to nanotechnology in fruits and vegetables. These techniques include active packaging, detection techniques, intelligent label systems, and combined preservation strategies. Research on nanotechnology-related intelligent labeling systems and combined preservation strategies are also briefly covered in this chapter.

Keywords Nanobiotechnology · Preservation · Shelf life · Food · Intelligent label

N. B. Nandi
Department of Chemistry, National Institute of Technology Agartala, Jirania, Tripura, India

J. Panda · P. Roy · R. Buragohain
Nano-biotechnology and Translational Knowledge Laboratory, Department of Applied Biology, University of Science and Technology Meghalaya (USTM), Ri-Bhoi, Meghalaya, India

K. Sarabandi
Department of Food Chemistry, Research Institute of Food Science and Technology (RIFST), Mashhad, Iran

P. C. Nath (✉)
Department of Food Technology, School of Applied and Life Sciences, Uttaranchal University, Dehradun, Uttarakhand, India

Y. K. Mohanta et al. (eds.), *Nano-microbiology for Sustainable Development*,
https://doi.org/10.1007/978-3-031-78845-1_6

6.1 Introduction

Nanotechnology has emerged as a major innovation with enormous promise to advance sustainability. To put it simply, nanotechnology is the study of materials or nanoparticles with one or more dimensions of less than 100 nm. It combines various applied science fields, including materials processing, environmental engineering, and biology, physics, and food technology (Fig. 6.1) (Baig et al., 2021; Mohanta et al., 2024). The structural elements, clusters, molecules, and crystallites of nanomaterials or nanostructure materials have ranging dimensions. These include zero dimensions for nanoparticles, nanoclusters, and nanotubes; one dimension for nanorods and nanotubes; two dimensions for nano-thin films; and three dimensions for nanomaterials in the 1–100 nm range. Besides, nanotechnology's success can be attributed to its promising outcomes, zero emission of pollutants, energy efficiency, and minimal space requirements. Along with these key success factors, nanotechnology has demonstrated a wide range of applications in the fields of food, agriculture, and environment (Nath et al., 2023a, b, c; Ray et al., 2023). A nanocomposite can be created by integrating nanostructure materials with different polymers, biomolecules, and other nanostructure substances, or by allowing the materials to exist in aggregate form and result in a substance with a somewhat larger size of particles (>100 nm) (Nath, 2022).

In addition to giving the body the calories it needs; fruits and vegetables are beneficial because they taste good and satisfy a variety of nutrient requirements. According to reports, a lack of appropriate shelf-life extension techniques causes

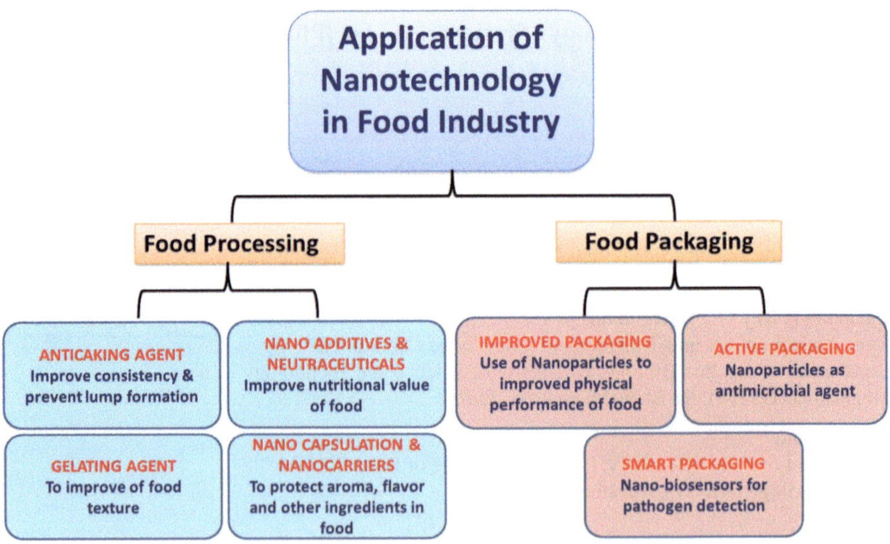

Fig. 6.1 A schematic depicting the function of nanotechnology in several food-related industries. (Figure adopted from Singh et al. (2017) and it is open article (copyright © 2023 by authors) distributed under the terms and condition of the Creative Commons Attribution (CC BY) License)

20–40% of food products fruits and vegetables to become tainted and deteriorate globally. Extension of life span of fruit and vegetable has thus grown in importance in the food industry. The main ways that freshness-keeping techniques increase the fruits and vegetables' shelf life are by postponing the ripening process, avoiding microbial contamination, and managing transpiration. Researchers have created numerous methods for extending shelf life thus far. One common physical-based technique for increasing the life span of fresh food products is cold storage, which involves keeping a low temperature in the storage environment. However, Table 6.1 demonstrated that the conventional methods of extending shelf life have certain inherent disadvantages. Recently, irradiation preservation which includes freshness of fruits and vegetables has received special attention. While irradiation preservation technology can offer a comparatively high level of disinfection without damaging the composition of fruits and vegetables, it has certain drawbacks, including a high initial investment cost, a high degree of sensitivity between radiation dosage and working time and the quality of the product, and a challenging operating condition control system (Bisht et al., 2021; Mishra et al., 2024). Chemical-based preservation techniques have the benefits of being inexpensive, highly effective, and simple to use. One such technique is the disinfection of fruits and vegetables with chlorine. However, consumers' trust in the harmlessness of food additives, particularly those that are synthetic or chemical, has diminished because of the recent breakout of food-borne diseases brought on by the excessive and prohibited insertion of food additives.

Table 6.1 Drawbacks of some traditional extension of shelf-life technology of fruits and vegetables

Techniques	Advantages	Drawbacks
Edible coating	The possibility of preserving fresh produce is growing. Not only do these coatings keep the nutritious characteristics intact, but they also make shipping from the production to merchants much easier	Poor moisture resistance, easy moisture absorption, inadequate shelf life, and challenging control over the concentration and thickness of the edible coating
Cold storage	Maintains freshness, enables bulk purchasing	Expensive investment; possible freezing or even chilling injury risks
Chemical preservation	Inhibit the factors causing spoilage. Prevent food from deterioration	Danger to safety; undesired residue
Heat treatment	Food refrigeration is eliminating the anti-nutritional elements (ANF), which include tannins, phytates, oxalates, enzyme inhibitors, thereby enhancing the nutrient availability and digestion	Readily cause vitamins, minerals, and other nutrients to be lost, which lowers the quality of food products
Irradiation	Increases shelf life and enhances safety	Expensive; current technology is still in its infancy; challenging to regulate radiation conditions

Innovative technology known as nanobiotechnology is gaining traction in the food industry. It leads to the creation of fresh food items with enhanced oral availability, more solubility, or thermal stability, with further advancements in food quality and safety. Utilizing biotechnological gadgets for food preservation has demonstrated encouraging outcomes in reducing food spoilage. The creation and advancement of exceptionally effective food preservatives is a crucial element for success in this particular field of application (Panda et al., 2024). It has been shown that using these tools has greatly reduced the amount of food deterioration, particularly for perishable foods (Lakshmayya et al., 2023). Overall, the use of nanotechnology in food preservation has produced some encouraging results in terms of extended shelf life without sacrificing the food's apparent safety concerns or organoleptic qualities.

6.2 Fundamental Theory of Nanotechnology-Based Preservations and Shelf Life Enhancement of Foods

In general, biological preservation, chemical preservation and physical preservation are the three main types of conventional methods used to preserve fruits and vegetables. While each preservation technique has a distinct focus, they all control three essential elements that are crucial to the preservation of quality:

(a) Managing the senescence process, which is typically accomplished by managing respiration
(b) Managing microorganisms, which is primarily accomplished by managing spoilage bacteria
(c) Managing internal water evaporation, this is primarily accomplished by managing the environment's relative humidity.

However, the conventional techniques for extending shelf life have certain inbuilt weaknesses including excessive investment, inadequate shelf life, High initial outlay, undesired residue etc. Materials with at least one structural dimension of less than 100 nm are commonly referred to as nanomaterials (Khan et al., 2020; Majumdar et al., 2020). As a result of materials shrinking from a macroscale to a nanoscale, nanomaterials differ from their macro-scopic state in a number of important ways, including exceptional hardness, barrier, and optical qualities (Mekuye & Abera, 2023). The fundamental ideas behind the introduction of nanoscience in the maintenance of fruits and vegetables freshness from these special material properties. Contamination and food deterioration are big problems that affect food safety and quality in the food production industry. While there are a number of physical, chemical, and biological techniques that can prolong food's shelf life, they are not without their drawbacks. A wide range of activities, including strong antioxidant and antibacterial properties, can be rendered by nanomaterial-based food shelf-life extension (Manikandan & Min, 2023; Shamim et al., 2024). The majority of

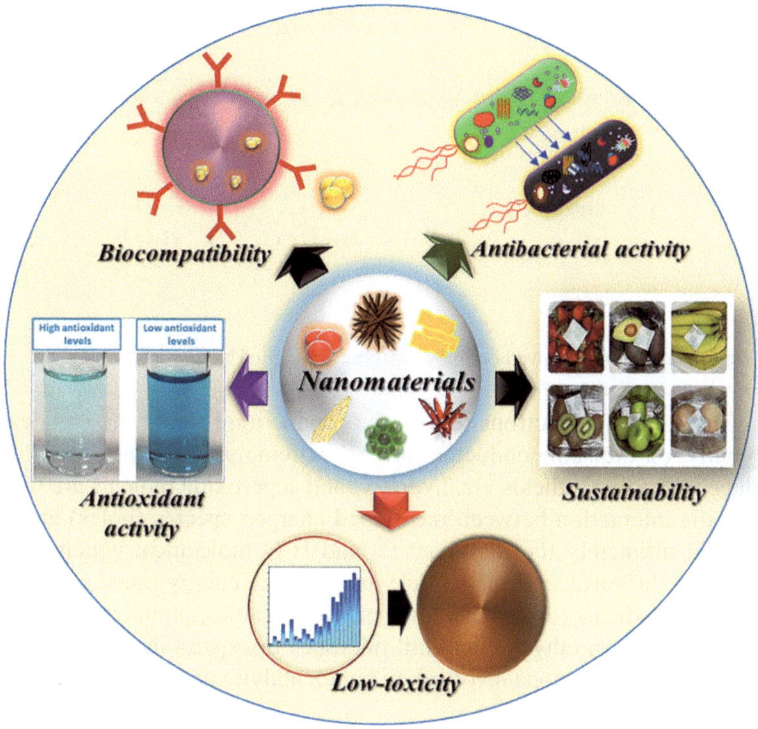

Fig. 6.2 The utilization of nanomaterials for the purpose of protecting food and extend the duration of shelf-life of food products. (Figure adopted from Manikandan and Min (2023) and it is a subscription based article (copyright © 2023 by authors) distributed under the terms and condition of the Creative Commons Attribution (CC BY) License)

research has shown that nanoparticles can elongate the longevity of fruits and vegetables and protect them from microbiological contamination (Fig. 6.2). The sections below go into detail about the fundamental ideas behind shelf-life extension techniques related to nanotechnology.

6.2.1 Photocatalytic Properties-Based NPs

The majority of photocatalytic nanomaterials are semiconductors with unique electrical structures, like metal oxides or sulfides. The energy band structure of semiconductor nanomaterials is typically composed of low energy valence bands and high energy conduction bands, with band gaps existing between the two, in contrast to metals that have a continuous band structure. Electrons on the valence bond are excited to the conduction band when exposed to light with energy greater than the bandgap. This results in creation of positively charged holes (h^+) on the valence

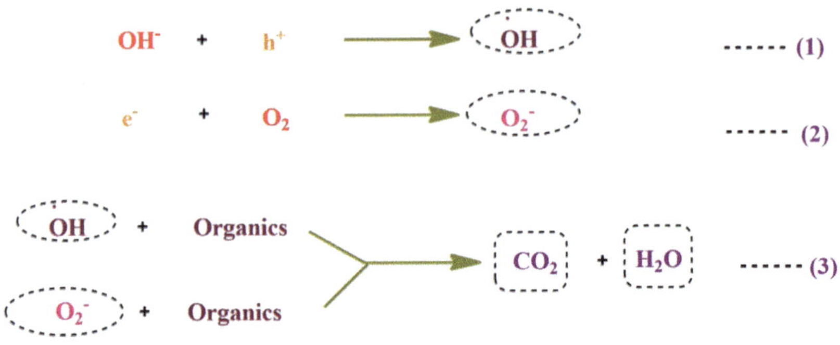

Fig. 6.3 Various reactions involved on photocatalytic activity of NPs

band as well as mobile electrons (e^-) on the conduction band at the identical time. Consequently, on the semiconductor surface, extremely active electron-hole pairs are produced. The two radicals viz. hydroxyl and superoxide radicals are formed as a result of the interaction between mentioned charged species (holes) and mobile electrons and ultimately the dissolved O_2 and H_2O molecules, which physically interacted with the surface of the particle and produce energy transfer (Yang et al., 2023). Because radicals formed can interact with organics, photocatalytic nanomaterials can break down ethylene and kill microbes to expand the life span of fruits and vegetables. The reactions involved on photocatalytic properties of NPs are portraited in Fig. 6.3.

The most popular nanomaterial used as photocatalyst for maintaining the quality of fruits and vegetables freshness is nano-TiO_2. It can slow down the deterioration of fruits and vegetables that have been harvested after harvesting and speed up the oxidative breakdown of ethylene and other gases produced during fruit and vegetable storage (Khan et al., 2023).

6.2.2 Antibacterial Properties of NPs

Antibacterial nanomaterials slow down the growth of microorganisms and eliminate individual ones, extending the life span of fruits and vegetables. Two categories of nano antibacterial materials can be distinguished based on their antibacterial active ingredients: oxide photocatalytic type and metal ion type. Metal ionic nano-antimicrobial particles are created by incorporating metal ions with antibacterial properties, like Al, Ni, Zn, Co, Cu, Fe, and Ag, into different types of natural or artificial mineral nano-carriers. Matrix materials which are porous, larger surface area, possess excellent properties of adsorption, non-toxic, and have chemical stability are commonly used as carriers of metallic ionic nano-antimicrobial materials (Dhar et al., 2023; Roy et al., 2024). Examples of these materials include phosphate and layered clay minerals. The metal ion dissolution theory states that when nano-antimicrobial materials are applied to preserve the nutritional value of vegetables and fruits, antimicrobial metal ions gradually dissolve . These dissolved metal ions then damage the energy the metabolism of bacterial cells and stop microorganisms from reproducing (Mohanta et al., 2023). Furthermore, functional groups of proteins and nucleic acids, such as amine (–NH) and/or sulfhydryl (–SH), can react with anti-microbial metal ions. Additionally, antimicrobial metal ions can penetrate bacterial cells and interact with both enzymes as well as DNA, impairing the physiological processes of microbes. The antibacterial component (positively charged component) and microbial cells (negatively charged component) interact with each other when they come into contact, according to the contact reaction theory. This allows them to effectively use charge transfer to breach the bacterial cell membrane and denature the protein. The bacterial cells can no longer breathe, cannot metabolize, and eventually vanish as a result.

The initial antimicrobial activity is preserved throughout the entire process, and none of the antimicrobial components are used up. As a result, the NPs with contact reaction properties suffer antibacterial activity over an extended period (Anvar et al., 2021).

6.2.3 Barrier Properties of NPs

Composites containing particulate nanomaterials typically exhibit superior "quantum mechanics" properties, such as improved barrier properties, when compared to conventional preservation techniques. Additionally, because the surface-to-volume ratio of the nanoparticles is larger, only a small amount of addition (roughly 5% weight by weight) is needed to produce an excellent interfacial interlinked with the polymer chain. This result concludes that the nanocomposite delivers a more effective shelf-life extension than the native polymer alone. Thermal treatment can effectively improve the barrier properties of nano cellulose fibril by decreasing porosity, raising crystallinity, and enhancing hydrophobicity. For example, they discovered that after thermal treatment by maintaining the temperature of 175 °C for about 3 h, the permeability of both H_2O vapor and O_2 was significantly decreased than it was before heat treatment (Sharma et al., 2014).

6.2.4 Thermal Properties of NPs

The energy of the polymeric matrix is improved by the advantageous interactions between the nanoparticles and matrix, which rearrange the polymer molecules into an arranged state that improves the thermal properties. As a result, the polymer crystallization temperature of the composite material was clearer, and its melting and heat distortion temperatures were higher. The thermal properties of the nanocomposite are crucial for post-processing treatments like pasteurizing fresh fruits and vegetables to increase shelf life.

6.3 Applications of Nanotechnology in Food Preservation and Shelf Life of Food

Nanoscience, which is based on nanomaterials, can preserve fruits and vegetables. In recent years, several studies have investigated the use of nanomaterials to preserve fruits and vegetables. Several books provide insightful summaries of the potential applications of nanoparticles in fresh food preservation, including methods for nano-preserving fruits and vegetables. Here is a rundown of the most common applications of nano-sized materials in fruit and vegetable preservation methods, as shown in Table 6.2. This chapter covers a wide range of strategies related to nanotechnology for extending the shelf life of fruits and vegetables.

6.3.1 Detection Methods Using Nanotechnology

Comparing nanomaterials to their macroscale counterparts, useful "new" properties can be observed. Since new detection technologies involving nanomaterials have higher sensitivity, selectivity, and/or detection rates than conventional detection technologies, there has been an increase in interest in the use of nanomaterials in the field of fruit and vegetable quality detection in recent years. The literatures that were retrieved indicate that the use of detection technologies involving nanomaterials to expand the life span of fruits and vegetables was primarily focused on the identification of food-borne pathogens and food allergens.

6.3.1.1 Nanoparticles for the Detection of Foodborne Pathogens

The presence of pathogens in food items can provide a route to foodborne illnesses. This point is reflected a serious threat to public health, especially for infants, everywhere in the world. Studies show that parasites, viruses and particularly bacteria are the most common pathogens responsible for foodborne illnesses. Toxins can remain in food even after food processing methods have eradicated the original pathogen, which makes it crucial to comprehend the important role they play in the emergence of foodborne illnesses. Common foodborne pathogens include bacteria, viruses, and parasites, as was previously mentioned. Foodborne pathogenic bacteria include *Shigella* species, *Escherichia coli* O157:H7, *Campylobacter jejuni*, *Listeria monocytogenes*, *Samonella* species, *Clostridium botulinum*, *Staphylococcus aureus*, *Bacillus cereus*, *Clostridium perfringens*, and *Vibrio cholera*. The Hepatitis E virus (HEV) is a common foodborne virus that should be especially concerned about in developing countries. When it comes to meat products in particular, poor food preparation and sanitation practices can result in a HEV infection that causes liver disease. Velusamy et al. (2010) divided the primary conventional testing techniques

Table 6.2 Some selected nanomaterials in the preservation of fruits and vegetables

Food item	Nano-materials used	Consequences	References
Carrot	ZnO	Nano-ZnO could increase the life span of carrot to almost 40 days while reducing the total colony producing units while storage	Xu et al. (2017)
Chineses Yam	$CaCO_3$	Innovative nano-packaging could considerably slow down the ripening process by preventing microbial growth, reducing browning, and preserving the activity of useful compounds	Luo et al. (2015)
Strawberries	ZnO	Nano sized ZnO could stop the microorganism's growth, postpone weight loss, and keep strawberries' useful nutrients intact	Sogvar et al. (2016)
Longan Fruit	Silica	Nanosilica was able to stop the activities of peroxidase and polyphenol oxidase, which are linked to the deterioration of storage product quality, in comparison to the control treatment. The longan fruit's nutritional content, including its total soluble solids and vitamin C content, had also been better preserved	Shi et al. (2013)
Apple	ZnO	Freshly cutted apples wraped in nanomaterials packaging had a 6-day longer shelf life than samples packed in conventional polyvinyl chloride film	Li et al. (2011)
Green Soybean	ZnO	Yeast, molds, coliforms, and other common microorganisms could all be considerably inhibited from growing by nano-ZnO	Yu et al. (2015)
Cucumber	Chitosan	A cucumber coated with chitosan nanoparticles may have increased antioxidant activity and prolonged shelf life	Mohammadi et al. (2016)
Chili, Radish, Lettuce	Ag	An efficient method of food packing that stops pathogen ingress	Seray et al. (2021)
Tomato	Se	Alterations to the tomato plant's enzymatic activity and enhance the yield of Tomato crops	Hernández-Hernández et al. (2019)
Green Lettuce	TiO_2	TiO_2 packaging material worked better for storing green vegetables	Yemmireddy and Hung (2015)
Grape	SiO_2	cut down on the phases of grape development	Zahedi et al. (2020)
Solid foods	Lipid	Drug and nutritional routes of administration	Ban et al. (2020)
Dairy foods	Protein	Great potential in catalysis, synthesis, bio imaging Found in foods in the form of casein	Samadarsi et al. (2020)
Wine and hone making industry	Carbon nanotube	The electronic conductivity of carbon nanotubes is recommended, and they exhibit strong stability and environmental friendliness	Cao et al. (2021)

into three groups: polymerase chain reaction (PCR), immunology-based assays, and culture-and-colony counting methods. Additionally, DNA loop-mediated isothermal amplification (LAMP) has attracted interest lately. These methods can be very successful, but they also have disadvantages. They could be costly, labor-intensive, and time-consuming (two to ten days). Though no single approach can meet all requirements, efforts are still being made to develop alternative detection techniques that may be less expensive, faster, or more sensitive than current methods. Au nanoparticles (GNPs) are extensively employed in the field of biomedical because of their high physio-chemical stability, biocompatibility, ease of functionalization with bioactive molecules, and distinctive optical properties obtained from the oscillations of electrons reside in the bands of conduction. Several GNP-based techniques can be used to identify a single pathogen, as we will discuss below.

GNPs in Solution-Based Colorimetric Assays

Food contamination can be identified early and without the need for costly equipment by applying a solution-based colorimetric assay that utilizes GNPs. Their colorimetric changes are often visible to the unaided eye and show high sensitivity. GNPs can be surface-conjugated with targeting moieties, like aptamers, oligonucleotides, or antibodies, to achieve specificity against target foodborne pathogens. According to Du et al. (2018), *Listeria monocytogenes* can be identified by utilizing oligonucleotide-functionalized AuNPs (OLG-GNPs) as a detection probe. For this colorimetric assay, bacterial DNA must be extracted and the hlyA gene from *L. monocytogenes* (which corresponds to the OLG probe) amplified by PCR before analysis can begin. Negative samples produced aggregation and a colorless solution that was easily visible to the unaided eye, while OLG-GNPs caused GNP aggregation to be inhibited by the presence of either the target genomic DNA or the PCR product hlyA gene. 48.4 and 100.4 ng were the corresponding limits of detection (Du et al., 2018). In order to detect *Salmonella typhimurium*, a common root of food-related illnesses, Ma et al. (2017) designed a GNP aptasensor. Aptamers can be short, targeted oligonucleotide sequences or single-stranded DNA. These can be taken up by the GNP surface and so stop the GNP from aggregating. In contrast, aptamers preferentially detached from AuNPs and attach to their particular targets when GNPs are present, causing GNP to aggregate. This implies that 56 CFU mL^{-1} of *S. typhimurium* may be found (Ma et al., 2017). In contrast to GNP colorimetric assays, which usually need sample enrichment, one study have developed a competitive binding assay that does not require it. Using the strong electrostatic interactions between bacteria, negatively charged enzymes (β-galactosidase), and positively charged polyethyleneimine-coated GNPs (PEI-GNPs), they have achieved this. PEI-GNPs bind to the enzymes and inhibit them; however, in an abundance of bacteria, the PEI-GNPs will bind especially to the teichoic acids of gram-positive microorganisms (*S. aureus*) or the LPS of gram-negative microorganisms (e.g., enterotoxicogenic *E. coli*), which will restore the function of the enzymes. Afterward, colorimetric monitoring of the enzyme activity can be achieved by hydrolyzing the substrate, chlorophenol red β-Dgalactopyranoside (Thiramanas & Laocharoensuk, 2016).

Surface-Enhanced Raman Scattering

Bozkurt et al. (2018) reported a relatively complex "sandwich assay" for the identification of *E. coli* by surface-enhanced Raman scattering (SERS) utilizing gold nanorods (GNRs) and spherical-shaped GNPs. In that case, alkaline phosphatase (ALP) produces hydrolyzed SERS substrate (5-bromo-4-chloro-3-indolyl phosphate; BCIP), which was used to label the SERS. There is a direct correlation between the quantity of ALP and the number of bacteria. To accomplish enrichment, *E. coli* antibody-conjugated gold-coated magnetic nanoparticles were employed. Following the separation procedure, GNRs coated with both an anti-*E. coli* antibody and ALP were included. This permitted for extremely particular interaction between ALP and *E. coli*. Then, the BCIP substrate was combined with the mixture of nanoparticles and *E. coli*. The BCIP could then be hydrolyzed by the surface ALP of any attached GNRs, yielding 5-bromo-4-chloro-3-indole (BCI). The method showed a high level of sensitivity, with a 10 CFU mL^{-1} detection limit (Bozkurt et al., 2018). A polymer SERS substrate was developed by Usitalo et al. utilizing a roll-to-roll technique. This was fused with GNPs and immunomagnetic separation to produce an easily adaptable platform for the identification of *Listeria innocua*. It was possible to reach an approximate detection limit of 1×10^4 CFU mL^{-1} (Uusitalo et al., 2016).

Refractrometric Sensor

An LPSR experiences a redshift in frequency due to a rise in the dielectric operation of the surrounding. This transfer can be measured with a spectrophotometer and has been the basis for many "refractometric" sensing schemes depending on propagated surface plasmon polaritons or localized surface plasmon resonances on discrete nanoparticles. The above phenomenon can be utilized to develop a sensor that can quickly identify *S. typhimurium*. This was accomplished by conjugating an aptamer and GNPs with a portable LSPR sensing ship and then incubating the sensor for thirty minutes in a species containing *S. typhimurium*. The LSPR peak intensity was then analyzed using UV-Vis spectroscopy. The presence of the bacteria is indicated by the rise in absorbance at the unknown wavelength. The limit of detection was 1×10^4 CFU mL^{-1}, according to reports (Oh et al., 2017).

Electrochemical Techniques

It is also possible to create electrochemical biosensors of pathogens utilizing a AuNP platform. For example, the addition of AuNPs to a biosensor for detecting *E. Coli* 0157:H7 was described by Wang and his coworkers (2015). Two different kinds of nanoparticles viz. polymer-coated MNPs and carbohydrate-coated AuNPs, were used in their investigation. The detector *E. coli* 0157:H7 was separated using MNPs capped with polyaniline, and the anti-*E. coli* 0157:H7 monoclonal antibody was then conjugated with the help of an electrostatic interaction. Using MNPs capped with polyaniline as a separator, target *E. coli* 0157:H7 was isolated from the analyte matrix. The other particle, AuNPs coated with glucose, was used as a detector by generating electrochemical signals when it was coupled to an anti-*E. coli* 0157:H7 polyclonal antibody. A potentiostat was used to measure the amount of *E. coli* by means of an electrochemical signal generated after the detector *E. coli* 0157:H7 complex of MNPs and GNPs was magnetically separated. There was a linear relationship between the signal and the amount of *E. coli* 0157:H7, andcould be identified using this method with concentrations of 10–10^6 CFU mL^{-1} (Wang & Alocilja, 2015).

Lateral Flow Assay

Lateral flow assays, also known as LFAs, are one kind of immunochromatographic method that have been widely utilized to identify target specimens from sample matrices. These analyses are quick, simple, and don't require expensive or complex equipment to find the signal. Additionally, the assay shows excellent portability. This technique was developed without GNPs at first, but their later addition increased the sensitivity of target analyte detection. Nucleic acid sequences or antibodies are commonly used as molecules to detect foodborne pathogens. Liu and his colleagues (2013) conjugated 16S rDNA/rRNA to GNPs to create nucleic acid-based LFAs for the purpose of detecting Salmonella in food. The recognition probe was made by adjusting a –SH group on an oligonucleotide, which is a sequence of DNA attached with residues of ten adenosine and –SH group. First, the Salmonella (target specimen) was amalgamated with a capture probe, followed by a detection probe. Together with the Ag enhancement at the test membrane, this hybridized analyte was dipped into the membrane. A sample containing salmonella was visible at the detection limit of 10^4 CFU mL^{-1} and formed a change in color within the membrane (Liu et al., 2013). It is important to keep in mind that, even though nucleic acid sequences can be used as a molecular ligand to target pathogens with great sensitivity, preparing the nucleic acid sequence usually requires PCR preparation. Regretfully, there are certain requirements when using PCR, including costly reagents, specialized equipment, and highly experienced specialists. This has elevated curiosity in isothermal amplification of nucleic acid sequences, a more recent technique for preparing nucleic acid sequences. This type of amplification uses a uniform temperature to magnify sequences of nucleic acid without the use of thermal cycling equipment.

6.3.1.2 Nanoparticles for Protection from Allergens

Food allergies are an immunological response triggered by consuming or coming into contact with a specific food. Food allergens are particular foodstuffs, usually proteins but also occasionally chemical substances that certain immune cells recognize and trigger an immunological reaction that can result in symptoms that range from minor to fatal. There is a general interest in the influence of food allergies on human health as well as the social and economic ramifications. The ability to quickly, accurately, affordably, and sensitively detect and quantify allergenic residues in food is essential to the food industry's judgment of allergen labeling and risk management programs. This can be a difficult analytical piece of work because of the generally low concentration of these compounds within food matrices and the possibility that matrix effects will obscure their detection. The unique optical and electrical characteristics of nanomaterials, along with current selective bio- as well as biomimetic receptors, have made it feasible to create novel nanosensors that could greatly enhance a number of fields, including quality control and food safety. Reduced reagent volumes, faster analysis times, low detection limits (single molecule detection), and multianalyte detection are just a few advantages of using nanosensors for allergen detection. The ability to closely integrate nanomaterials with the biological recognition element is one advantage of using them in biosensing (Das et al., 2023; Nandi et al., 2022; Nandi et al., 2023a, b). To create composite nanoscale materials that merge the functionalities of biomolecules with the precisely controlled sensing properties of nanomaterials, biological essence including proteins, peptides, enzymes, antibodies, carbohydrates, or oligonucleotides (DNA or RNA) can be combined with nanostructures. When it comes to using nanomaterials with biosensors to identify food allergens, metal nanoparticles (like Au and Ag), nanomaterials from carbon (like carbon nanotubes and graphene), magnetic nanoparticles, and semiconductor quantum dots are the most frequently used types. The type of NP material, its dimensions, surface characteristics, and physical ligand, as well as the biological molecule chosen, all affect the bioconjugation chemistry that is best suited for developing a biosensor (Feng et al., 2023).

Optical or electrochemical signal transductions are the most widely used techniques by biosensors for food analysis. An optical phenomenon produced by the interaction of the bio-recognition element and the analyte is converted into an electrical signal by optical transducers. Optical biosensors are generally inexpensive, small in size, highly sensitive, and highly specific. Fluorescence, surface-enhanced Raman scattering (SERS), chemiluminescence, UV-vis absorption, surface plasmon resonance (SPR) and interferometry are the most often employed transduction methods in optical biosensors based on nanoparticles for analysis of food. Plasmonic metal nanomaterials (PMNMs), particularly silver and gold, are essential for developing novel color-based biosensors for allergen detection. Any change to the target compound's size, shape, or presence in the NPs will have an impact on the local electron confinement and change the color of the colloidal solution and SPR absorption maxima. Pollet and his research groups reported on the first fiber optic SPR biosensor that identified Ara h 1 in food matrices using an enhanced SPR signal from nanobeads. The researchers used this platform to compare three different immune sensing methods: a sandwich assay enhanced by nanobeads, a label-free assay, and a sandwich assay with a secondary antibody. They found that the SPR sensor's detection limit was increased by two orders of magnitude, from 9 to 0.09 μg/mL, by using 19 nm-size superparamagnetic NPs coated with the detection antibody, in comparison to other methods. After 35 regeneration cycles, there was no sensitivity loss (Pollet et al., 2011). Quantum dots or QDs are the often-utilized nanoparticles in biosensing using fluorescence spectroscopy. They possess large absorption spectra, efficient resistance to photobleaching, size-tunable emission obtained in a narrow and symmetrical spectrum, and high photoluminescence quantum yields. The major drawback of QDs is their extreme toxicity, which is caused by the heavy metals that comprise them. However, a variety of coatings have been applied lately to lessen this risk and improve their appropriateness for aqueous media (Misra & Misra, 2023). One of the most widely utilized techniques in fluorescence biosensing for allergen detection is the combination of aptamers with carbon nanostructures, or graphene oxide (GO) or AuNPs, and noble metals. For the detection of shrimp allergen tropomyosin, Zhang and his groups (2017) reported a sensitive biosensor by modifying a selective aptamer on a graphene oxide substrate through hydrophobic and van der Waals interactions. The aptamer's shape changed upon the addition of tropomyosin, resulting in its release from the graphene oxide surface. The aptamer-tropomyosin complex in solution was identified using fluorescence spectroscopy ($\lambda_{em} = 480$ nm; $\lambda_{em} = 522$ nm) with the existence of the OliGreen ssDNA reagent. This technique allowed for a dynamic value range of 0.5–50 μg/mL and a limit of detection 4.2 nM (Zhang et al., 2017). Bioassays and electrochemical biosensors are widely used in food and beverage analysis. The sensors can be categorized according to the electrochemical principle that is applied as conductometric, voltammetric, potentiometric, amperometric, or impedimetric. Among all biomolecular electronics devices, amperometric biosensors are perhaps the most widely used, abundant, and profitable. Due to their affordability, sensitivity, easy to fabricate, portability, rapidity, uniform response, electrochemical biosensors may be more economical than other types of equipment for the allergens analysis in complex samples of food. Gold nanoparticles are the most broadly utilized nanoparticles for amplification of signal because of their high stability, ease of functionalization, and versatility in size and format synthesis. Numerous techniques for analyzing food allergens with gold nanoparticles in collaboration with electrochemical biosensors have been reported in the literature. Liu and his group created electrochemical immunosensors with glutathione-decorated AuNPs (GSH-AuNP) functionalized with 28 amino acid peptide fragments of the main Ara h 2

IgE-binding epitope and covered on the surface of pyrolytic graphite electrodes for the detection of antibodies against Ara h 2. They estimated the efficacy of different detection methods, including non-faradaic impedance, single-label amperometry, and faradaic electrochemical impedance spectroscopy (EIS) to detect model chicken anti-peanut antibodies (IgY) in serum. Horseradish peroxidase (HRP) amplification of cells was used to measure the non-faradaic impedance of the sensor to determine its best linear range and detection limits (5 μg/mL) (Liu et al., 2010).

6.3.2 *Active Packaging Based on Nanotechnology*

Due to the active ingredients in materials like ethylene removers, antioxidants, and antimicrobial compounds, active packaging is a useful method for expanding the life span of fruits and vegetables. Consumers' demands for convenient, high preservation effect, eco-friendly, and reusable freshness-keeping packaging can be met by incorporating nanotechnology into active packaging (Nath et al., 2023a, b, c). There are several uses for nanoparticles based on silver in the food packaging industry. Edible nanocoating can maintain and safeguard food freshness, functionalized packaging can improve physicochemical and antibacterial qualities, and thermal-resistant high-performance packages are among these uses Fig. 6.4 and Table 6.3 (Mathew & Radhakrishnan, 2021).

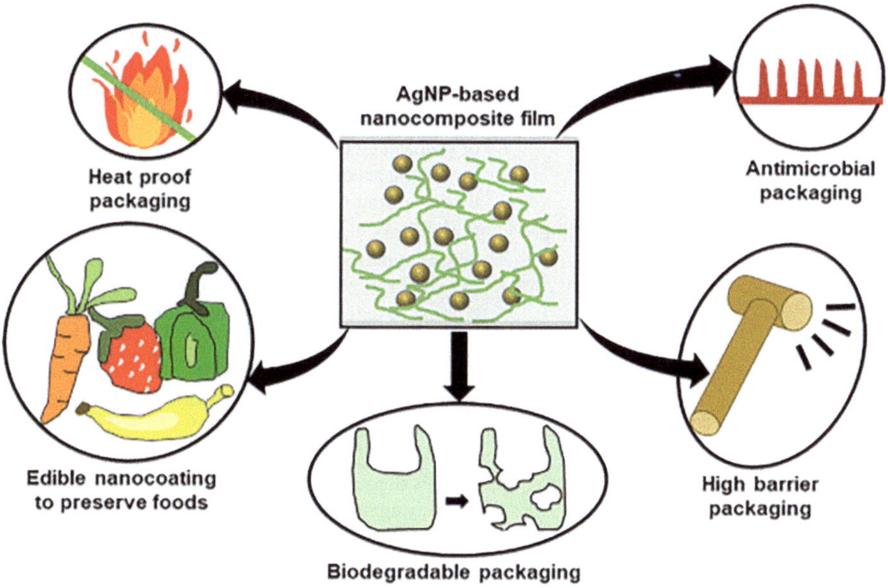

Fig. 6.4 Nanoparticles (Ag NPs) in food packaging and their potentiality. (Figure adapted from Mathew and Radhakrishnan (2021) and it is an open article (copyright © 2021 by authors) distributed under the terms and conditions of the Creative Commons Attribution (CC BY) License)

Table 6.3 The efficiency of edible coating used in food

Foods	Edible coating used in food	Efficiency	References
Fresh cut apple	Sodium Alginate	When compared to traditional emulsions at the same dosage, edible coatings based on nanoemulsions demonstrated slower growth of psychrophilic bacteria and higher *E. Coli* inactivation	Zambrano-Zaragoza et al. (2018)
Grape berry	Chitosan	The application of the nanoemulsion demonstrated antioxidant capacity retention while successfully inhibiting the early swelling of yeasts, *S. Typhimurium*, aerobic mesophiles, and molds	Zambrano-Zaragoza et al. (2018)
Fresh cut apple	Nopal mucilage	Compared to traditional emulsions, the coatings made with the nano-emulsion significantly inhibited PME and PPO activity	Zambrano-Zaragoza et al. (2018)
Guava	Solid lipid nanoparticles, Candeubawa S wax (carnauba wax and candelilla wax) Poloxamer 407	It would be simple to apply the possible usage of SLNs in edible coatings to reduce the senescence of various items	Zambrano-Zaragoza et al. (2018)
Guava	Poloxamer 40	Although it does so more slowly, the application of candeuba waxes (SLN) aids in maintaining the natural maturation process	Zambrano-Zaragoza et al. (2018)
Apples	Nano-SiO_2	Reduced respiration rate, preservation of firmness, and increased shelf life are the outcomes of preparing edible coating by ultrasonic processing and incorporating it into an SPI matrix	Liu et al. (2017)
Papaya	pear Montmorillonite (MMT)	Weight loss was seen in Papaya by adding montmorillonite and oregano essential oil; additionally, the edible coating assisted in slowing microbial growth	Cortez-Vega et al. (2014)
Strawberries	Montmorillonite (MMT)	The fresh covered strawberries' freshness was preserved, and the growth of mold was effectively inhibited by this edible coating	Zambrano-Zaragoza et al. (2018)
Citrus fruit	Silver nanoparticles (AgNPs)	AgNPs demonstrated noteworthy efficacy against *E. coli* and *S. aureus*, exhibiting advantageous outcomes for the conservation of citrus fruits	Koushesh Saba and Amini (2017)
Kinnow (Citrus reticulata)	Silver nanoparticles (AgNPs)	Ag nanoparticles and guar gum or CMC were introduced to a coating emulsion base	R. Liu et al. (2017)

(continued)

Table 6.3 (continued)

Foods	Edible coating used in food	Efficiency	References
Strawberries	Limonene	limonene liposomes dramatically reduced fungal growth	Shah et al. (2016)
Cantaloupe	Chitosan/nano-silica/nisin	It was discovered that their combination was ideal for extending shelf life since it preserved color, vitamin C, and peroxidase activity for up to days during storage	Dhital et al. (2017)
Blueberries	Nano-TiO_2/chitosan	At freezing temperatures, they could retain the nutritional makeup without sacrificing quality	Sami et al. (2021)
Blackberry	Chitosan	exhibited activity of antifungal against Racemosus	Xing et al. (2019)
Fresh Fruit (Redberry)	Arbutus unedo Sodium alginate	It shows the best results for the preservations	Miteluţ et al. (2021)
Fresh cut pineapples	Sodium Alginate	Citral nano-emulsion edible covering enhanced physico-chemical properties and inhibited microbiological development	Miteluţ et al. (2021)

6.3.2.1 Ripening Control

Reduced ethylene levels in storage environments can postpone the ripening process of fruits and vegetables, as ethylene is a key plant hormone in this process. The utilization of nanocomposite packaging materials, such as nano-Ag, nano-TiO_2, and montmorillonite, effectively prolonged the shelf life of ethylene-matured kiwifruit. The impact of a novel nano-packaging material, which consisted of nano-Ag, nano-silica, nano-TiO_2, and attapulgite, on the quality and shelf-life of *Flammulina velutipes* have been demonstrated. Researchers discovered that the presence of *F. velutipes* could potentially inhibit the generation of free radicals when using nano-packaging material. Additionally, it eliminated ethylene, controlled the air within the storage micro-environment, and inhibited microbial growth (Donglu et al., 2016).

6.3.2.2 Antioxidant Film

Numerous studies in the field of antioxidant packaging with nanomaterials have been conducted recently. Shi and his coworkers demonstrated that the activities of peroxidase and polyphenol oxidase, which are linked to the degradation of storage product quality, could be inhibited by chitosan/nano-silica coating. They ascribed the inhibition effect to the chelating action of the chitosan molecule and the adsorption of positively charged polymers of nano-silica. In comparison to control treatment products, the levels of malondialdehyde in chitosan/nano-SiO_2 coated samples were predominantly decreased. This indicated that the generation of reactive oxygen species and lipid peroxidation could be effectively controlled by nanohybrid film, greatly extending the shelf life of longan fruit (Nongbet et al., 2024; Shi et al., 2013).

6.3.2.3 Antimicrobial Film

Fruit and vegetable shelf life extension research has always focused on high-efficiency antimicrobial agents during storage because contamination due to microbes is the primary cause of rotting in fruits and vegetables (Nath et al., 2023a, b, c, 2024a, b; Sharma et al., 2023). To fully capitalize on the ability of Ag nanoparticles to be uniformly spread out in the films of alginate while retaining their inherent properties of antibacterial. Jiang \and team created a novel alginate/nano-silver coating material that enhances the quality of preservation, extends storage time, and inhibits the growth of microorganisms in cold storage shiitake mushrooms. According to their experimental findings, the shiitake mushroom cold storage with alginate/nano-silver coating material had a better freshness-keeping effect than the control group, which was only refrigerated for 16 days at 4 °C (Jiang et al., 2013). It is commonly known that silver nanoparticles have broad-spectrum antimicrobial activity. Costa and his research groups synthesized an Ag-montmorillonite film using an ion exchange mechanism which ultimately create a nano-Ag antimicrobial active packaging film having excellent mechanical properties. They discovered that the Ag-montmorillonite nanoparticles could increase the longevity of carrots shelf-life approximately more than 2 months through the estimation of sensory and microbiological quality (Costa et al., 2012).

6.3.3 Intelligent Label System Based on Nanotechnology

Intelligent label systems are acknowledged as an efficient way to further improve the fine management of the supply chain for fruits and vegetables, lower the consumption of perishable goods, and increase the safety of edible materials. These systems can precisely "identify" and "record" the wrapped environmental variables of fruits and vegetables and then supply the particulars to the purchaser intuitively. The most common intelligent labels used to extend the fruit and vegetables shelf life are gas, time-temperature, and freshness indicators. One example is the preparation of specific selective ethylene detection sensors using a copper complex and single-walled carbon nanotubes. Because copper can bind to ethylene molecules and is an essential cofactor of the ethylene (C_2H_4) receptors gene, stable copper-ethylene complexes are produced. Consequently, when the ethylene (C_2H_4) from wrapped fruits and vegetables propagates into the sealed microenvironment, the electrical surroundings of single-walled carbon nanotubes will alter. This will change the resistance of the sensors and disrupt the uniform state of the voltage between the electrodes. This carbon nanotube-based sensor was employed by Esser and his groups to identify the C_2H_4 emission of pear, banana, apple, avocado and orange. The findings showed that the reported sensor system could identify ethylene concentrations as low as ppm, and that the highest amount of ethylene was produced during storage in bananas. As a result, there was a good chance that this sensor system could determine whether fruits and vegetables were fit for human consumption (Esser et al., 2012). The intelligent label with time-temperature can be applied to analyze the real-time condition of food product freshness because temperature fluctuation and time variation have a direct impact on the freshness of fruits and vegetables, and improper storage conditions will severely damage the freshness of packaged products. To locate air leakage in packed products utilizing visualization approach, Suman and his coworkers interconnected an optoelectrical O_2 sensor that contained ion-pairing with nano-TiO_2, methylene blue, dodecyl sulfate to an electrical circuit (Suman et al., 2012). Despite the many benefits of nano-based intelligent labeling, intelligent label systems remain expensive projects. Since the fruit and vegetable industry has low-profit margins and lengthy investment return cycles, the use of nanotechnology in intelligent label systems for the preservation of fruits and vegetables is primarily in the laboratory research stage. Furthermore, barriers to advancing intelligent labels based on nanotechnology systems from research in the lab to broad-scale industrial production may include safety and legal issues.

6.4 Combined Preservation Techniques Based on Nanotechnology

Fruits and vegetables will have a longer shelf life when nanotechnologies are combined with other fresh-keeping treatments. The use of combined preservation strategies related to nanotechnology is becoming more and more popular as a creative and effective way to extend the shelf life of fruits and vegetables. To regulate the total number of bacteria and enhance the standard of vacuum-packaged Caixin, Liu et al. (2014), merge nano-zinc oxide with microwave heating. As anticipated, the vacuum-packed Caixin produced a sustainable product standard with the fewest bacterial colonies under treatment of 0.02 g/kg nano-ZnO microwave heating (2450 MHz) (Liu et al., 2014). Based on work published by Liu et al. (2014) for the preservation of carrots, Xu and team (2017), replaced the process of heating from the original microwave heating with radio frequency heating (RFH) because radio frequency heating technology experiences better heating condition, deeper depth to penetrate and uniform heating stability in contrast to microwave heating treatment. After analyzing the prepared carrots' overall colony-forming units and quality indicators, the authors concluded that the combined treatment could increase the shelf life of the product to 60 days by preventing microbial growth and improving product quality (Xu et al., 2017).

6.5 Commercialization of Nanobiotechnology-Based Products

There are two primary approaches to synthesizing nano-based packaging material to preserve the standard and increase the life span of food goods. One approach entails integrating nanoparticles into traditional food packaging materials such as films and containers. This second method involves using nanocomposites to make multi-layer packaging materials and coating them with organic, inorganic, or hybrid nanoparticles by immersing, spraying, or rubbing them on. Nanotechnology enables the creation of improved packaging materials that possess superior physical and mechanical properties. It also allows for the development of active packaging that exhibits antibacterial, antioxidative, and UV absorption properties. In addition, nanotechnology allows for the development of intelligent packaging that is capable of monitoring and regulating food conditions. The user's input is insufficient and lacks substantive information (Primožič et al., 2021). Table 6.4 displays the current available packaging options. Statistics MRC predicts that the global smart packaging market will be valued at $32.7 billion by 2022, growing at a compound annual rate of 11%. Researchers predict that intelligent packaging, with its unique, interactive, and customer-friendly features at a reduced cost, will grow at the highest rate compared to other types of smart packaging (Sahani & Sharma, 2021).

6.6 Conclusions and Future Scope

This chapter describes the latest advancements in fruit and vegetable shelf-life prolongation strategies relating to nanotechnology including active packaging, detection methods, intelligent label systems, and merge preservation techniques. It was observed that preservation techniques involving nanotechnology can significantly increase the fruits and vegetables shelf life and improve product quality while they are being stored. Furthermore, researchers can explore the mechanism of fruits and vegetables' quality degradation behavior during storage with the help of nanoscale detection. It should be noted that there are still many obstacles to be solved before there are widespread industrial trials for nanotechnology-related preservation techniques for extending the fruits and vegetables shelf life.

Table 6.4 Commercial availability of different nanomaterials-based packaging materials

Product and company	Materials and forms	Properties	References
Aegis HFX Resin and OXCE Resin Honeywell International Inc. USA	Nylon 6-nanoclay composite; Barrier of nylon resins	O_2 Scavenger	https://product.statnano.com/
Ageless®E Mitsubishi Gas Chemical Inc. Japan	Sodium carbonate; Sachet & labels	CO_2 Scavenging	https://www.mgc.co.jp/eng/products/sc/ageless/
Biomaster; Corporation, USA Addmaster Limited US	Nanosilver; Spray & bag	Anti-microbial	http://www.biomasterusa.com/
Carnation Instant food Carnation Breakfast Essential Switzerland	Titanium Dioxide; Spray	Anticaking agent	https://product.statnano.com/
Cryovac® OS System Cryovac Div., Sealed Air Corporation USA	Polymer; Tray, films	O_2 Scavenger	https://www.sealedair.com/
Ethysorb® Stay Fresh Ltd	PE-Nanoclay composite		https://magazines.marinelink.com/
Microspheres Bernard Technologies. Inc. USA	Chlorine dioxide	Microbiological Infestation	https://www.ssgca.com/
N-coat Multifilm Packing	Nano clay; Films	Act as a barrier of Gas	https://product.statnano.com/product/9780/n-coat
OMAC® Imperm® Mitsubishi Gas Chemical Inc. Japan	Cerium oxide; Film	O_2 Scavenger	https://www.mgc.co.jp/eng/products/sc/ageless-omac.html
OXy Guard® Clariant Ltd. Swaziland	Iron; Sachet & films	O_2 Scavenger	https://www.clariant.com/
RipeSense™ Sensor Ripesense limited New Zealand	Aromatic compounds; Stickers	Indicators	https://product.statnano.com/
TimeStrip® TimeStrip UK Ltd UK	TTI-based on enzyme, lipase, and pH; Stickers	Freshness (determined by color)	https://timestrip.com/
Tip Top bread George Weston Foods	Nanosized self-assembled liquid structure; Bag	Tuna fish oil combined with ethylene scavenger and nano capsule	https://www.tiptop.com.au/
UltraZap R Xtenda Pak pads Paper Pak Industries Canada	Allyl isothiocyanate (AIT) or scavenging molecular O_2; Tray pads	Have an anti-microbial pad CO_2 emitting properties	https://www.novipax.com/

References

Anvar, A. A., Ahari, H., & Ataee, M. (2021). Antimicrobial properties of food nanopackaging: A new focus on foodborne pathogens. *Frontiers in Microbiology, 12*. https://doi.org/10.3389/fmicb.2021.690706

Baig, N., Kammakakam, I., & Falath, W. (2021). Nanomaterials: A review of synthesis methods, properties, recent progress, and challenges. *Materials Advances, 2*(6), 1821–1871. https://doi.org/10.1039/D0MA00807A

Ban, C., Jo, M., Park, Y. H., Kim, J. H., Han, J. Y., Lee, K. W., Kweon, D.-H., & Choi, Y. J. (2020). Enhancing the oral bioavailability of curcumin using solid lipid nanoparticles. *Food Chemistry, 302*, 125328. https://doi.org/10.1016/j.foodchem.2019.125328

Bisht, B., Bhatnagar, P., Gururani, P., Kumar, V., Tomar, M. S., Sinhmar, R., Rathi, N., & Kumar, S. (2021). Food irradiation: Effect of ionizing and non-ionizing radiations on preservation of fruits and vegetables—A review. *Trends in Food Science & Technology, 114*, 372–385. https://doi.org/10.1016/j.tifs.2021.06.002

Bozkurt, A. G., Buyukgoz, G. G., Soforoglu, M., Tamer, U., Suludere, Z., & Boyaci, I. H. (2018). Alkaline phosphatase labeled SERS active sandwich immunoassay for detection of Escherichia coli. *Spectrochimica Acta Part A: Molecular and Biomolecular Spectroscopy, 194*, 8–13. https://doi.org/10.1016/j.saa.2017.12.057

Cao, B., Hu, M., Cheng, Y., Jing, P., Liu, B., Zhou, B., Wang, X., Gao, R., Sun, X., Du, Y., & Zhang, J. (2021). Tailoring the d-band center of N-doped carbon nanotube arrays with Co4N nanoparticles and single-atom Co for a superior hydrogen evolution reaction. *NPG Asia Materials, 13*(1), 1. https://doi.org/10.1038/s41427-020-00264-x

Cortez-Vega, W. R., Pizato, S., de Souza, J. T. A., & Prentice, C. (2014). Using edible coatings from Whitemouth croaker (Micropogonias furnieri) protein isolate and organo-clay nanocomposite for improve the conservation properties of fresh-cut 'Formosa' papaya. *Innovative Food Science & Emerging Technologies, 22*, 197–202. https://doi.org/10.1016/j.ifset.2013.12.007

Costa, C., Conte, A., Buonocore, G. G., Lavorgna, M., & Del Nobile, M. A. (2012). Calcium-alginate coating loaded with silver-montmorillonite nanoparticles to prolong the shelf-life of fresh-cut carrots. *Food Research International, 48*(1), 164–169. https://doi.org/10.1016/j.foodres.2012.03.001

Das, N., Debnath, P., Nandi, N. B., Purkayastha, A., Ghanta, S., & Misra, T. K. (2023). Combined experimental and computational investigation of a Schiff base derived from N-(1-naphthyl)ethylenediamine for fluoride recognition. *Monatshefte Für Chemie - Chemical Monthly, 154*(10), 1101–1114. https://doi.org/10.1007/s00706-023-03111-1

Dhar, P., Neog, U., Roy, B., Nandi, N. B., Deka, S. C., & Nath, P. C. (2023). A comprehensive guide to essential oil determination methods. In *Essential oils* (pp. 583–601). Wiley. https://doi.org/10.1002/9781119829614.ch26

Dhital, R., Joshi, P., Becerra-Mora, N., Umagiliyage, A., Chai, T., Kohli, P., & Choudhary, R. (2017). Integrity of edible nano-coatings and its effects on quality of strawberries subjected to simulated in-transit vibrations. *LWT, 80*, 257–264. https://doi.org/10.1016/j.lwt.2017.02.033

Donglu, F., Wenjian, Y., Kimatu, B. M., Xinxin, A., Qiuhui, H., & Liyan, Z. (2016). Effect of nanocomposite packaging on postharvest quality and reactive oxygen species metabolism of mushrooms (Flammulina velutipes). *Postharvest Biology and Technology, 119*, 49–57. https://doi.org/10.1016/j.postharvbio.2016.04.012

Du, J., Singh, H., Dong, W., Bai, Y., & Yi, T.-H. (2018). Colorimetric detection of Listeria monocytogenes using one-pot biosynthesized flower-shaped gold nanoparticles. *Sensors and Actuators B: Chemical, 265*, 285–292. https://doi.org/10.1016/j.snb.2018.03.067

Esser, B., Schnorr, J. M., & Swager, T. M. (2012). Selective detection of ethylene gas using carbon nanotube-based devices: Utility in determination of fruit ripeness. *Angewandte Chemie International Edition, 51*(23), 5752–5756. https://doi.org/10.1002/anie.201201042

Feng, L., Song, S., Li, H., He, R., Chen, S., Wang, J., Zhao, G., & Zhao, X. (2023). Nanobiosensors based on noble metal and semiconductor materials: Emerging trends and future prospects. *Metals, 13*(4), 792. https://doi.org/10.3390/met13040792

Hernández-Hernández, H., Quiterio-Gutiérrez, T., Cadenas-Pliego, G., Ortega-Ortiz, H., Hernández-Fuentes, A. D., Cabrera de la Fuente, M., Valdés-Reyna, J., & Juárez-Maldonado, A. (2019). Impact of selenium and copper nanoparticles on yield, antioxidant system, and fruit quality of tomato plants. *Plants, 8*(10), 355. https://doi.org/10.3390/plants8100355

Jiang, T., Feng, L., & Wang, Y. (2013). Effect of alginate/nano-Ag coating on microbial and physicochemical characteristics of shiitake mushroom (Lentinus edodes) during cold storage. *Food Chemistry, 141*(2), 954–960. https://doi.org/10.1016/j.foodchem.2013.03.093

Khan, S. A., Choudhury, R., Majumdar, M., Nandi, N. B., Roy, S., & Misra, T. K. (2020). Gluconate-stabilized silver nanoparticles as pH dependent dual-nanosensor for quantitative evaluation of methionine and cysteine. *ChemistrySelect, 5*(23), 7076–7085. https://doi.org/10.1002/slct.202001654

Khan, O. A., Zaidi, S., Islam, R. U., Naseem, S., & Junaid, P. M. (2023). Enhanced shelf-life of peach fruit in alginate based edible coating loaded with TiO_2 nanoparticles. *Progress in Organic Coatings, 182*, 107688. https://doi.org/10.1016/j.porgcoat.2023.107688

Koushesh Saba, M., & Amini, R. (2017). Nano-ZnO/carboxymethyl cellulose-based active coating impact on ready-to-use pomegranate during cold storage. *Food Chemistry, 232*, 721–726. https://doi.org/10.1016/j.foodchem.2017.04.076

Lakshmayya, N. S. V., Mishra, A. K., Mohanta, Y. K., Panda, J., Naik, B., Mishra, B., & Varma, R. S. (2023). Essential oils-based nano-emulsion system for food safety and preservation: Current status and future prospects. *Biocatalysis and Agricultural Biotechnology, 53*, 102897. https://doi.org/10.1016/j.bcab.2023.102897

Li, X., Li, W., Jiang, Y., Ding, Y., Yun, J., Tang, Y., & Zhang, P. (2011). Effect of nano-ZnO-coated active packaging on quality of fresh-cut 'Fuji' apple. *International Journal of Food Science & Technology, 46*(9), 1947–1955. https://doi.org/10.1111/j.1365-2621.2011.02706.x

Liu, H., Malhotra, R., Peczuh, M. W., & Rusling, J. F. (2010). Electrochemical immunosensors for antibodies to peanut allergen Ara h2 using gold nanoparticle–Peptide films. *Analytical Chemistry, 82*(13), 5865–5871. https://doi.org/10.1021/ac101110q

Liu, C.-C., Yeung, C.-Y., Chen, P.-H., Yeh, M.-K., & Hou, S.-Y. (2013). Salmonella detection using 16S ribosomal DNA/RNA probe-gold nanoparticles and lateral flow immunoassay. *Food Chemistry, 141*(3), 2526–2532. https://doi.org/10.1016/j.foodchem.2013.05.089

Liu, Q., Zhang, M., Fang, Z., & Rong, X. (2014). Effects of ZnO nanoparticles and microwave heating on the sterilization and product quality of vacuum-packaged Caixin. *Journal of the Science of Food and Agriculture, 94*(12), 2547–2554. https://doi.org/10.1002/jsfa.6594

Liu, R., Liu, D., Liu, Y., Song, Y., Wu, T., & Zhang, M. (2017). Using soy protein SiOx nanocomposite film coating to extend the shelf life of apple fruit. *International Journal of Food Science & Technology, 52*. https://doi.org/10.1111/ijfs.13478

Luo, Z., Wang, Y., Jiang, L., & Xu, X. (2015). Effect of nano-$CaCO_3$-LDPE packaging on quality and browning of fresh-cut yam. *LWT - Food Science and Technology, 60*(2), 1155–1161. https://doi.org/10.1016/j.lwt.2014.09.021

Ma, X., Song, L., Zhou, N., Xia, Y., & Wang, Z. (2017). A novel aptasensor for the colorimetric detection of S. typhimurium based on gold nanoparticles. *International Journal of Food Microbiology, 245*, 1–5. https://doi.org/10.1016/j.ijfoodmicro.2016.12.024

Majumdar, M., Khan, S. A., Nandi, N. B., Roy, S., Panja, A. S., Roy, D. N., & Misra, T. K. (2020). Green synthesis of iron nanoparticles for investigation of biofilm inhibition property. *ChemistrySelect, 5*(43), 13575–13583. https://doi.org/10.1002/slct.202003033

Manikandan, V., & Min, S. C. (2023). Roles of polysaccharides-based nanomaterials in food preservation and extension of shelf-life of food products: A review. *International Journal of Biological Macromolecules, 252*, 126381. https://doi.org/10.1016/j.ijbiomac.2023.126381

Mathew, S., & Radhakrishnan, E. K. (2021). Chapter 6 - Silver-based nanomaterials for food packaging applications. In A.-F. A. Abd-Elsalam (Ed.), *Nanobiotechnology for plant protection* (pp. 125–146). Elsevier. https://doi.org/10.1016/B978-0-12-823528-7.00003-2

Mekuye, B., & Abera, B. (2023). Nanomaterials: An overview of synthesis, classification, characterization, and applications. *Nano Select, 4*(8), 486–501. https://doi.org/10.1002/nano.202300038

Mishra, B., Mishra, A. K., Mohanta, Y. K., Yadavalli, R., Agrawal, D. C., Reddy, H. P., Gorrepati, R., Reddy, C. N., Mandal, S. K., Shamim, M. Z., & Panda, J. (2024). Postbiotics: the new horizons of microbial functional bioactive compounds in food preservation and security. *Food Production, Processing and Nutrition, 6*(1), 28. https://doi.org/10.1186/s43014-023-00200-w

Misra, K. P., & Misra, R. D. K. (2023). ZnO-based quantum dots for biosensing, cancer imaging and therapy: An overview. *Biomedical Materials & Devices, 1*(1), 99–107. https://doi.org/10.1007/s44174-022-00033-0

Miteluţ, A. C., Popa, E. E., Drăghici, M. C., Popescu, P. A., Popa, V. I., Bujor, O.-C., Ion, V. A., & Popa, M. E. (2021). Latest developments in edible coatings on minimally processed fruits and vegetables: A review. *Foods (Basel, Switzerland), 10*(11). https://doi.org/10.3390/foods10112821

Mohammadi, A., Hashemi, M., & Hosseini, S. M. (2016). Postharvest treatment of nanochitosan-based coating loaded with Zataria multiflora essential oil improves antioxidant activity and extends shelf-life of cucumber. *Innovative Food Science & Emerging Technologies, 33*, 580–588. https://doi.org/10.1016/j.ifset.2015.10.015

Mohanta, Y. K., Mishra, A. K., Panda, J., Chakrabartty, I., Sarma, B., Panda, S. K., Chopra, H., Zengin, G., Moloney, M. G., & Sharifi-Rad, M. (2023). Promising applications of phyto-fabricated silver nanoparticles: Recent trends in biomedicine. *Biochemical and Biophysical Research Communications, 688*, 149126. https://doi.org/10.1016/j.bbrc.2023.149126

Mohanta, Y. K., Nongbet, A., Panda, J., Chakrabartty, I., Mohanta, T. K., & Mishra, B. (2024). Chapter 10 - Nanobiotechnology in genetic engineering for abiotic stress resistance in crops. In R. N. Pudake, R. M. Tripathi, & C. P. Gill (Eds.), *Nanotechnology for abiotic stress tolerance and management in crop plants* (pp. 153–164). Academic. https://doi.org/10.1016/B978-0-443-18500-7.00010-7

Nandi, N. B., Purkayastha, A., Kłak, J., Ganguly, R., Ghanta, S., & Misra, T. K. (2022). Copper(II) complexes of 1,3-dimethyl-5-(4′/3′-pyridylazo)-6-aminouracil: Structures, redox, magnetic and protein binding properties. *Journal of Molecular Structure, 1252*, 132164. https://doi.org/10.1016/j.molstruc.2021.132164

Nandi, N. B., Das, N., Ghanta, S., Puzari, K. R., Dutta, P., Kłak, J., Sieroń, L., Maniukiewicz, W., & Misra, T. K. (2023a). Atropisomers and a copper(II) complex derived from 1,3-dimethyl-5-(8′-quinolinylazo)-6-aminouracil: structures, magnetism and biological properties. *New Journal of Chemistry, 47*(47), 21633–21647. https://doi.org/10.1039/D3NJ04598F

Nandi, N. B., Ghanta, S., Kłak, J., Sieroń, L., Maniukiewicz, W., & Kumar Misra, T. (2023b). Pseudohalide coligands guided structural motifs, magnetism and BSA-interactions of copper(II) complexes derived from 1,3-dimethy-6-aminouracil. *Polyhedron, 230*, 116215. https://doi.org/10.1016/j.poly.2022.116215

Nath, P. C. (2022). *Polyethylene terephthalate-clay nanocomposites* (pp. 183–202). https://doi.org/10.21741/9781644902035-8

Nath, P. C., Nandi, N. B., Khan, S. A., Bhunia, B., Bandyopadhyay, T. K., & Roy, B. (2023a). Smart packaging for commercial food products. In *Green sustainable process for chemical and environmental engineering and science* (pp. 197–209). Elsevier. https://doi.org/10.1016/B978-0-323-95644-4.00016-4

Nath, P. C., Nandi, N. B., Tiwari, A., Das, J., & Roy, B. (2023b). Applications of nanotechnology in food sensing and food packaging. In *Nanotechnology applications for food safety and quality monitoring* (pp. 321–340). Elsevier. https://doi.org/10.1016/B978-0-323-85791-8.00006-9

Nath, P. C., Ojha, A., Debnath, S., Sharma, M., Sridhar, K., Nayak, P. K., & Inbaraj, B. S. (2023c). Biogeneration of valuable nanomaterials from agro-wastes: A comprehensive review. *Agronomy, 13*(2), 561. https://doi.org/10.3390/agronomy13020561

Nath, P. C., Mishra, A. K., Sharma, R., Bhunia, B., Mishra, B., Tiwari, A., Nayak, P. K., Sharma, M., Bhuyan, T., Kaushal, S., Mohanta, Y. K., & Sridhar, K. (2024a). Recent advances in artificial intelligence towards the sustainable future of agri-food industry. *Food Chemistry, 447*, 138945. https://doi.org/10.1016/j.foodchem.2024.138945

Nath, P. C., Sharma, R., Debnath, S., Nayak, P. K., Roy, R., Sharma, M., Inbaraj, B. S., & Sridhar, K. (2024b). Recent advances in production of sustainable and biodegradable polymers from agro-food waste: Applications in tissue engineering and regenerative medicines. *International Journal of Biological Macromolecules, 259*, 129129. https://doi.org/10.1016/j.ijbiomac.2023.129129

Nongbet, A., Panda, J., Mohanta, Y. K., Chakrabartty, I., Shamim, M. Z., & Mohanta, T. K. (2024). Chapter 6—Role of nanoparticles to protect plants from abiotic stress by scavenging reactive oxygen species. In R. N. Pudake, R. M. Tripathi, & C. P. Gill (Eds.), *Nanotechnology for abiotic stress tolerance and management in crop plants* (pp. 95–114). Academic. https://doi.org/10.1016/B978-0-443-18500-7.00006-5

Oh, S. Y., Heo, N. S., Shukla, S., Cho, H.-J., Vilian, A. T. E., Kim, J., Lee, S. Y., Han, Y.-K., Yoo, S. M., & Huh, Y. S. (2017). Development of gold nanoparticle-aptamer-based LSPR sensing chips for the rapid detection of Salmonella typhimurium in pork meat. *Scientific Reports, 7*(1), 10130. https://doi.org/10.1038/s41598-017-10188-2

Panda, J., Mishra, A. K., Mohanta, Y. K., Patowary, K., Rauta, P. R., & Mishra, B. (2024). Exploring biopolymer for food and pharmaceuticals application in the circular bioeconomy: An agro-food waste-to-wealth approach. *Waste and Biomass Valorization*, 1–31. https://doi.org/10.1007/s12649-024-02452-0

Pollet, J., Delport, F., Janssen, K. P. F., Tran, D. T., Wouters, J., Verbiest, T., & Lammertyn, J. (2011). Fast and accurate peanut allergen detection with nanobead enhanced optical fiber SPR biosensor. *Talanta, 83*(5), 1436–1441. https://doi.org/10.1016/j.talanta.2010.11.032

Primožič, M., Knez, Ž., & Leitgeb, M. (2021). (Bio)nanotechnology in food science—Food packaging. *Nanomaterials, 11*(2), 292. https://doi.org/10.3390/nano11020292

Ray, M. K., Mishra, A. K., Mohanta, Y. K., Mahanta, S., Chakrabartty, I., Kungwani, N. A., Avula, S. K., Panda, J., & Pudake, R. N. (2023). Nanotechnology as a promising tool against phytopathogens: A futuristic approach to agriculture. *Agriculture, 13*(9), 1856. https://doi.org/10.3390/agriculture13091856

Roy, B., Bhunia, B., Bandyopadhyay, T. K., Khan, S. A., Nandi, N. B., & Nath, P. C. (2024). Water pollution by heavy metals and their impact on human health. In *Handbook of water pollution* (pp. 333–352). Wiley. https://doi.org/10.1002/9781119904991.ch10

Sahani, S., & Sharma, Y. C. (2021). Advancements in applications of nanotechnology in global food industry. *Food Chemistry, 342*, 128318. https://doi.org/10.1016/j.foodchem.2020.128318

Samadarsi, R., Mishra, D., & Dutta, D. (2020). Mangiferin nanoparticles fortified dairy beverage as a low glycemic food product: its quality attributes and antioxidant properties. *International Journal of Food Science & Technology, 55*(2), 589–600. https://doi.org/10.1111/ijfs.14310

Sami, R., Almatrafi, M., Elhakem, A., Alharbi, M., Benajiba, N., & Helal, M. (2021). Effect of nano silicon dioxide coating films on the quality characteristics of fresh-cut cantaloupe. *Membranes, 11*(2). https://doi.org/10.3390/membranes11020140

Seray, M., Skender, A., & Hadj-Hamou, A. S. (2021). Kinetics and mechanisms of Zn^{2+} release from antimicrobial food packaging based on poly (butylene adipate-co-terephthalate) and zinc oxide nanoparticles. *Polymer Bulletin, 78*(2), 1021–1040. https://doi.org/10.1007/s00289-020-03145-z

Shah, S., Qaisar, M., Jahangir, M., Abbasi, K., Khan, S., Ali, N., & Liaquat, M. (2016). Influence of CMC- and guar gum-based silver nanoparticle coatings combined with low temperature on major aroma volatile components and the sensory quality of kinnow (Citrus reticulata). *International Journal of Food Science & Technology, 51*, 2345–2352. https://doi.org/10.1111/ijfs.13213

Shamim, M. Z., Deka, P., Mohanta, Y. K., Mishra, B., Biswas, K., & Mishra, A. K. (2024). Chapter 5—Implications of caged molecular structure of buckminster fullerenes in food sciences and industry applications. In K. Biswas, Y. K. Mohanta, T. K. Mohanta, & B. Saravanan (Eds.), *Progress in biochemistry and biotechnology* (pp. 123–161). Academic. https://doi.org/10.1016/B978-0-443-15508-6.00008-7

Sharma, S., Zhang, X., Nair, S. S., Ragauskas, A., Zhu, J., & Deng, Y. (2014). Thermally enhanced high performance cellulose nano fibril barrier membranes. *RSC Advances, 4*(85), 45136–45142. https://doi.org/10.1039/C4RA07469F

Sharma, R., Nath, P. C., Bhunia, B., & Bandyopadhyay, T. K. (2023). Microbial exopolysaccharides: Production, properties, and food applications. In *Food process engineering and technology* (pp. 441–469). Springer Nature. https://doi.org/10.1007/978-981-99-6831-2_23

Shi, S., Wang, W., Liu, L., Wu, S., Wei, Y., & Li, W. (2013). Effect of chitosan/nano-silica coating on the physicochemical characteristics of longan fruit under ambient temperature. *Journal of Food Engineering, 118*(1), 125–131. https://doi.org/10.1016/j.jfoodeng.2013.03.029

Singh, T., Shukla, S., Kumar, P., Wahla, V., Bajpai, V. K., & Rather, I. A. (2017). Application of nanotechnology in food science: Perception and overview. *Frontiers in Microbiology, 8*, 01501. https://doi.org/10.3389/fmicb.2017.01501

Sogvar, O. B., Koushesh Saba, M., Emamifar, A., & Hallaj, R. (2016). Influence of nano-ZnO on microbial growth, bioactive content and postharvest quality of strawberries during storage. *Innovative Food Science & Emerging Technologies, 35*, 168–176. https://doi.org/10.1016/j.ifset.2016.05.005

Suman, G. V., Kumar, P., & Jain, V. K. (2012). Nanomaterial-based opto-electrical oxygen sensor for detecting air leakage in packed items and storage plants. *Journal of Experimental Nanoscience, 7*(6), 608–615. https://doi.org/10.1080/17458080.2011.551895

Thiramanas, R., & Laocharoensuk, R. (2016). Competitive binding of polyethyleneimine-coated gold nanoparticles to enzymes and bacteria: A key mechanism for low-level colorimetric detection of gram-positive and gram-negative bacteria. *Microchimica Acta, 183*(1), 389–396. https://doi.org/10.1007/s00604-015-1657-7

Uusitalo, S., Kögler, M., Välimaa, A.-L., Popov, A., Ryabchikov, Y., Kontturi, V., Siitonen, S., Petäjä, J., Virtanen, T., Laitinen, R., Kinnunen, M., Meglinski, I., Kabashin, A., Bunker, A., Viitala, T., & Hiltunen, J. (2016). Detection of Listeria innocua on roll-to-roll produced SERS substrates with gold nanoparticles. *RSC Advances, 6*(67), 62981–62989. https://doi.org/10.1039/C6RA08313G

Velusamy, V., Arshak, K., Korostynska, O., Oliwa, K., & Adley, C. (2010). An overview of foodborne pathogen detection: In the perspective of biosensors. *Biotechnology Advances, 28*(2), 232–254.

Wang, Y., & Alocilja, E. C. (2015). Gold nanoparticle-labeled biosensor for rapid and sensitive detection of bacterial pathogens. *Journal of Biological Engineering, 9*(1), 16. https://doi.org/10.1186/s13036-015-0014-z

Xing, Y., Li, W., Wang, Q., Li, X., Xu, Q., Guo, X., Bi, X., Liu, X., Shui, Y., Lin, H., & Yang, H. (2019). Antimicrobial nanoparticles incorporated in edible coatings and films for the preservation of fruits and vegetables. *Molecules, 24*(9), 1695. https://doi.org/10.3390/molecules24091695

Xu, J., Zhang, M., Bhandari, B., & Kachele, R. (2017). ZnO nanoparticles combined radio frequency heating: A novel method to control microorganism and improve product quality of prepared carrots. *Innovative Food Science & Emerging Technologies, 44*, 46–53. https://doi.org/10.1016/j.ifset.2017.07.025

Yang, F., Hu, P., Yang, F., Hua, X.-J., Chen, B., Gao, L., & Wang, K.-S. (2023). Photocatalytic applications and modification methods of two-dimensional nanomaterials: A review. *Tungsten*. https://doi.org/10.1007/s42864-023-00229-x

Yemmireddy, V. K., & Hung, Y.-C. (2015). Effect of binder on the physical stability and bactericidal property of titanium dioxide (TiO_2) nanocoatings on food contact surfaces. *Food Control, 57*, 82–88. https://doi.org/10.1016/j.foodcont.2015.04.009

Yu, N., Zhang, M., Islam, M. N., Lu, L., Liu, Q., & Cheng, X. (2015). Combined sterilizing effects of nano-ZnO and ultraviolet on convenient vegetable dishes. *LWT - Food Science and Technology, 61*(2), 638–643. https://doi.org/10.1016/j.lwt.2014.12.042

Zahedi, S. M., Karimi, M., & Teixeira da Silva, J. A. (2020). The use of nanotechnology to increase quality and yield of fruit crops. *Journal of the Science of Food and Agriculture, 100*(1), 25–31. https://doi.org/10.1002/jsfa.10004

Zambrano-Zaragoza, M. L., González-Reza, R., Mendoza-Muñoz, N., Miranda-Linares, V., Bernal-Couoh, T. F., Mendoza-Elvira, S., & Quintanar-Guerrero, D. (2018). Nanosystems in edible coatings: A novel strategy for food preservation. *International Journal of Molecular Sciences, 19*(3). https://doi.org/10.3390/ijms19030705

Zhang, Y., Wu, Q., Wei, X., Zhang, J., & Mo, S. (2017). DNA aptamer for use in a fluorescent assay for the shrimp allergen tropomyosin. *Microchimica Acta, 184*(2), 633–639. https://doi.org/10.1007/s00604-016-2042-x

Chapter 7
Eco-Friendly Production of Antimicrobial Nanoparticles for Sustainability

Arunima Biswas, Jibanjyoti Panda, Amilia Nongbet, and Renald Blundell

Abstract One of the primary causes of worldwide morbidity and mortality is infectious diseases. Antibiotics represent a major landmark discovery of the twentieth century, having saved millions of lives from deadly pathogens. Unfortunately, microbes are increasingly acquiring significant antimicrobial resistance. Such a crisis is fast becoming the most pressing issue in public health due to the inefficacy of existing drugs and the lack of new antimicrobials. Therefore, sustainable development ensures that present demands are met without risking future generations' ability to do the same. Focusing on Sustainable Development Goal 3, which states that by 2030, improve mental health and well-being while reducing premature mortality from non-communicable disorders by one-third through prevention and treatment, the authors of this chapter have made their way to that goal. Global efforts are underway to develop new, more effective antibacterial chemicals and delivery and targeting methods. Nanotechnology may help fight antibiotic resistance by creating innovative antimicrobials that disrupt microbial processes and increase biofilm penetration. Nanotechnology can improve nanomedicine applications of numerous natural substances. The classical nanotechnology physical and chemical synthesis techniques employing 'top-down' or 'bottom-up' approaches commonly use

A. Biswas (✉)
Department of Microbiology, Raidighi College, Raidighi, West Bengal, India

J. Panda
Nano-Biotechnology and Translational Knowledge Laboratory, Department of Applied Biology, University of Science and Technology Meghalaya (USTM), Ri-Bhoi, Meghalaya, India

A. Nongbet
Department of Botany, School of Biological Sciences, University of Science and Technology Meghalaya, Ri-Bhoi, Meghalaya, India

R. Blundell
Centre for Molecular Medicine, Biomedical Sciences Building, University of Malta, Msida, Malta

Y. K. Mohanta et al. (eds.), *Nano-microbiology for Sustainable Development*,
https://doi.org/10.1007/978-3-031-78845-1_7

poisonous and hazardous substances and extreme physical parameters that may harm humans, ecosystems, and the environment. These are complex, expensive, time-consuming, and arduous techniques. Scientists worldwide are concentrating on eco-friendly, safer, and easier production processes employing natural resources for sustainability and antibacterial efficacy. In green production of more advanced antibacterial nanoparticles, plant-based natural resources and microbial biological material are promising.

Keywords Green nanotechnology · Antimicrobial nanoparticles · Synthesis · Natural resources · Eco-friendly · Sustainable development

7.1 Introduction

Antibiotics, regarded as the epitome of last century's "wonder drugs", transformed once mortally feared deadly bacterial diseases into easily curable infections. They have served as a blessing to mankind in the battle against pathogenic and harmful microbes. Since the discovery of the first and the iconic antibiotic penicillin, numerous other antimicrobial agents have been developed and marketed. As early as 1945, Sir Alexander Fleming raised the alarm regarding antibiotic overuse and the evolution of antimicrobial resistance or AMR (Kungwani et al., 2024; Pariente, 2022). This is a menace to the global population as it has the potential to cause large-scale health crises since resistant microbes can proliferate and spread rapidly and cause regular common infections that are no longer treatable with standard drugs. Inhibition of crucial enzyme activity, target modification, and new improved, or altered functioning of efflux pumps are a few of the most important defensive measures used by bacteria to acquire resistance against specific antibiotics (Muteeb, 2023; Nayak et al., 2024). The unavailability of newer, improved drugs has further enhanced the problem. This sad scenario is mostly attributable to a lack of proper regulatory requirements and public awareness and also due to reduced financial incentives and research grants (Fig. 7.1). The "golden era" of antibiotics ended because researchers could not maintain the pace of the discovery of new antimicrobials in the face of emerging resistant pathogens. Since treatment options for infections caused by antimicrobial-resistant bacteria are limited, AMR is the next big threat that is slowly developing into a silent pandemic and may result in considerable morbidity and mortality with a huge economic impact, thus necessitating urgent intervention. According to the World Health Organization (WHO) report of 2019, AMR has been responsible for the deaths of 700,000 people, while it is estimated that by 2050 the figure will reach an alarming 20 million, costing over a staggering $2.9 trillion (Salam et al., 2023). Sustainable development entails the implementation of strategies that satisfy the present requirements of society while safeguarding the capacity of future generations to meet their own needs. Here, the authors have emphasized sustainable goal 3, which seeks to decrease premature mortality caused

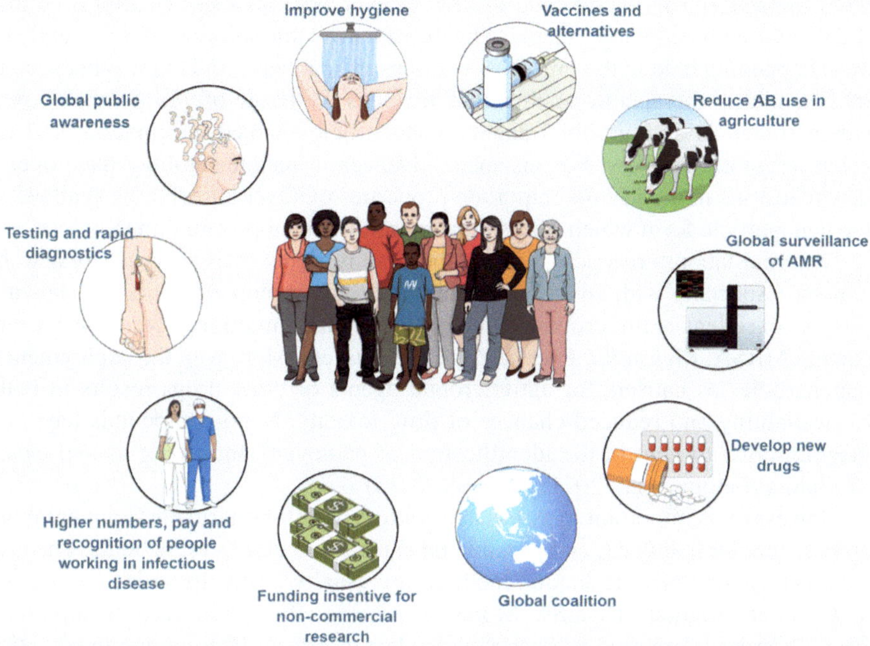

Fig. 7.1 Depicts a few major factors associated with the solution of AMR. (Figure adapted from Stanley et al. (2022) and it is open article (copyright © 2022 by authors) distributed under the terms and conditions of the Creative Commons Attribution (CC BY) License)

by non-communicable illnesses by 33% through prevention and treatment. Additionally, it aspires to enhance mental health and well-being by the year 2030. This chapter demonstrates the efficient and eco-friendly use of nanoparticles obtained from various sources for sustainable antibacterial treatment.

Microbes have also become resistant to other forms of antimicrobial agents. An excellent example is that of silver. For many years, different silver-based products have been used topically as antimicrobial compounds. Silver nitrate and silver sulfadiazine are common instances. They can exert broad-spectrum bactericidal activity as they can bind with vital prokaryotic macromolecules, like proteins, at their thiol (–SH) group resulting in complete inhibition of crucial pathways such as respiration, electron transfer, transport systems, DNA replication, transcription etc., ultimately leading to cell death. These properties have also rendered silver compounds extremely effective in the treatment of chronic wounds and burns, which are often poly-microbial with both Gram-positive cocci and Gram-negative bacteria causing severe harm. Moreover, alteration in normal blood flow patterns in the vicinity of these wounds minimizes the effect of systemic antibiotics thus making silver compounds the medicine of choice. Unfortunately, overuse of such silver compounds has led to the emergence of bacterial silver resistance and consequently has compromised its therapeutic utility. To overcome all these problems of drug resistance, nanotechnology seems to be the best available solution. It is currently the most happening field with numerous functions in healthcare and many other sectors. Its potential in therapeutic applications includes the usage of various metal oxide nanoparticles as antimicrobial agents, nanoparticle-enabled vaccination, and targeted antibiotic delivery. For instance, advances in nanotechnology have opened new platforms to modify the important properties of silver allowing its synthesis in the nanoparticle form which offers greater antimicrobial potential than before.

Different laboratories across the globe have reported their success story in this context, especially with silver, gold, and zinc oxide nanoparticles being shown to possess excellent antimicrobial efficacy even against multidrug-resistant bacteria such as MRSA, *Klebsiella*, *E. coli*, *Pseudomonas*, etc. Moreover, the deployment of nanoparticles as carriers for antimicrobial agents or other drugs results in better bioavailability and reduced chances of drug toxicity. Nano-compounds have also been reported to help in the identification of pathogens and to act as anti-cancer mediators (Fatima et al., 2021; Xu et al., 2023)

However it is important to assess and address their long-term environmental outcomes, economic impact, and possible unwanted side effects on patients. The classical nanotechnology production methods are usually costly, hazardous, laborious, and not environment-friendly. Moreover, chemically synthesized nanoparticles have often been reported to have adverse health effects like cardiac dysfunction, immune suppression, lung inflammation, liver problems, and accumulation in non-target organs resulting in altered metabolism and organ-specific toxicity (Borehalli et al., 2023). Eco-friendly, affordable, reliable, pragmatic, and secure synthesis methods using natural resources can be used to rule out such menaces. Since antimicrobial nanoparticles are gradually becoming the need of the hour due to their high potential as a timely remedy to address the menaces of AMR, it is imperative that they be synthesized for that purpose using sustainable greener technologies (Biswas, 2023). Over the past several years, scientists have focussed on using algae, fungi, bacteria, viruses, and plant–based resources for the production of efficient and non-toxic metallic nanoparticles. Such environmentally safe methods, often denoted as 'green nanotechnology', are proving to be extremely viable alternatives to traditional approaches. Figure 7.2 demonstrates the overall idea of different techniques for nanomaterial synthesis.

7.2 Nanoparticles as Antimicrobial Agents: Uses and Demands

In the past couple of decades, nanomaterials have emerged as valuable and convenient tools in diverse fields ranging from medicine to agriculture (Ray et al., 2023). They have immense and extensive applications. Due to their nanoscale sizes, they present large surface/volume ratios, distinct characteristic morphology, and

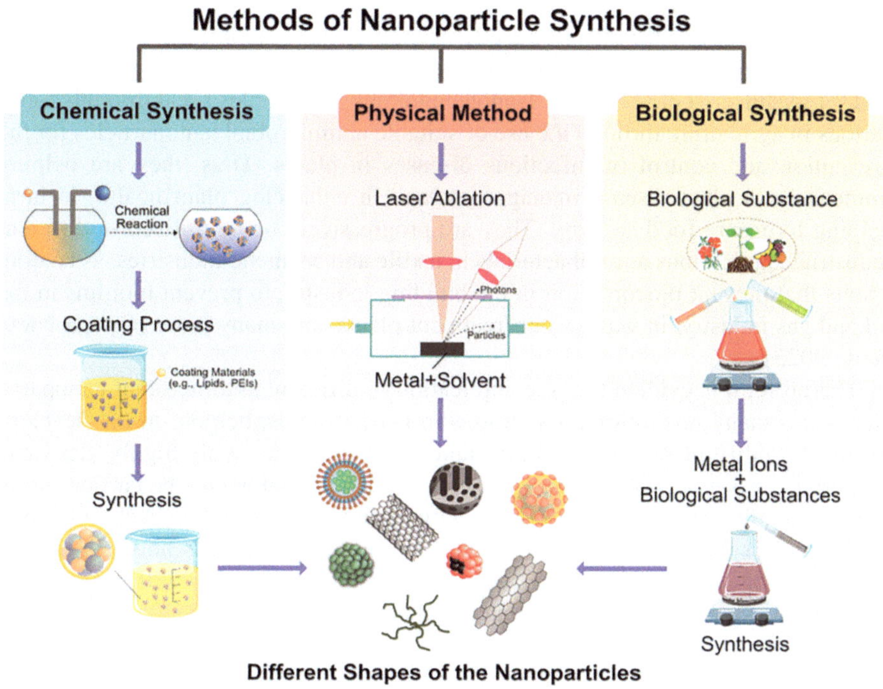

Fig. 7.2 Schematic representation of the different methods of nanoparticle synthesis. (Figure adapted from Jeyaraj et al. (2019) and it is an open article (copyright © 2019 by authors) distributed under the terms and conditions of the Creative Commons Attribution (CC BY) License)

dimensions that match the scale of natural biomolecules. Such unique properties have allowed scientists to design and develop various types of multi-functional nanoparticles for disease diagnosis, targeted drug delivery, treatment of resistant infections, and effective management of deadly diseases such as cancer. The most exciting application is its great potential in addressing the AMR crisis, resulting in intense research focus on its antimicrobial therapeutic properties. Nanoparticles of gold, silver, titanium oxide, etc. have also been successfully used against deadly viruses like Ebola, HIV, influenza, and HBV. The designing of sophisticated nano-vaccines is under intense investigation and is being considered to be of prime importance especially after the Covid-19 pandemic, to induce strong, reproducible immune responses against difficult and fatal pathogens (Fatima et al., 2021; Muteeb, 2023).

Apart from bio-medicine, antimicrobial nanoparticles also have increasing demand in different other fields due to their unique properties and applications. For example, certain nanoparticles can control microbial bio-deterioration and degradation of cellulose-containing cultural heritage like archaeological manuscripts (Fouda et al., 2019). In addition, many antimicrobial nanoparticles can be used for enhancing food safety, security, and quality via increased shelf life and food preservation properties and have consequently opened up new possibilities for innovation in the food preparation, processing, packaging, and storage sectors (Jadhav et al., 2023; Lakshmayya et al., 2023). Nanotechnology also enables breakthrough applications in agriculture through the use of suitable antimicrobial nanoparticles for the prevention and control of infectious diseases in plants. Thus, they are helping immensely in plant disease management and in enhancing plant health, in turn, helping to ensure food security. They are progressively being utilized in different industrial applications as antibacterials in textile and cosmetic industries, as coating agents that prevent biocorrosion in the building industry, to prevent biofilms in the oil and gas industry, in wastewater treatment plants, and many more (Ortega-Nieto et al., 2023).

Therefore, it is evident that the application of different antimicrobial nanoparticles offers many revolutionary solutions in myriad fields, helping in the development of multi-functionalized drugs and other products with highly desirable properties. It is essential that the synthesis of such crucial agents be carried out in an eco-friendly and sustainable manner with minimum risks or drawbacks. Greener, eco-friendly synthesis techniques of antimicrobial nanomaterials have become an active area of interest because of numerous advantages coupled with a variety of applications, often offering better functional benefits than before (Huq et al., 2023).

7.3 Sustainable Development Goals and Nanotechnology

Every nation envisions building a society where living conditions and resources can nourish and support common people's needs without disrupting the stability of the natural ecosystem or planetary integrity. Any long-lasting sustainable development policy aims to find a balance between social well-being, environmental safety, and financial growth, in a lasting, viable manner. The mission is to satisfy the needs of the present generation without compromising the security and welfare of future generations. In 2015, the United Nations General Assembly adopted 17 Sustainable Development Goals that address different pressing global challenges, including climate change and environmental degradation (Purvis et al., 2019). Nanotechnology is a key tool enabling modern technologies of the twenty-first century and its endless capabilities align with the United Nations' Sustainable Development Goals. Most of countries have consciously started making significant investments in nanotechnology as they have recognized and appreciated its radical, game-changing breakthrough prospects that can immensely help mankind, the environment, and the planet (Pokrajac et al., 2021). Thus, it is vital to ensure the responsible development of nanotechnology itself. Scientists and policymakers must ascertain that health and environmental risks associated with the production and use of nanoparticles are nil or negligible.

7.4 Green Nanotechnology, Environment and Sustainability

According to the National Nanotechnology Initiative, nanotechnology is defined as the understanding, manipulation, and control of matter at a scale between 1 and 100 nanometres (nm). Scientists can take advantage of the unique physical, chemical, optical, biological, and mechanical properties of nanomaterials to revolutionize technology and industry and, in turn, benefit society. Nanomaterials are traditionally produced using Top-down and Bottom-up approaches. The top-down approach is the step-wise breaking down of bulk material to get nanostructures, while the bottom-up approach is the build-up of nanostructures via atom-by-atom or molecule-by-molecule self-assembly (Bayda et al., 2020). The physical production processes employ methods like spray pyrolysis, high-energy irradiation, laser ablation, and ion implantation, and the chemical approaches include the use of the sonochemical method, microemulsion method, chemical reduction technique, sol-gel process, and electrochemistry. But, as mentioned above, these approaches have severe drawbacks, especially with respect to environmental safety and sustainability (Biswas, 2023). The physical production methods are mostly energy intensive and usually require special costly devices while the chemical pathways often use high temperature and toxic reagents and produce hazardous by-products. These key factors negatively influence the commercial-level scale-up of production process, both economically as well as environmentally. Furthermore, the clinical use of nanoparticles generated through chemical techniques has been restricted due to concerns of adverse biocompatibility, issues of stability and reports of toxicity. Serious concerns have been raised regarding the ecological safety and sustainability of chemically synthesized nanoparticles which have already put an enormous burden on the environment and economy because of their harmful effects on plant, animal, human and their ecosystems (Bhardwaj et al., 2020; Mohanta et al., 2024).

Such limitations of physical and chemical production methods compelled researchers to look for other means of nanoparticle synthesis and they turned towards the biological world. This initiated the epoch of green nanotechnology where scientists aimed to use environmentally safer, simpler methods using natural resources for sustainability (Fig. 7.3). Initial success was obtained by many laboratories with a variety of biologically active macromolecules such as proteins, DNA, enzymes, and polysaccharides where the molecules often served both as templates and stabilizing agents, easily yielding highly stable antimicrobial metal nanoparticle suspensions on a large scale. However, there were quite a few challenges in using such biological molecules due to their easy degradation, overall high cost, contamination risk, and requirement of special storage conditions. On the other hand, nanoparticles synthesized with the help of plant and microbial resources are relatively non-hazardous, more biocompatible, easily water-soluble, quite stable, considerably beneficial, and reasonably cost-effective. Biological cells have the inherent ability to take up and accumulate inorganic metal ions from their surrounding ecological niche and can further process those to synthesize extracellular and

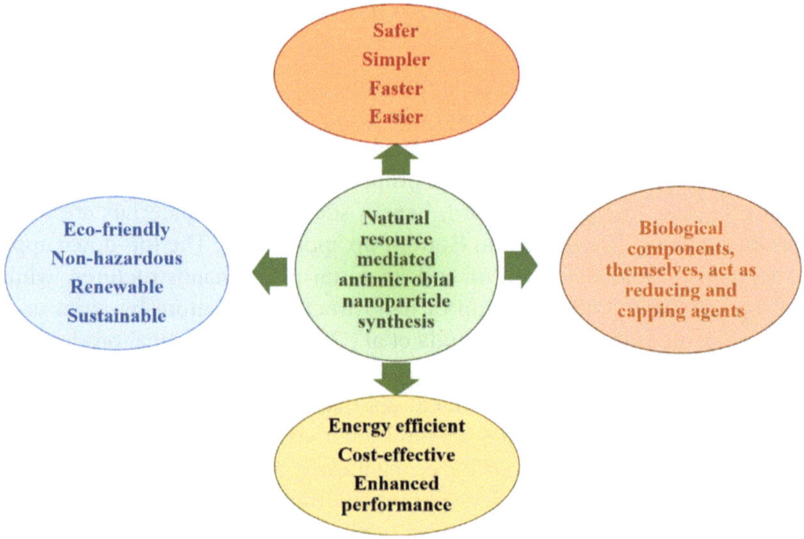

Fig. 7.3 A summary of the key merits of nanoparticle synthesis using natural resources

intracellular nanoparticles, thus offering a great convenient opportunity for safety and sustainability. Such protocols also have the advantage of reproducibility along with ease of scaling up with well-defined morphology (Borehalli et al., 2023). Renewable and easily available natural resources (like honey) are being used by some nations for novel green synthetic approaches, which is proving to be safer, easier, non-toxic with the nanoparticles often showing improved therapeutic efficacy (Biswas, 2023). As a result, the manufacturing of nanoparticles by biological methods has emerged as a new sustainable trend in this sector.

The current primary goal of all nanotechnologists is to develop a superior large-scale production method that will be reliable and viable and would ensure desired chemical composition, morphology, and better functions of the synthesized nanomaterials. It is a dynamic and evolving research process. But, at present, the green synthetic strategy is the most favorable choice. Antimicrobial nanoparticles are in great demand in various industries and the health sector. Such nanoparticles interfere with bacterial systems through various pathways like Reactive Oxygen Species (ROS)-induced oxidative stress, protein degradation, enzyme inhibition, alterations in gene expression, cell wall damage, cell membrane disruption, metal release, etc., all leading to dysfunction of cellular processes and ultimately cell death (Nongbet et al., 2024; Wahab et al., 2023). Figure 7.4 represents the overall mechanism for antimicrobial treatment through nanoparticles. Laboratories across the globe are working on different greener routes of antimicrobial nanoparticle production, using biological systems like bacteria, fungi, algae, mushrooms, naturally obtained

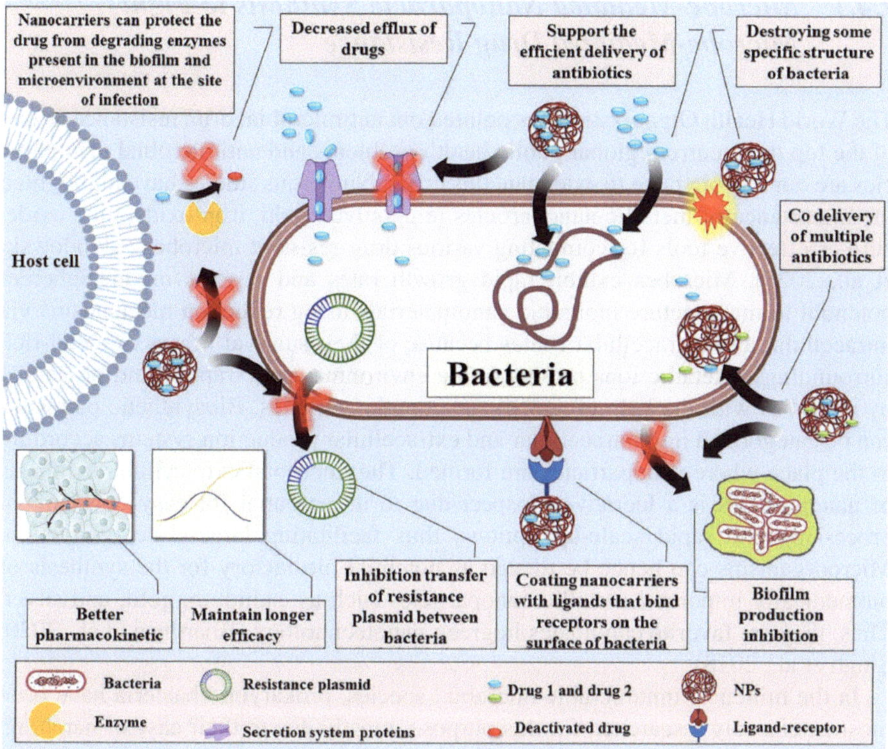

Fig. 7.4 Illustration of nanocarriers used for antibacterial treatments. The usage of nanocarriers can boost the activity and cellular absorption of drugs. Nanocarriers can be modified or filled with molecules that can interfere with the process of quorum sensing or alter the arrangement of biofilms. In addition, they have been utilized to enhance the pharmacokinetic profile of drugs and augment their cellular absorption. Nanocarriers can shield the medicine from the degrading enzymes present in the biofilm. Furthermore, the concurrent administration of medications using nanocarriers yields superior results compared to the simultaneous administration of free drugs. Furthermore, directing drugs specifically to infected areas can enhance effectiveness and minimize toxicity in comparison to delivering drugs without targeting. (Figure adapted from Aflakian et al., (2023) and it is an open article (copyright © 2023 by authors) distributed under the terms and conditions of the Creative Commons Attribution (CC BY) License)

polysaccharides, enzymes, and plant-based resources including different plant parts or extracts from various fruit, seed, leaf, flower, bark, honey, etc. Not only are the production methods economic and non-hazardous, but the synthesized nanoparticles, in addition, usually exhibit special enhanced features including better catalytic activity, higher stability, greater biocompatibility and biodegradability and significantly lower toxicity profile (Biswas, 2023; Ortega-Nieto et al., 2023).

7.4.1 Microbe-Mediated Nanoparticle Synthesis to Fight Microbe-Mediated Drug Resistance

The World Health Organization has pointed out antimicrobial drug resistance as one of the top three current global public health problems and antimicrobial nanoparticles are our one big hope to extenuate this crisis. Numerous studies have highlighted the importance of metallic nanoparticles (e.g., silver, gold, iron oxide, zinc oxide, etc.) as effective tools for combating various drug-resistant microbes (Skłodowski et al., 2023). Microbes exhibit rapid growth rates and possess unique inherent potential to manufacture inorganic nanomaterials using reduction mechanisms via intracellular and extracellular routes because of their survival fitness in metal-rich surroundings. Metallic ions present in the environment are trapped and processed by microbes with the help of specific enzymatic activities. Biosynthetic pathways can be categorized into intracellular and extracellular production systems according to the place where nanoparticles are formed. The microbial extracellular synthesis of nanoparticles is a lucrative prospect due to its potential for easy downstream processing and rapid scale-up options, thus facilitating large-scale production. Microorganisms can hence be treated as potential bio-factory for the synthesis of biomedically important metallic nanoparticles such as cadmium, gold, and silver. Thus, they are favorite candidates in green nanotechnology (Bhardwaj et al., 2020; Singh et al., 2018).

In the milieu of innumerable microbial species, prokaryotic bacteria have been most extensively researched for this purpose, mostly due to their ease of handling, low cost of cultivation, and wide options for genetic manipulation. Common bacteria like *E. coli, Bacillus subtilis,* and *Lactobacillus*, filamentous cyanobacterium like *Plectonema boryanum*, mesophilic bacterium like *Shewanella algae*, endophytic bacterium like *Bacillus cereus,* extremophilic actinomycetes like *Thermomonospora* species and many more have been extensively used to produce metal nanoparticles (Velusamy et al., 2016). Biomolecules secreted by bacteria can be used as capping as well as stabilizing agents during nanoparticle synthesis, so that there is no need to add such agents from outside.

Medical use of silver can be traced back to ancient times. Antimicrobial potential of this noble metal has been known since ages. Hence the exploration of the antibacterial capacity of silver nanoparticles was an obvious path of choice during the early developmental days of nanotechnology. Silver nanoparticles have indeed been reported to be the maximum effective antimicrobial agent amongst all other nanomaterials and they are usually quite stable, adding to their efficacy (Skłodowski et al., 2023). Their antibacterial competence covers both Gram-negative and Gram-positive bacteria, including multidrug resistant strains and they also possess antifungal potential against *C. albicans* and other fungal infections. More importantly, they exhibit significant synergistic effect against pathogenic bacteria including Extended Spectrum Beta-Lactamase (ESBL)-producing Enterobacteriaceae like *Escherichia coli* and *Klebsiella pneumoniae*, Methicillin-resistant *Staphylococcus aureus,* multi drug resistant *Salmonella enterica, Vibrio species, Bacillus anthracis* etc., in combination with antibacterial organic compounds or antibiotics like nitrofurantoin, carbenicillin, vancomycin, rifampicin, ciprofloxacin with increment in efficacy rising up to 50% (Bhardwaj et al., 2020). Since different silver-tolerant bacterial species are naturally capable of reducing silver ions with ease, they are widely used as standard candidates in preparation of silver nanoparticle with distinct size, shape and morphology by laboratories across the world. Some examples of bacterial strains that have been extensively used for the purpose include: *Escherichia coli, Pseudomonas proteolytica, Lactobacillus casei, Aeromonas sp. SH10, Bacillus cereus, Bacillus cecembensis, Bacillus licheniformis, Bacillus amyloliquefaciens, Bacillus indicus, Enterobacter cloacae, Arthrobacter gangotriensis* and many more (Singh et al., 2018).

A popular method of bacteria-mediated silver nanoparticle emphasizes the simplicity of the green production process, especially since the involved biological molecules, themselves, perform dual functions of formation and stabilization of such nanoparticles in the aqueous medium. As AMR is a global threat, bacteria-mediated synthesis of silver nanoparticles offers an excellent eco-friendly alternative as the process can be effectively and safely used to prevent and combat infections, decontaminate medical supplies, sterilize industrial products, and develop new bactericidal products for disinfection or purification purposes. The attention captured by them is reflected in the steady growth of their market, the ascending figures of high demand in industrial sectors, and increased research funding related to them (Bruna et al., 2021).

Likewise, for the preparation of gold nanoparticles, several bacterial species have been used, such as *E. coli DH5a, Brevibacterium casei, Pseudomonas aeruginosa, Bacillus subtilis 168, Bacillus megaterium D01, Plectonema boryanum UTEX 485, Desulfovibrio desulfuricans, Rhodopseudomonas capsulata, Shewanella alga,* etc. (Biswas, 2023). Gold nanoparticles also have good stability, excellent biocompatibility, significant antibacterial properties and can be easily modified for functional enhancement. They are first-rate drug carriers that can augment the effects of loaded antibacterial drugs thus acting as excellent nano-arsenal against ESBL producing resistant pathogens (Rizvi et al., 2023).

The use of fungal systems is equally effective and convenient. For example, during the early 2000, silver nanoparticles in the size range of 2–5 nm could be produced extracellularly by silver-tolerant fungal species like yeast strain MKY3, when challenged with 1 mM soluble silver during the log phase of cell growth (Kowshik et al., 2003). Such extracellular synthesis of nanoparticles is highly profitable and viable from the point of view of large-scale production and easy downstream processing. In fact, compared to bacterial system, many laboratories have obtained higher yield of nanoparticles with different shapes, sizes and therapeutic activity, because of the larger biomass of fungal systems. Many fungal species have since been used to produce antimicrobial nanoparticles, such as *Saccharomyces cerevisiae, Aspergillus oryzae, Alternaria alternata, Penicillium fellutanum, Fusarium oxysporum, Trichoderma viride, Verticillium luteoalbum, Colletotrichum* sp., *Trichothecium sp.*, etc. Many of them show various ranges of antimicrobial efficacy, including antifungal activity against fluconazole-susceptible and fluconazole-resistant *Candida albicans* isolates, and against pathogenic bacteria like *Streptococcus pneumoniae, Klebsiella pneumoniae, Vibrio cholerae, Bacillus anthracis*, etc (Abdel-Hadi et al., 2023; Elshafei et al., 2021).

It is worth mentioning that microbe-made green metal nanoparticles are also known for their wide spectrum anti-viral prospective against many viruses, like hepatitis B virus, herpes simplex virus type I, II, HIV, respiratory syncytial virus, monkey pox virus, etc. They can act by targeting processes of virus binding to the host cell or virus penetration or replication within the host. This enables myriad applications in the health sector and bio-remedy, food and textile industries, smart agriculture, air filtration, and wastewater treatment (Luceri et al., 2023). On the other hand, virus-mediated synthesis of nanoparticles has also been attempted by many groups, especially by employing plant viruses and bacteriophages to treat infections and diseases in humans and animals (Ghosh et al., 2021). Viral bodies, themselves, can be viewed as naturally occurring nanoparticles due to their 20–500 nm dimensions and have been explored for drug delivery, targeted vaccines, and immunotherapeutics. The major concern of this technique is its safety issue which needs to be addressed with utmost precaution and priority.

Actinomycetes need special mention in the context of green nanotechnology. They are considered by many to be the superior groups among microbial species of commercial importance because of the various bioactive components and extracellular enzymes present, which are crucial in metal nanoparticle synthesis. Species of *Streptomyces, Thermomonospora, Rhodococcus, Nocardia* have been identified from actinomycetes to be most effective in producing antimicrobial nanoparticles that work great against *E. coli, S. aureus, Enterococcus faecalis, P. aeruginosa, C. albicans, C. glabrata, Alternaria alternata and Fusarium oxysporum,* as determined by standard protocols like agar well diffusion technique, MIC and MBC analysis (Ghosh et al., 2021). Thus, the use of microbial bio-factories is a viable eco-friendly strategy for the sustainable production of anti-microbial nanoparticles. However, they have their limitations. Synthetic methods involving fungus, and bacteria become challenging as they involve strain separation, proper preservation of pure culture, and maintenance of optimum growth parameters and culture media. Hence, many researchers are opting for plant-based systems primarily because they are simple to extract, plentiful, and easier to work with (Parmar et al., 2022).

7.4.2 Plant-Based Eco-Friendly Synthesis of Antimicrobial Nanoparticles

Eco-friendly plant-derived nanoparticles are regarded as decisive components of future green chemistry and medicinal chemistry to control various life-threatening infections and fatal diseases of man and animal, with zero or negligible adverse side effects. They are expected to accomplish this feat without threatening our ecosystem and with concomitant protection of our environment.

Plants have various biomolecules and are the source of many primary and secondary plant metabolites (carbohydrates, proteins, enzymes, co-enzymes, phenolics, tannins, flavonoids, flavones, alkaloids, terpenoids, terpenes, saponins, etc.). These are capable of converting metal salt to a zero-oxidation state, thus generating stable nanoparticles. Antimicrobial metal nanoparticles have been synthesized by using several plant parts, including zinc nanoparticles from leaf extract of coriander (*Coriandrum sativum*) and essential oil of lemongrass (*Cymbopogon sp*), gold and silver nanoparticles from biomass and leaf extract of oat (*Avena sativa*), silver nanoparticles by plant extract of alfalfa (*Medicago sativa*), leaf extract of aloe vera (*Aloe barbadensis* Miller) neem (*Azadirachta indica*), cabbage (*Brassica oleracea*) and tulsi (*Ocimum sanctum*), seed extract of Mustard (*Brassica juncea*) and seed aqueous extract of *Mangifera indica*, stem extract of Mangosteen (*Garcinia mangostana*), extract of green tea (*Camellia sinensis*) and zest extract of lemon (*Citrus limon*), cupric oxide nanoparticles by leaf and natural fruit extract of tamarind (*Tamarindus indica L.*), and many more (Bhardwaj et al., 2020; Khane et al., 2022; Mohammed et al., 2023; Parmar et al., 2022; Perveen et al., 2021). A great advantage is that plant extracts can serve as both reducing and stabilizing agents during the production process due to the reducing functional groups present in different phytochemicals (Mohanta et al., 2023). Thus, they can be used to replace chemical-reducing agents like sodium borohydride, sodium citrate, and ascorbate, which are not only toxic and expensive but in many cases, detrimental to our environment. This property of phytochemicals facilitates easy, safe, single-step synthesis of desired nanoparticles. However, a few challenges commonly faced with these techniques include disadvantages such as a lack of control over morphology and size, and the issue of lower stability in aqueous solutions due to the presence of active biomolecules in plant extracts (Tabasum et al., 2022).

Plant-made nanoparticles have been widely reported to protect from *Escherichia coli, Klebsiella pneumoniae, Staphylococcus aureus,* and different *Candida species* and are also effective against biofilms (Ali et al., 2022; Tabasum et al., 2022). Phytonanotechnology has also unlocked the possibility of personalized medicine, which is an individualized approach to making decisions regarding disease prevention, diagnosis, and treatment of a specific individual. This has become possible due to the precise targeting capability and low toxicity of eco-friendly antimicrobial nanoparticles which can be used to customize therapeutic management depending on the personal attributes of each patient to outweigh blanket treatment (Alghamdi et al., 2022). Nanoparticles can be designed to minimize both local and systemic adverse side effects, increase the bioavailability of oral drugs, and enhance the half-life and overall pharmacokinetic and pharmacodynamic properties. Moreover, nanoparticles can also help reduce the frequency of drug administration or usage, resulting in better compliance thus upgrading clinical outcomes, especially for chronic infections. The emergence of medicinal plant-based green synthesis of metallic nanoparticles is believed to play a pivotal role in the field of theragnostic, which is a multipurpose portmanteau of synchronized diagnostics and therapeutics to simultaneously detect and treat infections and ailments (Xulu et al., 2022). While the therapeutic effects of metal nanoparticles in various infections and diseases are well documented and well established, their diagnostic use, especially those of plant-derived nanoparticles, in general, is still in its infancy. It has been reported that the fluorescence properties of metal nanoparticles can be exploited in the detection of microorganisms including resistant bacteria while their magnetic traits are used for cancer diagnosis. For example, nanotheranostics can perform the dual function of diagnosis and therapy in the case of methicillin-resistant *S. aureus* (MRSA) using the same metal nanoparticle for both purposes at the same time. Diagnosis can be via visual detection, and/or by fluorescence microscopy techniques. Simultaneous treatment can be carried out either via nanoparticle-mediated photothermal therapy inducing thermal injury or by photodynamic therapy releasing reactive oxygen species to induce prokaryotic cell damage (Panigrahi et al., 2022; Xulu et al., 2022). Purified compounds or phytochemicals like curcumin from turmeric (*Curcuma longa*), quercetin and kaempferol from Indian pennywort (*Centella asiatica*), resveratrol from common grape vine (*Vitis vinifera*), thymol from thyme (*Thymus vulgaris*), chlorogenic acid from coffee (*Coffea canephora, Coffea arabica*), cinnamaldehyde from cinnamon (*Cinnamomum zeylanicum*) have been reported to yield antimicrobial gold and silver nanoparticles which can act as excellent plant-based nanotheranostic agents (Xulu et al., 2022). Nanotheranostics are being referred to as "game-changers" that can protect against highly resistant bacteria (notorious superbugs like MRSA) and their tolerant biofilms through rapid diagnosis and multimodal therapeutic mechanisms at one go. Moreover, nanotheranostic approach using green nanotechnology is also predicted to be highly effective and safe to combat several viral diseases (Mosselhy et al., 2021).

Plant-derived nanoparticles have also shown significant efficacy against parasitic infectious protozoa including *Plasmodium falciparum* that causes malaria, a life-threatening vector borne common tropical disease. Current anti-malarial treatment depends mainly on chemotherapy, which is limited by serious toxic adverse side effects and development of drug-resistant parasites leading to treatment failure. Quite a few laboratories have reported fruitful and efficient green synthesis of metal nanoparticles (like silver and titanium oxide nanoparticles) with high anti-plasmodial activity using plant-based materials like aqueous leaf and/or fruit extract of *Catharanthus roseus, Salvia officinalis, Momordica charantia, Crataegus ambigua* (Gujjari et al., 2022). To put it in a nutshell, plant-based eco-friendly synthesis of nanoparticles has emerged as a sustainable and profitable green chemistry route for large-scale rapid synthesis of stable nanoparticles with extensive and diverse antimicrobial properties (Bagheri et al., 2023).

7.4.3 Honey-Mediated Eco-Friendly Synthesis of Antimicrobial Nanoparticles

Honey is a precious, renewable natural resource that is available globally and whose nutritional and medicinal properties have been appreciated since early civilizations. Ancient literature and medical manuscripts cite the use of honey by their physicians to heal injured or infected patients (Palma-Morales et al., 2023). In recent times, there has been renewed interest in harnessing the therapeutic powers of natural honey. Various laboratories in different countries have reported their findings about the antimicrobial, antioxidant, wound-healing anti-inflammation, anti-tumor, and other pharmaceutical effects of their native honey. However, the scientific community is still mostly skeptical of its curative impact, and, thereby, its use in modern medicine is unfortunately limited (Biswas, 2023). Its most popular use in the industry is only as a sweetening agent. The 'World Charter for Nature', formed by the United Nations back in 1982, and the 'Sustainable Development Goals' set up in 2015, all emphasize the need for optimum, responsible, sensible, sustainable usage of all global natural resources for their various commercial, pharmacological, aesthetic, social and cultural values. Our planet is rich in diverse, precious and unique biotic and abiotic natural products. Regrettably, many of these have not yet been studied and characterized systematically and scientifically for their various valuable properties. This has left a huge gap in recognizing and appreciating their prospect and is delaying their acceptance by the scientific community and common public. Honey is one such neglected bio-resource that has not been optimally used to its full potential. It happens mainly due to lack of awareness about various utility of honey on a bigger scale.

Honey generally contains 80–85% carbohydrate (about 40% fructose, 35% glucose, 5% sucrose), 0.1–0.4% protein, 0.2% ash, 15–17% water, small quantities of vitamins, enzymes, amino acids along with other valuable ingredients such as phenolic antioxidants. The precise complex chemical composition, physical features, and beneficial properties of honey vary with its floral origin on which honey bees foraged, species of the honey bees, geographical region, differences in climatic parameters, and nature of vegetation (Balasooriya et al., 2017; Keskin et al., 2022). Notably, the high sugar and low water content, acidic pH, osmolarity, hydrogen peroxide, and diverse phytochemical content confer honey its ability to kill a wide spectrum of bacteria.

With the advent of the 'green nanotechnology' concept, eco-friendly synthesis of nanoparticles has commenced a new era, in which honey can play a promising role. Honey-mediated sustainable production is a relatively novel concept that is being used by many nanotechnologists to synthesize antimicrobial silver, gold, iron oxide, and copper nanoparticles that are effective against both gram-positive and gram-negative bacteria including *Enterococcus faecalis*, *Escherichia coli*, *Bacillus* sp., *Staphylococcus aureus* etc and can help in wound healing (Bahari et al., 2022; Ismail et al., 2019; Neupane et al., 2019). A great contributing factor is that honey acts as both a stabilizing and reducing agent, although further studies are needed to identify the actual ingredients that are responsible for the reduction of metal ions and bestowing stability. This synthesis method can be carried out safely and conveniently at room temperature and does not produce any toxic by-products. Figure 7.5 represents how natural honey provides a simple, rapid, cost-effective, biocompatible, safe method of the sustainable nanoparticle synthesis process. However, a general drawback of this process is that the reduction of silver ions using honey occurs best at alkaline pH, where some of the inherent valuable properties of honey may get lost.

At room temperature 20 gm of honey is dissolved in 80 ml of deionized water

↓

15 ml of this honey is added to 20 ml aqueous solution of 1mM silver nitrate

↓

Stirred well for a few minutes

↓

pH is adjusted to 5.5 using NaOH to initiate the reduction of Ag ions

↓

Rapid reduction is indicated by the colour change of the solution to brown-yellow

↓

Decreased particle size is obtained with increasing pH and increasing honey concentration

↓

Silver nanoparticles can be analyzed using X-ray diffraction and electron microscopy

↓

Assayed for antimicrobial activity using standard protocols

Fig. 7.5 The figure represents a simplified protocol of honey-mediated antimicrobial silver nanoparticle synthesis

A few laboratories have also reported a faster and easier method of sunlight-induced, honey-mediated synthesis of antimicrobial silver nanoparticles using only natural honey and silver nitrate as ingredients (Kumar et al., 2020). In this method, equal volumes of 3% honey solution and 0.1 M silver nitrate solution were mixed together. The mixture was continuously stirred with the help of a magnetic stirrer to obtain a completely miscible solution which was then exposed to bright sunlight for 10 minutes. The colour change of the mixture to nearly a yellowish-brown suggested the formation of silver nanoparticles as honey reduced silver nitrate forming the metal nanoparticles. An increase in color intensity with increasing time of sunlight exposure, indicated an increase in the concentration of nanoparticles formed. Successful synthesis was confirmed by UV-visible spectrophotometry and electron microscopy analysis. Interestingly, a follow-up study to find out the relation between silver nanoparticle production and the wavelength of the incident light indicated that blue and yellow lights are best for the production process. Another group has reported successful synthesis of highly stable, pure silver nanoparticles using 30gm of honey as a capping and reducing agent in combination with 5 kGy dose of gamma radiation, yielding significant antimicrobial activity against standard bacterial strains as well as silver-resistant clinical isolates from infected wounds and burns (Hosny et al., 2017; Kumar et al., 2020). Copper is another multifunctional antibacterial agent. Copper nanoparticles can effectively bind and conjugate with biomolecules like proteins, enzymes, and nucleic acid, exhibit potent inhibition toward various bacterial species, and can be used as nano-drugs. But copper is highly unstable as it can easily oxidize at atmospheric conditions to produce cuprous or cupric oxides. Thereby, it is essential to have a surface-protecting stabilizing agent that can form complexes with the copper nanoparticles and honey can easily play that role. So, honey-mediated antimicrobial copper nanoparticles have been synthesized using ultrasonic irradiation (Ismail et al., 2019). Similarly, other nanoparticles, such as chromium oxide and iron oxide nanoparticles, have also been synthesized using *different varieties of* honey (Neupane et al., 2019; Shahid et al., 2023).

Such nanoparticles often exhibit enhanced superior antimicrobial potential than nanoparticles made by classical methods, which can be attributed to additive effects conferred by honey due to its own medicinal properties (Biswas, 2023; Neupane et al., 2019). Thus, they can be seriously considered as an excellent alternative to standard antibiotics. However, it is absolutely crucial to recognize precisely the correct type of honey for a specific nanomaterial, according to its intended applications. This is still a budding prospective field that remains to be fully optimized and needs thorough research. Sadly, as of now, on a global scale, honey-mediated eco-friendly nano-synthesis has a very small footprint since only a handful of laboratories are interested in exploring the immense valuable prospects of different varieties of honey available in nature, for use in green nanotechnology. For example, in India, in spite of the government initiative of the 'Sweet Revolution" scheme to boost apiculture and optimum usage of honey, every year only 30–40% of India's natural honey is harvested for sale and export. The rest simply dries up, leading to a hitherto unrecognised huge waste of a precious research resource that holds great promise in fighting AMR and in sustainable health management.

7.4.4 Algae-Mediated Eco-Friendly Synthesis of Antimicrobial Nanoparticles

Algae form a remarkable natural resource with multiple applications in various industries including the health sector and pharmaceutical industry. They are highly versatile and sustainable organisms. They are capable of photosynthesis, can grow quickly in different environments, and produce valuable bioactive compounds with diverse interesting properties. Algae are currently classified into ten major phyla (namely, Cyanobacteria, Bacillariophyta, Charophyta, Chlorophyta, Glaucophyta, Cryptophyta, Ochrophyta, Rhodophyta, Euglenozoa, Miozoa), which can be grouped into macroalgae and microalgae according to significant differences in their shape and size (Dai et al., 2022). Both of them have made great impact in the field of medicine and provide excellent platforms for the biosynthesis of high-value products, including various nanoparticles, thus giving rise to the cost-effective and environmentally friendly discipline of Phyco-nanotechnology, which holds great promise (El-Sheekh et al., 2022; Mandal et al., 2023). Their rich content of secondary metabolites is especially beneficial as these can act as capping and reducing agents during the synthesis process.

Antimicrobial metal nanoparticles have been synthesized using extracts of different marine algae species, including *Cystoseira myrica* (brown alga), *Ulva rigida* (green alga), and *Gracilaria foliifera* (red alga), freshwater microalgae *Scenedesmus* species, *Arthrospira platensis, Chlorella vulgaris*, cyanobacteria *Chroococcus minutus* and many more (El-Sheekh et al., 2022; Sidorowicz et al., 2023). They can protect against bacterial infections by *Staphylococcus aureus, Bacillus subtilis, Pseudomonas aeruginosa, Escherichia coli., Serratia marcescens, K. pneumoniae* and are also effective as anti-fungal agents against *Candida* species, *Aspergillus* species, *Fusarium* species etc (Sidorowicz et al., 2023). This is a new field and needs further research to elucidate the exact mechanism involving the action of algal metabolites in the green synthesis process. This in turn can help optimize growth parameters and production conditions to get maximum yield in minimum time.

7.5 Clinical Trial and Commercialization of Nanotechnology-Based Antimicrobial Products

The nanomedicine field is rapidly evolving due to the emergence of novel delivery systems, technologies, treatment methods, pharmacological authorizations, and even the clinical setbacks of current therapies. Table 7.1 displays a collection of the antibacterial applications of nanoparticles that have been recorded on the website-www.clinicaltrials.gov. Unique attributes of nanoparticles enhance the biological, physical, and chemical aspects under study for a variety of purposes. The FDA-approved materials are the optimal options for constructing these nanocarriers. Clinical settings have previously used FDA-approved antibiotic-loaded nanoparticles (Table 7.2). It is anticipated that further antimicrobial nanoparticles will become available for utilization in medical therapy. Due to recent advancements in industrial production and scaling up, it is anticipated that the availability of commercially viable antibacterial nanoparticles will increase shortly (Yeh et al., 2020).

7.6 Challenges

Despite the wide range of potential benefits and applications of nanoparticles, there are significant concerns about long-term risks to the environment and also about possible health hazards from widespread and continued exposure, especially during therapy. This has led to the emergence of new scientific disciplines like nanotoxicology.

Green nanotechnology has its additional challenges. These include selecting the correct raw material(s), setting up optimum reaction conditions, controlling product yield and quality, and applicability, which are all crucial deciding factors for production scale-up (Ying et al., 2022). For example, the shape and size of

Table 7.1 Study on the applicability of antibacterial nanoparticles through clinical trials

Nanoparticles	Strain/diseases	Invention	Status of trail	Phase of trail	Reference no.
Ag	*Streptococcus mutans*, Dental biofilms	Fluoride	Completed	1	NCT01950546
AgNP	*Serratia marcescens*	Alcohol-based hand gel	Unknown	3	NCT00659204
AgNP	The prevention of the proliferation of bacterial strains in the face or other regions of the body.	Topical approved anti-microbial gel	Unknown	1	NCT03752424
AgNP and Chitosan	*Enterococcus faecalis*	Bioceramic sealer	Completed	4	NCT04481945
Alkylated polyethyleneimine (PEI) nanoparticles	*S. mutans*	Flowable and hybrid composite resin disks	Completed	2	NCT00299598
Au-Ag-Cu_2O	Drug-resistant Microbial Keratitis	Hollow gold and silver alloy cuprous oxide shell nano-shell hydrogel, Voriconazole eye drops	Recruiting	1	NCT05268718
Chitosan	*E. faecalis*	Sodium Hypochlorite	Unknown	2	NCT03719261
Insoluble anti-bacterial nanoparticles (IABN)	–	Composite resin	Unknown	2	NCT00389714
Nanosilver fluoride	*S. mutans*	Chitosan and fluoride	Completed	3	NCT03186261
PLGA	*E. faecalis*, Endodontic infections	Ciprofloxacin	Completed	1	NCT05442736
Polyethyleneimine nanoparticles	*S. mutans* and *E. faecalis*	–	Unknown	2	NCT01167985
Quaternary Ammonium Polyethyleneimine nanoparticles	–	Silicon	Unknown	1	NCT01007240
TiO_2 and Chitosan	*S. mutans*	–	Completed	3	NCT04365270

Table 7.2 Some of the authorized nanoparticles and their antimicrobial properties

Nanoparticles	Application	Approved agency	Mode of action	References
ZnO	Antibacterial, Food additive	European Food Safety Authority, Food and Drug Administration	Zn^{2+} ion release, intracellular ROS generation, mechanical cell wall damage from bacterial surface contact	Vassallo et al. (2020)
AgNPs	Wound-dressing applications	Food and Drug Administration	Alteration of permeability, dissension in the proton gradient, ROS production, cytochromes and the electron transport chain's inactivation, interactions between sulfhydryl groups of biomolecules, denaturation of the thirty-second ribosomal subunit, Interaction with microbial DNA under a blocking of replication	Burduşel et al. (2018) and Vassallo et al. (2020)
AuNPs	Therapeutic drug delivery; contrast agent; photothermal therapy and radiosensitizer; gene transfection.	Food and Drug Administration	Inhibiting ATPase activity, collapse membrane potential, and tRNA interaction with a ribosomal subunit	Sibuyi et al. (2021)
Dendrimer based NPs (VivaGel® BV)	Anti-infective to stop recurring bacterial vaginosis.	Food and Drug Administration	Disturbance of membranes	Halwani (2022) and Pellett Madan et al. (2015)

nanoparticles produced by different extracts of the same biological species may vary widely depending on geographical distribution, and their efficacy may also fluctuate seasonally. Seasonal availability of some natural resources, time restrictions, additional upstream or downstream processing steps, and possible requirement of greater energy input are also factors that may hinder large-scale production. Sometimes it is also difficult to standardize purity and yield due to the heterogenous nature of the starting materials. In addition, the biocompatibility of metallic nanoparticles is also influenced by the method of synthesis and the conditions used (Chakroborty et al., 2022). Microbial pure cultures are difficult to maintain, and contamination risk is usually high. Vital active biomolecules are susceptible to degradation with changes in temperature or pH. Another worrisome aspect of repeated use of metallic nanoparticles is whether pathogens can become resistant to them, and how rapidly they might do so. Moreover, nanoparticles may potentially interfere with the host intestinal microbiome thus compromising gut health.

Gut dysbiosis can lead to a lot of health hazards (Skłodowski et al., 2023). More in-depth research is needed to address all these concerns. However, in an era of increasing antimicrobial resistance and climate crisis, eco-friendly nanoparticles appear to hold the most sustainable promise to improve current therapies, especially to combat AMR.

7.7 Conclusion

The society is living in an age threatened by viral pandemic, drug-resistant microbes, and the silent peril of biofilms. We are also facing climate crisis and energy crisis. Post-lockdown, many nations have suffered from economic deficits. As human civilization continues to confront new unprecedented challenges, the advent of new technologies is needed now more than ever to address such issues. Designed judiciously, technology can play a crucial role in decoupling development and climate degradation, making mankind's progress sustainable. Nature has its machinery of delivering vital services to sustain life on the planet and this expertise has been perfected over hundreds of millions of years. Unfortunately, most of the previous technological revolutions of man had not only been expensive and/or energy-intensive but had also largely taken for granted all healthy natural systems, consuming and depleting biological resources irresponsibly, without an appropriate appreciation of their immense power and value. Green nanotechnology, on the other hand, aims to harness the power of our natural resources sustainably. Eco-friendly production of antimicrobial nanoparticles is one such example in which renewable biological resources (including plants, algae, bacteria, fungi, actinomycetes, viruses, etc.) are used to synthesize valuable nanoparticles with a wide range of pharmaceutical and industrial applications via simple, cost-effective, safer, greener schemes with minimum energy input and almost zero hazardous waste (Borehalli et al., 2023; Yılmaz et al., 2023). They can help mankind to step up its battle against life-threatening drug-resistant infections. Hence, this is an area of active research. However, to date, most green nanotechnologies have been conducted successfully only on a laboratory scale. Initiatives are needed by policymakers to evaluate the feasibility of adapting these to commercial-scale synthesis as some might turn out to be expensive initially to move into production. Financial and regulatory barriers often become the greatest hurdles. Collaboration between academia and industry with relevant regulations from government, can help overcome such challenges. The immense potential of eco-friendly antimicrobial nanoparticles for expanded use in health sector and extended applicability in other areas (like the industrial sector) puts us on track for a sustainable, better and healthier future.

Acknowledgement The author would like to thank Dr. Sasabindu Jana, Principal, Raidighi College, West Bengal, India, and the Department of Microbiology, Raidighi College for their support. The author would also like to thank the editors of the book.

References

Abdel-Hadi, A., Iqbal, D., Alharbi, R., Jahan, S., Darwish, O., Alshehri, B., Banawas, S., Palanisamy, M., Ismail, A., Aldosari, S., Alsaweed, M., Madkhali, Y., Kamal, M., & Fatima, F. (2023). Myco-synthesis of silver nanoparticles and their bioactive role against pathogenic microbes. *Biology, 12*(5), 661. https://doi.org/10.3390/biology12050661

Aflakian, F., Mirzavi, F., Aiyelabegan, H. T., Soleimani, A., Gholizadeh Navashenaq, J., Karimi-Sani, I., Rafati Zomorodi, A., & Vakili-Ghartavol, R. (2023). Nanoparticles-based therapeutics for the management of bacterial infections: A special emphasis on FDA approved products and clinical trials. *European Journal of Pharmaceutical Sciences, 188*, 106515. https://doi.org/10.1016/j.ejps.2023.106515

Alghamdi, M. A., Fallica, A. N., Virzì, N., Kesharwani, P., Pittalà, V., & Greish, K. (2022). The promise of nanotechnology in personalized medicine. *Journal of Personalized Medicine, 12*(5), 673. https://doi.org/10.3390/jpm12050673

Ali, S. G., Jalal, M., Ahmad, H., Sharma, D., Ahmad, A., Umar, K., & Khan, H. M. (2022). Green synthesis of silver nanoparticles from Camellia sinensis and its antimicrobial and antibiofilm effect against clinical isolates. *Materials, 15*(19), 6978. https://doi.org/10.3390/ma15196978

Bagheri, A. R., Aramesh, N., Hasnain, M. S., Nayak, A. K., & Varma, R. S. (2023). Greener fabrication of metal nanoparticles using plant materials: A review. *Chemical Physics Impact, 7*, 100255. https://doi.org/10.1016/j.chphi.2023.100255

Bahari, N., Hashim, N., Md Akim, A., & Maringgal, B. (2022). Recent advances in honey-based nanoparticles for wound dressing: A review. *Nanomaterials, 12*(15), 2560. https://doi.org/10.3390/nano12152560

Balasooriya, E. R., Jayasinghe, C. D., Jayawardena, U. A., Ruwanthika, R. W. D., De Silva, R. M., & Udagama, P. V. (2017). Honey mediated green synthesis of nanoparticles: New era of safe nanotechnology. *Journal of Nanomaterials, 2017*, 5919836. https://doi.org/10.1155/2017/5919836

Bayda, S., Adeel, M., Tuccinardi, T., Cordani, M., & Rizzolio, F. (2020). The history of nanoscience and nanotechnology: From chemical-physical applications to nanomedicine. *Molecules, 25*(1), 112. https://doi.org/10.3390/molecules25010112

Bhardwaj, B., Singh, P., Kumar, A., Kumar, S., & Budhwar, V. (2020). Eco-friendly greener synthesis of nanoparticles. *Advanced Pharmaceutical Bulletin*, *10*(4), 566–576. https://doi.org/10.34172/apb.2020.067

Biswas, A. (2023). Potential of honey in sustainable health management. *Ecology, Environment and Conservation*, *29*, S81–S87. https://doi.org/10.53550/eec.2023.v29i01s.013

Borehalli Mayegowda, S., Roy, A., Manjula, N. G., Pandit, S., Alghamdi, S., Almehmadi, M., Allahyani, M., Awwad, N. S., & Sharma, R. (2023). Eco-friendly synthesized nanoparticles as antimicrobial agents: An updated review. *Frontiers in Cellular and Infection Microbiology, 13*, 1224778. https://doi.org/10.3389/fcimb.2023.1224778

Bruna, T., Maldonado-Bravo, F., Jara, P., & Caro, N. (2021). Silver nanoparticles and their antibacterial applications. *International Journal of Molecular Sciences, 22*(13), 7202. https://doi.org/10.3390/ijms22137202

Burdușel, A.-C., Gherasim, O., Grumezescu, A. M., Mogoantă, L., Ficai, A., & Andronescu, E. (2018). Biomedical applications of silver nanoparticles: An up-to-date overview. *Nanomaterials (Basel, Switzerland), 8*(9), 681. https://doi.org/10.3390/nano8090681

Chakroborty, S., Nath, N., & Pal, K. (2022). Limitations of nanomaterials insights in green chemistry sustainable route: Review on novel applications. *Green Processing and Synthesis, 11*(1), 951–964. https://doi.org/10.1515/gps-2022-0081

Dai, N., Wang, Q., Xu, B., & Chen, H. (2022). Remarkable natural biological resource of algae for medical applications. *Frontiers in Marine Science, 9*, 912924. https://doi.org/10.3389/fmars.2022.912924

Elshafei, A. M., Othman, A. M., Elsayed, M. A., Al-Balakocy, N. G., & Hassan, M. M. (2021). Green synthesis of silver nanoparticles using Aspergillus oryzae NRRL447 exogenous proteins Optimization via central composite design, characterization and biological

applications. *Environmental Nanotechnology, Monitoring & Management, 16*, 100553. https://doi.org/10.1016/j.enmm.2021.100553
El-Sheekh, M. M., Morsi, H. H., Hassan, L. H. S., & Ali, S. S. (2022). The efficient role of algae as green factories for nanotechnology and their vital applications. *Microbiological Research, 263*, 127111. https://doi.org/10.1016/j.micres.2022.127111
Fatima, F., Siddiqui, S., & Khan, W. A. (2021). Nanoparticles as novel emerging therapeutic antibacterial agents in the antibiotics resistant era. *Biological Trace Element Research, 199*(7), 2552–2564. https://doi.org/10.1007/s12011-020-02394-3
Fouda, A., Abdel-Maksoud, G., Abdel-Rahman, M. A., Salem, S. S., Hassan, S. E. D., & El-Sadany, M. A. H. (2019). Eco-friendly approach utilizing green synthesized nanoparticles for paper conservation against microbes involved in biodeterioration of archaeological manuscript. *International Biodeterioration and Biodegradation, 142*, 160–169. https://doi.org/10.1016/j.ibiod.2019.05.012
Ghosh, S., Ahmad, R., Zeyaullah, M., & Khare, S. K. (2021). Microbial nano-factories: Synthesis and biomedical applications. *Frontiers in Chemistry, 9*, 626834. https://doi.org/10.3389/fchem.2021.626834
Gujjari, L., Kalani, H., Pindiprolu, S. K., Arakareddy, B. P., & Yadagiri, G. (2022). Current challenges and nanotechnology-based pharmaceutical strategies for the treatment and control of malaria. *Parasite Epidemiology and Control, 17*, e00244. https://doi.org/10.1016/j.parepi.2022.e00244
Halwani, A. A. (2022). Development of pharmaceutical nanomedicines: From the bench to the market. *Pharmaceutics, 14*(1), 106. https://doi.org/10.3390/pharmaceutics14010106
Hosny, A. M. S., Kashef, M. T., Rasmy, S. A., Aboul-Magd, D. S., & El-Bazza, Z. E. (2017). Antimicrobial activity of silver nanoparticles synthesized using honey and gamma radiation against silver-resistant bacteria from wounds and burns. *Advances in Natural Sciences: Nanoscience and Nanotechnology, 8*(4), 045009. https://doi.org/10.1088/2043-6254/aa8b44
Huq, M. A., Rahman, M. M., Singh, P., & Akter, S. (2023). Editorial: Eco-friendly synthesis & antibacterial applications of metal nanoparticles. *Frontiers in Bioengineering and Biotechnology, 11*, 1157374. https://doi.org/10.3389/fbioe.2023.1157374
Ismail, N. A., Shameli, K., Wong, M. M. T., Teow, S. Y., Chew, J., & Sukri, S. N. A. M. (2019). Antibacterial and cytotoxic effect of honey mediated copper nanoparticles synthesized using ultrasonic assistance. *Materials Science and Engineering C, 104*, 109899. https://doi.org/10.1016/j.msec.2019.109899
Jadhav, R., Pawar, P., Choudhari, V., Topare, N., Raut-Jadhav, S., Bokil, S., & Khan, A. (2023). An overview of antimicrobial nanoparticles for food preservation. *Materials Today Proceedings, 72*, 204–216. https://doi.org/10.1016/j.matpr.2022.07.045
Jeyaraj, M., Gurunathan, S., Qasim, M., Kang, M. H., & Kim, J. H. (2019). A comprehensive review on the synthesis, characterization, and biomedical application of platinum nanoparticles. *Nanomaterials, 9*(12), 1719. https://doi.org/10.3390/nano9121719
Keskin, M., Kaya, G., & Keskin, S. (2022). Nanotechnology in honey. In *Nanotechnology in functional foods* (pp. 87–101). https://doi.org/10.1002/9781119905059.ch4
Khane, Y., Benouis, K., Albukhaty, S., Sulaiman, G. M., Abomughaid, M. M., Al Ali, A., Aouf, D., Fenniche, F., Khane, S., Chaibi, W., Henni, A., Bouras, H. D., & Dizge, N. (2022). Green synthesis of silver nanoparticles using aqueous Citrus limon zest extract: Characterization and evaluation of their antioxidant and antimicrobial properties. *Nanomaterials (Basel, Switzerland), 12*(12). https://doi.org/10.3390/nano12122013
Kowshik, M., Ashtaputre, S., Kharrazi, S., Vogel, W., Urban, J., Kulkarni, S. K., & Paknikar, K. M. (2003). Extracellular synthesis of silver nanoparticles by a silver-tolerant yeast strain MKY3. *Nanotechnology, 14*(1), 95–100. https://doi.org/10.1088/0957-4484/14/1/321
Kumar, A. S., Madhu, G., John, E., Kuttinarayanan, S. V., & Nair, S. K. (2020). Optical and antimicrobial properties of silver nanoparticles synthesized via green route using honey. *Green Processing and Synthesis, 9*(1), 268–274. https://doi.org/10.1515/gps-2020-0029

Kungwani, N. A., Panda, J., Mishra, A. K., Chavda, N., Shukla, S., Vikhe, K., Sharma, G., Mohanta, Y. K., & Sharifi-Rad, M. (2024). Combating bacterial biofilms and related drug resistance: Role of phyto-derived adjuvant and nanomaterials. *Microbial Pathogenesis*, 106874. https://doi.org/10.1016/j.micpath.2024.106874

Lakshmayya, N. S. V., Mishra, A. K., Mohanta, Y. K., Panda, J., Naik, B., Mishra, B., & Varma, R. S. (2023). Essential oils-based nano-emulsion system for food safety and preservation: Current status and future prospects. *Biocatalysis and Agricultural Biotechnology, 53*, 102897. https://doi.org/10.1016/j.bcab.2023.102897

Luceri, A., Francese, R., Lembo, D., Ferraris, M., & Balagna, C. (2023). Silver nanoparticles: Review of antiviral properties, mechanism of action and applications. *Microorganisms, 11*(3), 629. https://doi.org/10.3390/microorganisms11030629

Mandal, A. K., Nayak, R., Pradhan, B., Behera, C., Behera, A. K., Parida, S., Patra, S., Hembram, P., & Jena, M. (2023). Algal-derived nanoparticles and their antibacterial potential: Current evidence and future prospectives. *Journal of Microbiological Methods, 211*, 106790. https://doi.org/10.1016/j.mimet.2023.106790

Mohammed, Y. H. I., Alghamdi, S., Jabbar, B., Marghani, D., Beigh, S., Abouzied, A. S., Khalifa, N. E., Khojali, W. M. A., Huwaimel, B., Alkhalifah, D. H. M., & Hozzein, W. N. (2023). Green synthesis of zinc oxide nanoparticles using cymbopogon citratus extract and its antibacterial activity. *ACS Omega, 8*(35), 32027–32042. https://doi.org/10.1021/acsomega.3c03908

Mohanta, Y. K., Mishra, A. K., Panda, J., Chakrabartty, I., Sarma, B., Panda, S. K., Chopra, H., Zengin, G., Moloney, M. G., & Sharifi-Rad, M. (2023). Promising applications of phyto-fabricated silver nanoparticles: Recent trends in biomedicine. *Biochemical and Biophysical Research Communications, 688*, 149126. https://doi.org/10.1016/j.bbrc.2023.149126

Mohanta, Y. K., Nongbet, A., Panda, J., Chakrabartty, I., Mohanta, T. K., & Mishra, B. (2024). Chapter 10—Nanobiotechnology in genetic engineering for abiotic stress resistance in crops. In R. N. Pudake, R. M. Tripathi, & C. P. Gill (Eds.), *Nanotechnology for abiotic stress tolerance and management in crop plants* (pp. 153–164). Academic. https://doi.org/10.1016/B978-0-443-18500-7.00010-7

Mosselhy, D. A., Assad, M., Sironen, T., & Elbahri, M. (2021). Nanotheranostics: A possible solution for drug-resistant staphylococcus aureus and their biofilms? *Nanomaterials, 11*(1), 1–36. https://doi.org/10.3390/nano11010082

Muteeb, G. (2023). Nanotechnology—A light of hope for combating antibiotic resistance. *Microorganisms, 11*(6), 1489. https://doi.org/10.3390/microorganisms11061489

Nayak, D., Mishra, A. K., Malla, S., Swain, U., Panigrahi, K., Panda, J., Rustagi, S., Rauta, P. R., & Mohanta, Y. K. (2024). Eco-friendly mycogenic synthesis of iron oxide nanoparticles using Volvariella volvacea and unveiling its potential as antibiofilm and anticancer agents. *ECS Journal of Solid State Science and Technology, 13*(8), 087003. https://doi.org/10.1149/2162-8777/ad658b

Neupane, B. P., Chaudhary, D., Paudel, S., Timsina, S., Chapagain, B., Jamarkattel, N., & Tiwari, B. R. (2019). Himalayan honey loaded iron oxide nanoparticles: Synthesis, characterization and study of antioxidant and antimicrobial activities. *International Journal of Nanomedicine, 14*, 3533–3541. https://doi.org/10.2147/IJN.S196671

Nongbet, A., Panda, J., Mohanta, Y. K., Chakrabartty, I., Shamim, M. Z., & Mohanta, T. K. (2024). Chapter 6—Role of nanoparticles to protect plants from abiotic stress by scavenging reactive oxygen species. In R. N. Pudake, R. M. Tripathi, & C. P. Gill (Eds.), *Nanotechnology for abiotic stress tolerance and management in crop plants* (pp. 95–114). Academic. https://doi.org/10.1016/B978-0-443-18500-7.00006-5

Ortega-Nieto, C., Losada-Garcia, N., Prodan, D., Furtos, G., & Palomo, J. M. (2023). Recent advances on the design and applications of antimicrobial nanomaterials. *Nanomaterials, 13*(17), 2406. https://doi.org/10.3390/nano13172406

Palma-Morales, M., Huertas, J. R., & Rodríguez-Pérez, C. (2023). A comprehensive review of the effect of honey on human health. *Nutrients, 15*(13), 3056. https://doi.org/10.3390/nu15133056

Panigrahi, L. L., Sahoo, B., & Arakha, M. (2022). Nanotheranostics and its role in diagnosis, treatment and prevention of COVID-19. *Frontiers of Materials Science, 16*(2), 220611. https://doi.org/10.1007/s11706-022-0611-y

Pariente, N. (2022). The antimicrobial resistance crisis needs action now. *PLoS Biology, 20*(11), e3001918. https://doi.org/10.1371/journal.pbio.3001918

Parmar, S., Kaur, H., Singh, J., Matharu, A. S., Ramakrishna, S., & Bechelany, M. (2022). Recent advances in green synthesis of Ag NPs for extenuating antimicrobial resistance. *Nanomaterials, 12*(7), 1115. https://doi.org/10.3390/nano12071115

Pellett Madan, R., Dezzutti, C. S., Rabe, L., Hillier, S. L., Marrazzo, J., McGowan, I., Richardson, B. A., & Herold, B. C. (2015). Soluble immune mediators and vaginal bacteria impact innate genital mucosal antimicrobial activity in young women. *American Journal of Reproductive Immunology (New York, N.Y. : 1989), 74*(4), 323–332. https://doi.org/10.1111/aji.12412

Perveen, R., Shujaat, S., Naz, M., Qureshi, M. Z., Nawaz, S., Shahzad, K., & Ikram, M. (2021). Green synthesis of antimicrobial silver nanoparticles with Brassicaceae seeds. *Materials Research Express, 8*(5), 055007. https://doi.org/10.1088/2053-1591/ac006b

Pokrajac, L., Abbas, A., Chrzanowski, W., Dias, G. M., Eggleton, B. J., Maguire, S., Maine, E., Malloy, T., Nathwani, J., Nazar, L., Sips, A., Sone, J., van den Berg, A., Weiss, P. S., & Mitra, S. (2021). Nanotechnology for a sustainable future: addressing global challenges with the International Network Sustainable Nanotechnology. *ACS Nano, 15*(12), 18608–18623. https://doi.org/10.1021/acsnano.1c10919

Purvis, B., Mao, Y., & Robinson, D. (2019). Three pillars of sustainability: In search of conceptual origins. *Sustainability Science, 14*(3), 681–695. https://doi.org/10.1007/s11625-018-0627-5

Ray, M. K., Mishra, A. K., Mohanta, Y. K., Mahanta, S., Chakrabartty, I., Kungwani, N. A., Avula, S. K., Panda, J., & Pudake, R. N. (2023). Nanotechnology as a promising tool against phytopathogens: A futuristic approach to agriculture. *Agriculture, 13*(9), 1856. https://doi.org/10.3390/agriculture13091856

Rizvi, S. M. D., Lila, A. S. A., Moin, A., Hussain, T., Kamal, M. A., Sonbol, H., & Khafagy, E. S. (2023). Antibiotic-loaded gold nanoparticles: A nano-arsenal against ESBL producer-resistant pathogens. *Pharmaceutics, 15*(2), 430. https://doi.org/10.3390/pharmaceutics15020430

Salam, M. A., Al-Amin, M. Y., Salam, M. T., Pawar, J. S., Akhter, N., Rabaan, A. A., & Alqumber, M. A. A. (2023). Antimicrobial resistance: A growing serious threat for global public health. *Healthcare (Basel, Switzerland), 11*(13), 1946. https://doi.org/10.3390/healthcare11131946

Shahid, H., Shah, A. A., Shah Bukhari, S. N. U., Naqvi, A. Z., Arooj, I., Javeed, M., Aslam, M., Chandio, A. D., Farooq, M., Gilani, S. J., & Bin Jumah, M. N. (2023). Synthesis, characterization, and biological properties of iron oxide nanoparticles synthesized from Apis mellifera honey. *Molecules, 28*(18), 6504. https://doi.org/10.3390/molecules28186504

Sibuyi, N. R. S., Moabelo, K. L., Fadaka, A. O., Meyer, S., Onani, M. O., Madiehe, A. M., & Meyer, M. (2021). Multifunctional gold nanoparticles for improved diagnostic and therapeutic applications: A review. *Nanoscale Research Letters, 16*(1), 174. https://doi.org/10.1186/s11671-021-03632-w

Sidorowicz, A., Fais, G., Casula, M., Borselli, M., Giannaccare, G., Locci, A. M., Lai, N., Orrù, R., Cao, G., & Concas, A. (2023). Nanoparticles from microalgae and their biomedical applications. *Marine Drugs, 21*(6), 352. https://doi.org/10.3390/md21060352

Singh, J., Dutta, T., Kim, K. H., Rawat, M., Samddar, P., & Kumar, P. (2018). "Green" synthesis of metals and their oxide nanoparticles: Applications for environmental remediation. *Journal of Nanobiotechnology, 16*(1), 84. https://doi.org/10.1186/s12951-018-0408-4

Skłodowski, K., Chmielewska-Deptuła, S. J., Piktel, E., Wolak, P., Wollny, T., & Bucki, R. (2023). Metallic nanosystems in the development of antimicrobial strategies with high antimicrobial activity and high biocompatibility. *International Journal of Molecular Sciences, 24*(3), 2104. https://doi.org/10.3390/ijms24032104

Stanley, D., Batacan, R., & Bajagai, Y. S. (2022). Rapid growth of antimicrobial resistance: The role of agriculture in the problem and the solutions. *Applied Microbiology and Biotechnology, 106*(21), 6953–6962. https://doi.org/10.1007/s00253-022-12193-6

Tabasum, H., Bhat, B. A., Sheikh, B. A., Mehta, V. N., & Rohit, J. V. (2022). Emerging perspectives of plant-derived nanoparticles as effective antimicrobial agents. *Inorganic Chemistry Communications, 145*, 110015. https://doi.org/10.1016/j.inoche.2022.110015

Vassallo, A., Silletti, M. F., Faraone, I., & Milella, L. (2020). Nanoparticulate antibiotic systems as antibacterial agents and antibiotic delivery platforms to fight infections. *Journal of Nanomaterials, 2020*(1), 6905631. https://doi.org/10.1155/2020/6905631

Velusamy, P., Kumar, G. V., Jeyanthi, V., Das, J., & Pachaiappan, R. (2016). Bio-inspired green nanoparticles: Synthesis, mechanism, and antibacterial application. *Toxicological Research, 32*(2), 95–102. https://doi.org/10.5487/TR.2016.32.2.095

Wahab, S., Salman, A., Khan, Z., Khan, S., Krishnaraj, C., & Yun, S. I. (2023). Metallic nanoparticles: A promising arsenal against antimicrobial resistance—Unraveling mechanisms and enhancing medication efficacy. *International Journal of Molecular Sciences, 24*(19), 14897. https://doi.org/10.3390/ijms241914897

Xu, Y., Li, H., Li, X., & Liu, W. (2023). What happens when nanoparticles encounter bacterial antibiotic resistance? *Science of the Total Environment, 876*, 162856. https://doi.org/10.1016/j.scitotenv.2023.162856

Xulu, J. H., Ndongwe, T., Ezealisiji, K. M., Tembu, V. J., Mncwangi, N. P., Witika, B. A., & Siwe-Noundou, X. (2022). The use of medicinal plant-derived metallic nanoparticles in theranostics. *Pharmaceutics, 14*(11), 2437. https://doi.org/10.3390/pharmaceutics14112437

Yeh, Y. C., Huang, T. H., Yang, S. C., Chen, C. C., & Fang, J. Y. (2020). Nano-based drug delivery or targeting to eradicate bacteria for infection mitigation: A review of recent advances. *Frontiers in Chemistry, 8*, 00286. https://doi.org/10.3389/fchem.2020.00286

Yılmaz, G. E., Göktürk, I., Ovezova, M., Yılmaz, F., Kılıç, S., & Denizli, A. (2023). Antimicrobial nanomaterials: A review. *Hygiene, 3*(3), 269–290. https://doi.org/10.3390/hygiene3030020

Ying, S., Guan, Z., Ofoegbu, P. C., Clubb, P., Rico, C., He, F., & Hong, J. (2022). Green synthesis of nanoparticles: Current developments and limitations. *Environmental Technology & Innovation, 26*, 102336. https://doi.org/10.1016/j.eti.2022.102336

Chapter 8
Nano-biosensors for Detecting Microbial Pathogens in Agriculture

Sudip Mondal, Thi Thuy Truong, Vu Hoang Minh Doan, Thi Tuong Vy Phan, Jaeyeop Choi, Byeongil Lee, and Junghwan Oh

Abstract The growing world population requires a big increase in farming, but harmful germs remain a major problem. Traditional methods for detecting pathogens have limitations in sensitivity, specificity, and speed. Nano-biosensors are revolutionizing pathogen identification in agriculture with their precision and efficiency. This chapter offers a detailed overview of nano-biosensors, covering their principles, materials, and applications for detecting microbial pathogens. It emphasizes the crucial role of pathogen detection in safeguarding agricultural productivity and food security. The basics of nano-biosensors, including their characteristics and operational principles, are explored in depth. Nano-biosensors are categorized based on design, composition, and intended use, highlighting the role of nanomaterials in enhancing sensor performance. Components like nanoparticles, nanocomposites, nanowires, and nanotubes are examined for their ability to detect pathogens

S. Mondal
Digital Healthcare Research Center, Pukyong National University, Busan, Republic of Korea

Smart Gym-Based Translational Research Center for Active Senior's Healthcare, Pukyong National University, Busan, Republic of Korea

T. T. Truong
Industry 4.0 Convergence Bionics Engineering, Department of Biomedical Engineering, Pukyong National University, Busan, Republic of Korea

V. H. M. Doan · J. Choi
Smart Gym-Based Translational Research Center for Active Senior's Healthcare, Pukyong National University, Busan, Republic of Korea

T. T. V. Phan
Center for Advanced Chemistry, Institute of Research and Development, Duy Tan University, Danang, VietNam

Department of Environmental and Chemical Engineering, Duy Tan University, Danang, VietNam

Y. K. Mohanta et al. (eds.), *Nano-microbiology for Sustainable Development*,
https://doi.org/10.1007/978-3-031-78845-1_8

in complex agricultural environments. Biological recognition elements such as antibodies, antigens, enzymes, and nucleic acids are crucial for the accuracy of nano-biosensors in identifying different microbial species. The chapter discusses optical methods, electrochemical techniques, and Surface Plasmon Resonance (SPR) as effective approaches for rapid pathogen identification. Applications of Nano-biosensors in agriculture cover the detection of bacterial, fungal, and viral pathogens. Despite their promise, challenges in adoption, ethical considerations, and environmental impacts are addressed, offering a balanced view. Future trends and innovations in nano-biosensors are highlighted, aiming for more effective and sustainable biosensing solutions in agriculture.

Keywords Nano-biosensors · Pathogen detection · Agriculture · Nanomaterials · Biosensor technology

8.1 Introduction

The global population is projected to reach 8.5 billion by 2030 and is anticipated to grow exponentially to 9.8 billion by 2050, posing challenges in natural resource availability and food production (Khan et al., 2021; Oluwole et al., 2023). Issues such as diminishing land space, food scarcity, competition for resources, and the need to enhance crop yield in adverse environmental conditions exacerbate these challenges. Microbial pathogens like bacteria, fungi, and viruses can inflict severe diseases on plants, leading to significant crop losses if not detected early. Prompt detection allows for timely intervention, preventing widespread infections and economic setbacks for farmers and the agricultural sector. Hence, there is a pressing need for sustainable methods to enhance resource efficiency in agriculture and food industries, including the use of plant hormones, various pathogens, herbicides,

B. Lee
Digital Healthcare Research Center, Pukyong National University, Busan, Republic of Korea

Industry 4.0 Convergence Bionics Engineering, Department of Biomedical Engineering, Pukyong National University, Busan, Republic of Korea

Department of Smart Healthcare, Pukyong National University, Busan, Republic of Korea

J. Oh (✉)
Digital Healthcare Research Center, Pukyong National University, Busan, Republic of Korea

Industry 4.0 Convergence Bionics Engineering, Department of Biomedical Engineering, Pukyong National University, Busan, Republic of Korea

Smart Gym-Based Translational Research Center for Active Senior's Healthcare, Pukyong National University, Busan, Republic of Korea

Ohlabs Corp., Busan, Republic of Korea
e-mail: jungoh@pknu.ac.kr

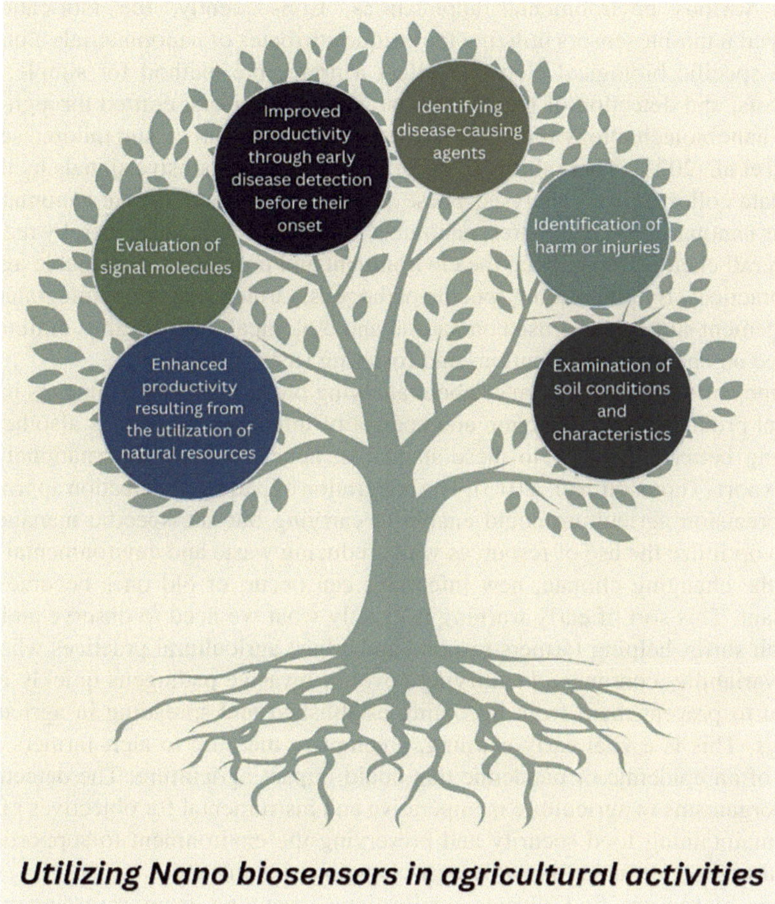

Fig. 8.1 Nano-biosensors used for agricultural activities

fertilizers, and metal ions (Maitra et al., 2021). Early detection and mitigation of these pathogens before harvest are crucial for maintaining food safety and public health. Some pathogens can persist in soil, affecting future crops, underscoring the importance of targeted treatment strategies to minimize long-term impacts on soil quality and ecosystem health. Nanotechnology has emerged as a promising solution, facilitates functional materials to augment conventional practices in the food sector (Fig. 8.1). Various synthesis methods, including physical, chemical, and biological routes, are employed to produce nanomaterials, leveraging their unique properties to enhance existing technologies.

Early identification facilitates prompt and cost-effective interventions, mitigating financial losses. The swift and precise detection of pathogens enables more targeted utilization of pesticides and fungicides. Nano-biosensors, which integrate nanomaterials with biological components, have significantly enhanced sensing capabilities across various environmental applications. Consequently, the fabrication of advanced nano-biosensors utilizing the unique attributes of nanomaterials alongside highly specific biological elements offers a preferable method for simple, early diagnosis, and detection of plant diseases. Nano-biosensors, crafted through innovative nanobiotechnology methods, are non-invasive, sensitive, and tailored sensors (Singh et al., 2023). Their ability to generate real-time responsive signals facilitates easy data collection and analysis. These sensors incorporate diverse nanomaterials such as nanoparticles, nanowires, nanotubes, and nanocomposites, thereby reducing the overall chemical impact on the environment and promoting sustainable agricultural practices. By pinpointing specific pathogens, farmers can adopt integrated pest management strategies focused on natural and biological controls, thus diminishing reliance on chemical interventions and fostering sustainability.

Some nations have strict rules about allowing presence of the pathogens in agricultural products. This detection and control of infectious pathogens also helps in allowing better adherence to these standards, hence enabling international trade with export (Thomson et al., 2013). The integration of pathogen detection approaches with precision agriculture could enable to carrying out site-specific management and so optimize the use of resources while reducing waste and environmental risks. With the changing climate, new infections can occur, or old ones become more abundant. This sort of early warning is exactly what we need to observe and react on such shifts, helping farmers respond and adjust agricultural practices when climate variability continues. Identifying novel or invasive pathogens quickly is also critical to prevent them from becoming established and spreading in agricultural settings. This is a vital early warning, preemptive measure to alert farmers in the event of an epidemic or pandemic that could impact agriculture. The detection of microorganisms in agriculture is imperative and instrumental for objectives ranging from maintaining food security and preserving the environment to supporting an agricultural sector under threat by ever-changing circumstances.

Nano-biosensors find diverse applications, spanning from assessing natural resource abundance in ecosystems like soil quality and groundwater availability. In agriculture, these compact, portable devices enable farmers to monitor and manage soil conditions directly on-site. They have proven effective in analyzing soil fertility, pH levels, moisture content, and mineral concentrations, as well as detecting pests and mineral deficiencies before disease onset (Ghorbani et al., 2008). The biosensor market reached a value of US $15.5 billion in 2015 and was projected to expand to approximately US$ 26.9 billion by 2020 (Thakur et al., 2022). Networks of nano-biosensors can forecast crop yields and detect various diseases in field crops. Recent initiatives aim to integrate nano-biosensors with information technology (IT) to disseminate results to remote areas, benefiting farmers in isolated regions. Nano-biosensors are gaining traction in the food sector, particularly in ensuring food safety during production and packaging by detecting allergens and pathogens. The food industry faces numerous challenges, including food security, safety, availability, and utilization. Nano-biosensors offer distinct advantages such as ease of operation, high sensitivity, and fast detection of pesticide remains compared to conventional methods like high-performance liquid chromatography (HPLC) and gas chromatography-mass spectrometry (GC–MS) (Christopher et al., 2020; Pudake et al., 2021; Zia et al., 2022).

8.2 Types of Nano-biosensors

Different types of nanosensors are employed for detecting agricultural pathogens, each utilizing distinct mechanisms for precise and sensitive identification. These diverse nano-biosensor platforms offer high sensitivity, selectivity, and rapid detection capabilities, making them valuable tools for agricultural pathogen detection and crop disease management.

8.2.1 *Multiple Channel Nanobiosensor*

This type of sensor integrates multiple channels or sensing elements to detect different analytes simultaneously. It offers high throughput and enables the detection of various contaminants, including pathogens, toxins, and chemical residues, in food samples (Srivastava et al., 2018).

8.2.2 *Array-Based Sensor*

Array-based sensors utilize arrays of receptors or probes immobilized on a solid substrate. Each receptor specifically binds to a target analyte, allowing for the simultaneous detection of multiple contaminants in a single sample. These sensors provide rapid and multiplexed detection capabilities (Ranbir et al., 2023).

8.2.3 *Aptameric Sensor*

Aptameric sensors utilize aptamers, which are short, single-stranded DNA or RNA molecules, as recognition elements. Aptamers exhibit high affinity and specificity toward target molecules, enabling sensitive detection of contaminants, such as pathogens, allergens, and toxins, in food samples. Their appeal has grown significantly due to their advantageous properties, such as high stability across a wide range of pH and temperatures, along with their cost-effectiveness. Typically consisting of 15 to 40 bases, aptamers adopt three-dimensional structures that enable them to specifically bind to their complementary regions through mechanisms like van der Waals forces, hydrogen bonding, and electrostatic interactions (Mondal et al., 2022).

8.2.4 Chemosynthetic Mimotope Peptide Sensor

Chemosynthetic mimotope peptide sensors utilize synthetic peptides that mimic the structure of natural antigens or toxins. They offer high specificity and sensitivity for the detection of specific contaminants, pathogens, or allergens in food samples (You et al., 2020).

8.2.5 Biofilm and Toxin Nanobiosensor

These sensors target biofilms formed by bacteria on food surfaces, as well as toxins produced by microbial contamination. They employ specialized detection strategies to monitor and control biofilm formation and toxin production, ensuring food safety and quality (Wang et al., 2020).

8.2.6 Nanoparticle-Based Biosensors

These biosensors utilize nanoparticles functionalized with specific ligands or antibodies that bind to target pathogens, resulting in measurable signals. Quantum dots, gold nanoparticles, and magnetic nanoparticles are commonly employed due to their unique optical and magnetic properties (Koedrith et al., 2015).

8.2.6.1 Nanowire Biosensors

Nanowires offer high surface-to-volume ratios, enhancing sensitivity. Functionalized nanowires can detect pathogens by measuring changes in electrical conductivity or impedance upon pathogen binding. Silicon nanowires and carbon nanotubes are frequently used in this approach (Song et al., 2017).

8.2.6.2 Nanoporous Biosensors

These sensors feature nanoporous materials with tailored pore sizes to capture pathogens selectively. Binding events alter the electrical, optical, or mass transport properties of the nanopores, enabling detection. Materials like porous silicon and aluminum oxide are commonly utilized (Wu et al., 2022).

8.2.6.3 Graphene-Based Biosensors

Graphene's exceptional electrical conductivity and large surface area make it suitable for biosensing applications. Functionalized graphene or graphene oxide can bind to target pathogens, leading to changes in electrical conductivity or fluorescence, which are quantified for detection (Bobrinetskiy & Knezevic, 2018; Mathur et al., 2018).

8.2.6.4 Plasmonic Biosensors

Plasmonic nanoparticles exhibit unique optical properties due to surface plasmon resonance (SPR), enabling label-free detection of pathogens. Binding events cause shifts in SPR spectra, providing real-time detection. Gold and silver nanoparticles are commonly used in plasmonic biosensors (Balbinot et al., 2021).

8.2.6.5 Carbon-Based Biosensors

Carbon-based nanomaterials, such as carbon nanotubes and graphene, are utilized in biosensors for their excellent electrical conductivity and biocompatibility. Functionalized carbon nanomaterials selectively bind to target pathogens, resulting in measurable changes in electrical properties for detection (Smart et al., 2020).

8.2.6.6 Quantum Dot Biosensors

Quantum dots are semiconductor nanoparticles with tunable optical properties. Functionalized quantum dots can selectively bind to target pathogens, emitting fluorescent signals upon binding, which are quantified for detection. Their narrow emission spectra enable multiplexed detection of multiple pathogens simultaneously (Hong & Lee, 2018).

8.3 Materials Used for Nano-biosensor Fabrication

A nano-biosensor consists of three essential components: a biological probe, a transducer, and a data recording unit. The biological probe typically employs affinity-based materials such as antibody-antigen interactions, enzyme-substrate interactions, nucleic acid interactions, and cell-based interactions (Verma & Rani, 2021). These probes can be composed of DNA, enzymes, antibodies, biomimetic materials, or aptamers. Synthetic receptors have replaced natural biological components to mimic functions with greater speed and specificity. Analytes are detected by these biological probes in combination with various nanoparticles like metallic,

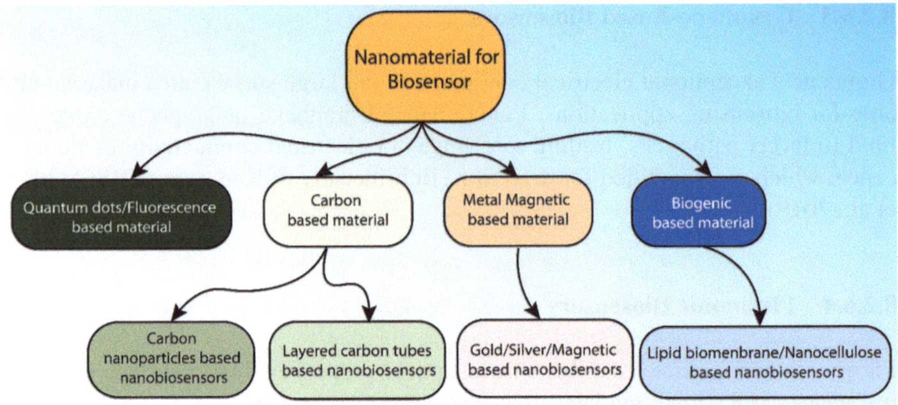

Fig. 8.2 Different types of nanomaterials used for agricultural pathogenic biosensor development

magnetic, quantum dots, graphene oxide, and carbon nanotubes (Fig. 8.2) (Neethirajan et al., 2018b). Transducers convert biological signals into digital ones. They can utilize electrochemical signals such as amperometric, voltammetric, and potentiometric signals, as well as optical signals including colorimetric, surface plasmon resonance, metallic fluorescence, and optical fibers (Shawon et al., 2020). Advanced materials are employed for nanosensing applications in the development of nano-biosensors. Carbon nanotube-based nanomaterials, including single-walled and multiwalled carbon nanotubes (SWCNTs, MWCNTs), are commonly used due to their superior surface properties for immobilizing biological components (Verma et al., 2020). However, a significant limitation is their low solubility in aqueous environments. Additional materials employed comprise fullerenes, titanium, and silicon oxides, while polymers such as zeolites, chitosan, and polyacrylic acid are employed for encapsulation. Metallic nanoparticles, notably gold, and silver, are favored for nano-biosensors due to their enhanced safety characteristics, particularly in food applications for detecting contaminants and pathogens. Detection of pathogenic contaminants can be achieved with metallic components-based nano-biosensors, providing safer alternatives for food items. Another common material for nano-biosensor fabrication relies on fluorescence measurement. Fluorescence-based nano-biosensors utilize two methodologies for sensing: quantum dots sensors based on semiconducting properties and fluorescence resonance energy transfer (FRET) intermolecular charge transfer (ICT) (Girigoswami & Akhtar, 2019). Although fluorescence-based detection offers easier alternatives, it is associated with high costs that can be mitigated using metal-enhanced fluorescence (MEF) strategies. Zinc oxide (ZnO) is another crucial metallic counterpart used in nanosensor fabrication. Nano-biosensors often utilize light-emitting diodes (LEDs) for the detection of free radicals in food packaging materials. ZnO possesses salient properties such as high catalytic power, high isoelectric point, and strong adsorption capability (Malhotra & Mandal, 2019). Table 8.1 represents the different types of nanomaterials used for biosensor fabrication with their senility level and target molecules.

Table 8.1 Different categories of nano-biosensors utilized in agriculture

Nano-biosensor	Detection mechanism	Applications	Sensitivity	References
Quantum dots	Fluorescence	Detection of *F. oxysporum*	25 mg/mL	Khaledian et al. (2017)
Multiwalled carbon nanotubes	Surface Plasmon Resonance	Detection of Cymbidium Mosaic virus	–	Antonacci et al., 2018)
Carbon nanomaterials	Volatile organic compound profile	Detection of pathogens	80–90% accuracy	Cui et al. (2019)
DNA-based biosensor	Electrochemical	Detection of *Phytophthorapulmivora* causing black pod rot in cacao pod	–	Franco et al. (2019)
Si NP-Rubpy-IgG	Fluorescence	Detection of *X. campestris*	3×10^2 CFU	Yüksel et al. (2015)
Calcium phosphate	Electrochemical	Pathogenic detection	0.30 ng/mL	Manjunatha et al. (2016)
Graphene	Electrochemical	Pesticide detection of chlorothlonil and chlorpyrifos methyl	0.13 mg/L	Xu et al. (2020)
Carbon dots CDs@SBA15@ MIP	Electrochemical	Detection of kaempferol polyphenol in vegetables	14 μg/L	He et al. (2020)
Green fluorescent protein	Fluorescent based	Detection of strigolactones for plant growth and parasitism	–	Chesterfield et al. (2020)
Zinc oxide and Copper	Electrochemical	Enhance the germination of tomato chili and cucurbits in Mexico	50 ppm	Negrete (2020)
TiO_2 and SnO_2 NP	Electrochemical	Detection of p-Ethylguaiacol	35–62 nM	Li et al. (2019)
Cholinergic enzyme	Square wave voltammetry	Detection of malathion	0.31 ng/L	Bao et al. (2019)
Cholinergic enzyme	Differential pulse voltammetry	Detection of dichlorvos	6.4 μg/mL	Mishra et al. (2021)
Cholinergic enzyme	Amperometric	Detection of chlorpyrifos	0.01 μg/L	Hou et al. (2019)
Cu(I) tris(pyrazolyl) borate-SWCNT	Chemiresistor	Detection of Ethylene	0.5 ppm	Cellini et al. (2017)
Enzyme	Amperometric	Detection of Malathion	0.001 μg/L	Zhang et al. (2019)
CdTe QD-Rd	FRET	Detection of *Citrus tristeza*	220 ng/mL	Chen et al. (2018)

(continued)

Table 8.1 (continued)

Nano-biosensor	Detection mechanism	Applications	Sensitivity	References
Enzyme	Differential pulse voltammetry	Detection of paraoxon	1.1 ng/mL	Mahmoudi et al. (2019)
Enzyme	Coloured reaction	Detection of chlorpyrifos	3.3 μg/L	Fu et al. (2019)
Au Nanoparticles	Electrochemistry	*P. syringae*	214 pM	Darr et al. (2017)
Au Nanoparticles	PCR mediated AuNPs based sensing assay	*Phytophthora infestans* (causing potato late blight)	0.1 pg/μL	Zhan et al. (2018)
Silver Nanoparticles	SERS	Aflatoxins	5×10^{-5} M	Lau et al. (2017)

8.4 Applications of Nano-biosensors in Agriculture

8.4.1 Detection of Bacterial Pathogens

Nano-biosensors employed for bacterial pathogen detection often utilize nanomaterials such as nanoparticles, nanowires, or nanotubes functionalized with specific receptors, antibodies, or aptamers that can selectively bind to target bacterial species. These sensors can detect the presence of bacterial pathogens in soil, water, or plant tissues with high sensitivity and specificity. In the identification process, nano-biosensors interact with bacterial pathogens, triggering a measurable signal that indicates their presence (Fig. 8.3). This signal can be optical, electrical, or chemical in nature, depending on the sensing mechanism of the nano-biosensor. For example, in optical nano-biosensors, changes in light absorption or emission properties upon bacterial binding are detected and quantified. Several aptamer-based sensors have been developed for detecting pathogen biomarkers such as *S. aureus*, *E. coli*, and *C. jejuni* (Li et al., 2021). For instance, researchers designed an ultra-sensitive magnetic fluorescence aptasensor for detecting *S. aureus* by utilizing fluorescence-resonance energy transfer. In this approach, aptamers were immobilized on the surface of Fe_3O_4 and modified carbon dots (CDs), with CDs serving as the fluorescence donor and Fe_3O_4 as the receptor for the "off-on" sensor. Upon binding with the target bacteria, the Fe_3O_4 /CDs aptasensor disassembled, leading to the recovery of CDs fluorescence. The detection range achieved was between 50×10^7 CFU mL^{-1} and a detection limit of 8 CFU mL^{-1} (Cui et al., 2019).

Another instance entails the identification of *E. coli* utilizing graphene oxide (GO)-modified gold nanoparticles (Au-NPs) augmented with aptamers tailored for *E. coli*. This nanosensor, exhibiting a detection threshold of 10 cells/mL in water and coconut water-enriched samples, showcased exceptional specificity towards its

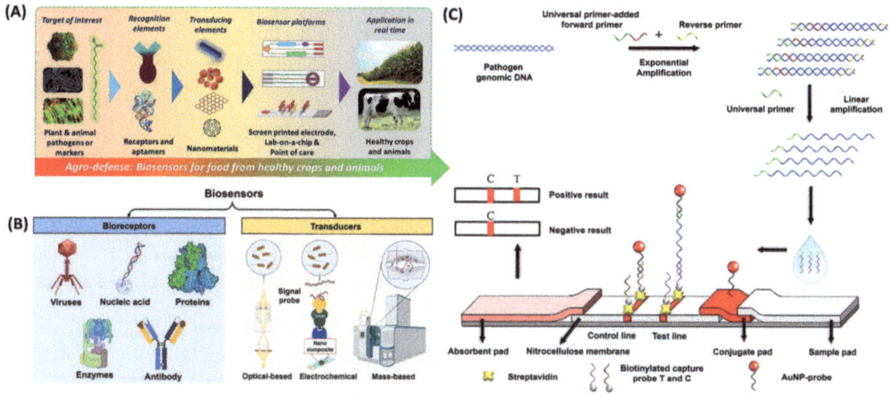

Fig. 8.3 (**a, b**) Progress in different biosensing technologies for the detection of phytopathogens. (Represented with permission from Neethirajan et al. 2018a, Elsevier (Neethirajan et al., 2018a), and Ali et al. 2022, Elsevier (Ali et al., 2022)) (**c**) Gold nanoparticle-enabled lateral flow biosensor for precise visual identification of *Phytophthora infestans*, the causative agent of potato late blight. (Represented with permission from Zhan et al. 2018, Elsevier, (Zhan et al., 2018))

target, exhibiting no cross-reactivity with other bacterial species. The aggregation-induced color change from red to blue, observable by the naked eye, further facilitates detection (Gupta et al., 2021). Similarly, a nano-biosensor utilizing quantum dots (QDs) functionalized with aptamer II and coated with magnetic nanoparticles (NPs) showed promise for *E. coli* detection in water samples. The fluorescence values recorded for various bacterial concentrations demonstrated the capability of the CdTe-MPA QDs conjugated with aptamer II to selectively capture and detect *E. coli* (Pandit et al., 2022).

8.4.2 Identification of Fungal and Viral Pathogens

Fungal infections in economically valuable plants cause substantial declines in both yield and quality, frequently resulting in the complete loss of affected plants. To mitigate these losses, it is crucial to promptly detect and identify the pathogens. Early detection and precise identification of fungi can effectively curb the spread of infection (Fig. 8.4). Traditional methods, considered the benchmark in fungal detection, rely on symptom assessment, culture-based techniques, morphological observation, and biochemical identification (Ray et al., 2017). The Food and Agriculture Organization (FAO) reports that globally, the average postharvest losses of food amount to approximately 29% in highly developed nations and around 38% in developing countries across Asia, Africa, Latin America, and Southeast Asia (Gustavsson et al., 2011). These losses are notably more pronounced for horticultural crops and exotic fruits (Alkan & Fortes, 2015). Microbial infections stand out as the primary cause of postharvest losses on a worldwide scale. Among the most

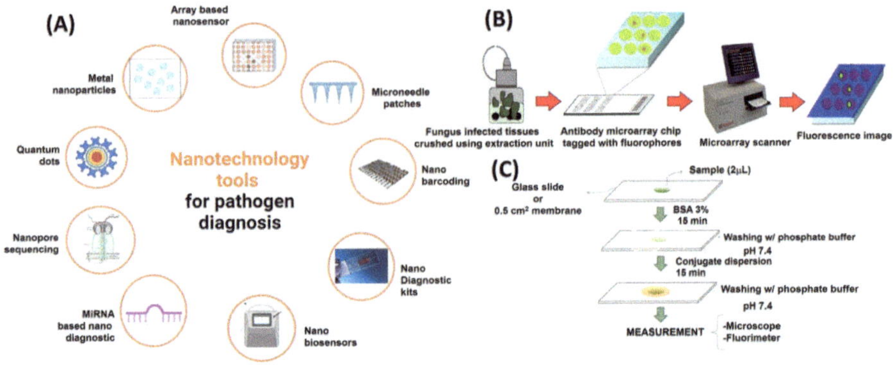

Fig. 8.4 (**a**) Diagrammatic representation of different nanotechnology tools for phytopathogen detection. (Represented from Ray et al. 2023, MDPI (Ray et al., 2023)), (**b**) The fluorescence detection process involves tagging antibodies with fluorophores and fixing them onto a microarray chip. Crushed fungus-infected tissues are stained with a fluorescent dye and dispersed over the antibody microarray chip, forming a monolayer of fungal antigens. Detection occurs using a microarray scanner equipped with a fiber optic diode laser system, quantitatively analyzing the target infection based on the fluorescence intensity, reflecting the level of plant infection. (Represented with permission from Ray et al. 2017, Elsevier (Ray et al., 2017)). (**c**) The paper-based immunoassay for detecting soybean rust spores begins with applying vegetal extract to a nitrocellulose membrane, which is then blocked with a protein solution. Detection follows using fluorescent nanoparticles coated with polyclonal anti-spore antibodies targeting Asian soybean rust. (Represented with permission from Miranda et al. 2013, Elsevier (Miranda et al., 2013)).

prevalent post-harvest diseases are green and blue mold, attributed to Penicillium digitatum and *P. italicum*, respectively (Perez et al., 2016), brown rot, caused by *Monilinia fructicola*, Alternaria black spot, typically induced by *Alternaria alternata, Rhizopus rot*, associated with *Rhizopus stolonifer* (Perez et al., 2016), and anthracnose, brought about by *Colletotrichum brevisporum, C. boninense*, and *C. brevisporum* (Palou et al., 2015). A paper-based immunosensor with high sensitivity was utilized to detect the fungus *Phakopsora pachyrhizi*, the causative agent of Asian soybean rust, enabling early disease diagnosis. Polyclonal antibodies, tagged with fluorescent nanoparticles, were used as the recognition elements. The fluorescence signal exhibited a linear response range spanning from 0.0032 to 28 μg/mL. The immunosensor achieved a detection limit of 2.2 ng/mL, roughly equivalent to 8–12 spores/mL (Miranda et al., 2013). Malecka et al. (2014) recently demonstrated a voltammetric method for detecting nucleic acid from Plum pox virus (PPV) at picomolar concentrations using glassy carbon electrodes (Malecka et al., 2014). This label-free approach involved monitoring the detection of pathogen-related DNA immobilized on the electrode using Osteryoung square wave voltammetry. By measuring changes in electron transfer resulting from the hybridization reaction, the voltammetric technique enabled the detection of 22-mer and 42-mer complementary target DNA sequences of PPV within a concentration range of 10–50 pg/ml. The research demonstrated successful differentiation between diseased and unaffected leaf samples, reaching a detection threshold of 12.8 pg/ml. Yet, the method's specificity towards other plant pathogens was not assessed. Lin et al. (2014) pioneered a label-free surface plasmon resonance (SPR) immunosensor employing gold nanorods for detecting two orchid viruses: Cymbidium mosaic virus and Odontoglossum ringspot virus (Lin et al., 2014). In this approach, antibodies specific to these orchid viruses were modified with AuNRs to serve as sensing layers. This modification helped to broaden the spectral region, mitigating the color interference issue often encountered in sample matrices. The developed technology achieved detection limits of 48 pg/ml for CymMV and 42 pg/ml for ORSV. While the study did not report the stability of the established SPR biosensing system, it assessed specificity by using a mixture of target and non-target viral antigens. The performance was evaluated by observing signal changes resulting from the binding of viral antigens to antibodies on the surface of AuNRs.

8.4.3 Impacts on Crop Yield and Quality

Nano-biosensors have the potential to significantly impact crop yield and quality in agriculture. By enabling precise and rapid detection of various factors affecting plant health, such as pathogens, pests, nutrient levels, and environmental conditions, nano-biosensors contribute to early identification and mitigation of issues that can hamper crop growth and development (Fig. 8.5). This early detection allows for timely interventions, including targeted application of fertilizers, pesticides, and irrigation, thereby optimizing crop yield and quality. Additionally, nano-biosensors facilitate real-time monitoring of crop health parameters, enabling farmers to make data-driven decisions to enhance productivity and minimize losses.

Disease Detection Nano-biosensors are employed for the detection of pathogens, including bacteria, viruses, and fungi, responsible for various crop diseases. By detecting these pathogens at early stages, nano-biosensors help prevent disease outbreaks and minimize crop losses.

Nutrient Monitoring Nano-biosensors enable the precise monitoring of nutrient levels in soil and plants. By providing real-time data on nutrient availability, these sensors help optimize fertilization practices, ensuring that crops receive the necessary nutrients for healthy growth and development.

Environmental Monitoring Nano-biosensors can monitor environmental parameters such as temperature, humidity, and pH levels in soil and water. This information allows farmers to assess environmental conditions and make informed decisions regarding irrigation, drainage, and soil management.

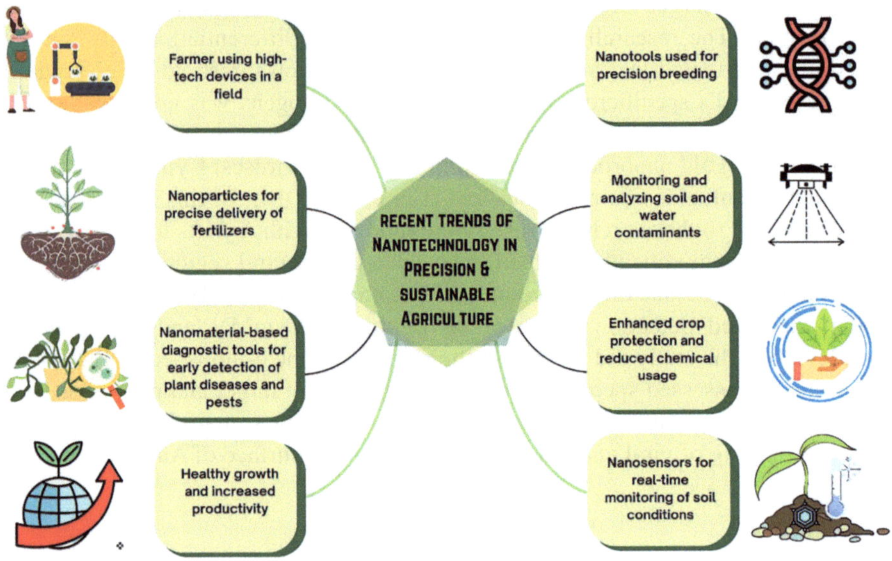

Fig. 8.5 Impacts of nanotechnology on crop yield and quality. (Represented from Tyagi et al., 2023, Frontiers in plant science (Tyagi et al., 2023))

Pesticide Residue Detection Nano-biosensors are utilized to detect pesticide residues in crops and soil. By accurately measuring pesticide levels, these sensors help ensure compliance with regulatory standards and minimize the risk of pesticide contamination in food products.

Quality Control Nano-biosensors contribute to quality control in agriculture by detecting contaminants, allergens, and pathogens in food products. By ensuring the safety and quality of agricultural products, these sensors enhance consumer confidence and facilitate trade in global markets.

8.4.4 Challenges and Future Directions

(A) ***Sensitivity and Specificity***: One of the primary challenges in nano-biosensors for detecting microbial pathogens in agriculture is achieving high sensitivity and specificity. While nanomaterials offer unique properties for enhancing sensor performance, ensuring selective detection of target pathogens amidst complex environmental matrices remains a challenge.

(B) ***Stability and Reliability***: Nano-biosensors must maintain stability and reliability under varying environmental conditions encountered in agricultural settings, including temperature fluctuations, humidity, and soil composition. Ensuring sensor robustness and longevity is essential for practical field deployment.

(C) ***Cost and Accessibility***: The development and fabrication of nano-biosensors can be costly, limiting their widespread adoption, especially in resource-constrained agricultural regions. Addressing cost-effectiveness and scalability challenges is essential to democratize access to these technologies for farmers.

(D) ***Standardization and Regulation***: Lack of standardized protocols and regulations for nano-biosensor development and deployment pose challenges in ensuring consistency and reliability across different sensor platforms. Establishing guidelines and regulatory frameworks is necessary to validate sensor performance and ensure data integrity.

8.4.5 Environmental and Ethical Considerations

The utilization of nanomaterials in biosensors raises concerns about their potential environmental impact when introduced into agricultural ecosystems (Table 8.2). It is crucial to understand how nanomaterials behave in soil and water systems to prevent negative effects on environmental health. The introduction of nano-biosensors in agricultural settings could potentially disturb ecological balances by altering microbial communities or inadvertently affecting non-pathogenic organisms. Therefore, it is essential to assess and minimize any unintended ecological consequences to preserve ecosystem stability. The use of nano-biosensors for data

Table 8.2 Different commercially available nano-biosensors

Analytes	Product invention Country	Company
Glucose	Japan	Fuji Electric Co.
Glucose, saccharose, ethanol, methanol, lactose	Sweden	Chemel AB
Glucose, lactate, malate	Italy	Tectronik
Sucrose, L-glutamate, alcohol	UK	Gwent Sensor Ltd.
Lactate	France	Biosentec
Ethanol, methanol, glucose, lactate, glycerol	UK, USA	Analox Instruments
Glucose, L-lactate, L-glutamate, L-glutamine, alchol, sucrose, methanol, ammonia	USA	Nova Biomedical
Folic acid, biotin, antibiotics in cereals, meat, milk, infant food, honey	Worldwide	Biacore GE
Salmonella and Enterococcus	USA	Ambri Ltd
NutriChip, DNA detection with array technology	Germany	GeneScan Europe AG/Scil Diagnostic GmbH
Sensors and components for microorganism detection	UK	Don Whitley Scientific Ltd.
Degradation products of ATP	Canada	Pegasus Biotechnology
Online monitoring of the concentration of total microorganisms in water samples in 15 min	Spain	Biosensores S.L

collection and analysis raises ethical issues related to data privacy, ownership, and consent. Protecting the privacy of farmers and consumers and ensuring transparent data governance practices are essential for maintaining trust and upholding ethical standards.

8.4.6 *Innovations in Nano-Biosensor Technology*

Future developments in nano-biosensors aim to incorporate multiple functions like pathogen detection, environmental monitoring, and assessing plant health within a unified platform. These multifunctional sensors provide comprehensive insights into agricultural systems, enabling thorough monitoring and effective management. Integration of nano-biosensors with IoT platforms allows for real-time data collection, analysis, and decision-making in precision agriculture. IoT-enabled sensor networks facilitate remote monitoring and control of agricultural processes, optimizing resource utilization and boosting productivity. Current research focuses on creating sustainable nanomaterials that minimize environmental impact and enhance biocompatibility. Methods such as green synthesis and biodegradable materials offer eco-friendly options for constructing nano-biosensors. Advances in miniaturization technologies enable the development of compact, portable nano-biosensors suitable for on-site detection in agricultural fields. Handheld or smartphone-based sensor devices empower farmers with quick and accessible pathogen monitoring capabilities. Additionally, integrating machine learning and artificial intelligence algorithms improves the analysis and interpretation of nano-biosensor data. This enables predictive modeling of disease outbreaks and provides optimal decision support for agricultural management practices. AI-driven sensor systems offer intelligent insights for proactive disease management and strategies to mitigate risks effectively.

8.5 Conclusion

The transformative potential of nano-biosensors in agriculture addresses the critical need for enhanced pathogen detection to support global food security. These advanced sensors offer unique precision and efficiency, surpassing the limitations of traditional methods in sensitivity, specificity, and speed. Applying diverse nanomaterials and innovative biological recognition elements, Nano-biosensors excel in detecting minute pathogen concentrations within complex agricultural environments. Despite current challenges and considerations, their ongoing development promises increasingly effective and sustainable solutions for protecting agricultural yields worldwide. Global collaboration is needed to speed up the use and development of Nano-Biosensor technologies, strengthening agriculture against future microbial threats.

References

Ali, Q., Zheng, H., Rao, M. J., Ali, M., Hussain, A., Saleem, M. H., Nehela, Y., Sohail, M. A., Ahmed, A. M., Kubar, K. A., Ali, S., Usman, K., Manghwar, H., & Zhou, L. (2022). Advances, limitations, and prospects of biosensing technology for detecting phytopathogenic bacteria. *Chemosphere, 296*, 133773.

Alkan, N., & Fortes, A. M. (2015). Insights into molecular and metabolic events associated with fruit response to post-harvest fungal pathogens. *Frontiers in Plant Science, 6*, 889.

Antonacci, A., Arduini, F., Moscone, D., Palleschi, G., & Scognamiglio, V. (2018). Nanostructured (Bio) sensors for smart agriculture. *TrAC Trends in Analytical Chemistry, 98*, 95–103.

Balbinot, S., Srivastav, A. M., Vidic, J., Abdulhalim, I., & Manzano, M. (2021). Plasmonic biosensors for food control. *Trends in Food Science & Technology, 111*, 128–140.

Bao, J., Huang, T., Wang, Z., Yang, H., Geng, X., Xu, G., Samalo, M., Sakinati, M., Huo, D., & Hou, C. (2019). 3D graphene/copper oxide nano-flowers based acetylcholinesterase biosensor for sensitive detection of organophosphate pesticides. *Sensors and Actuators B: Chemical, 279*, 95–101.

Bobrinetskiy, I. I., & Knezevic, N. Z. (2018). Graphene-based biosensors for on-site detection of contaminants in food. *Analytical Methods, 10*(42), 5061–5070.

Cellini, A., Blasioli, S., Biondi, E., Bertaccini, A., Braschi, I., & Spinelli, F. (2017). Potential applications and limitations of electronic nose devices for plant disease diagnosis. *Sensors, 17*(11), 2596.

Chen, B. B., Liu, M. L., Zhan, L., Li, C. M., & Huang, C. Z. (2018). Terbium(III) modified fluorescent carbon dots for highly selective and sensitive ratiometry of stringent. *Analytical Chemistry, 90*(6), 4003–4009.

Chesterfield, R. J., Whitfield, J. H., Pouvreau, B., Cao, D., Alexandrov, K., Beveridge, C. A., & Vickers, C. E. (2020). Rational design of novel fluorescent enzyme biosensors for direct detection of strigolactones. *ACS Synthetic Biology, 9*(8), 2107–2118.

Christopher, F. C., Kumar, P. S., Christopher, F. J., Joshiba, G. J., & Madhesh, P. (2020). Recent advancements in rapid analysis of pesticides using nano biosensors: A present and future perspective. *Journal of Cleaner Production, 269*, 122356.

Cui, F., Sun, J., de Dieu Habimana, J., Yang, X., Ji, J., Zhang, Y., Lei, H., Li, Z., Zheng, J., & Fan, M. (2019). Ultrasensitive fluorometric angling determination of Staphylococcus aureus in vitro and fluorescence imaging in vivo using carbon dots with full-color emission. *Analytical Chemistry, 91*(22), 14681–14690.

Darr, J. A., Zhang, J., Makwana, N. M., & Weng, X. (2017). Continuous hydrothermal synthesis of inorganic nanoparticles: Applications and future directions. *Chemical Reviews, 117*(17), 11125–11238.

Franco, A. J. D., Merca, F. E., Rodriguez, M. S., Balidion, J. F., Migo, V. P., Amalin, D. M., Alocilja, E. C., & Fernando, L. M. (2019). DNA-based electrochemical nanobiosensor for the detection of Phytophthora palmivora (Butler) Butler, causing black pod rot in cacao (Theobroma cacao L.) pods. *Physiological and Molecular Plant Pathology, 107*, 14–20.

Fu, Q., Zhang, C., Xie, J., Li, Z., Qu, L., Cai, X., Ouyang, H., Song, Y., Du, D., & Lin, Y. (2019). Ambient light sensor based colorimetric dipstick reader for rapid monitoring organophosphate pesticides on a smart phone. *Analytica Chimica Acta, 1092*, 126–131.

Ghorbani, R., Wilcockson, S., Koocheki, A., & Leifert, C. (2008). Soil management for sustainable crop disease control: A review. *Environmental Chemistry Letters, 6*, 149–162.

Girigoswami, K., & Akhtar, N. (2019). Nanobiosensors and fluorescence based biosensors: An overview. *International Journal of Nano Dimension, 10*(1), 1–17.

Gupta, R., Kumar, A., Kumar, S., Pinnaka, A. K., & Singhal, N. K. (2021). Naked eye colorimetric detection of Escherichia coli using aptamer conjugated graphene oxide enclosed Gold nanoparticles. *Sensors and Actuators B: Chemical, 329*, 129100.

Gustavsson, J., Cederberg, C., Sonesson, U., Van Otterdijk, R., & Meybeck, A. (2011). *Global food losses and food waste*. FAO.

He, J., Zhang, L., Xu, L., Kong, F., & Xu, Z. (2020). Development of nanozyme-labeled biomimetic immunoassay for determination of sulfadiazine residue in foods. *Advances in Polymer Technology, 2020*, 1–8.

Hong, S., & Lee, C. (2018). The current status and future outlook of quantum dot-based biosensors for plant virus detection. *The Plant Pathology Journal, 34*(2), 85.

Hou, W., Zhang, Q., Dong, H., Li, F., Zhang, Y., Guo, Y., & Sun, X. (2019). Acetylcholinesterase biosensor modified with ATO/OMC for detecting organophosphorus pesticides. *New Journal of Chemistry, 43*(2), 946–952.

Khaledian, S., Nikkhah, M., Shams-bakhsh, M., & Hoseinzadeh, S. (2017). A sensitive biosensor based on gold nanoparticles to detect Ralstonia solanacearum in soil. *Journal of General Plant Pathology, 83*, 231–239.

Khan, N., Ray, R. L., Sargani, G. R., Ihtisham, M., Khayyam, M., & Ismail, S. (2021). Current progress and future prospects of agriculture technology: Gateway to sustainable agriculture. *Sustainability, 13*(9), 4883.

Koedrith, P., Thasiphu, T., Weon, J.-I., Boonprasert, R., Tuitemwong, K., & Tuitemwong, P. (2015). Recent trends in rapid environmental monitoring of pathogens and toxicants: Potential of nanoparticle-based biosensor and applications. *The Scientific World Journal*, 2015.

Lau, H. Y., Wu, H., Wee, E. J. H., Trau, M., Wang, Y., & Botella, J. R. (2017). Specific and sensitive isothermal electrochemical biosensor for plant pathogen DNA detection with colloidal gold nanoparticles as probes. *Scientific Reports, 7*(1), 38896.

Li, Z., Askim, J. R., & Suslick, K. S. (2019). The optoelectronic nose: Colorimetric and fluorometric sensor arrays. *Chemical Reviews, 119*(1), 231–292.

Li, D., Liu, L., Huang, Q., Tong, T., Zhou, Y., Li, Z., Bai, Q., Liang, H., & Chen, L. (2021). Recent advances on aptamer-based biosensors for detection of pathogenic bacteria. *World Journal of Microbiology and Biotechnology, 37*, 1–20.

Lin, H.-Y., Huang, C.-H., Lu, S.-H., Kuo, I. T., & Chau, L.-K. (2014). Direct detection of orchid viruses using nanorod-based fiber optic particle plasmon resonance immunosensor. *Biosensors and Bioelectronics, 51*, 371–378.

Mahmoudi, E., Fakhri, H., Hajian, A., Afkhami, A., & Bagheri, H. (2019). High-performance electrochemical enzyme sensor for organophosphate pesticide detection using modified metal-organic framework sensing platforms. *Bioelectrochemistry, 130*, 107348.

Maitra, S., Brestic, M., Bhadra, P., Shankar, T., Praharaj, S., Palai, J. B., Shah, M. M. R., Barek, V., Ondrisik, P., & Skalický, M. (2021). Bioinoculants—Natural biological resources for sustainable plant production. *Microorganisms, 10*(1), 51.

Malecka, K., Michalczuk, L., Radecka, H., & Radecki, J. (2014). Ion-channel genosensor for the detection of specific DNA sequences derived from Plum pox virus in plant extracts. *Sensors, 14*(10), 18611–18624.

Malhotra, S. P. K., & Mandal, T. (2019). Zinc oxide nanostructure and its application as agricultural and industrial material. *Contaminants in Agriculture and Environment: Health Risks and Remediation, 1*, 216.

Manjunatha, S., Biradar, D., & Aladakatti, Y. R. (2016). Nanotechnology and its applications in agriculture: A review. *Journal of Farm Sciences, 29*(1), 1–13.

Mathur, A., Gupta, R., Kondal, S., Wadhwa, S., Pudake, R. N., Kansal, R., Pundir, C. S., & Narang, J. (2018). A new tactics for the detection of S. aureus via paper based geno-interface incorporated with graphene nano dots and zeolites. *International Journal of Biological Macromolecules, 112*, 364–370.

Miranda, B. S., Linares, E. M., Thalhammer, S., & Kubota, L. T. (2013). Development of a disposable and highly sensitive paper-based immunosensor for early diagnosis of Asian soybean rust. *Biosensors and Bioelectronics, 45*, 123–128.

Mishra, A., Kumar, J., Melo, J. S., & Sandaka, B. P. (2021). Progressive development in biosensors for detection of dichlorvos pesticide: A review. *Journal of Environmental Chemical Engineering, 9*(2), 105067.

Mondal, R., Dam, P., Chakraborty, J., Paret, M. L., Katı, A., Altuntas, S., Sarkar, R., Ghorai, S., Gangopadhyay, D., Mandal, A. K., & Husen, A. (2022). Potential of nanobiosensor in sustainable agriculture: the state-of-art. *Heliyon, 8*(12), e12207.

Neethirajan, S., Ragavan, K. V., & Weng, X. (2018a). Agro-defense: Biosensors for food from healthy crops and animals. *Trends in Food Science & Technology, 73*, 25–44.

Neethirajan, S., Ragavan, V., Weng, X., & Chand, R. (2018b). Biosensors for sustainable food engineering: challenges and perspectives. *Biosensors, 8*(1), 23.

Negrete, J. C. (2020). Nanotechnology an option in Mexican agriculture. *Journal of Biotechnology & Bioinformatics Research, 2*(2), 1–3.

Oluwole, O., Ibidapo, O., Arowosola, T., Raji, F., Zandonadi, R. P., Alasqah, I., Lho, L. H., Han, H., & Raposo, A. (2023). Sustainable transformation agenda for enhanced global food and nutrition security: A narrative review. *Frontiers in Nutrition, 10*.

Palou, L., Montesinos-Herrero, C., Tarazona, I., Besada, C., & Taberner, V. (2015). Incidence and etiology of postharvest fungal diseases of persimmon (Diospyros kaki Thunb. cv. Rojo Brillante) in Spain. *Plant Disease, 99*(10), 1416–1425.

Pandit, C., Alajangi, H. K., Singh, J., Khajuria, A., Sharma, A., Hassan, M. S., Parida, M., Semwal, A. D., Gopalan, N., & Sharma, R. K. (2022). Development of magnetic nanoparticle assisted aptamer-quantum dot based biosensor for the detection of Escherichia coli in water samples. *Science of the Total Environment, 831*, 154857.

Perez, M. F., Contreras, L., Garnica, N. M., Fernández-Zenoff, M. V., Farías, M. E., Sepulveda, M., Ramallo, J., & Dib, J. R. (2016). Native killer yeasts as biocontrol agents of postharvest fungal diseases in lemons. *PLoS One, 11*(10), e0165590.

Pudake, R. N., Jain, U., & Kole, C. (Eds.). (2021). *Biosensors in agriculture: Recent trends and future perspectives*. Springer.

Ranbir, S. G., Singh, H., Kaur, N., & Singh, N. (2023). Portable sensor array for on-site detection and discrimination of pesticides and herbicides using multivariate analysis. *Analytical Chemistry, 95*(39), 14533–14540.

Ray, M., Ray, A., Dash, S., Mishra, A., Achary, K. G., Nayak, S., & Singh, S. (2017). Fungal disease detection in plants: Traditional assays, novel diagnostic techniques and biosensors. *Biosensors and Bioelectronics, 87*, 708–723.

Ray, M. K., Mishra, A. K., Mohanta, Y. K., Mahanta, S., Chakrabartty, I., Kungwani, N. A., Avula, S. K., Panda, J., & Pudake, R. N. (2023). Nanotechnology as a promising tool against phytopathogens: A futuristic approach to agriculture. *Agriculture, 13*(9), 1856.

Shawon, Z. B. Z., Hoque, M. E., & Chowdhury, S. R. (2020). Nanosensors and nanobiosensors: Agricultural and food technology aspects. In *Nanofabrication for smart nanosensor applications* (pp. 135–161). Elsevier.

Singh, N. A., Narang, J., Garg, D., Jain, V., Payasi, D., Suleman, S., & Swami, R. K. (2023). Nanoparticles synthesis via microorganisms and their prospective applications in agriculture. *Plant Nano Biology*, 100047.

Smart, A., Crew, A., Pemberton, R., Hughes, G., Doran, O., & Hart, J. (2020). Screen-printed carbon based biosensors and their applications in agri-food safety. *TrAC Trends in Analytical Chemistry, 127*, 115898.

Song, D., Li, Y., Lu, X., Sun, M., Liu, H., Yu, G., & Gao, F. (2017). Palladium-copper nanowires-based biosensor for the ultrasensitive detection of organophosphate pesticides. *Analytica Chimica Acta, 982*, 168–175.

Srivastava, A. K., Dev, A., & Karmakar, S. (2018). Nanosensors and nanobiosensors in food and agriculture. *Environmental Chemistry Letters, 16*, 161–182.

Thakur, M., Wang, B., & Verma, M. L. (2022). Development and applications of nanobiosensors for sustainable agricultural and food industries: Recent developments, challenges and perspectives. *Environmental Technology & Innovation, 26*, 102371.

Thomson, G., Penrith, M. L., Atkinson, M., Thalwitzer, S., Mancuso, A., Atkinson, S., & Osofsky, S. (2013). International trade standards for commodities and products derived from animals: The need for a system that integrates food safety and animal disease risk management. *Transboundary and Emerging Diseases, 60*(6), 507–515.

Tyagi, P. K., Arya, A., Ramniwas, S., & Tyagi, S. (2023). *Recent trends in nanotechnology in precision and sustainable agriculture* (Vol. 14, p. 1256319). Frontiers Media SA.

Verma, M. L., & Rani, V. (2021). Biosensors for toxic metals, polychlorinated biphenyls, biological oxygen demand, endocrine disruptors, hormones, dioxin, phenolic and organophosphorus compounds: A review. *Environmental Chemistry Letters, 19*(2), 1657–1666.

Verma, M. L., Kumar, S., Das, A., Randhawa, J. S., & Chamundeeswari, M. (2020). Chitin and chitosan-based support materials for enzyme immobilization and biotechnological applications. *Environmental Chemistry Letters, 18*, 315–323.

Wang, L., Huo, X., Qi, W., Xia, Z., Li, Y., & Lin, J. (2020). Rapid and sensitive detection of Salmonella Typhimurium using nickel nanowire bridge for electrochemical impedance amplification. *Talanta, 211*, 120715.

Wu, B., Xu, H., Shi, Y., Yao, Z., Yu, J., Zhou, H., Li, Y., Chen, Q., & Long, Y. (2022). Microelectrode glucose biosensor based on nanoporous platinum/graphene oxide nanostructure for rapid glucose detection of tomato and cucumber fruits. *Food Quality and Safety, 6*, fyab030.

Xu, Y., Dhaouadi, Y., Stoodley, P., & Ren, D. (2020). Sensing the unreachable: Challenges and opportunities in biofilm detection. *Current Opinion in Biotechnology, 64*, 79–84.

You, K., Luo, X., Hu, W., Xu, Y., Guo, J., & He, Q. (2020). Environmental-friendly gold nanoparticle immunochromatographic assay for ochratoxin A based on biosynthetic mimetic mycotoxin-conjugates. *World Mycotoxin Journal, 13*(2), 267–276.

Yüksel, S., Schwenkbier, L., Pollok, S., Weber, K., Cialla-May, D., & Popp, J. (2015). Label-free detection of Phytophthora ramorum using surface-enhanced Raman spectroscopy. *Analyst, 140*(21), 7254–7262.

Zhan, F., Wang, T., Iradukunda, L., & Zhan, J. (2018). A gold nanoparticle-based lateral flow biosensor for sensitive visual detection of the potato late blight pathogen, Phytophthora infestans. *Analytica Chimica Acta, 1036*, 153–161.

Zhang, P., Sun, T., Rong, S., Zeng, D., Yu, H., Zhang, Z., Chang, D., & Pan, H. (2019). A sensitive amperometric AChE-biosensor for organophosphate pesticides detection based on conjugated polymer and Ag-rGO-NH_2 nanocomposite. *Bioelectrochemistry, 127*, 163–170.

Zia, R., Taj, A., Younis, S., Bukhari, S. Z., Latif, F., Feroz, Y., Fatima, K., Imran, A., & Bajwa, S. Z. (2022). Application of nanosensors for pesticide detection. In *Nanosensors for smart agriculture* (pp. 259–302). Elsevier.

Chapter 9
Revolutions in Biotic Stress Management and Sustainable Agriculture Through Microbial-Mediated Nanoformulation

Riya Sharma, Shrishty Pandey, Ramesh Namdeo Pudake, Navya Tyagi, Arti Mishra, and Jaagriti Tyagi

Abstract Biotic stress threatens sustainable agriculture by causing economic losses, reduced yields, and environmental harm. Conventional methods like chemical pesticides worsen these issues by fostering pest resistance, harming non-target species, and endangering pollinators. Nanoformulations techniques, like nanoprecipitation, nanoparticles, and nanocapsules, offer precise and targeted biotic stress management. Microbial nanoformulation, which uses microbes to synthesize nanoparticles, represents a significant advancement, enabling environmentally friendly production. These microbial-assisted nanoparticles enhance the stability and effectiveness of nanoformulations, providing a natural means to control pests and diseases. Despite challenges in maintaining formulation stability and scaling up production, nanoparticle-based technology holds promise for sustainable agriculture. Achieving targeted delivery and controlled release of nanoscale products is crucial for mitigating biotic stress with minimal environmental impact, ensuring food security. Integrating nanoparticle formulations with integrated pest management (IPM) strategies enhances efficacy and reduces reliance on traditional pesti-

R. Sharma · S. Pandey
Amity Institute of Microbial Technology, Amity University Uttar Pradesh, Noida, Uttar Pradesh, India

R. N. Pudake
Amity Institute of Nanotechnology, Amity University Uttar Pradesh, Noida, Uttar Pradesh, India

N. Tyagi
Dr DY Patil Biotechnology and Bioinformatics Institute, Pune, Maharashtra, India

A. Mishra
Umeå Plant Science Center, Department of Plant Physiology, Umeå University, Umeå, Sweden

J. Tyagi (✉)
Amity Institute of Microbial Technology, Amity University Uttar Pradesh, Noida, Uttar Pradesh, India

Dr MPS Group of Institutions, Agra, India

Y. K. Mohanta et al. (eds.), *Nano-microbiology for Sustainable Development*,
https://doi.org/10.1007/978-3-031-78845-1_9

cides. Nanoformulated pesticides, applied at lower rates and with reduced off-target effects, support environmental sustainability and lower agriculture's ecological footprint. Microbial nanoformulation techniques further improve outcomes by leveraging microbes' natural abilities to produce and stabilize nanoparticles, fostering resilient and sustainable agricultural systems. This chapter highlights the potential of nanoformulation, including microbial-assisted approaches, to enhance global food security, promote sustainable practices, and reduce environmental degradation, guiding agriculture toward a resilient and eco-friendly future.

Keywords Microbial nano formulation · Biotic stress · Nanoparticles · Integrated pest management

9.1 Introduction

Nano microbiology, an interdisciplinary field that merges nanotechnology and microbiology, focuses on studying and manipulating microbial cells at the nanoscale (1–100 nanometers). This field has the potential to address biotic stress comprises unfavorable effects on plants caused by living organisms, including pathogens (e.g., bacteria, viruses, fungi, and nematodes), pests (e.g., insects, herbivores), and weeds. Pathogens harm plants through feeding or infection, eventually decreasing growth, yield, and ultimately the internal structure of plants. It largely affects agriculture, forestry, and natural ecosystems, resulting in high economic losses and substantial environmental degradation (Tyagi et al., 2022). Management of biotic problems often involves the reduction of damage through pest and pathogen control using pesticides, resistant cultivars, cultural methods, and biological control (Singh et al., 2023).

Fertilizers are vital in managing biotic stress in food crops by enhancing plant health and resilience to pests and diseases. Essential nutrients like nitrogen, phosphorus, and potassium bolster plants' natural defence mechanisms, stimulating the production of protective compounds. Fertilizers also aid in rapid crop recovery post-damage, minimizing yield losses and promoting sustainability. However, fertilizer use poses challenges. Environmental contamination is a significant concern, with excessive application leading to nutrient runoff and water contamination, causing eutrophication and harmful algal blooms (Chaudhary et al., 2023). Continuous use without proper soil management can degrade its quality, resulting in nutrient imbalances, acidification, and reduced microbial diversity.

Degraded soils require increased fertilizer inputs, potentially stimulating weed growth and pest proliferation, complicating stress management. High fertilizer costs burden farmers economically, particularly in intensive agricultural systems. Dependence on synthetic fertilizers can degrade soil fertility, emphasizing the need for sustainable soil management practices like organic amendments and crop rotation. Addressing these challenges is vital for sustainable biotic stress management in food crop production.

Biofertilizers enhance soil health by promoting nutrient cycling, improving soil structure, and fostering microbial diversity. Also, beneficial microorganisms in biofertilizers can suppress plant pathogens by competing or producing antimicrobial compounds, activating plants' natural defense mechanisms like systemic acquired resistance (SAR) (Bidyarani et al., 2022; Eevera et al., 2023). They also enhance nutrient uptake by solubilizing essential nutrients, improving plant nutrition and vigor, and reducing susceptibility to biotic stresses (Chaudhary et al., 2023). Biofertilizers offer a sustainable approach to plant management, promoting productivity and environmental stewardship. The viability of biofertilizers in managing biotic stress depends on factors such as soil conditions, climate, crop type, and microorganism strain. However, their effectiveness may vary and not always be readily observable. Some biofertilizers target specific pathogens or pests, limiting their applicability. Compatibility issues with other inputs like chemical fertilizers or pesticides can complicate their use. Regulatory constraints and certification requirements may hinder widespread adoption. Moreover, higher costs and limited availability compared to conventional inputs pose barriers to their use (Singh et al., 2021). In contemporary agricultural strategies, nanofertilizers offer an innovative approach to combat biotic stress in crops. Nanoformulations precisely target stress mitigation through nutrient delivery and activation of resilient enzymes, complementing the role of biofertilizers in sustainable farming (Das & Das, 2019; Rawat, 2023).

9.2 Microbial- Based Nanoformulation

9.2.1 Synthesis of Microbial-Based Nanoformulations

Microbial-based nanoformulations are synthesized through various strategies involving microorganisms such as bacteria, fungi, yeast, and microalgae. These strategies include biological synthesis, extracellular synthesis, fungal, and yeast synthesis. For instance, silver nanoparticles have been synthesized using the culture supernatant of *Streptomyces griseoruber*, while *Rhizopus stolonifer* and white rot fungi are used to synthesize nanoparticles extracellularly. In the case of biosynthesis, microorganisms like *Klebsiella pneumoniae* employ biological processes and microbial enzymes to reduce metal ions and form nanoparticles. Fungi, such as thermophilic filamentous fungi, utilize a green synthesis process to create gold nanoparticles. Yeast also plays a role in synthesizing nanoparticles like silver through biological processes. The main steps involved in the microbial synthesis of nanoparticles include selecting suitable microorganisms based on their ability to synthesize nanoparticles. These microorganisms have specific enzymes and mechanisms facilitating nanoparticle synthesis. Metal ions are then reduced by microbial enzymes and proteins, either intracellularly or extracellularly. Intracellular synthesis involves transporting metal ions into the microbial cell, where they are reduced by enzymes, while extracellular synthesis occurs on the cell surface or in the culture medium (Nongbet et al., 2024; Mohanta et al., 2024).

Microbial enzymes, such as oxidoreductases, play a crucial role in reducing metal ions to form nanoparticles. To prevent nanoparticle aggregation and enhance stability, microorganisms secreted proteins, carbohydrates, and other organic molecules act as capping agents. In intracellular synthesis, metal ions are transported into the microbial cell through various mechanisms, including electrostatic attraction and cellular transporters. The reduced metal ions are converted into nanoparticles through biochemical mechanisms involving microbial enzymes and cofactors. Finally, the synthesized nanoparticles are characterized and purified to ensure their stability, size, and shape. These steps provide a sustainable, eco-friendly, and biocompatible approach to nanoparticle fabrication, offering significant advantages over traditional chemical processes.

9.2.2 Encapsulation Techniques for Active Ingredients

Encapsulation techniques play a pivotal role in nano-formulating active ingredients for managing biotic stress in plants, safeguarding them from degradation, and facilitating targeted delivery to pests and pathogens. This process involves incorporating the active core into a matrix structure through physical means and stabilizing it using materials like chitosan, alginate, and other biopolymers. By encapsulating active ingredients, such as pesticides and phytohormones, these techniques ensure stability, controlled release, and precise delivery to target pests and pathogens, thereby enhancing the efficacy of biotic stress management in plants (Zehra et al., 2021; Sarkar et al., 2022). Encapsulation falls into three categories: chemical, physical-chemical, and physical-mechanical approaches, each enabling the controlled release of nano-encapsulated compounds like phytohormones, pesticides, and plant growth-promoting microorganisms (PGPMs), thereby bolstering plant stress tolerance. The encapsulation process preserves the activities of bioactive compounds within biodegradable polymers or other materials, promoting enhanced plant stress tolerance and growth promotion. Encapsulation systems offer numerous advantages, including controlled delivery, protection against harsh conditions, sustained release, and targeted delivery to pests and pathogens (Ur Rahim et al., 2021; Maity et al., 2022). These systems effectively mitigate the negative impacts of biotic stress factors, enhance plant resilience, and optimize crop yield by ensuring efficient and precise delivery of active compounds to plants requiring stress management (Bartolucci et al., 2020).

Among the biopolymers utilized for encapsulating active ingredients in nanoformulations aimed at managing biotic stress in plants, sodium alginate (ALG), chitosan, and gelatin stand out as highly effective options (Tamta et al., 2024). Extensively researched for their ability to safeguard active compounds like pesticides and phytohormones from degradation and facilitate targeted delivery to pests and pathogens, these biopolymers offer notable advantages (Danish et al., 2021).

Sodium alginate (ALG) is the prevailing biopolymer for encapsulation, lauded for its stability in both acidic and neutral pH environments, biocompatibility, biodegradability, and low accumulation in the human body. Its wide availability and cost-effectiveness further cement its popularity in encapsulation applications. Chitosan, another compelling biopolymer, produces smaller and more durable microcapsules under varied pH conditions, complemented by its biocompatibility, biodegradability, and non-toxic nature (Sneharani & Kumar, 2021; Humbal & Pathak, 2023). Gelatin, a natural polymer, offers affordability and biodegradability, with studies indicating enhanced encapsulation efficiency when combined with ALG (Riseh et al., 2022). These biopolymers play a crucial role in encapsulating diverse microbial agents, including plant growth-promoting rhizobacteria (PGPR). Utilizing nanoscale biopolymer coatings for these microbial agents holds promise for advancing sustainable biological control strategies and fostering plant growth under stress (Nag et al., 2021; Shafiei et al., 2023).

Chitosan-based nanoformulations have emerged as effective tools for managing biotic stress in plants by encapsulating bioactive molecules such as pesticides, biocontrol agents, and plant growth-promoting rhizobacteria (PGPR) (Rajput et al., 2024). These formulations enhance the stability, controlled release, and targeted delivery of active ingredients, thereby improving their efficacy and reducing environmental pollution (Singh et al., 2022; Khosropour et al., 2023).

The bio-adhesive properties of chitosan offer excellent protection to encapsulated agrochemicals, increasing their stability and bioavailability within plants. Controlled release behavior is crucial in mitigating the negative impacts associated with excessive pesticide use (Saleem & Zaidi, 2020; Tyagi et al., 2022; Delian et al., 2023; Tyagi et al., 2023). Agrochemical release from chitosan matrices can be triggered by biotic and abiotic stress factors, with mechanisms including pore diffusion, surface desorption, capsule swelling, and degradation (Elsharkawy et al., 2022; Nongbet et al., 2024). The diffusion-controlled mechanism relies on a rate gradient, while surface desorption involves release from the capsule's surface. Hydration-induced swelling of chitosan capsules facilitates agrochemical release. Enzymatic reactions or environmental factors can degrade the capsule matrix, effectively releasing agrochemicals (Rajput et al., 2024; Khosropour et al., 2023). These mechanisms underscore the potential of chitosan-based nanoformulations in sustainable agriculture.

9.2.3 Functionalization of Nanoparticles for Targeted Delivery

Nanoparticles, typically ranging from 1 to 100 nanometers in size, exhibit unique physical, chemical, and biological properties owing to their small dimensions, high surface area to volume ratio, and quantum effects (Pramanik et al., 2023). In the realm of managing biotic stress in food crops, nanoparticles have garnered considerable attention for their promising applications in agriculture. They present several advantages, including targeted delivery, enhanced bioavailability, and reduced environmental impact compared to conventional agrochemicals (Tortella et al., 2023).

Nanoparticles play a crucial role in combating biotic stress in food crops, with various types exhibiting significant efficacy (Mohanta et al., 2024). Metal-based nanoparticles, in particular, have garnered attention for their antimicrobial properties. For example, silver nanoparticles (AgNP) have demonstrated effectiveness against a wide range of plant pathogens, including bacteria, fungi, and viruses (Sultana et al., 2021; Almoneafy et al., 2023). They are commonly utilized in formulations such as nano sprays or coatings to combat infections such as bacterial blight, fungal leaf spots, and viral diseases in crops (Munir et al., 2023). Similarly, copper nanoparticles (CuNP) exhibit robust antimicrobial activity and find applications in agriculture for controlling fungal diseases such as powdery mildew, downy mildew, and anthracnose in crops (Nawaz et al., 2023). These nanoparticles, when incorporated into agricultural practices, contribute to enhancing crop health and productivity while minimizing the reliance on conventional chemical treatments.

Nanoparticles, spanning various categories such as carbon-based, polymeric, and metal oxide nanoparticles, offer innovative solutions in sustainable agriculture by effectively managing biotic stressors in food crops while minimizing environmental impacts associated with conventional agrochemicals. Carbon nanotubes (CNTs) possess unique mechanical and physicochemical properties, and research indicates their potential to enhance plant defense mechanisms against pathogens by bolstering plant immune responses (Priya et al., 2023). Additionally, CNTs are explored for their role in the controlled release of agrochemicals, thereby enhancing their efficacy (Ayub et al., 2022). Polymeric nanoparticles, exemplified by chitosan nanoparticles derived from chitin, a common polymer, exhibit inherent antifungal and antibacterial properties. These nanoparticles serve as carriers for encapsulating bioactive compounds and agrochemicals, facilitating controlled release and targeted delivery to plants for disease management (Mali et al., 2020). Metal oxide nanoparticles, such as zinc oxide nanoparticles (ZnO NPs), demonstrate antimicrobial properties and show promise in mitigating fungal and bacterial infections in crops. They find application in formulations for seed treatment or foliar application, enhancing plant resistance to pathogens.

While nanoparticles offer promising avenues for sustainable agriculture, it is crucial to ensure proper risk assessment and regulatory oversight to address potential environmental and health concerns associated with their use in agriculture (Upadhayay et al., 2022). By leveraging the capabilities of nanoparticles judiciously and responsibly, agriculture can progress towards more sustainable and resilient practices, safeguarding both crop productivity and environmental integrity (War et al., 2020).

9.3 Enhanced Delivery Systems

9.3.1 Controlled Release Mechanisms

Biotic stresses from insects and pathogens present significant challenges to global food security and agricultural sustainability. Traditional pest and pathogen management methods often involve the indiscriminate use of pesticides and fungicides, leading to environmental pollution, ecosystem disruption, and health risks. However, recent advancements in nanotechnology provide promising solutions for alleviating biotic stresses in food crops through improved delivery systems (Shoala, 2023). Controlled release mechanisms utilizing nanoparticles and nanoformulations play a pivotal role in mitigating biotic stress in plants by augmenting the expression of genes responsible for antioxidant enzymes, fostering plant growth, and facilitating stress alleviation (Chandel et al., 2022).

One exemplary technique involves employing mesoporous silica nanoparticles (MSN) modified with decanethiol, which are then loaded with salicylic acid (SA). Here, the mesopores of MSN serve as reservoirs for SA, while decane chains act as gatekeepers, preventing SA molecules from prematurely escaping. The introduction of glutathione (GSH) prompts the removal of decane chains, facilitating the gradual release of salicylic acid from the MSN, thus ensuring prolonged protection against biotic stress (Ding et al., 2023) Furthermore, cerium dioxide nanoparticles (CeO_2 -NPs) have emerged as a promising solution for enhancing plant growth and development amidst stressful conditions. Their application demonstrates a positive impact on plant stress tolerance, further underscoring their potential in agricultural practices (Du et al., 2022; Falsini et al., 2023; Civilini et al., 2024).

Nanoencapsulated systems, especially for controlled phytohormone release, attract attention for their regulated compound delivery. Advancements in encapsulation techniques enhance efficacy in bolstering plant stress tolerance (Valle-Garcia et al., 2023). Despite promising stress mitigation and vitality enhancement, challenges like field application constraints, potential side effects, and dosage complexity hinder widespread adoption. Controlled release via nanoparticles offers a viable approach to enhancing plant resilience against biotic stressors. Strategies with mesoporous silica nanoparticles, cerium dioxide nanoparticles, and nanoencapsulated systems show significant potential in agricultural biotic stress management (Dhakate et al., 2022; Nandini et al., 2023; Sales et al., 2024).

9.3.2 Targeted Delivery to Pests or Pathogens

Utilizing nanoparticles for precise delivery to pests or pathogens represents a cutting-edge strategy in combating bacterial, fungal, and viral infections in plants, along with controlling insect pests and nematodes. Nanoparticles exhibit various mechanisms to effectively target and eliminate microbes, including surface deposition, disruption of cell walls and membranes via released metal ions, and triggering toxic mechanisms through reactive oxygen species production (Abdel- Aziz et al., 2023).

Two primary nanoparticle types are utilized: inorganic-based (INPs) and bacteria-premised nanobiopesticides (B-NBPs), both meticulously crafted in terms of size, shape, composition, and surface characteristics. B-NBPs comprise nanostructures containing small particles of a biological control agent (BBCA) or engineered structures with antagonistic activity against plant pathogens/pests. They outperform conventional pesticides due to their smaller size, enabling better distribution on insect surfaces and deeper penetration into plant cells. Moreover, B-NBPs shield BBCAs from adverse environmental conditions, enhance their dispersibility and wettability, and offer a smart delivery platform with sustained release at the target site (Pan et al., 2023; Najitha et al., 2023).

On the contrary, INPs can be customized with specific structures to release ions, impacting toxicity against microbes within plants. Their effectiveness depends on size, shape, structure, and surface chemistry, influencing successful attachment and entry into microbial cells. Surface reactivity and coating provide appropriate charge, facilitating reaction and binding to the target (Otari et al., 2024). INPs' efficacy against phytopathogens varies depending on microbe sensitivities, incubation duration, and treatment stage in infected plants. The uptake pathway, whether foliar or soil sprayed, affects INPs' impact against microbes based on their effective dose. In summary, targeted nanoparticle delivery shows promise in combating plant infections and pest infestations. INPs and B-NBPs, with their tailored properties, offer a more efficient and eco-friendly alternative to traditional agrochemicals (Victoria et al., 2023; Loyal et al., 2023).

Utilizing silver nanoparticles (AgNPs) as a targeted nanoformulation against the bacterial pathogen *Xanthomonas axonopodis* pv. *vesicatoria* (Xav) represents a significant advancement in combating bacterial spot disease in tomato plants (Oliveira et al., 2022). AgNPs possess properties that make them effective against *Xav* by penetrating bacteria cell walls, disrupting membrane integrity, and triggering the release of reactive oxygen species (ROS), ultimately leading to bacterial demise. Moreover, AgNPs have shown efficacy in reducing disease severity, enhancing plant growth, and increasing crop yield (Padmakumar et al., 2023). This multifaceted effectiveness highlights AgNPs' potential as a holistic solution for managing plant diseases and pests (Tortella et al., 2023). The application of AgNPs demonstrates the promise of targeted nanoformulations in delivering microbe-targeting agents precisely where needed, offering eco-friendly and sustainable solutions to agricultural challenges while minimizing environmental impact (Sahu et al., 2023).

9.4 Nano-Formulated Pesticide

9.4.1 Increased Potency Against Pests and Diseases

Nano-encapsulated pesticides offer numerous advantages, foremost among them being the ability to diminish pesticide dosage and human exposure, thereby fostering environmental friendliness. By harnessing nanotechnology, these pesticides are engineered to address the inefficiencies of traditional counterparts. They achieve this by packaging pesticide molecules within nanoscale carriers, enhancing their absorption into pest tissues and efficacy at lower doses. Consequently, this approach minimizes pesticide release into the environment, a pivotal benefit for environmental preservation (Wu & Li, 2022). Moreover, nanoencapsulation elevates the assessment of insecticides by encapsulating active pesticide ingredients within thin-walled sacs at the nanometer scale, it also regulates their release and enhances efficiency. Particularly in the case of nano-herbicides, targeted delivery to weeds is facilitated, leaving minimal residual compounds in the soil and environment—a strategy aligned with eco-friendly practices (Satvekar et al., 2023; Adetuyi & Adetunji, 2024).

Nano-insecticides exhibit superiority over bulk substances on various fronts. They offer enhanced efficiency for both natural and chemical insecticides through controlled release, thereby reducing environmental contamination and ensuring easy and safe handling. Additionally, they are more susceptible to photodegradation and less toxic to non-target organisms compared to bulk substances. Polymer-based nanoformulations, predominantly utilizing polysaccharides such as chitosan, alginates, starch, and polyesters, have been pivotal in synthesizing nano-insecticides (Sarkar et al., 2024).

Apart from nanoencapsulation, nano pesticides, and nano insecticide formulation techniques, alternative methods exist for incorporating nano-formulated products (Rafiq et al., 2022; Sumithra & Ranjitha, 2023). These include nano-formulated, biosensors, nano-genetic approaches, nano-apparatus, nano herbicides, and an array of forthcoming innovations anticipated shortly, as depicted in Fig. 9.1.

Nanobiofungicides, a subset of nano-biopesticides, utilize engineered nanomaterials to enhance the effectiveness of biopesticides from natural sources. They exhibit superior solubility, mobility, and resilience compared to conventional pesticides, inhibiting pathogen growth and inducing neurotoxic effects on insects (Hassanisaasi et al., 2022; Sheokand et al., 2023). Production involves extracting active plant compounds (APCs), integrating them with nanoparticles, or embedding them within a polymer matrix. Their diminutive size and expansive surface area-to-volume ratio contribute to heightened efficacy, solubility, mobility, and minimal toxicity. Environmental friendly and biodegradable, nanobiofungicides aid in environmental cleanup by reducing the risk of harmful metal accumulation in plants (Ranjani & Hemalatha, 2023; Vyas et al., 2024). These formulations offer numerous

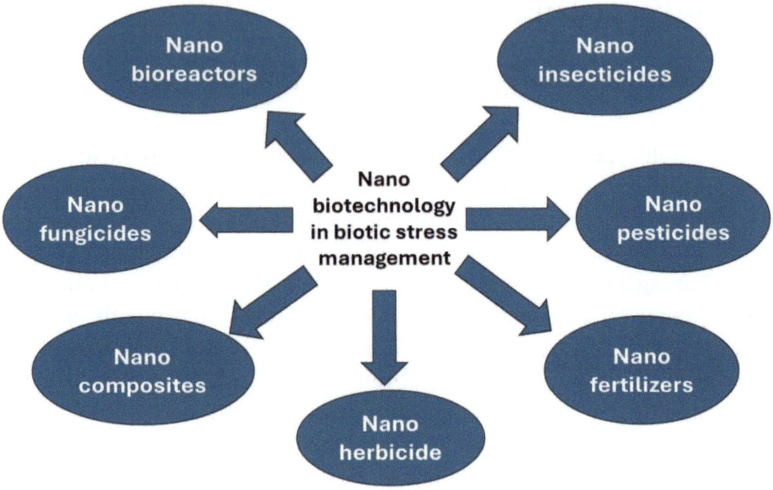

Fig. 9.1 Different forms of nanoparticles contribute to biotic stress management and sustainable development

advantages over traditional pesticides, including increased efficacy, durability, and reduced environmental impact. They also provide enhanced effectiveness and decreased likelihood of pesticide resistance.

Embracing nanobiofungicides could optimize agricultural resource utilization through precision farming techniques (Rahman & Upadhyaya, 2023). Integration of nano-barcodes facilitates monitoring of agricultural product quality. Available nanobiofungicides include nano-formulations of agro-nanofungicides, *Zataria multiflora*, and ginger essential oils, effectively combating plant pathogenic fungi like tomato root rot and Botrytis bunch rot (Patel et al., 2023; Sivalingam et al., 2024).

Nano biosensors are pivotal in combating biotic stress in plants by promptly detecting diseases and pests, thus safeguarding plant productivity. They excel in monitoring stress markers such as drought, diseases, and pest infestations, even at the individual plant level (Tyagi et al., 2017; Rathore et al., 2024). Their precision enables targeted interventions, minimizing unnecessary treatments. Comprising bioreceptors, transducers, and detectors, nano biosensors convert plant chemical signals into digital information with attributes like biocompatibility, portability, stability, specificity, and high signal detection accuracy (Kumar et al., 2021a, b). By utilizing bioreceptors to interact with stress-indicating analytes, transducers measure and transform these signals into quantifiable data, which detectors amplify and analyze, excelling in detecting specific stress markers with remarkable sensitivity and precision.

Various nano biosensors, including Optical nanosensors and nano-mechanical sensors, cater to diverse detection needs in agriculture. They monitor plant signaling molecules like hydrogen peroxide for stress response assessment and detect microbial contamination in soil. Utilizing materials such as gold nanoparticles and carbon nanotubes, these sensors can assess urea levels and urease activity in soil, crucial for optimizing crop quality and yield (Guleria et al., 2023). Quantum dots and plant viruses enable non-invasive monitoring of volatile substances on leaf surfaces, aiding in pest management strategies. Nano biosensors, merging nanoscience, biology, and electronics, promise to transform plant stress detection, ensuring sustainable agriculture and food security (Miguel-Rojas & Pérez-de-Luque, 2023).

Nano genetic apparatuses leverage nanotechnology to deliver RNA interference (RNAi) molecules, targeting specific genes in pests and pathogens. This disrupts their vital biological processes, effectively curbing pest and disease populations in crops like cotton and maize. Furthermore, RNAi-based plant-incorporated protectants (PIPs) offer highly targeted and eco-friendly solutions for insect pest control. These PIPs utilize RNAi, demonstrate exceptional specificity and environmental safety in managing pests. Delivery methods for PIPs vary and encompass foliar application, root uptake, and seed treatment.

Research has highlighted their efficacy in controlling pests across various crops, including soybean, corn, and cotton, showcasing their potential for sustainable pest management practices (Kaur et al., 2021). Similarly, nanoparticle-based delivery systems bolster plant antibodies, triggering systemic acquired resistance (SAR) against various pathogens. Nano-formulated plant immunizations, applied to crops such as tobacco and tomatoes, demonstrate promising disease prevention outcomes. In summary, these innovative technologies provide targeted solutions for crop protection, employing precise genetic targeting and immune response enhancement. Nano formulations offer an eco-friendly approach to combat biotic stress, decreasing environmental harm and residue levels. These advanced pesticides, utilizing nanotechnology, deliver active ingredients precisely, enhancing efficacy while minimizing pollution. Customizable for specific needs, they exhibit lower toxicity, quick degradation, and fewer emissions than conventional pesticides. Controlled release mechanisms ensure gradual, protective action, reducing pollution and residues (Hassan et al., 2024).

9.5 Biocontrol Agents in Nanoformulations

9.5.1 Utilization of Beneficial Microorganisms

Numerous types of microbes serve as biofertilizers, offering benefits such as insecticide and fungicide properties. These include *Bacillus thuringiensis var kurstaki* and *B. thuringiensis var tenebrionis, Bacillus subtilis, Talaromyces flavus, Clitoria ternatea (*butterfly pea*), Trichoderma,* plant growth promoting microbes, arbuscular mycorrhizal fungi, microalgae such as *Nostoc piscinale, Chlamydopodium fusiforme,* and *Chlorella vulgaris*. Additionally, nanomaterials derived from *Sargassum muticum, Caulerpa scalpelliformis,* and *Mesocyclops longisetus* exhibit biocompatibility, which is crucial for their integration into host systems without adverse effects

(Iqbal et al., 2019; Kumar et al., 2021a, b). Factors influencing this interaction include nanoparticle characteristics like size, shape, surface charge, and chemistry. Biocompatibility, the ability of nanomaterials to seamlessly integrate into a host system without causing harm, hinges on the interaction between nanoparticles and microbes. Various factors dictate this relationship, including the physicochemical traits of nanoparticles like size, shape, surface charge, and chemistry (Wang et al., 2020; Nawaz et al., 2023).

Microbes engage in synthesizing nanoparticles through metal capture, enzymatic reduction, and capping processes (Ahmed et al., 2023). Enzymatic reduction involves converting metal ions into elemental forms, aided by microbial enzymes. Subsequently, a protective layer forms around the nanoparticles through capping, preventing clumping and enhancing stability. The choice of microorganisms profoundly influences nanoparticle compatibility. For example, fungiexcel in synthesizing both intracellular and extracellular nanoparticles due to their specialized enzymes, resulting in well-dispersed and stable nanoparticles ideal for agricultural and plant disease management applications. Additionally, nanoparticle concentration and zeta potential play pivotal roles in compatibility (Khati et al., 2023). Larger nanoparticles demonstrate lower toxicity to plants compared to smaller ones, with toxicity also correlating with nanoparticle concentration. In essence, the interplay between nanoparticle physicochemical properties, microbial synthesis, and nanoparticle characteristics like concentration and zeta potential shapes their compatibility. Understanding these dynamics is paramount for tailoring nanoparticles for diverse applications in agriculture and plant disease management. Harnessing such microbes facilitates the development of an eco-friendly approach towards sustainability via green synthesis methods. This avenue of micro-nanotechnology holds promise in enhancing productivity and alleviating biotic stress in the future (Pudake et al., 2019; Kour et al., 2022).

9.5.2 Combination with Integrated Pest Management (IPM) Strategies

Integrated Pest Management (IPM) serves as a comprehensive strategy for pest control; integrating biological, cultural, physical, and chemical methods to manage pests while minimizing adverse environmental and human health effects. Nanoparticle-based pesticides, herbicides, and insecticides align with IPM principles, offering targeted delivery, enhanced efficacy, and reduced environmental impact. These solutions offer advantages over conventional pesticides, with their augmented surface area-to-volume ratio providing enhanced reactivity and effectiveness. Moreover, they can be customized to target specific pests, reducing reliance on broad-spectrum pesticides that may harm non-target organisms. Encapsulation of nanoparticle-based pesticides within polymers or other materials enables controlled release, reducing the need for repeated applications (Nazeer et al., 2023; Chaudhary et al., 2023). Nano-biopesticides, a subset of nanoparticle-based pesticides, are gaining attention for insect pest management. Derived from natural compounds like plant extracts or microorganisms, they offer environmental friendliness compared to synthetic pesticides. Nano-biopesticides provide stability, cost-effectiveness, and specificity, making them suitable for integration into IPM strategies (Yadav et al., 2023).

Nanoformulations integrated with IPM strategies offer significant advantages in pest management and agricultural sustainability. According to recent studies, nanopesticides demonstrate remarkable effectiveness, with a 31.5% increase in efficacy against target organisms compared to their non-nanoscale counterparts, as evidenced by an 18.9% improvement in field trials. Importantly, they also exhibit a 43.1% reduction in toxicity towards non-target organisms, thereby minimizing environmental harm. Moreover, nanopesticides contribute to the conservation of active ingredients (AIs), reducing premature loss by 41.4% and decreasing leaching potential in soils by 22.1%. Beyond their efficacy and environmental benefits, nanoformulations offer additional advantages. They enhance foliar adhesion, leading to improved crop yield and quality. Additionally, they serve as a responsive nanoscale delivery platform, effectively mitigating various biotic and abiotic stresses encountered in agricultural settings (Tyagi et al., 2017, 2021; Mahto et al., 2021). By promoting sustainable agriculture and bolstering global food security through higher yields and reduced environmental impacts, nanopesticides represent a promising advancement in IPM strategies.

9.5.3 Improvements in Soil Health and Crop Productivity

Nano particle-based fertilizers utilize essential nutrients like nitrogen (N), phosphorus (P), potassium (K), iron (Fe), and manganese (Mn), either alone or combined with nano-dimensional adsorbents, to achieve targeted delivery and controlled release (Abd-Elsalam & Alghuthaymi, 2023). This reduces nutrient loss through leaching and runoff while minimizing rapid degradation. Their high surface area to volume ratio and penetration ability makes nanofertilizers a viable alternative to chemical fertilizers, enhancing soil fertility and crop productivity over time. They also mitigate environmental risks by stimulating seed germination, nitrogen metabolism, photosynthesis, and stress tolerance. Nanofertilizers require smaller quantities for soil application, thus improving ease of use and lowering transportation costs. However, like all fertilizers, nanofertilizers have limitations and must be applied safely for sustainable agriculture (Khan et al., 2021). Therefore, their properties with microbe action trigger certain genes which regulate plant defense, resistance mechanism towards disease, signaling exchange for enzymatic release which led to the eradication of biotic stress antioxidant release both lead to improvement in soil health, crop productivity and other effects mentioned in Fig. 9.2 (Table 9.1).

In a study by Tiwari et al. (2024), significant impacts were observed in the productivity of chickpea *(Cicer arietinum)* suffering from biotic stress of Fusarium wilt

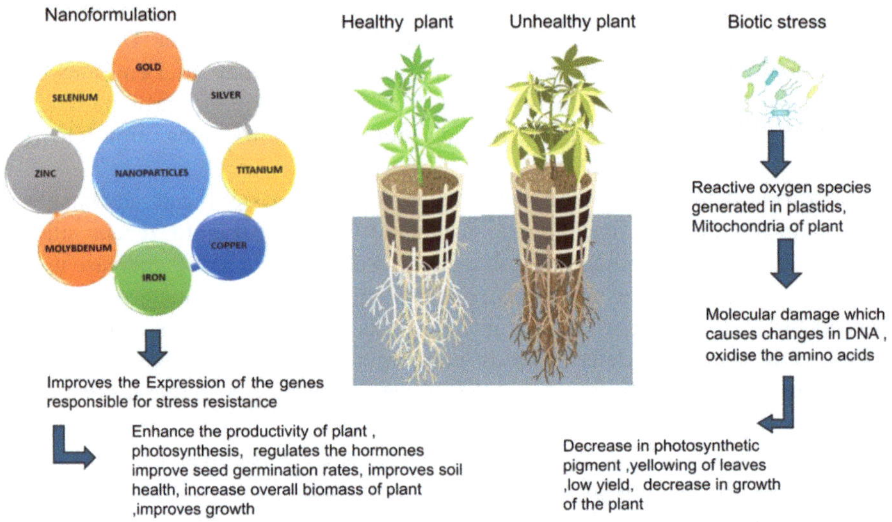

Fig. 9.2 Effects of nanoparticle-based nanoformulations on soil and crop productivity

when treated with copper oxide nanoparticles (CuO NPs) synthesized from coffee powder. This treatment led to enhanced growth and a reduction in disease incidence, accompanied by elevated levels of proteins and increased activity of enzymes such as phenylalanine ammonia-lyase (PAL), polyphenol oxidase (PPO), and nitrate reductase (NR) involved in defense responses against pathogens. Furthermore, the treatment regulated genes controlling oxidative stress, including those related to proline, hydrogen peroxide (H_2O_2), and malondialdehyde (MDA) levels.

9.6 Challenges and Future Perspective

In the realm of sustainable biotic stress management in food crops, we're faced with both challenges and exciting opportunities. One major challenge is ensuring that emerging technologies, such as nanoformulations, are not only effective but also scalable, affordable, and compliant with regulatory standards. These concerns are crucial because while nanotechnology holds immense promise for agriculture, its widespread adoption hinges on its economic viability and adherence to strict regulations. Collaborative efforts among researchers, industry players, and policymakers are essential to tackle these obstacles and facilitate the integration of nanotechnology into agricultural practices.

Looking ahead, the future of nanotechnology in sustainable biotic stress management is filled with potential. There's a vast scope for innovative research and development, ranging from exploring new applications to pushing the boundaries of

Table 9.1 Effects of nanoparticles in alleviating biotic stress in diseased crops

S. No.	Nanoparticle	Biotic agent and disease	Crops	Response/Effect	References
1.	Titanium dioxide nanoparticles (TiO_2 NPs)	*Puccinia striiformis f. sp. Tritici (Pst).* (Wheat yellow rust)	Wheat (*Triticum aestivum*)	Biosynthesis of NPs led to an increase in enzymatic activities like Superoxide dismutase (SOD), catalase (CAT), Peroxidase (POD), non-enzymatic activities like Total flavonoid content (TFC), Total proline content (TPC) in host, elevation of stress protein expression, defense mechanism, biochemical and osmotic changes led to inhibition of biotic stress	Satti et al. (2022)
2.	Silicon nanoparticles (SiNPs)	*Meloidogyne incognita* (Root-knot nematode)	Eggplant (*Solanum melongena*)	Biosynthesis of NPs showed 87.00–98.50% nematocidal, larvicidal effects in J2 juvenile phase, inhibited nematode reproduction, gall, egg masses on roots further treatment on host lead to increase plant morphology and physiology growth, root weight, leaves, reduction in Root gall index (RGI), Egg mass index (EGI) proves mitigation of root infection	El-Ashry et al. (2022)
3.	Selenium nanoparticles (SeNPs)	*Cochiobolus sativus* (Spot Blotch Disease)	Wheat (*Triticum aestivum*)	Green synthesis of NPs led to inhibition of phytopathogenic fungi, increased plant growth parameters (proline, sugar, flavonoids, phenolic, plant weight, Chlorophyll pigments, membrane stability fold were increased from 19 to 81 led to the eradication of biotic stress	Shahbaz et al. (2023)

(continued)

Table 9.1 (continued)

S. No.	Nanoparticle	Biotic agent and disease	Crops	Response/Effect	References
4.	Silicon nanoparticles (SiNPs)	*Rhizoctonia solani* (Root rot disease)	Oats (*Avena sativa L.*)	Increased protein content led to enhancement in plant growth, germination rate, water holding capacity, osmotic and membrane stability, tolerance index, reduced Disease Index (DI) 78.82% to 44.2% showed reduction in stress	Ahmad et al. (2023)
5.	Silver nanoparticles (AgNPs)	*Meloidogyne incognita* (Root-knot nematode)	Ajwain (*Trachyspermum ammi*)	Demonstrated nematocidal activity due to defense enzyme activities (peroxidase, catalase, superoxide dismutase, and ascorbate peroxidase), increased growth, nitrate reductase, carbonic anhydrase activity photosynthetic pigments seed oil with reduced root-knot index, egg, gall formation led to a reduction in stress causing component	Danish et al. (2021)
6.	Silica nanoparticles (SiO_2NPs)	*Magnaporthe oryzae* (Rice blast)	Rice (*Oryza sativa*)	Enhanced resistance, induced expression of salicylic acid genes promoted root development, drought resistance genes activation of systemic acquired resistance (SAR) leads to a reduction in biotic disease stress (70%), phytotoxicity, fungal inhibition, change in lesion patterns led to the protection of a host	Du et al. (2022)
7.	Manganese oxide nanoparticles (MnONPs)	*Aspergillus niger* (Black rot disease)	Tomato (*Solanum lycopersicum*)	Biosynthesized NPs posing stabilizing and reducing agent, the existence of manganese, oxygen increased sugar levels, and total soluble solids (TSS), and reducing sugars showed the most effective activity at an optimal concentration (2.5 mg/mL) lead to a reduction in rot disease stress	Anar et al. (2023)

8.	Chitosan- Salicylic acid (SA) NFs (CH-NFs) and Zinc sulphate NFs (ZnNFs)	*Puccinia striiformis. sp. tritici (Pst)* (Stripe rust)	Wheat (*Triticum aestivum*)	Green nano formulation increased enzymatic activities of phenylalanine ammonia lyase, tyrosine ammonia lyase and polyphenol oxidase. Nutrients, Total soluble sugars (TSS), pathogen resistance genes controlled disease effectively	Hembade et al. (2022)
9.	Chitosan nanoparticles + Thiamine	*Fusarium oxysporum* (Wilt disease)	Chickpea *(Cicer arietinum)*	Nano-encapsulation supplemented seed germination, plant growth, and increased phytohormone production Indole acetic acid (IAA), defense enzymes lead to suppression of disease	Muthukrishnan et al. (2019)
10.	Chitosan (CS) and chitosan nanoparticles (Ch NPs)	*Fusarium andiyazi* (*Wilt disease*)	Tomato (*Solanum lycopersicum*)	Inhibition of phytopathogen due to enzymatic activity, protein and antioxidant genes such as β-1,3-glucanase , chitinase, PR-1 and PR-10 genes, SOD fold increased up to 4.5 led to antifungal activity showed biocontrol efficiency to overcome biotic stress.	
11.	Zinc oxide nanoparticles (ZnONPs)	*Rhizoctonia solani Thanatephorus cucumeris* (Root rot or pocket rot disease)	Beet root (*Beta vulgaris*)	Increased plant physiological and biochemical parameter, caused reduction in disease index, root gall, nematode population, inhibited phytopathogens, Juveniles elevated enzymatic activities Superoxide Dismutase (SOD), Catalase (CAT), Phenylalanine Ammonia-Lyase (PAL), and Polyphenol Oxidase (PPO) led to disease suppression	Khan et al. (2021)

(continued)

Table 9.1 (continued)

S. No.	Nanoparticle	Biotic agent and disease	Crops	Response/Effect	References
12.	Silicon dioxide nanoparticles (SiO_2NPs)	*Meloidogyne incognita, Pectobacterium betavasculorum* and *Rhizoctonia solani* (Root rot or pocket rot disease)	Beet root *(Beta vulgaris*)	Enhanced chlorophyll, shoot, root weight, defense enzymes Superoxide Dismutase SOD, Catalase (CAT), Phenylalanine Ammonia-Lyase (PAL), and Polyphenol Oxidase (PPO) led to 84.2% and 91.9% reduction in Disease Index, root gall, and nematode reproduction with seed priming activity 200 mg L^{-1}	
13.	Silver nanoparticles (AgNPs)	*Aspergillus flavus*	Rice (*Oryza sativa*)	Biosynthesized NPs increased chlorophyll pigment, Relative Water Content, (RWC), Total Carotenoid Content (TCC), SOD (Superoxide Dismutase), POD (Peroxidase), CAT (Catalase), TFC (Total Flavonoid Content), TPC (Total Phenolic Content) lead to proliferation of phytopathogen and reduction in biotic stress	Sultana et al. (2021)
14.	Silver nanoparticles (AgNPs)	*Alternaria solani* (Early blight disease)	Tomato (*Solanum lycopersicum*)	Enhanced plant resistance, antioxidant enzymes SOD (Superoxide Dismutase), POD (Peroxidase), CAT (Catalase), balanced Reactive oxygen species (ROS), secondary metabolites, oxidative stress markers (callose, lignin, H_2O_2 and $O^-{}_2$, led to defense and resistance against diseased host thus eradiation of biotic stress	Narware et al. (2024)
15.	Chitosan fabricated biogenic silver nanoparticles (Ch@ BSNP)	*Xanthomonas campestris* (Bacterial spot of tomato)	Tomato (*Solanum lycopersicum*)	Biosynthesized NPs showed an increase in health compounds lycopene, carotene content, phenolics (49.10%), chlorophyll, carotenoids, water efficiency, transpiration rate, reduction in stress genes led to alleviation of biotic stress	Giri et al. (2023)

16.	Silicon dioxide nanoparticles (SiO_2NP)	*Rhizoctonia solani* (Rhizoctonia root rot)	Wheat (*Triticum aestivum*)	Biosynthesized Nps improved germination, Vegetative growth, photosynthetic components chlorophyll, carotenoid, defense hormones (Salicylic acid), stimulation of antioxidants, Peroxidase (POD), Superoxide Dismutase (SOD), Ascorbate Peroxidase (APX), Polyphenol Oxidase (PPO), Catalase (CAT), phenolics, led to complete reduction of phytopathogenic mycelial, root rot, counteract Reactive oxygen species (ROS), Methylglyoxal	Abdelrhim et al. (2021)
17.	Zinc oxide nanoparticles (ZnONPs)	*Fusarium oxysporum* (Fusarium wilt of chickpea)	Chick pea (*Cicer arietinum*)	Demonstrated suppression of mycelial growth, heightened antioxidant activity, elevated levels of sugars, phenols, total proteins, and superoxide dismutase (SOD), resulting in antifungal activity and reduction of disease incidence by 90%	Farhana et al. (2022)
18.	Chitosan nanoparticles + salicylic acid (SA) (CHNPs)	*Puccinia triticina* (Wheat Leaf Rust)	Wheat (*Triticum aestivum*)	Reduced germination of phytopathogenic spores, diminished symptoms, incubation, receptivity, condensed spore, lysis due to expression of PR (PR1-PR5 and PR10) defense genes, β-1, 4-glucanase, peroxidase, and chitinase, catalase enzyme led to spore condensation and lysis, causing the collapse of the fungal structure under biotic stress	Elsharkawy et al. (2022)

(continued)

Table 9.1 (continued)

S. No.	Nanoparticle	Biotic agent and disease	Crops	Response/Effect	References
19.	Chitosan nanoparticles + Thyme vulgaris essential oil (ThE-CsNPs)	*Fusarium solani* (Root rot disease)	Tomato (*Solanum lycopersicum)*	Upregulation of expression of WRKY transcription factors Increased phenolic, flavonoid, antioxidant enzyme 70% superoxide dismutase (SOD), 92.5% catalase (CAT), and 36.1% ascorbate peroxidase (APX) antioxidant enzymes, antifungal activity led to efficiency of Nps to control rot disease	Abd-Ellatif et al. (2022)
20)	Ferric oxide nanoparticles (Fe_2O_3)	*Fusarium oxysporum* (Fusarium wilt of tomato)	Tomato (*Solanum lycopersicum*)	Enhanced nutrient content, including chlorophyll pigments, carotenoids, total soluble proteins, phenols, and antioxidant enzyme activity (peroxidase, polyphenol oxidase, catalase, superoxide dismutase), facilitated growth, resulting in a significant decrease in disease index by 15.62% at a concentration of 20 μg/mL and 82.15% at 10 μg/mL, and by 25% with 71.42% protection	Elbasuney et al. (2022)
21.	Copper oxide nanoparticles (CuONPs)	*Meloidogyne incognita* (Root-knot nematodes)	Cowpea (*Vigna unguiculata*)	Biosynthesized NPs showed nematocidal activity due to improved biochemical mechanisms, chlorophyll pigment, protein content, promoted growth and yield, leading to apoptosis, necrosis, and distortion in the infected host, reduced gall size by 46.46%, 73.23%, and 80.30%, and egg mass from 61% to 96% at concentrations of 50 ppm, 100 ppm, and 200 ppm caused roughness and rupture of nematodes	Tauseef et al. (2021)

22.	Selenium Nanoparticles (SeNPs) + Copper	*Alternaria solani* (Early blight disease)	Tomato (*Solanum lycopersicum*)	NPs reduced the impact of pathogenic fungi by boosting enzyme activity such as superoxide dismutase, ascorbate peroxidase in leaves, increased (263–293%) glutathione peroxidase (GPX) and phenylalanine ammonia lyase led to enhancement in non-enzymatic antioxidants like vitamin C, glutathione, phenols, and flavonoids in fruits, effectively alleviating biotic stress	
23.	Copper nanoparticles (CuNPs) potassium silicate	*Clavibacter michiganensis* (Bacterial wilt and canker)	Tomato (*Solanum lycopersicum*)	Reduction in pathogen severity resulted from the bolstered activity of defense enzymes (SOD, PAL, GPX, APX) and antioxidant compounds (reduced glutathione, total phenols), along with increased concentrations of lycopene and β-carotene in the fruits, fostering host tolerance against the biotic stress-causing component	
24.	Silver nanoparticles (AgNPs)	*Meloidogyne javanica* (Root knot nematode)	Tomato (*Solanum lycopersicum*)	Green synthesized Nps showed highest nematicide activity reduced 89.5–100% egg masses, mortality, gall J2s juvenile phase due to increased expression of Phenylalanine Ammonia-Lyase (PAL), Poly Phenol Oxidase (PPO) led to enhancement in plant immunity against root knot nematode	Ghareeb et al. (2020)

nanoscience. For instance, developing sophisticated nanoencapsulation techniques can enable targeted delivery of bioactive compounds, enhancing crop resilience against pests and diseases. Additionally, leveraging RNA interference for eco-friendly pest control showcases another avenue for innovation in this field.

In this dynamic landscape, interdisciplinary collaboration is key. Bringing together experts from various fields fosters fresh perspectives and catalyzes innovation. Forward-thinking initiatives that promote collaboration between scientists, engineers, agronomists, and policymakers will be instrumental in driving progress and unlocking the full potential of nanotechnology in agriculture. By embracing these challenges and forging new paths for innovation, we can pave the way for a more sustainable and resilient future in food crop production.

9.7 Conclusion and Future Perspectives

Nanoformulations offer a promising avenue for administering biotic stressors in food crops, presenting innovative solutions to combat pests, diseases, and other biotic challenges while promoting sustainable agriculture. The small size and large surface area of nanoparticles enable precise delivery of bioactive compounds, enhancing their effectiveness and minimizing environmental impact. Nanoencapsulated biopesticides, antimicrobial agents, and RNA interference particles demonstrate improved stability, bioavailability, and targeted action against a wide range of pathogens and pests. Additionally, nanoparticle-based elicitors can modulate plant defense mechanisms, preparing crops for enhanced resistance against biotic stressors. Despite their potential, the widespread adoption of nanoformulations in agriculture requires addressing several challenges. Regulatory approval, environmental fate, and potential ecological effects necessitate rigorous risk assessment and standardized testing protocols to ensure the safety and efficacy of nano-enabled agrochemicals. Collaborative efforts between scientists, policymakers, and stakeholders are essential to navigate the ethical, social, and regulatory dimensions of nanotechnology in agriculture.

Looking ahead, continued research efforts, coupled with responsible innovation and effective governance, will be crucial for harnessing the full potential of nanoformulations in biotic stress management. Integrating nanotechnology with traditional agricultural practices can contribute to sustainable intensification, resilience, and food security in the face of evolving biotic stressors and global environmental challenges. By leveraging nanotechnology's precision and versatility, it is possible to develop tailored solutions to enhance crop protection, optimize resource utilization, and promote ecological balance in agricultural systems. Thus, nanoformulations represent a promising approach to address the complex challenges facing modern agriculture while advancing towards a more sustainable and adaptable food future.

References

Abdel-Aziz, S. M., Prasad, R., El Enshasy, H., & Sukmawati, D. (2023). Prospects of microbial nanotechnology for promoting climate resilient agriculture. In *Nanoparticles and plant-microbe interactions* (pp. 163–186). Academic.

Abd-Ellatif, S., Ibrahim, A. A., Safhi, F. A., Abdel Razik, E. S., Kabeil, S. S., Aloufi, S., Alyamani, A. A., Basuoni, M. M., Al Shamrani, S. M., & Elshafie, H. S. (2022). Green synthesized of Thymus vulgaris chitosan nanoparticles induce relative WRKY-genes expression in Solanum lycopersicum against Fusarium solani, the causal agent of root rot disease. *Plants, 11*(22), 3129.

Abdelrhim, A. S., Mazrou, Y. S., Nehela, Y., Atallah, O. O., El-Ashmony, R. M., & Dawood, M. F. (2021). Silicon dioxide nanoparticles induce innate immune responses and activate antioxidant machinery in wheat against Rhizoctonia solani. *Plants, 10*(12), 2758.

Abd-Elsalam, K.A., & Alghuthaymi, M.A. (Eds.). (2023). *Nanofertilizers for sustainable agroecosystems: Recent advances and future trends.*

Adetuyi, B. O., & Adetunji, C. O. (2024). Application of plant-based nanobiopesticides for mitigation of several biotic stress. In *Handbook of agricultural biotechnology, volume 1: Nanopesticides* (p. 239).

Ahmad, I., Younas, Z., Mashwani, Z. U. R., Raja, N. I., & Akram, A. (2023). Phytomediated selenium nanoparticles improved physio-morphological, antioxidant, and oil bioactive compounds of sesame under induced biotic stress. *ACS Omega, 8*(3), 3354–3366.

Almoneafy, A. A., Algam, S. A., Alhammadi, A. S., Moustafa-Farag, M., & Moghalles, M. A. (2023). Role of fungi-mediated nanoparticles in mitigation of biotic and abiotic stresses in plants. In *Fungal cell factories for sustainable nanomaterials productions and agricultural applications* (pp. 601–633). Elsevier.

Anar, M., Akbar, M., Tahir, K., Chaudhary, H. J., & Munis, M. F. H. (2023). Biosynthesized manganese oxide nanoparticles maintain firmness of tomato fruit by modulating soluble solids and reducing sugars under biotic stress. *Physiological and Molecular Plant Pathology, 127*, 102126.

Ayub, M. A., Naeem, A., Ur Rehman, M. Z., Farooqi, Z. U. R., Umar, W., Fatima, H., Nadeem, M., & Shabaan, M. (2022). Role of nanotechnology in enhancing crop production and produce quality. In *Sustainable nanotechnology for environmental remediation* (pp. 703–764).

Bartolucci, C., Antonacci, A., Arduini, F., Moscone, D., Fraceto, L., Campos, E., Attaallah, R., Amine, A., Zanardi, C., Cubillana-Aguilera, L. M., & Santander, J. M. P. (2020). Green nanomaterials fostering agrifood sustainability. *TrAC Trends in Analytical Chemistry, 125*, 115840.

Bidyarani, N., Vyas, G., Jaiswal, J., Deshmukh, S. K., & Kumar, U. (2022). Recent approaches in nanobioformulation for sustainable agriculture. In *Nanotechnology in agriculture and environmental science* (pp. 166–182). CRC Press.

Chandel, S. S., Shree, B., Mondal, K., & Mali, S. (2022). Mechanism of nanoparticles-mediated alleviating biotic and abiotic stresses in agricultural crops: Recent advances and future perspectives. In *The role of nanoparticles in plant nutrition under soil pollution: Nanoscience in nutrient use efficiency* (pp. 117–139).

Chaudhary, P., Chaudhary, A., Khati, P., Kumar, G., Tyagi, J., & Behera, M. (2023). Nanotechnology and omics approach in agrobiotechnology. In *Agricultural and environmental nanotechnology: Novel technologies and their ecological impact* (pp. 341–352). Springer Nature.

Civilini, M., Colautti, A., Brunello, A., Saccomanno, N., Marchiol, L., Foscari, A., & Iacumin, L. (2024). Impact of CeO_2 nanoparticles on the microbiota of the S. flos-cuculi L. (Caryophyllaceae) rhizosphere. *Environmental Science: Nano.*

Danish, M., Altaf, M., Robab, M. I., Shahid, M., Manoharadas, S., Hussain, S. A., & Shaikh, H. (2021). Green synthesized silver nanoparticles mitigate biotic stress induced by Meloidogyne incognita in Trachyspermum ammi (L.) by improving growth, biochemical, and antioxidant enzyme activities. *ACS Omega, 6*(17), 11389–11403.

Das, A., & Das, B. (2019). Nanotechnology a potential tool to mitigate abiotic stress in crop plants. In *Abiotic and biotic stress in plants.*

Delian, E., Bădulescu, L., Iacomi, B., & Lagunovschi-Luchian, V. (2023, September). The current sustainable tools for plant diseases management. In *IX South-Eastern Europe symposium on vegetables and potatoes 1391* (pp. 413–436).
Dhakate, P., Kandhol, N., Raturi, G., Ray, P., Bhardwaj, A., Srivastava, A., Kaushal, L., Singh, A., Pandey, S., Chauhan, D. K., & Dubey, N. K. (2022). Silicon nanoforms in crop improvement and stress management. *Chemosphere, 305*, 135165.
Ding, Y., Xiao, Z., Chen, F., Yue, L., Wang, C., Fan, N., Ji, H., & Wang, Z. (2023). A mesoporous silica nanocarrier pesticide delivery system for loading acetamiprid: Effectively manage aphids and reduce plant pesticide residue. *Science of the Total Environment, 863*, 160900.
Du, J., Liu, B., Zhao, T., Xu, X., Lin, H., Ji, Y., Li, Y., Li, Z., Lu, C., Li, P., & Zhao, H. (2022). Silica nanoparticles protect rice against biotic and abiotic stresses. *Journal of Nanobiotechnology, 20*(1), 197.
Eevera, T., Kumaran, S., Djanaguiraman, M., Thirumaran, T., Le, Q. H., & Pugazhendhi, A. (2023). Unleashing the potential of nanoparticles on seed treatment and enhancement for sustainable farming. *Environmental Research, 236*, 116849.
El-Ashry, R. M., El-Saadony, M. T., El-Sobki, A. E., El-Tahan, A. M., Al-Otaibi, S., El-Shehawi, A. M., Saad, A. M., & Elshaer, N. (2022). Biological silicon nanoparticles maximize the efficiency of nematicides against biotic stress induced by Meloidogyne incognita in eggplant. *Saudi Journal of Biological Sciences, 29*(2), 920–932.
Elbasuney, S., El-Sayyad, G. S., Attia, M. S., & Abdelaziz, A. M. (2022). Ferric oxide colloid: Towards green nano-fertilizer for tomato plant with enhanced vegetative growth and immune response against fusarium wilt disease. *Journal of Inorganic and Organometallic Polymers and Materials, 32*(11), 4270–4283.
Elsharkawy, M. M., Omara, R. I., Mostafa, Y. S., Alamri, S. A., Hashem, M., Alrumman, S. A., & Ahmad, A. A. (2022). Mechanism of wheat leaf rust control using chitosan nanoparticles and salicylic acid. *Journal of Fungi, 8*(3), 304.
Falsini, S., Nieri, T., Paolini, A., Schiff, S., Papini, A., Mugnai, L., Gonnelli, C., & Ristori, S. (2023). Tannins-lignin mixed nanoformulations for improving the potential of neem oil as fungicide agent. *Environmental Science and Pollution Research, 30*(13), 39131–39141.
Farhana, Munis, M. F. H., Alamer, K. H., Althobaiti, A. T., Kamal, A., Liaquat, F., Haroon, U., Ahmed, J., Chaudhary, H. J., & Attia, H. (2022). ZnO nanoparticle-mediated seed priming induces biochemical and antioxidant changes in chickpea to alleviate Fusarium wilt. *Journal of Fungi, 8*(7), 753.
Ghareeb, R. Y., Alfy, H., Fahmy, A. A., Ali, H. M., & Abdelsalam, N. R. (2020). Utilization of Cladophora glomerata extract nanoparticles as eco-nematicide and enhancing the defense responses of tomato plants infected by Meloidogyne javanica. *Scientific Reports, 10*(1), 19968.
Giri, V. P., Pandey, S., Srivastava, S., Shukla, P., Kumar, N., Kumari, M., Katiyar, R., Singh, S., & Mishra, A. (2023). Chitosan fabricated biogenic silver nanoparticles (Ch@ BSNP) protectively modulate the defense mechanism of tomato during bacterial leaf spot (BLS) disease. *Plant Physiology and Biochemistry, 197*, 107637.
Guleria, G., Thakur, S., Shandilya, M., Sharma, S., Thakur, S., & Kalia, S. (2023). Nanotechnology for sustainable agro-food systems: The need and role of nanoparticles in protecting plants and improving crop productivity. *Plant Physiology and Biochemistry, 194*, 533–549.
Hassan, S., Karaila, G. K., Singh, P., Meenatchi, R., Venkateswaran, A. S., Ahmed, T., Bansal, S., Kamalraj, R., Kiran, G. S., & Selvin, J. (2024). Implications of fungal nanotechnology for sustainable agriculture-applications and future perspectives. *Biocatalysis and Agricultural Biotechnology*, 103110.
Hassanisaadi, M., Barani, M., Rahdar, A., Heidary, M., Thysiadou, A., & Kyzas, G. Z. (2022). Role of agrochemical-based nanomaterials in plants: Biotic and abiotic stress with germination improvement of seeds. *Plant Growth Regulation, 97*(2), 375–418.
Hembade, V. L., Yashveer, S., Taunk, J., Sangwan, S., Tokas, J., Singh, V., Redhu, N. S., Grewal, S., Malhotra, S., & Kumar, M. (2022). Chitosan-Salicylic acid and Zinc sulphate nano-

formulations defend against yellow rust in wheat by activating pathogenesis-related genes and enzymes. *Plant Physiology and Biochemistry, 192*, 129–140.

Humbal, A., & Pathak, B. (2023). Application of nanotechnology in plant growth and diseases management: Tool for sustainable agriculture. In *Agricultural and environmental nanotechnology: Novel technologies and their ecological impact* (pp. 145–168). Springer Nature.

Iqbal, M., Umar, S., & Mahmooduzzafar (2019). Nano-fertilization to enhance nutrient use efficiency and productivity of crop plants. In *Nanomaterials and plant potential* (pp. 473–505).

Kaur, R., Choudhury, A., Chauhan, S., Ghosh, A., Tiwari, R., & Rajam, M. V. (2021). RNA interference and crop protection against biotic stresses. *Physiology and Molecular Biology of Plants, 27*(10), 2357–2377.

Khan, M., Khan, A. U., Hasan, M. A., Yadav, K. K., Pinto, M. M., Malik, N., Yadav, V. K., Khan, A. H., Islam, S., & Sharma, G. K. (2021). Agro-nanotechnology as an emerging field: A novel sustainable approach for improving plant growth by reducing biotic stress. *Applied Sciences, 11*(5), 2282.

Khati, P., Lohani, S., Raypa, P., Sharma, A., Kumari, A., Maura, J., Rani, V., & Maithani, D. (2023). Nanoparticles and Their Application as Nano-bioformulations in Agriculture. In *Advances in Nanotechnology for Smart Agriculture* (pp. 161–180). CRC Press.

Khosropour, E., Hakimi, L., & Weisany, W. (2023). Role of engineered nanomaterials in biotic stress managements. In *Engineered nanomaterials for sustainable agricultural production, soil improvement and stress management* (pp. 257–272). Academic.

Kour, H., Khan, S. S., Kour, D., Singh, S., Kumari, S., Kaur, M., Khan, R. T., & Yadav, A. N. (2022). Nanotechnologies for microbial inoculants as biofertilizers in the horticulture. In *Sustainable horticulture* (pp. 201–261). Academic.

Kumar, A., Choudhary, A., Kaur, H., Mehta, S., & Husen, A. (2021a). Smart nanomaterial and nanocomposite with advanced agrochemical activities. *Nanoscale Research Letters, 16*, 1–26.

Kumar, J., Ramlal, A., Mallick, D., & Mishra, V. (2021b). An overview of some biopesticides and their importance in plant protection for commercial acceptance. *Plants, 10*(6), 1185.

Loyal, A., Pahuja, S. K., Sharma, P., Malik, A., Srivastava, R. K., & Mehta, S. (2023). Potential environmental and human health implications of nanomaterials used in sustainable agriculture and soil improvement. In *Engineered nanomaterials for sustainable agricultural production, soil improvement and stress management* (pp. 387–412). Academic.

Mahto, R., Rani, P., Bhardwaj, R., Singh, R. K., Prasad, S. K., & Rakshit, A. (2021). Nanotechnology: A potential approach for abiotic stress management. In *Advances in nano-fertilizers and nanopesticides in agriculture* (pp. 249–264). Woodhead Publishing.

Maity, D., Gupta, U., & Saha, S. (2022). Biosynthesized metal oxide nanoparticles for sustainable agriculture: Next-generation nanotechnology for crop production, protection and management. *Nanoscale, 14*(38), 13950–13989.

Mali, S. C., Raj, S., & Trivedi, R. (2020). Nanotechnology a novel approach to enhance crop productivity. *Biochemistry and Biophysics Reports, 24*, 100821.

Miguel-Rojas, C., & Pérez-de-Luque, A. (2023). Nanobiosensors and nanoformulations in agriculture: New advances and challenges for sustainable agriculture. *Emerging Topics in Life Sciences, 7*(2), 229–238.

Mohanta, Y. K., Nongbet, A., Panda, J., Chakrabartty, I., Mohanta, T. K., & Mishra, B. (2024). Nanobiotechnology in genetic engineering for abiotic stress resistance in crops. In *Nanotechnology for abiotic stress tolerance and management in crop plants* (pp. 153–164). Academic.

Munir, N., Gulzar, W., Abideen, Z., Hancook, J. T., El-Keblawy, A., & Radicetti, E. (2023). Nanotechnology improves disease resistance in plants for food security: Applications and challenges. *Biocatalysis and Agricultural Biotechnology*, 102781.

Muthukrishnan, S., Murugan, I., & Selvaraj, M. (2019). Chitosan nanoparticles loaded with thiamine stimulate growth and enhances protection against wilt disease in Chickpea. *Carbohydrate Polymers, 212*, 169–177.

Nag, S., Ghosh, S., Dey, A., Majhi, S., Mondal, A., Mishra, A., Mitra, O., & Ahmad, F. (2021). *Plant biotic stress management.*

Najitha Banu, A., Rana, N., Kudesia, N., Sadaf, D., & Raut, A. M. (2023). Fabricated nanofertilizers: A clean and feasible substitute for conventional fertilizers. In *Nanofertilizers for sustainable agroecosystems: Recent advances and future trends* (pp. 35–59). Springer Nature.

Nandini, B., Mawale, K. S., & Giridhar, P. (2023). Nanomaterials in agriculture for plant health and food safety: A comprehensive review on the current state of agro-nanoscience. *3 Biotech, 13*(3), 73.

Narware, J., Singh, S. P., Ranjan, P., Behera, L., Das, P., Manzar, N., & Kashyap, A. S. (2024). Enhancing tomato growth and early blight disease resistance through green-synthesized silver nanoparticles: Insights into plant physiology. *South African Journal of Botany, 166*, 676–689.

Nawaz, A., Rehman, H. U., Usman, M., Wakeel, A., Shahid, M. S., Alam, S., Sanaullah, M., Atiq, M., & Farooq, M. (2023). Nanobiotechnology in crop stress management: An overview of novel applications. *Discover Nano, 18*(1), 74.

Nazeer, A., Ahmad, F., Verma, N., & Ahmad, S. (2023). Nanopesticides in agriculture: Some examples. In *Targeted delivery of nanopesticides and nanofertilizers in sustainable agricultural farming* (pp. 141–154). Springer Nature.

Nongbet, A., Panda, J., Mohanta, Y. K., Chakrabartty, I., Shamim, M. Z., & Mohanta, T. K. (2024). Role of nanoparticles to protect plants from abiotic stress by scavenging reactive oxygen species. In *Nanotechnology for abiotic stress tolerance and management in crop plants* (pp. 95–114). Academic.

Oliveira-Pinto, P. R., Mariz-Ponte, N., Torres, A., Tavares, F., Fernandes-Ferreira, M., Sousa, R. M., & Santos, C. (2022). Satureja montana L. essential oil, montmorillonite and nanoformulation reduce Xanthomonas euvesicatoria infection, modulating redox and hormonal pathways of tomato plants. *Scientia Horticulturae, 295*, 110861.

Otari, S., Bapat, V. A., Lakkakula, J., Kadam, U. S., & Suprasanna, P. (2024). Advancements in bionanotechnological applications for climate-smart agriculture and food production. *Biocatalysis and Agricultural Biotechnology*, 103117.

Padmakumar, A., Pavani, C., Eswar, K., Kong, L., Yang, W., Gopalakrishnan, S., Cahill, D. M., & Rengan, A. K. (2023). Bacteria-premised nanobiopesticides for the management of phytopathogens and pests. *ACS Agricultural Science & Technology, 3*(5), 370–388.

Pan, X., Guo, X., Zhai, T., Zhang, D., Rao, W., Cao, F., & Guan, X. (2023). Nanobiopesticides in sustainable agriculture: Developments, challenges, and perspectives. *Environmental Science: Nano, 10*(1), 41–61.

Patel, C., Singh, J., Karunakaran, A., & Ramakrishna, W. (2023). Evolution of nano-biofertilizer as a green technology for agriculture. *Agriculture, 13*(10), 1865.

Pramanik, B., Sar, P., Bharti, R., Gupta, R., Purkayastha, S., Sinha, S., Chattaraj, S., & Mitra, D. (2023). Multifactorial role of nanoparticles in alleviating environmental stresses for sustainable crop production and protection. *Plant Physiology and Biochemistry*, 107831.

Priya, S., Kumar, A., Upadhayay, V. K., Chaudhary, A., Parveen, H., & Kumar, G. (2023). Impact of nanoparticles on biotic stress tolerance. In *Advances in nanotechnology for smart agriculture* (pp. 197–220). CRC Press.

Pudake, R. N., Mittal, J., Tripathi, R. M., Tyagi, J., & Mohanta, T. K. (2019). Biochemical responses of maize seedlings exposed to SnNPs. *Micro & Nano Letters, 14*(6), 645–649.

Rafiq, S., Mantoo, M. A., Askary, T. H., Akbar, R., Jan, U., Akhter, A., & Bhadu, S. S. (2022). Nanopesticides: An emerging tool for pest management. *The Pharma Innovation Journal, 10*, 1659–1666.

Rahman, R., & Upadhyaya, H. (2023). Myconanoparticles for management of various biotic and abiotic stresses in plants. In *Fungal cell factories for sustainable nanomaterials productions and agricultural applications* (pp. 483–509). Elsevier.

Rajput, V. D., Singh, A., Ghazaryan, K., Minkina, T. M., & Al-Tawaha, A. R. M. (Eds.). (2024). *Sustainable agriculture: Nanotechnology and biotechnology for crop production and protection.* Walter de Gruyter GmbH & Co KG.

Ranjani, S., & Hemalatha, S. (2023). Influence of polyherbal nanoformulation on plant growth and biochemical constituents in legume seedlings. *Journal of Plant Growth Regulation, 42*(11), 7142–7150.
Rathore, A., Hasan, W., Pujar, K., Singh, R., Panotra, N., & Satapathy, S. N. (2024). Nanotech for crop protection: Utilizing nanoparticles for targeted pesticide delivery. *Uttar Pradesh Journal of Zoology, 45*(6), 46–71.
Rawat, S. (2023). Nanopriming technology for sustainable agriculture. In *Nanopriming approach to sustainable agriculture* (pp. 21–42). IGI Global.
Riseh, R. S., Hassanisaadi, M., Vatankhah, M., Soroush, F., & Varma, R. S. (2022). Nano/microencapsulation of plant biocontrol agents by chitosan, alginate, and other important biopolymers as a novel strategy for alleviating plant biotic stresses. *International Journal of Biological Macromolecules, 222*, 1589–1604.
Sahu, B., Choudhary, V. K., Sahu, M. P., Kumar, K. K., Sujayanand, G. K., Gopi, R., Prakasam, V., Sridhar, J., Mallikarjuna, J., Singh, H. K., & Sharma, K. C. (2023). Biotic Stress Management. In *Trajectory of 75 years of Indian Agriculture after Independence* (pp. 619–653). Springer Nature.
Saleem, H., & Zaidi, S. J. (2020). Recent developments in the application of nanomaterials in agroecosystems. *Nanomaterials, 10*(12), 2411.
Sales, H. B., de Carolino, A., de Nunes, R. Z., Macalia, C. M., Ruzo, C. M., da Pinto, C., de Bezerra, J., Campelo, P. H., Ţălu, Ş., de Souza, L. K., & Sanches, E.A. (2024). Advances in agricultural technology: A review of slow-release nanofertilizers and innovative carriers. In *Communications in soil science and plant analysis* (pp. 1–34).
Sarkar, M. R., Rashid, M. H. O., Rahman, A., Kafi, M. A., Hosen, M. I., Rahman, M. S., & Khan, M. N. (2022). Recent advances in nanomaterials based sustainable agriculture: An overview. *Environmental Nanotechnology, Monitoring & Management, 18*, 100687.
Sarkar, M. M., Ghosh, R., & Roy, S. (2024). Engineering native nanoparticles for enhancing functional attributes and their prospects in the synthesis of nanoformulations for the betterment of agriculture. In *Food production, diversity, and safety under climate change* (pp. 249–261). Springer Nature.
Satti, S. H., Raja, N. I., Ikram, M., Oraby, H. F., Mashwani, Z. U. R., Mohamed, A. H., Singh, A., & Omar, A. A. (2022). Plant-based titanium dioxide nanoparticles trigger biochemical and proteome modifications in Triticum aestivum L. under biotic stress of Puccinia striiformis. *Molecules, 27*(13), 4274.
Satvekar, R., Chavan, Y., Sahoo, A., & Nandre, V. S. (2023). Nanoformulations of natural compounds for herbicide and agri-food application. In *New horizons in natural compound research* (pp. 427–443). Academic.
Shafiei-Masouleh, S. S., Khankahdani, H. H., & Kumar, S. (2023). Natural biopolymer nanomaterials in biotic stress management. In *Biotic stress management of crop plants using nanomaterials* (pp. 53–70). CRC Press.
Shahbaz, M., Akram, A., Mehak, A., Haq, E. U., Fatima, N., Wareen, G., Fitriatin, B. N., Sayyed, R. Z., Ilyas, N., & Sabullah, M. K. (2023). Evaluation of selenium nanoparticles in inducing disease resistance against spot blotch disease and promoting growth in wheat under biotic stress. *Plants, 12*(4), 761.
Sheokand, M., Jain, K., Rana, V., Dhaka, S., Rana, A., Singh, K. P., & Dhaka, R. K. (2023). Nanobiochar-based formulations for sustained release of agrochemicals in precision agriculture practices. In *Handbook of green and sustainable nanotechnology: Fundamentals, developments and applications* (pp. 2413–2438). Springer International Publishing.
Shoala, T. (2023). The dual role of nanotechnology in the management of biotic and abiotic stresses in plants. In *Nanotechnology in agriculture and agroecosystems* (pp. 51–70). Elsevier.
Singh, R. P., Handa, R., & Manchanda, G. (2021). Nanoparticles in sustainable agriculture: An emerging opportunity. *Journal of Controlled Release, 329*, 1234–1248.
Singh, A., Rajput, V. D., Rawat, S., Sharma, R., Singh, A. K., Kumar, P., Singh, A. K., Minkina, T., Singh, R. P., & Singh, S. (2022). Geoinformatics and nanotechnological approaches for coping

up abiotic and biotic stress in crop plants. In *Sustainable agriculture systems and technologies* (pp. 337–359).
Singh, A. K., Mishra, K. K., Jeevan, B., Paschapur, A. U., Verma, G., & Kant, L. (2023). Nanotechnology in biotic stress management: Future challenges and opportunities. In *Biotic stress management of crop plants using nanomaterials* (pp. 173–192).
Sivalingam, S., Golla, G., Arunachalam, L., Singh, T., Karthikeyan, G., Shanthi, A., & Malaichamy, K. (2024). Encapsulation of essential oil to prepare environment friendly nanobio-fungicide against Fusarium oxysporum f. sp. lycopersici: An experimental and molecular dynamics approach. *Colloids and Surfaces A: Physicochemical and Engineering Aspects, 681*, 132681.
Sneharani, A. H., & Kumar, M. P. (2021). Nanoformulations in soil health and agriculture. In *Nanotechnology in sustainable agriculture* (pp. 211–224). CRC Press.
Sultana, T., Javed, B., Raja, N. I., & Mashwani, Z. U. R. (2021). Silver nanoparticles elicited physiological, biochemical, and antioxidant modifications in rice plants to control Aspergillus flavus. *Green Processing and Synthesis, 10*(1), 314–324.
Sumithra, B. S., & Ranjitha, G. (2023). *Nanotechnology and their applications in insect pest control* (p. 161). Integrated Publications.
Tamta, S., Vimal, V., Verma, S., Gupta, D., Verma, D., & Nangan, S. (2024). Recent development of nanobiomaterials in sustainable agriculture and agrowaste management. *Biocatalysis and Agricultural Biotechnology*, 103050.
Tauseef, A., Gupta, J., Rehman, A., & Uddin, I. (2021). Differential response of cowpea towards the CuO nanoparticles under Meloidogyne incognita stress. *South African Journal of Botany, 139*, 175–182.
Tiwari, V., Bambharoliya, K. S., Bhatt, M. D., Nath, M., Arora, S., Dobriyal, A. K., & Bhatt, D. (2024). Application of green synthesized copper oxide nanoparticles for effective mitigation of Fusarium wilt disease in roots of Cicer arietinum. *Physiological and Molecular Plant Pathology*, 102244.
Tortella, G., Rubilar, O., Pieretti, J. C., Fincheira, P., de Melo Santana, B., Fernández-Baldo, M. A., Benavides-Mendoza, A., & Seabra, A. B. (2023). Nanoparticles as a promising strategy to mitigate biotic stress in agriculture. *Antibiotics, 12*(2), 338.
Tyagi, J., Varma, A., & Pudake, R. N. (2017). Evaluation of comparative effects of arbuscular mycorrhiza (*Rhizophagus intraradices*) and endophyte (*Piriformospora indica*) association with finger millet (*Eleusine coracana*) under drought stress. *European Journal of Soil Biology, 81*, 1–10.
Tyagi, J., Shrivastava, N., Sharma, A. K., Varma, A., & Pudake, R. (2021). Effect of *Rhizophagus intraradices* on growth and physiological performance of Finger Millet (*Eleusine coracana* L.) under drought stress. *Plant Science Today, 8*(4), 912–923.
Tyagi, J., Chaudhary, P., Jyotsana, U. B., Bhandari, G., & Chaudhary, A., (2022). Impact of endophytic fungi in biotic stress management. In *Plant protection: From chemicals to biologicals* (pp. 447–462).
Tyagi, J., Mishra, A., Kumari, S., Singh, S., Agarwal, H., Pudake, R. N., Varma, A., & Joshi, N. C. (2023). Deploying a microbial consortium of *Serendipita indica, Rhizophagus intraradices*, and *Azotobacter chroococcum* to boost drought tolerance in maize. *Environmental and Experimental Botany, 206*, 105142.
Upadhyay, T. K., Kumar, V. V., Sharangi, A. B., Upadhye, V. J., Khan, F., Pandey, P., Kamal, M. A., Baba, A. Y., & Hakeem, K. R. (2022). Nanotechnology-based advancements in postharvest management of horticultural crops. *Phyton, 91*(3), 471.
Ur Rahim, H., Qaswar, M., Uddin, M., Giannini, C., Herrera, M. L., & Rea, G. (2021). Nanoenable materials promoting sustainability and resilience in modern agriculture. *Nanomaterials, 11*(8), 2068.
Valle-García, J. D., Ali, A., Patra, J. K., Kerry, R. G., Das, G., & Fernández-Luqueño, F. (2023). Integration of eco-friendly biological and nanotechnological strategies for better agriculture: A sustainable approach. In *Agricultural and environmental nanotechnology: Novel technologies and their ecological impact* (pp. 647–674).

Victoria, J., Tripathi, S., Prakash, V., Tiwari, K., Mahra, S., Sharma, A., Rana, S., Kandhol, N., Sahi, S., Tripathi, D. K., & Sharma, S. (2023). Encapsulated nanopesticides application in plant protection: Quo vadis? *Plant Physiology and Biochemistry*, 108225.

Vyas, T. K., More, B., & Mehta, M. P. (2024). Improving stress resilience in plants by nanoparticles. In *Improving stress resilience in plants* (pp. 73–96). Academic.

Wang, Z., Yue, L., Dhankher, O. P., & Xing, B. (2020). Nano-enabled improvements of growth and nutritional quality in food plants driven by rhizosphere processes. *Environment International, 142*, 105831.

War, J. M., Fazili, M. A., Mushtaq, W., Wani, A. H., & Bhat, M. Y. (2020). Role of nanotechnology in crop improvement. In *Nanobiotechnology in agriculture: An approach towards sustainability* (pp. 63–97).

Wu, H., & Li, Z. (2022). Recent advances in nano-enabled agriculture for improving plant performance. *The Crop Journal, 10*(1), 1–12.

Yadav, A., Yadav, K., Ahmad, R., & Abd-Elsalam, K. A. (2023). Emerging frontiers in nanotechnology for precision agriculture: Advancements, hurdles and prospects. *Agrochemicals, 2*(2), 220–256.

Zehra, A., Rai, A., Singh, S. K., Aamir, M., Ansari, W. A., & Upadhyay, R. S. (2021). An overview of nanotechnology in plant disease management, food safety, and sustainable agriculture. In *Food security and plant disease management* (pp. 193–219).

Chapter 10
From Lab to Landfill: Exploring Nanobiotechnological Approaches for E-Waste Remediation and Environmental Sustainability

Arnab Chakraborty, Rajshree Amrit, Mrunal S. Wagh, Manjit Kumar Ray, Bibhu Prasad Panda, and Guy Raymond Feuya Tchouya

Abstract The swift spike in electronic waste (e-waste) in modern society urges for novel sustainable management methods. Worldwide microbial-aided nanobiotechnology is now considered as a potential key for managing e-waste and promoting environmental sustainability. This approach harnesses microbial consortia and nanomaterials to remediate, recycle, and recover e-waste. The chapter shows that microbial-aided nanobiotechnology could potentially break down and detoxify various e-waste materials, reducing environmental pollution and resource depletion. Microbial processes utilizing nanomaterial platforms could concentrate on recovering important metals and components from electronic gadgets, supporting circular economy principles and resource conservation. Its potential is intriguing, but scalability, cost-effectiveness, and technical optimization need further research. Bioreactor design, novel microbial strains, and microorganism-nanomaterial synergy are potential research areas awaiting to be delved into. Academic, industrial, and regulatory collaboration is needed to successfully use microbial-aided nanobio-

A. Chakraborty · R. Amrit
School of Biosciences and Technology, Vellore Institute of Technology, Vellore, Tamil Nadu, India

M. S. Wagh (✉)
Research and Development, Organica Biotech Private Limited, Mumbai, Maharashtra, India

M. K. Ray
Department of Applied Biology, School of Biological Sciences, University of Science and Technology Meghalaya, Ri-Bhoi, Meghalaya, India

B. P. Panda
Environmental Sciences, Department of Chemistry, ITER, Siksha' O' Anusandhan (Deemed to be University), Bhubaneswar, Odisha, India

G. R. F. Tchouya
Department of Chemistry, Faculty of Science, Université des Sciences et Techniques de Masuku, Franceville, Gabon

Y. K. Mohanta et al. (eds.), *Nano-microbiology for Sustainable Development*,
https://doi.org/10.1007/978-3-031-78845-1_10

technology in e-waste disposal. Safe and sustainable worldwide deployment of these technologies needs strong legislative frameworks and regulatory norms. This chapter meticulously highlights the pivotal role of microbial nanotechnology in optimizing treatment methods for maximum efficiency and sustainability. It aims to showcase the synergistic effects of integrating microbiology and nanotechnology for sustainable treatment, scientifically illustrating how microbial nanobiocomposites amplify efficiency in e-waste recycling processes through enhanced eco-friendly material recovery.

Keywords E-waste management · Nanobiotechnology · Environmental sustainability · Resource recovery · Circular economy · Bioremediation

10.1 Introduction

In the contemporary age of technology, the growing reliance on electronic devices has resulted in an alarming rise in electronic waste (e-waste) output, thereby presenting formidable impediments to the attainment of environmental sustainability objectives (Yadav et al., 2022; Varsha et al., 2024). Unsafe disposal of e-waste may have devastating impacts on ecosystems and human well-being due to its complex mixing of toxic and non-toxic elements. Landfilling and incineration the two simplest and most conventional methods of e-waste disposal, exacerbate environmental degradation and resource depletion (Ramanayaka et al., 2019). As a result, investigating the possibilities of microbial-assisted nanobiotechnology to improve e-waste management procedures and address these urgent concerns is becoming more important. Taking into account an academic and scientific standpoint, e-waste is a treasure trove of various metals, polymers, and other substances that microbes may use for biotransformation and metabolic processes. The intricate composition of e-waste, encompassing heavy metals, rare earth elements, and organic contaminants, presents a dual spectrum of challenges and opportunities for microbial-mediated processes (Haghighatjoo & Mohammadi, 2020).

Microorganisms have distinct metabolic abilities that can be utilized for various processes, including bioleaching, biomineralization, and bioaccumulation (Ray et al., 2023). These processes enable the extraction of valuable metals from electronic components while minimizing adverse effects on biodiversity. In addition, the synthesis of nanomaterials using microbial enzymes and biomolecules allows for the exploration of innovative methods in e-waste recycling and resource recovery (Rienzie et al., 2019; Rana et al., 2020). Progress in biotechnology has opened up new possibilities for applying microbial nanobiotechnology to manage e-waste in innovative ways. By implementing genetic engineering techniques, microbial strains have the potential to be enhanced to possess improved abilities in metal mobilization, detoxification, and biodegradation. In addition, the incorporation of nanomaterial-based sensors and biosensors allows for the continuous monitoring of e-waste treatment processes, resulting in improved efficiency and control (Rienzie

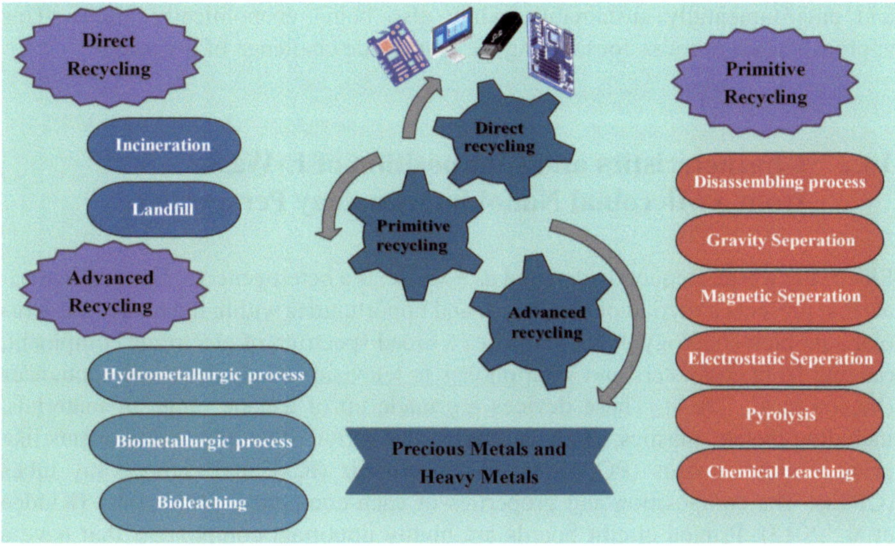

Fig. 10.1 Various conventional and non-conventional modes of E-waste recycling

et al., 2019). Incorporating microbial nanobiotechnology into e-waste management strategies shows potential for meeting sustainability goals (Pant et al., 2012).

Through the integration of biological processes and nanomaterial-based technologies, it is feasible to enhance resource recovery rates, mitigate environmental pollution, and minimize energy consumption in comparison to various conventional methods (Fig. 10.1). In addition, the utilization of microbial consortia and biofilms can improve the resilience and adaptability of e-waste treatment systems in various environmental conditions. The utilization of microbial nanobiotechnology for e-waste management has substantial environmental and economic ramifications (Yadav et al., 2022). This approach helps in controlling greenhouse gas emissions, conserving natural resources, and creating economic opportunities through the circular economy by reducing reliance on environmentally harmful practices and valorizing recovered materials. In the future, numerous challenges and opportunities await in the realm of microbial nanobiotechnology in managing e-waste (Pant et al., 2012). Factors like microbial diversity, process scalability, and regulatory frameworks will play a vital role in the widespread acceptance of these technologies. In order to drive innovation and tackle emerging challenges in this rapidly evolving field, it is crucial to foster interdisciplinary collaboration among microbiologists, nanotechnologists, environmental engineers, and policymakers (Ghulam & Abushammala, 2023). Ultimately, the utilization of microbial-assisted nanobiotechnology has the power to completely transform the way we are committed to addressing the issue of e-waste management, working towards a better tomorrow. Through the utilization of microorganisms and the exploitation of nanomaterials' distinct characteristics, researchers may contribute towards a solution that is both efficient and environmentally sustainable, while also being economically viable. This approach holds promise for tackling the worldwide challenge of e-waste.

10.2 Characteristics and Composition of E-Waste from a Microbial Nanobiotechnology Perspective

Electronic waste, frequently denoted as e-waste, is a heterogeneous blend of materials that presents a myriad of challenges and opportunities within the realm of microbial nanobiotechnology. It encompasses a broad spectrum of electronic equipment, ranging from computers and smartphones to televisions and household appliances (Varsha et al., 2024). These devices are made up of a wide range of materials, including metals, plastics, glass, ceramics, and various electronic components like printed circuit boards (PCBs), integrated circuits (ICs), and cathode ray tubes (CRTs). The composition and properties of each component are distinct (Kiddee et al., 2013). Printed circuit boards are highly important components that have a significant impact. They contain valuable metals like gold, silver, and copper, as well as potentially harmful substances like lead and mercury (Işıldar et al., 2018). The presence of heavy metals like cadmium, lead, and lithium-ion batteries commonly found in e-waste can pose a significant environmental risk if not handled with utmost care and caution. The release of heavy metals into the soil, water, and air can result in contamination, causing harmful effects on human health and wildlife. In addition, it is important to note that electronic waste may contain brominated flame retardants, such as polybrominated diphenyl ethers (PBDEs). These compounds are considered persistent organic pollutants because they can build up in the environment, posing risks to ecosystems and human health over time (Lithner et al., 2011). Informal recycling practices are common in many developing countries, where methods such as incineration or acid leaching are used to extract valuable materials from e-waste (Ramanayaka et al., 2019). These practices pose significant risks to workers, exposing them to harmful fumes and chemicals. Additionally, they contribute to the release of pollutants into the environment, worsening environmental degradation and endangering public health. Additionally, electronic waste frequently retains hazardous substances such as toxic chemicals, plasticizers, and long-chain hydrocarbons generated during the manufacturing process.

Viewed through the lens of microbial nanobiotechnology, the composition of e-waste reveals intriguing characteristics that capture attention. E-waste is deemed as a treasured repository of metals that can be efficiently extracted using microbial-driven mechanisms like bioleaching and biomineralization (Bindschedler et al., 2017). Microorganisms possess the enzymatic capability to mobilize metals from solid substrates, converting insoluble metal compounds into soluble forms that are readily extractable (Rana et al., 2020). This capability holds significant promise in resource recovery, particularly in the extraction of precious metals like gold, silver, copper, and palladium from electronic components, thereby reducing the reliance on environmentally harmful mining practices and promoting environmental sustainability (Cayumil et al., 2016; Sethurajan et al., 2019). Furthermore, the organic segment of e-waste serves as a substrate for microbial metabolism and biotransformation. Microorganisms harbor a diverse array of metabolic pathways that enable them to break down different types of organic pollutants found in electronic waste, including polychlorinated biphenyls (PCBs), plasticizers, and polycyclic aromatic hydrocarbons (PAHs) (Pant et al., 2012). Microbes take part in a pivotal setup in the degradation and remediation of complex biochemical compounds found in e-waste-contaminated environments. Through processes such as biodegradation and bioremediation, these microorganisms break down these compounds into simpler, less toxic forms, effectively detoxifying the environment (Zhang et al., 2010). Moreover, the heterogeneous nature of electronic waste provides conducive conditions for the proliferation and cooperative interactions of microbial communities. Microbial consortia comprise diverse populations of microorganisms with complementary metabolic abilities, allowing them to collectively degrade complex mixtures of pollutants and withstand fluctuating environmental conditions (Haghighatjoo and Mohammadi, 2020). By harnessing the intricate interplay among various microbial species, one can optimize the efficiency and resilience of microbial-based mechanisms for e-waste management. In a nutshell, when viewed from a microbial nanobiotechnology perspective, the attributes and composition of electronic waste underscore its potential as a valuable resource for sustainable material retrieval and environmental restoration. Through the utilization of microorganisms' metabolic capabilities and the unique properties of nanomaterials, microbial nanobiotechnology offers innovative solutions to address the complexities associated with e-waste, all while advocating for environmental sustainability and resource conservation (Ghulam and Abushammala, 2023). From Table 10.1, a clear view of the various probable bioremediation targets, which are generally toxic if left untreated in the environment, can be observed.

10.3 Biotechnological Applications in Microbial Nanobiotechnology for E-Waste Management

Biotechnological applications in microbial nanobiotechnology offer potential solutions for addressing electronic waste (e-waste), which is becoming an increasingly pressing global issue. E-waste, as we have already discussed, contains valuable metals and toxic substances that call for innovative and sustainable management strategies. By leveraging the power of genetic engineering and nanobiotechnology, scientists hope to significantly improve the efficiency and sustainability of metal recovery in microbial processes. Important biotechnological approaches hold the key in this field, these include biometallurgy along with its specific counterparts like biosorption, bioleaching, and bioaccumulation, a brief idea of which can be seen in Fig. 10.2.

Table 10.1 Various bioremediation targets in E-waste and their health impacts

Probable remediation targets	Common sources	Characteristics	Uses	Adverse health impacts	References
Plasticizers and long chain hydrocarbons					
Brominated flame retardants (BFRs)	Printed circuit boards and plastic housings	Generally classified into polybrominated biphenyls (PBBs), (TBBPA) tetrabromobisphenols, and polybrominated diphenyl ethers (PBDEs)	Used in the form of an insulating material to reduce flammability	This leads to risks of hormonal disorders caused by the inhalation of toxic vapours released because of its combustion	Kiddee et al. (2013)
Polyoxymethylene (POM)	High-performance engineering components	Has a high dielectric strength and a low dissipation factor	Provides electrical insulation and prevents electrical leakage	A notable carcinogen with skin corrosion potency	Lithner et al. (2011)
Chlorofluorocarbons (CFCs)	Refrigerators and air conditioners	It is chemically inert, has a high thermal stability, low toxicity, and non-flammable nature	As a refrigerant, aerosol, blowing agent, dielectric fluid, etc.	Heavily impacts ozone layer depletion which therefore leads to skin cancer. It can also lead to cases of cardiac sensitization	Tsai (2014)
PCB (Polychlorinated Biphenyls)	Transformers, Printed circuit boards, Heat transfer fluids and capacitors	Organochlorine compounds ($C_{12}H_{10}$−xClx)	Used as dielectric and coolant fluids for electrical apparatus	A probable carcinogenic and bioaccumulating agent with detrimental effects on the endocrine, reproductive, and immune systems of various animals	Ghulam and Abushammala (2023)
Polyacrylonitrile (PAN)	Water and air purifiers, super capacitors etc.	The powdered form has a conductivity of <10–10 S cm^{-1}	Widely used insulating material	A notable carcinogen and skin sensitization agent	Lithner et al. (2011), Khayyam et al. (2020), and Koutsonas (2017)
PVC (Polyvinyl chloride)	Computer accessories, cabling, and plastic housings	The latex of polyvinyl chloride is a viscous, white liquid with a % by mass emulsion of PVC in water	Used as a very high-quality sheathing, insulating, and bedding material	Often regarded as the most toxic plasticizer with high carcinogenicity and flammability	Kaczorek-Chrobak and Fangrat (2020) and Kiddee et al. (2013)

Heavy metals					
Chromium (Cr)	Metal housings, untreated galvanized steel plates, data tapes, floppy disks, decorative hardeners	Prevents surface oxidation **Weight (%):** –0.0063 **Recycling (%):** –0	For coating and plating e-appliances	Exposure to high amounts of it leads to neurodegenerative diseases, various types of cancers and ulcers, chromosomal abnormalities, multiple organ failures	Lunk (2015), Needhidasan et al. (2014), and Priya and Hait (2017)
Antimony (Sb)	CRT glass, plastic computer housings, infrared detectors, diodes and cables	When alloyed with lead or other metals it improves their hardness and strength	As a melting agent and solder alloy	It is a notable carcinogen. Prolonged Sb vapour exposure causes ulcers and stomach disorders	Kiddee et al. (2013) and Barragan et al. (2020)
Lead (Pb)	Acid batteries, solders, ceramic components, CRT, fluorescent tubes etc.	**Weight (%):** –6.2988 **Recycling (%):** –5	It is widely used for X-ray shielding, damp proofing, etc.	High levels of bioaccumulation lead to acute and chronic effects leading to neurodegenerative and reproductive disorders	Needhidasan et al. (2014)
Mercury (Hg)	Flat panel displays, thermostats, lamps, batteries etc.	It possesses remarkable properties, such as its density, uniform expansion and contraction with temperature variations, and impressive electrical conductivity.	Through the interaction of electric current and gas, the mercury produces ultraviolet (UV) light	Psychological disorders, affects the CNS, kidney etc. Bioaccumulation often has a fatal effect on foetuses	Pant et al. (2012), Rustagi and Singh (2010), and Priya and Hait (2017)
Cadmium (Cd)	Chip resistors, Rechargeable Ni-Cd batteries, printers, semiconductors etc.	The battery exhibits low internal resistance, allowing for high discharge rates without compromising its capacity. Additionally, it can be operated within a wide temperature range **Weight (%):** –0.0094 **Recycling (%):** –0	Also used in electroplating and in nuclear reactors where it acts as a neutron absorber.	Profound acute and chronic exposure effects ranging from oxidative damage, kidney failure to respiratory diseases	Genchi et al. (2020), Ghulam and Abushammala (2023), and Needhidasan et al. (2014)

(continued)

Table 10.1 (continued)

Probable remediation targets	Common sources	Characteristics	Uses	Adverse health impacts	References
Nickel (Ni)	CRT, Ni-Cd batteries, computer housing and printed circuit boards	Hasa the ability to improve storage capacity at minimal costs	Used in miniaturizing of batteries, armour plating, making of austenitic stainless steel etc.	A potent carcinogen with severe allergic reaction inducing capabilities	Padiyar (2011) and Pant et al. (2012)
Coinage metals					
Gold (Au)	PBCs of mobile phones, Computers, Transistors, etc.	Highly efficient conductor of electricity, remains corrosion-free free and effectively carries small voltages	Utilized in electroplating chemicals, manufacturing of gold bonding wire and sputter targets	High concentrations may lead to a decrease in body weight, reduced RBC count, speel index, etc.	Kim et al. (2011), Goodman (2002), and Zhang et al. (2010)
Copper (Cu)	TV board scraps, PBCs of mobile phones, Computers etc.	Recognized for its superior electrical conductivity and low level of resistivity	Widely utilized in the generation, transmission, and distribution of electrical and thermal energy, telecommunications, electronics circuitry manufacturing etc.	High doses lead to hepatic toxicity, acute gastrointestinal symptoms, adverse effects on cholesterol and glucose metabolism	Ramanayaka et al. (2019), Taylor et al. (2020), and Collins and Klevay (2011)

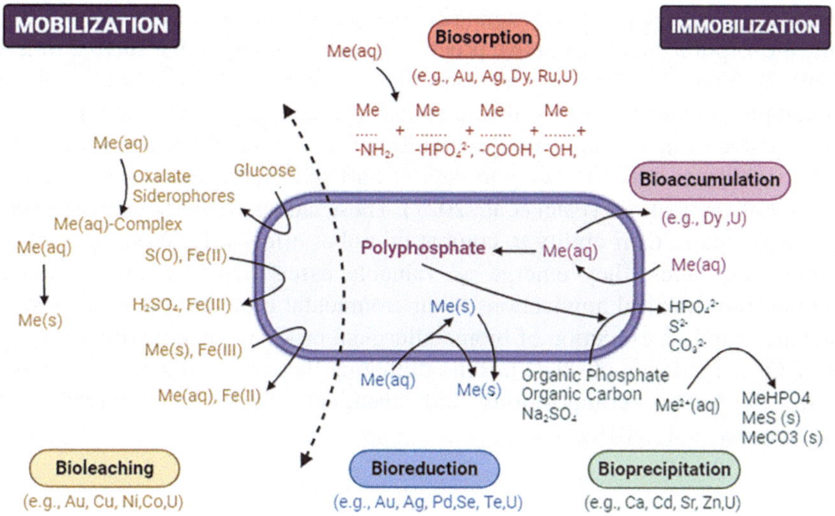

Fig. 10.2 Systematic representation of various microbial approaches for metal waste recovery

10.3.1 Biometallurgy

Biometallurgy seamlessly merges the intricacies of biological processes with the precision of metallurgical techniques (Hennebel et al., 2015). At its core, this interdisciplinary field harmonizes the innate abilities of microorganisms with the rich tapestry of metals and metal-bearing minerals, unlocking an array of promising avenues for metal extraction, recovery, and environmental remediation. Biomining, a foundational pillar of biometallurgy, orchestrates the remediation and extraction of metals from waste ore materials through bio-recovery utilizing the metabolic prowess of microbes (Han et al., 2022). Within microbial consortia, comprising a diverse array of bacteria and fungi, these organisms serve as potent biocatalysts, catalyzing the dissolution of metals from solid sources by transforming insoluble compounds into soluble forms (Krebs et al., 1997; Beolchini et al., 2012). Noteworthy among these microbial marvels are microalgae, demonstrating remarkable proficiency in both biosorption and bioaccumulation mechanisms, thus enhancing metal recovery efficiency, particularly in the context of challenging matrices such as electronic waste (Mukherjee et al., 2023). Another pivotal facet of biometallurgy lies in bioremediation, a noble endeavor aimed at addressing the remediation and mitigation of hazardous contaminants, including heavy metals and radionuclides, from polluted sites (Gadd, 2000). Microbial-mediated processes, exemplified by biosorption, furnish rapid and efficacious methodologies for metal removal from aqueous solutions, utilizing biological materials such as microorganisms and agricultural waste (Fomina and Gadd, 2014). Distinguishing between biosorption and bioaccumulation is paramount in unraveling the nuanced mechanisms underlying metal uptake by organisms over time (Zabochnicka-Świątek & Krzywonos, 2014). While biosorption entails a passive process of metal binding to the surface of living organisms, bioaccumulation demands metabolic energy for the active uptake and accumulation of substances, occurring either intracellularly or extracellularly.

The advent of metal nanoparticles resulting from biometallurgical processes has ignited significant interest, owing to their remarkable characteristics and the broad spectrum of applications (Patel et al., 2021). These nanoparticles boast unique properties, particularly their ability to combat microbes effectively. With inherent antimicrobial attributes, they emerge as valuable assets across multiple domains, spanning from medical applications to environmental protection efforts. However, achieving seamless execution of biometallurgical processes is not without its challenges. One significant obstacle lies in optimizing these processes to accommodate the diverse matrix compositions and chemical species of targeted metals (Ramanayaka et al., 2019).

10.3.2 Bioleaching

Bioleaching stands as a remarkable biohydrometallurgical process utilized in mining for the bio-recovery and remediation of economically important metals from substandard ores, offering a sustainable alternative to traditional extraction methods (Priya & Hait, 2017). By harnessing the metabolic activities of microorganisms, bioleaching converts insoluble metal compounds into soluble forms, facilitating their extraction from ores. As an example, the process of CuS oxidation to $CuSO_4$ emphasizes the bioleaching process, in which microorganisms play a vital role in aiding the dissolution of metallic elements (Rawlings, 2002). Within the field of electronic waste (e-waste) management, bioleaching has emerged as a promising approach for extracting economically valuable metals such as gold, silver, copper, lead, nickel, and more from discarded electronic devices. Several types of bacteria and fungi, such as *Chromobacterium violaceum*, *Pseudomonas species*, *Aspergillus species*, and *Penicillium simplicissimum* have been found to extract metals from e-waste by using organic acids or enzymes (Li et al., 2015; Dusengemungu et al., 2021). This approach to synthesis not only helps in recovering metals but also supports the sustainable use of resources. Nevertheless, the diverse makeup of e-waste presents obstacles to effective metal extraction through individual microorganisms. To address this limitation, scientists are investigating the potential of mixed cultures or microbial consortia, taking advantage of the beneficial interactions between different species of microorganisms (Valix, 2017; Jiang et al., 2017). Co-cultivating different types of microorganisms has yielded encouraging outcomes in improving the effectiveness of extracting metals from electronic waste. As shown in a study by Işıldar et al. (2016), a specific combination of microorganisms was found to be highly effective in extracting copper and gold from electronic waste. This combination included acidophilic bacteria like *Acidithiobacillus ferrivorans* and *Acidithiobacillus thiooxidans*, as well as cyanide-producing bacteria such as *Pseudomonas fluorescens* and *Pseudomonas putida*. In a recent study, researchers found a way to enhance the process of chromium recovery by using a combination of biosurfactant-producing bacteria and sulfur-oxidizing bacteria (Kaksonen et al., 2018).

Beyond metal recovery, bioleaching offers several advantages over conventional extraction methods, including the utilization of low-grade ores left behind by previous mining activities (Rawlings, 2002). Bacteria, owing to their small size, rapid growth, and adaptability, serve as effective agents for metal absorption. Fungal hyphae, with their extensive surface area-to-volume ratio, act as efficient accumulators of heavy metals (Fu et al., 2012). Moreover, microbes produce enzymatic complexes like laccase, which play a crucial role in detoxifying toxic compounds present in e-waste (Igiri et al., 2018).

In the field of microbial nanobiotechnology, microbial biocatalytic agents show great potential for efficient and environmentally friendly removal of heavy metal ions from e-waste, offering a cost-effective solution. These agents facilitate the biosorption of metal species inside cells, aiding in the leaching of different metals from e-waste. In addition, microbial nanomaterials have been implemented to neutralize environmental toxins, demonstrating the wide range of applications and cost-effectiveness of microbial nanobiotechnology in nanomanufacturing (Verma, 2017; Verma et al., 2020).

10.3.3 Biosorption

Biosorption emerges as an environmentally sustainable and economically viable mechanism for the removal of pollutants from effluents, especially those resistant to biodegradation, utilizing a diverse array of dead and living cells, alongside agricultural waste, as effective biosorbents. In the realm of e-waste degradation, the integration of microbial nanobiotechnology amplifies biosorption efficacy, achieved through immobilizing microbial cells onto nanoparticles. Nanoparticles exhibit distinctive physicochemical properties rendering them ideal for immobilizing microbes. This immobilization significantly enhances microbial activity, affording protection and fostering biomass degradation (Azeez et al., 2022). By serving as biomass support, nanoparticles elevate biosorption efficacy and selectivity, marking a paradigm shift in bioremediation strategies. The advantages of employing biosorption for e-waste degradation through microbial nanobiotechnology are manifold. Notably, biosorption demonstrates rapid contaminant removal, boasting high binding capacity across diverse pH and temperature conditions (Kiliç et al., 2008). Microorganisms can absorb heavy metals through different pathways, depending on the characteristics of their cellular compounds, contingent upon the uptake site as can be seen in Fig. 10.3. Moreover, immobilizing microbial cells onto nanoparticles confers mechanical robustness, shields against environmental rigors, and enhances stability and reusability (Gadd, 2009).

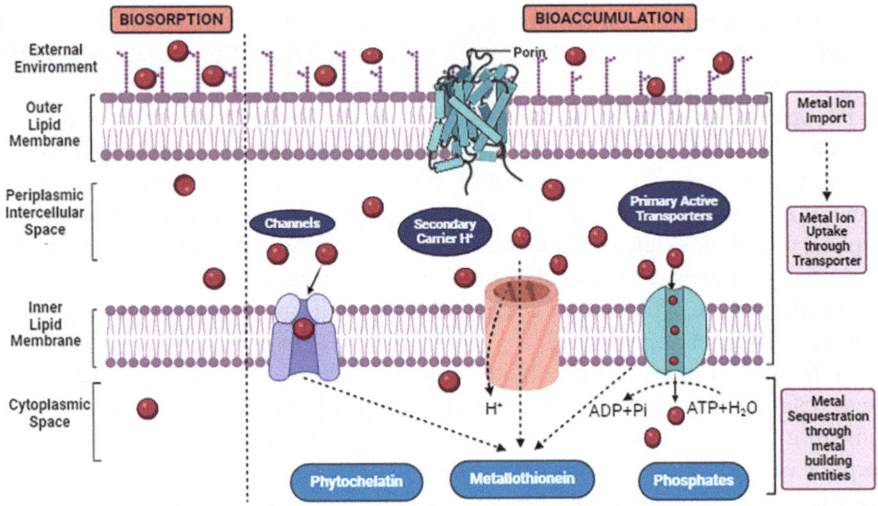

Fig. 10.3 Mechanism of biosorption and bioaccumulation

In a study conducted by Cai et al. (2011), *Magnetospirillum gryphiswaldense* MSR-1 was found to demonstrate an exceptional ability in the bio-transformation of gold from its ionic state to a zero-valency metal and then bio-accumulating it as spherical nanoparticles through biosorption. This groundbreaking research underscores the potential of microbial nanobiotechnology in mediating the transformation of heavy metal pollutants, offering a sustainable approach to e-waste management. Another study by Yang et al., (2021); introduced an innovative solution in the form of algal-bacterial aerobic granular sludge (AGS), demonstrating its effectiveness in chromium (VI) reduction and total removal under specific operational conditions. This novel tool highlights the versatility of microbial systems and underscores the importance of tailored approaches in addressing e-waste challenges. Furthermore, the study by Manasi et al. (2014); reported discovering a new bacterial strain, *Halomonas*, from electronic industry effluent. This bacterial strain exhibited significant utility in adsorbing cadmium, highlighting the potential of indigenous microbial communities in mitigating heavy metal pollution associated with e-waste disposal. Such findings underscore the importance of exploring microbial biodiversity and harnessing their innate capabilities for sustainable waste management practices. On the other hand, fungi have also emerged as promising biosorbents. A study elucidated the remarkable prowess of fungi in sequestering metal ions from aqueous solutions, suggesting a potential avenue for bioremediation strategies targeting e-waste contaminants. This research contributes to expanding our understanding of microbial diversity and their applications in environmental remediation (Bindschedler et al., 2017)

These seminal studies collectively underscore the pivotal role of microbial nanobiotechnology in e-waste degradation. By harnessing the unique capabilities of microorganisms and nanoparticles, researchers pave the way for innovative and sustainable solutions to mitigate the environmental impacts of e-waste accumulation.

10.3.4 Bioaccumulation

Bioaccumulation encompasses the active or passive absorption and accumulation of various substances, including heavy metals, within living organisms. Microbes exhibit distinctive metabolic pathways and surface characteristics that facilitate their specific metal accumulation (Fig. 10.3) from the surrounding environment (Pant et al., 2012). Biomass commonly contains metals as a protective mechanism against the harmful effects of metal toxicity. Certain intracellular proteins, specifically sulfur-containing peptides like metallothionein and phytochelatins, can form complexes with metals, and immobilize them (Bindschedler et al., 2017). This inherent capacity of microorganisms serves as the foundation for harnessing the potential of microbial-assisted nanobiotechnology in e-waste management. Here, nanomaterials play an essential role as carriers or enhancers for microbial bioaccumulation processes. Nanoparticles, such as iron oxide or carbon-based materials, have the potential to undergo alteration to optimize their ability to bind with metals or foster the attachment of microorganisms. Bioaccumulation mechanisms work better when nanomaterials are used as substrates for microbes to colonize and make it easier for metals to be absorbed. As instance, it has been frequently observed that elevated concentrations of Pb, Cd, Zn, and Hg are present in the carpophores of macrofungi in polluted areas (Bindschedler et al., 2017; Cayumil et al., 2016). Identifying and optimizing microbial strains with a significant capacity for metal accumulation is of utmost importance in scientific research. This entails the examination of microbial isolates from various environments, including locations affected by metal contamination, where they can then be subjected to genetic modification or adaptive evolution techniques to enhance their ability to absorb metals.

Controlled experiments have demonstrated the bioaccumulation of Ag in *Agaricus bisporus* fruiting bodies. Furthermore, certain fungal species have shown elevated levels of Au and Sb, suggesting a species-specific accumulation (Bindschedler et al., 2017). Thus, it becomes crucial to optimize the characteristics of nanomaterials to enhance their interaction with both microorganisms and electronic waste constituents. The surface modification of nanoparticles using ligands or biomolecules has the potential to improve selectivity for specific metals and facilitate microorganism attachment. Optimizing the bioaccumulation efficiency necessitates meticulous adjustment of process parameters, including pH, temperature, and nutrient availability. Bioreactor systems are utilized to regulate environmental conditions and expand bioaccumulation processes for real-world implementation. The application of microbial-assisted nanobiotechnology in e-waste management presents numerous advantages in terms of environmental sustainability (Yadav et al., 2022). First and foremost, it offers a cost-effective and environmentally conscious alternative to traditional approaches like incineration or landfilling. Utilizing natural biological processes, effectively mitigates the production of secondary pollutants and reduces energy consumption. In addition, the metals obtained from electronic waste can be recycled, thereby aiding in the preservation of limited resources and promoting a circular economy (Kiddee et al., 2013).

10.4 Enzymatic Interactions and Optimization Strategies Involving E-Waste Degradation

Enzymes serve as pivotal catalysts in the intricate process of enzymatic degradation, a crucial approach in managing the ever-growing challenge of e-waste. This degradation process unfolds through several sequential steps, beginning with the interaction between enzyme molecules and the polymer surfaces prevalent in e-waste. This interaction is finely regulated by various environmental factors such as temperature, pH levels, and the concentrations of enzymes and substrates in the surrounding medium (Mohanan et al., 2020). Once enzymes contact the polymer surface, they undergo adsorption, anchoring themselves onto the surface. This adsorption process, influenced by diverse factors including the polymer's surface properties and enzyme affinity, marks the initial stage of enzymatic degradation. Following adsorption, enzymes catalyze the breakdown of the polymer through a series of catalytic reactions. It is noteworthy that the efficiency of the adsorption process can be optimized by adjusting parameters such as enzyme concentration and polymer surface area, thereby enhancing the overall degradation of e-waste. The interaction between enzymes and nanomaterials constitutes another critical facet of e-waste degradation. This interaction profoundly influences the chemical and physical properties of both enzymes and nanomaterials (Castagnola et al., 2017; Zhao et al., 2015). Depending on the method of enzyme immobilization employed, whether physical or chemical bonding, distinct alterations may occur. For instance, physical bonding can induce conformational transitions in enzymes, impacting their catalytic activity. Enzyme orientation, influenced by factors like enzyme and nanomaterial properties, pH, and temperature, also plays a pivotal role in enzymatic reactions (Polyák et al., 2019).

Various pathways of interaction between enzymes and nanoparticles exist, with enzymes encapsulated or combined with nanoparticles for diverse functions, including nanozymes and biodegradation (Chen et al., 2017). Studies suggest that enzymes immobilized on nanoparticles with smaller diameters exhibit enhanced activity, as higher curvatures prevent unnecessary interactions and decrease enzyme denaturation (Asuri et al., 2006). Despite the vast potential of microbial nanobiotechnology in e-waste management, its commercialization remains a daunting challenge, with only a fraction realized thus far. To scale up microorganism-assisted nanotechnology for efficient e-waste management, continuous support from researchers and government funding is imperative (Shukla, 2020). In addition to enzyme-mediated processes, the fabrication of silver nanoparticles (AgNPs) from electronic waste offers a sustainable approach. This involves preparing an $AgNO_3$ solution from leaching agents like nitric acid and collecting silver from e-waste, providing an environmentally friendly method for nanoparticle synthesis. Recycled Ag NPs-coated fabric has demonstrated comparable antimicrobial properties to pure Ag NPs-coated fabrics, indicating the potential for green-synthesized Ag NPs as alternative antimicrobial agents in various industries (Mondal et al., 2023). E-waste recycling is paramount for mitigating environmental pollution arising from discarded electronic devices. Enzymes, widely utilized in industrial processes, have garnered attention for their applications in environmental remediation, particularly in the development of enzyme-loaded nanomaterials (Qamar et al., 2022). These advancements hold promise to revolutionize the way of regulating e-waste, leading to a more sustainable and environmentally friendly approach, thereby minimizing its detrimental impact on the environment. Furthermore, specific enzyme pathways contribute significantly to e-waste degradation. For instance, PETase facilitates the hydrolysis of polyethylene terephthalate (PET), a common material found in e-waste, while proteases play a role in breaking down protein residues resistant to degradation, such as insoluble proteins. These enzyme-mediated processes contribute to reducing the toxicity of e-waste and facilitating its recycling (Janeeshma et al., 2024; Bittencourt et al., 2023).

10.5 Environmental and Economic Implications of Microbial Nanobiotechnology in E-Waste Management

Potentially, microbial nanobiotechnology can help address the growing issue of e-waste worldwide. The number of electronics that are thrown away is expected to rise due to the exponential expansion of electronic gadgets and is predicted to reach 44.4 million metric tons in the next 5 years. Effective e-waste recycling is therefore becoming more and more necessary, and analysts project that the global market will reach $65.8 billion by 2026 (Seif et al., 2024). Traditional e-waste recycling methods often involve hazardous chemicals and high energy consumption, contributing to environmental degradation and health risks. In contrast, microbial nanobiotechnology provides an economical and environmentally friendly substitute by using genetically modified microorganisms to extract useful components from electronic trash (Han et al., 2022). The distinctive properties of nanoparticles, whose nanostructures range in size from 0.1 to 100 nm, are utilized by these technologies. As a result, their surface area to volume ratio is large and their materials acquire many advantageous qualities (Mondal et al., 2023). Mint Innovation, a biometallurgy firm based in Auckland, New Zealand, has undertaken pioneering research on applying microbial nanobiotechnology in e-waste recycling. The methodology used by Mint Innovation incorporates biological and chemical cleanup processes within their procedure. Whereby treating grounded e-waste with acids and oxidants is the first step in the process (Hubau & Bryan, 2022). While rich metals like gold and palladium are left behind in the solid residue, this technique aids in the extraction of less valuable metals like copper and tin. After extracting gold ions through another cycle of leaching, the bacteria *Cupriavidus metallidurans* is introduced to the extracted gold ions.

Moreover, researchers demonstrated the efficacy of biotic-abiotic co-assemblies in e-waste remediation. These assemblies comprise synthetic bacterial cells engineered to display heavy-metal-capturing proteins on their surfaces, coupled with magnetic nanoparticles. This novel method allows for the effective extraction of harmful heavy metals, like cadmium and lead, from electronic waste, thus reducing the risk of environmental pollution (Zhu et al., 2020). Although there are still some obstacles, the field of microbial nanobiotechnology shows great promise for e-waste recycling. One significant challenge lies in a lack of studies that explore the transition from successful laboratory experiments to large-scale industrial applications. This transition is crucial for effectively implementing e-waste bioremediation on a commercial level. In addition, it is essential to prioritize the development of molecular toolkits that can assist in genetically modifying less prevalent microorganisms that have the potential for e-waste recycling (Han et al., 2022). A comprehensive examination has been conducted on e-waste management, delving into various approaches including Life Cycle Assessment (LCA), Material Flow Analysis (MFA), Multi-Criteria Analysis (MCA), and Extended Producer Responsibility (EPR). Developed countries have made significant progress in managing e-waste by implementing directives like the Waste Electrical and Electronic Equipment (WEEE) directive. This regulation emphasizes the importance of reducing waste disposal and improving environmental standards (EU, 2002). Research efforts focus on component separation for recycling and recovery of rare and precious metals, contributing to sustainable e-waste management practices. These abovementioned approaches are further described in detail in Table 10.2.

10.6 Recent Updates and Findings on the Nano-bioengineering Front About E-Waste Remediation

According to contemporary research, the quantity of electronic waste produced in 2019 exceeded 50 million metric tons and predicted that this number will rise to 74 million metric tons by 2030 (Al-Salem et al., 2022). The rise in electronic waste generation demands urgent and effective waste management solutions, considering the potential risks to the environment and public health (Wu et al., 2018; Wang et al., 2021). When it comes to the bioremediation of e-waste, certain microorganisms possess a natural ability to break down specific components of electronic waste. Through the utilization of high-throughput genetic tools in synthetic biology,

Table 10.2 Notable approaches dealing with sustainable eco-friendly e-waste management

Approaches	Characteristic features	Target domain	Regulatory principles	Practical applications	References
Life Cycle Assessment (LCA)	Reveals the environmental footprint of products from inception to disposal.	Environmental sustainability	Drives policy innovation by exposing eco-impact hotspots	Guides decisions on green product design and sustainable production methods.	Withanage and Habib (2021) and Wagh et al. (2024)
Material Flow Analysis (MFA)	Maps the intricate journey of materials, spotlighting resource circulation.	Resource efficiency	Propels policy evolution by identifying material flow optimizations	Identifies avenues for resource recovery and minimizes waste in e-waste recycling.	Azizi et al. (2023)
Multi-Criteria Analysis (MCA)	Weighs diverse factors, from economic to ecological, for informed decision-making.	Decision optimization	Sparks policy creativity by integrating varied stakeholder perspectives	Prioritizes e-waste interventions based on a spectrum of criteria.	Pažek et al. (2020) and Kiddee et al. (2013)
Extended Producer Responsibility (EPR)	Mandates producers to own the lifecycle impact of their creations.	Producer responsibility	Fuels policy creativity by incentivizing sustainable product stewardship	Catalyses eco-conscious product design and fosters robust recycling infrastructure.	Patra and Mishra (2024)

there is potential to improve these capabilities and make significant progress in bioremediation and resource recovery (Han et al., 2022). In a recent study, Saad et al. (2022) successfully devised a highly effective approach to produce selenium nanoparticles. They utilized the bacterium *Bacillus subtilis* AS12 to demonstrate the promising capabilities of bacterial-mediated biosynthesis in stabilizing and evaluating the nanoparticles' resistance to heavy metal accumulation.

Recent studies have shown that bacteria and fungi have shown exceptional promise in tackling the issues presented by e-waste, thanks to their impressive abilities and distinct physiological characteristics. Certain types of bacteria, like *Pseudomonas aeruginosa* and *Streptomyces spp.*, have gained recognition for their ability to effectively absorb substances. This is due to their small size, fast growth in controlled environments, and their ability to adapt to different environmental conditions. As an example, bacterial microbes have a wide range of physiological characteristics and often rely on metal ions as cofactors. This makes them highly effective in absorbing metals from electronic waste. Because of their special qualities, these bacteria can break down oxidoreductase enzyme complexes, such as the laccase enzyme complex. Prokaryotic bacteria (such as *Pseudomonas aeruginosa* and *Streptomyces spp.*) and eukaryotic fungi (like *Pleurotus florida*, *Trametes versicolor*, and *Aspergillus niger*) both include this enzyme system. Because of its substrate non-specificity, the laccase enzyme system can help in the detoxification of hazardous substances by oxidative coupling an enzyme called laccase, which is essential for removing and recovering metals from waste printed circuit boards (WPCBs), is produced more readily when electronic waste containing metals is introduced (Kaur et al., 2022; Igiri et al., 2018).

On the other hand, fungi provide a lot of benefits when it comes to cleaning up e-waste. They can withstand high metal concentrations, have a variety of survival strategies, work well with biomass, and produce more nanoparticles from waste printed circuit boards (WPCBs) with greater productivity (Šebesta et al., 2022). Notably, because of their remarkable surface area-to-volume ratio, fungal hyphae can effectively deposit heavy metals (Fu et al., 2012). Consequently, these microbes play a critical role in the biosorption process, which is used to remove different metals from e-waste. This entails the internal absorption of metal species within the cell, assisting in the economical and environmentally responsible removal of heavy metal ions from e-waste. Although microbial nanotechnology is still in its initial stages, it has opened up a world of possibilities for future research and development. For example, studies show how microorganism-driven nanotechnology might enhance agri-food industry operations by providing green solutions for a sustainable environment by neutralizing and detoxifying harmful waste materials (Verma et al., 2023). Algae are another species that is well-known for their contribution to the bio-reduction of metals, which makes it easier to produce nanoparticles of silver and gold. Microbe-gold interactions are influenced by the polarity of the microbial surface, which is determined by the presence of specific functional groups like carboxyl, hydroxyl, and amine groups. This ultimately determines how any bioleaching process turns out. By slightly harming the microbial cells, the integration of biological processes with a chemical agent such as polyvinylpyrrolidone was discovered to reduce these interactions and, consequently, the diffusion control between microorganisms and metal ions (Pant et al., 2012). By applying bioremediation techniques, the field of microbial nanobiotechnology has significant potential in resolving the problem of managing e-waste. This technique reduces the ecological footprint of electronic waste by using microbes to break down or transform potentially harmful components of the waste into less harmful materials (Ayilara & Babalola, 2023). However, because of its large volume, intricate componentry, and related toxicity, managing e-waste poses significant hurdles. Before emerging laboratory-scale bioremediation successes can be expanded for commercial use in recycling e-waste, a number of issues must be resolved. The complexity and toxicity of e-waste, which can impede the growth and activity of microorganisms, provide one of the main obstacles to e-waste bioremediation. Recent studies have, however, improved the ability of microorganisms to withstand the toxicity of e-waste and increased the effectiveness of bioleaching for the recovery of metals. In terms of managing e-waste, the combination of nanotechnology and microbes has demonstrated potential. Enzymes and microorganisms can be made more active by nanomaterials, which increases their capacity to degrade contaminants and extract useful materials from e-waste (Malik et al., 2022).

10.7 Conclusion and Future Prospects

In a nutshell, the integration of microbial-assisted nanobiotechnology offers a promising solution to address the continuously expanding problem of electronic waste management, all while promoting a greener and more sustainable future. This approach leverages the intricate interplay between microbial consortia and nanomaterials to offer a holistic solution for mitigating the adverse environmental impacts of electronic waste disposal. This chapter highlights the effectiveness of using microbial-assisted nanobiotechnology to efficiently break down and remediate complex e-waste materials, reducing the release of hazardous substances into the environment. In addition, the combination of microbial processes and nanomaterial-based platforms allows for the extraction of precious metals and components from electronic devices, in line with the principles of conserving resources and promoting a circular economy. Although there are exciting possibilities, there are still obstacles and constraints that require more research and improvement. Certain challenges need to be addressed to fully implement microbial-assisted nanobiotechnology in industrial environments, including scalability and cost-effectiveness. In addition, a thorough evaluation and optimization of microbial strains, coupled with the meticulous planning of bioreactor systems, are vital to attain optimal performance and reproducibility. To move forward, it ought to be a priority for future research to focus on tackling these obstacles while also exploring new possibilities for improving the effectiveness and practicality of microbial-assisted nanobiotechnology in the discipline of e-waste management. It involves the design and development of innovative bioreactor designs, the study of resilient microbial communities, and the investigation of how microbial and nanomaterial components can work together harmoniously. Furthermore, the effective application of microbial-assisted nanobiotechnology relies on the cooperation of academia, industry, and regulatory agencies. Developing effective legislative frameworks and regulatory guidelines is crucial to ensuring the safe and sustainable implementation of these technologies worldwide. Proper scientific education with public awareness initiatives plays an essential role in fostering an attitude of responsible consumption and waste management practices. By involving stakeholders and increasing awareness about the ecological consequences of electronic waste, people have the potential to instigate a paradigm shift in favour of more sustainable patterns of consumption and approaches to waste management. Ultimately, the utilization of microbial-assisted nanobiotechnology presents a remarkable opportunity to revolutionize e-waste management, paving the way for a more sustainable environment and the preservation of valuable resources. By employing a combination of scientific exploration, imaginative thinking, and cooperative efforts, we can fully unlock the capabilities of this technology to tackle one of the foremost critical environmental issues of our era.

References

Al-Salem, S. M., Leeke, G. A., El-Eskandarany, M. S., Van Haute, M., Constantinou, A., Dewil, R., & Baeyens, J. (2022). On the implementation of the circular economy route for e-waste management: A critical review and analysis for the case of the State of Kuwait. *Journal of Environmental Management, 323*, 116181.

Asuri, P., Bale, S. S., Karajanagi, S. S., & Kane, R. S. (2006). The protein-nanomaterial interface. *Current Opinion in Biotechnology, 17*(6), 562–568. https://doi.org/10.1016/j.copbio.2006.09.002

Ayilara, M. S., & Babalola, O. O. (2023). Bioremediation of environmental wastes: The role of microorganisms. *Frontiers in Agronomy, 5*. https://doi.org/10.3389/fagro.2023.1183691. Frontiers Media S.A.

Azeez, R., Hasan, M. M., Alawi, N. M., & Nguyen, H. M. (2022). A review study of immobilized microbial-nanoparticles: Techniques and biotechnology applications. *Iraqi Journal of Oil and Gas Research (IJOGR), 2*(2), 87–101. https://doi.org/10.55699/ijogr.2022.0202.1030

Azizi, D. D. S., Hanafiah, M. M., & Woon, K. S. (2023). Material flow analysis in WEEE management for circular economy: A content review on applications, limitations, and future outlook. *Sustainability (Switzerland), 15*(4). https://doi.org/10.3390/su15043505. MDPI.

Barragan, J. A., Ponce De León, C., Alemán Castro, J. R., Peregrina-Lucano, A., Gómez-Zamudio, F., & Larios-Durán, E. R. (2020). Copper and antimony recovery from electronic waste by hydrometallurgical and electrochemical techniques. *ACS Omega, 5*(21), 12355–12363. https://doi.org/10.1021/acsomega.0c01100

Beolchini, F., Fonti, V., Dell'Anno, A., Rocchetti, L., & Vegliò, F. (2012). Assessment of biotechnological strategies for the valorization of metal bearing wastes. *Waste Management, 32*(5), 949–956. https://doi.org/10.1016/j.wasman.2011.10.014

Bindschedler, S., Vu Bouquet, T. Q. T., Job, D., Joseph, E., & Junier, P. (2017). Fungal biorecovery of gold from e-waste. *Advances in Applied Microbiology, 99*, 53–81. https://doi.org/10.1016/bs.aambs.2017.02.002

Bittencourt, G. A., Porto De Souza Vandenberghe, L., José Martínez-Burgos, W., Kley Valladares-Diestra, K., Fá Tima Murawski De Mello, A., Maske, B. L., Brar, S. K., Varjani, S., Vinicius, G., Pereira, M., & Soccol, C. R. (2023). Emerging contaminants bioremediation by enzyme and nanozyme-based processes-A review. https://doi.org/10.1016/j.isci

Cai, F., Li, J., Sun, J., & Ji, Y. (2011). Biosynthesis of gold nanoparticles by biosorption using Magnetospirillum gryphiswaldense MSR-1. *Chemical Engineering Journal, 175*(1), 70–75. https://doi.org/10.1016/j.cej.2011.09.041

Castagnola, V., Cookman, J., de Araújo, J. M., Polo, E., Cai, Q., Silveira, C. P., Krpetić, Yan, Y., Boselli, L., & Dawson, K. A. (2017). Towards a classification strategy for complex nanostructures. *Nanoscale Horizons, 2*(4), 187–198. https://doi.org/10.1039/c6nh00219f. Royal Society of Chemistry.

Cayumil, R., Khanna, R., Rajarao, R., Mukherjee, P. S., & Sahajwalla, V. (2016). Concentration of precious metals during their recovery from electronic waste. *Waste Management, 57*, 121–130. https://doi.org/10.1016/j.wasman.2015.12.004

Chen, M., Zeng, G., Xu, P., Lai, C., & Tang, L. (2017). How do enzymes 'meet' nanoparticles and nanomaterials? *Trends in Biochemical Sciences, 42*(11), 914–930. https://doi.org/10.1016/j.tibs.2017.08.008. Elsevier Ltd.

Collins, J. F., & Klevay, L. M. (2011). Copper. *Advances in Nutrition, 2*(6), 520–522. https://doi.org/10.3945/an.111.001222

Dusengemungu, L., Kasali, G., Gwanama, C., & Mubemba, B. (2021). Overview of fungal bioleaching of metals. *Environmental Advances, 5*, 100083. https://doi.org/10.1016/j.envadv.2021.100083

EU. (2002). Directive 2002/96/EC of the European parliament and of the council of 27 January 2003 on waste electrical and electronic equipment (WEEE). *Official Journal of the European Union (Ed.), L037*, 0024-39.

Fomina, M., & Gadd, G. M. (2014). Biosorption: Current perspectives on concept, definition and application. *Bioresource Technology, 160*, 3–14. https://doi.org/10.1016/j.biortech.2013.12.102

Fu, Y. Q., Li, S., Zhu, H. Y., Jiang, R., & Yin, L. F. (2012). Biosorption of copper (II) from aqueous solution by mycelial pellets of Rhizopus oryzae. *African Journal of Biotechnology, 11*, 1403–1411.

Gadd, G. M. (2000). Bioremedial potential of microbial mechanisms of metal mobilization and immobilization. *Current Opinion in Biotechnology, 11*(3), 271–279. https://doi.org/10.1016/s0958-1669(00)00095-1

Gadd, G. M. (2009). Biosorption: Critical review of scientific rationale, environmental importance and significance for pollution treatment. *Journal of Chemical Technology and Biotechnology, 84*, 13–28.

Genchi, G., Sinicropi, M. S., Lauria, G., Carocci, A., & Catalano, A. (2020). The effects of cadmium toxicity. *International Journal of Environmental Research and Public Health, 17*(11). https://doi.org/10.3390/ijerph17113782. MDPI AG.

Ghulam, S. T., & Abushammala, H. (2023). Challenges and opportunities in the management of electronic waste and its impact on human health and environment. *Sustainability, 15*(3), 1837.

Goodman, P. (2002). Current and future uses of gold in electronics. *Gold Bulletin, 35*.

Haghighatjoo, M., & Mohammadi, A. (2020). Bioleaching for nanoparticle production as a solution for environmental pollution. *Journal of Biosafety, 13*(3), 61–70.

Han, P., Teo, W. Z., & Yew, W. S. (2022). Biologically engineered microbes for bioremediation of electronic waste: Wayposts, challenges and future directions. *Engineering Biology, 6*(1), 23–34. https://doi.org/10.1049/enb2.12020

Hennebel, T., Boon, N., Maes, S., & Lenz, M. (2015). Biotechnologies for critical raw material recovery from primary and secondary sources: R&D priorities and future perspectives. *New Biotechnology, 32*(1), 121–127. https://doi.org/10.1016/j.nbt.2013.08.004

Hubau, A., & Bryan, C. G. (2022). Metal recovery from e-wastes. In *Biomining technologies: Extracting and recovering metals from ores and wastes* (pp. 239–259). Springer.

Igiri, B. E., Okoduwa, S. I. R., Idoko, G. O., Akabuogu, E. P., Adeyi, A. O., & Ejiogu, I. K. (2018). Toxicity and bioremediation of heavy metals contaminated ecosystem from tannery wastewater: A review. *Journal of Toxicology, 2018*. https://doi.org/10.1155/2018/2568038. Hindawi Limited.

Işıldar, A., van de Vossenberg, J., Rene, E. R., van Hullebusch, E. D., & Lens, P. N. L. (2016). Two-step bioleaching of copper and gold from discarded printed circuit boards (PCB). *Waste Management, 57*, 149–157. https://doi.org/10.1016/j.wasman.2015.11.033

Işıldar, A., Rene, E. R., van Hullebusch, E. D., & Lens, P. N. L. (2018). Electronic waste as a secondary source of critical metals: Management and recovery technologies. *Resources, Conservation and Recycling, 135*, 296–312. https://doi.org/10.1016/j.resconrec.2017.07.031

Janeeshma, E., Habeeb, H., Sinha, S., Arora, P., Chattaraj, S., das Mohapatra, P. K., Panneerselvam, P., & Mitra, D. (2024). Enzymes-mediated solid waste management: A sustainable practice for recycling. *Waste Management Bulletin, 1*(4), 104–113. https://doi.org/10.1016/j.wmb.2023.10.007

Jiang, L. L., Zhou, J. J., Quan, C. S., & Xiu, Z. L. (2017). Advances in industrial microbiome based on microbial consortium for biorefinery. *Bioresources and Bioprocessing, 4*(1). https://doi.org/10.1186/s40643-017-0141-0. Springer Science and Business Media Deutschland GmbH.

Kaczorek-Chrobak, K., & Fangrat, J. (2020). PVC-based copper electric wires under various fire conditions: Toxicity of fire effluents. *Materials, 13*(5). https://doi.org/10.3390/ma13051111

Kaksonen, A. H., Boxall, N. J., Gumulya, Y., Khaleque, H. N., Morris, C., Bohu, T., Cheng, K. Y., Usher, K. M., & Lakaniemi, A. M. (2018). Recent progress in biohydrometallurgy and microbial characterisation. *Hydrometallurgy, 180*, 7–25. https://doi.org/10.1016/j.hydromet.2018.06.018

Kaur, P., Sharma, S., Albarakaty, F. M., Kalia, A., Hassan, M. M., & Abd-Elsalam, K. A. (2022). Biosorption and bioleaching of heavy metals from electronic waste varied with microbial genera. *Sustainability (Switzerland), 14*(2). https://doi.org/10.3390/su14020935

Khayyam, H., Jazar, R. N., Nunna, S., Golkarnarenji, G., Badii, K., Fakhrhoseini, S. M., Kumar, S., & Naebe, M. (2020). PAN precursor fabrication, applications and thermal stabilization process in carbon fiber production: Experimental and mathematical modelling. *Progress in Materials Science, 107*. https://doi.org/10.1016/j.pmatsci.2019.100575. Elsevier Ltd.

Kiddee, P., Naidu, R., & Wong, M. H. (2013). Electronic waste management approaches: An overview. *Waste Management, 33*(5), 1237–1250. https://doi.org/10.1016/j.wasman.2013.01.006

Kiliç, M., Keskin, M. E., Mazlum, S., & Mazlum, N. (2008). Effect of conditioning for Pb(II) and hg(II) biosorption on waste activated sludge. *Chemical Engineering and Processing: Process Intensification, 47*(1), 31–40. https://doi.org/10.1016/j.cep.2007.07.019

Kim, E. Y., Kim, M. S., Lee, J. C., & Pandey, B. D. (2011). Selective recovery of gold from waste mobile phone PCBs by hydrometallurgical process. *Journal of Hazardous Materials, 198*, 206–215. https://doi.org/10.1016/j.jhazmat.2011.10.034

Koutsonas, S. (2017). Electrical conductivity of degraded polyacrylonitrile powder by microwave irradiation for supercapacitor devices or other mobile applications. *Materials Letters, 193*, 203–205. https://doi.org/10.1016/j.matlet.2017.02.001

Krebs, W., Brombacher, C., Bosshard, P. P., Bachofen, R., & Brandl, H. (1997). Microbial recovery of metals from solids. *FEMS Microbiology Reviews, 20*(3–4), 605–617. https://doi.org/10.1111/j.1574-6976.1997.tb00341.x

Li, J., Liang, C., & Ma, C. (2015). Bioleaching of gold from waste printed circuit boards by Chromobacterium violaceum. *Journal of Material Cycles and Waste Management, 17*(3), 529–539. https://doi.org/10.1007/s10163-014-0276-4

Lithner, D., Larsson, A., & Dave, G. (2011). Environmental and health hazard ranking and assessment of plastic polymers based on chemical composition. *Science of the Total Environment, 409*(18), 3309–3324. 10.1016/j.scitotenv.2011.04.038

Lunk, H. J. (2015). Discovery, properties and applications of chromium and its compounds. *ChemTexts, 1*(1). https://doi.org/10.1007/s40828-015-0007-z

Malik, S., Dhasmana, A., Preetam, S., Mishra, Y. K., Chaudhary, V., Bera, S. P., Ranjan, A., Bora, J., Kaushik, A., Minkina, T., Jatav, H. S., Singh, R. K., & Rajput, V. D. (2022). Exploring microbial-based green nanobiotechnology for wastewater remediation: A sustainable strategy. *Nanomaterials, 12*(23). https://doi.org/10.3390/nano12234187. MDPI.

Manasi, Rajesh, V., Santhana, A., Kumar, K., & Rajesh, N. (2014). Biosorption of cadmium using a novel bacterium isolated from an electronic industry effluent. *Chemical Engineering Journal, 235*, 176–185. https://doi.org/10.1016/j.cej.2013.09.016

Mohanan, N., Montazer, Z., Sharma, P. K., & Levin, D. B. (2020). Microbial and enzymatic degradation of synthetic plastics. *Frontiers in Microbiology, 11*. https://doi.org/10.3389/fmicb.2020.580709. Frontiers Media S.A.

Mondal, M. S., Paul, A., & Rhaman, M. (2023). Recycling of silver nanoparticles from electronic waste via green synthesis and application of AgNPs-chitosan based nanocomposite on textile material. *Scientific Reports, 13*(1). https://doi.org/10.1038/s41598-023-40668-7

Mukherjee, S., Paul, S., Bhattacharjee, S., Nath, S., Sharma, U., & Paul, S. (2023). Bioleaching of critical metals using microalgae. *AIMS Environmental Science, 10*(2), 226–244. https://doi.org/10.3934/environsci.2023013. American Institute of Mathematical Sciences.

Needhidasan, S., Samuel, M., & Chidambaram, R. (2014). Electronic waste—An emerging threat to the environment of urban India. *Journal of Environmental Health Science and Engineering, 12*(1). https://doi.org/10.1186/2052-336X-12-36. BioMed Central Ltd.

Padiyar, N. (2011). Nickel allergy - is it a cause of concern in everyday dental practice? *International Journal of Contemporary Dentistry, 2*(1), 12.

Pant, D., Joshi, D., Upreti, M. K., & Kotnala, R. K. (2012). Chemical and biological extraction of metals present in E waste: A hybrid technology. *Waste Management, 32*(5), 979–990. https://doi.org/10.1016/j.wasman.2011.12.002

Patel, A., Enman, J., Gulkova, A., Guntoro, P. I., Dutkiewicz, A., Ghorbani, Y., Rova, U., Christakopoulos, P., & Matsakas, L. (2021). Integrating biometallurgical recovery of metals with biogenic synthesis of nanoparticles. *Chemosphere, 263*. https://doi.org/10.1016/j.chemosphere.2020.128306

Patra, B. C., & Mishra, D. N. (2024). Extended producer responsibility (EPR) and E-waste management: A roadmap for entrepreneurs. In *Sustainable solutions for E-waste and development* (pp. 136–145). IGI Global.

Pažek, K., Prišenk, J., Bukovski, S., Prevolšek, B., & Rozman, C. (2020). Multicriteria assessment of the quality of waste sorting centers—A case study. *Sustainability (Switzerland), 12*(9). https://doi.org/10.3390/su12093780

Polyák, P., Urbán, E., Nagy, G. N., Vértessy, B. G., & Pukánszky, B. (2019). The role of enzyme adsorption in the enzymatic degradation of an aliphatic polyester. *Enzyme and Microbial Technology, 120*, 110–116. https://doi.org/10.1016/j.enzmictec.2018.10.005

Priya, A., & Hait, S. (2017). Comparative assessment of metallurgical recovery of metals from electronic waste with special emphasis on bioleaching. *Environmental Science and Pollution Research, 24*(8), 6989–7008. https://doi.org/10.1007/s11356-016-8313-6

Qamar, M., Basharat, A., Qamar, S. A., Bilal, M., Franco, M., & Iqbal, H. M. (2022). Enzyme-loaded nanostructured materials for the degradation of environmental pollutants. *Current Opinion in Environmental Science & Health, 30*, 100400. https://doi.org/10.1016/j.coesh.2022.100400

Ramanayaka, S., Keerthanan, S., & Vithanage, M. (2019). Urban mining of E-waste: Treasure hunting for precious nanometals. In *Handbook of electronic waste management: International best practices and case studies* (pp. 19–54). Elsevier. https://doi.org/10.1016/B978-0-12-817030-4.00023-1

Rana, S., Mishra, P., Wahid, Z., Thakur, S., Pant, D., & Singh, L. (2020). Microbe-mediated sustainable bio-recovery of gold from low-grade precious solid waste: A microbiological overview. *Journal of Environmental Sciences (China), 89*, 47–64. https://doi.org/10.1016/j.jes.2019.09.023. Chinese Academy of Sciences.

Rawlings, D. E. (2002). Heavy metal mining using microbes. *Annual Review of Microbiology, 56*, 65–91. https://doi.org/10.1146/annurev.micro.56.012302.161052

Ray, M. K., Panda, J., Panda, B. P., Mohanta, T. K., & Mohanta, Y. K. (2023). *Mycoremediation of heavy metals and/or metalloids in soil BT—Land remediation and management: bioengineering strategies* (H. Sarma & S. Joshi, Eds.) (pp. 161–190). Springer Nature. https://doi.org/10.1007/978-981-99-4221-3_8

Rienzie, R., Perera, A. T. D., & Adassooriya, N. M. (2019). Biorecovery of precious metal nanoparticles from waste electrical and electronic equipments. In *Electronic waste management and treatment technology* (pp. 133–152). Elsevier. https://doi.org/10.1016/B978-0-12-816190-6.00006-6

Rustagi, N., & Singh, R. (2010). Mercury and health care. *Indian Journal of Occupational and Environmental Medicine, 14*(2), 45–48. https://doi.org/10.4103/0019-5278.72240

Saad, A. M., Sitohy, M. Z., Sultan-Alolama, M. I., El-Tarabily, K. A., & El-Saadony, M. T. (2022). Green nanotechnology for controlling bacterial load and heavy metal accumulation in Nile tilapia fish using biological selenium nanoparticles biosynthesized by Bacillus subtilis AS12. *Frontiers in Microbiology, 13*. https://doi.org/10.3389/fmicb.2022.1015613

Šebesta, M., Vojtková, H., Cyprichová, V., Ingle, A. P., Urík, M., & Kolenčík, M. (2022). Mycosynthesis of metal-containing nanoparticles-synthesis by ascomycetes and basidiomycetes and their application. *International Journal of Molecular Sciences, 24*(1), 304. https://doi.org/10.3390/ijms24010304

Seif, R., Salem, F. Z., & Allam, N. K. (2024). E-waste recycled materials as efficient catalysts for renewable energy technologies and better environmental sustainability. *Environment, Development and Sustainability, 26*, 5473–5508. https://doi.org/10.1007/s10668-023-02925-7

Sethurajan, M., van Hullebusch, E. D., Fontana, D., Akcil, A., Deveci, H., Batinic, B., Leal, J. P., Gasche, T. A., Ali Kucuker, M., Kuchta, K., Neto, I. F. F., Soares, H. M. V. M., & Chmielarz, A. (2019). Recent advances on hydrometallurgical recovery of critical and precious elements from end of life electronic wastes—A review. *Critical Reviews in Environmental Science and Technology, 49*(3), 212–275. https://doi.org/10.1080/10643389.2018.1540760

Shukla, P. (2020). Microbial nanotechnology for bioremediation of industrial wastewater. *Frontiers in Microbiology, 11*. https://doi.org/10.3389/fmicb.2020.590631

Taylor, A. A., Tsuji, J. S., Garry, M. R., McArdle, M. E., Goodfellow, W. L., Adams, W. J., & Menzie, C. A. (2020). Critical review of exposure and effects: Implications for setting regulatory health criteria for ingested copper. *Environmental Management, 65*(1), 131–159. https://doi.org/10.1007/s00267-019-01234-y

Tsai, W. T. (2014). Chlorofluorocarbons. In *Encyclopedia of toxicology: Third edition* (pp. 883–884). Elsevier. https://doi.org/10.1016/B978-0-12-386454-3.01118-0

Valix, M. (2017). Bioleaching of electronic waste: Milestones and challenges. In *Current developments in biotechnology and bioengineering: Solid waste management* (pp. 407–442). Elsevier Inc. https://doi.org/10.1016/B978-0-444-63664-5.00018-6

Varsha, S., Thomas, B. S., Kundu, C., Vuppaladadiyam, A. K., Duan, H., & Bhattacharya, S. (2024). Can e-waste recycling provide a solution to the scarcity of rare earth metals? An overview of e-waste recycling methods. *Science of the Total Environment, 924*, 171453. https://doi.org/10.1016/j.scitotenv.2024.171453

Verma, M. L. (2017). Nanobiotechnology advances in enzymatic biosensors for the agri-food industry. In Environmental Chemistry Letters (Vol. 15, Issue 4, pp. 555–560). Springer Verlag. https://doi.org/10.1007/s10311-017-0640-4

Verma, M. L., Joseph, P. J., & Sen Karaman, D. (2023). Microbial nanotechnology: a new frontier in microbiology. *Frontiers in Microbiology, 14*, 1296573

Verma, M. L., Dhanya, B. S., Sukriti, Rani, V., Thakur, M., Jeslin, J., & Kushwaha, R. (2020). Carbohydrate and protein based biopolymeric nanoparticles: Current status and biotechnological applications. *International Journal of Biological Macromolecules, 154*, 390–412. https://doi.org/10.1016/j.ijbiomac.2020.03.105

Wagh, M. S., Sowjanya, S., Nath, P. C., Chakraborty, A., Amrit, R., Mishra, B., Mishra, A. K., & Mohanta, Y. K. (2024). Valorisation of agro-industrial wastes: Circular bioeconomy and biorefinery process—A sustainable symphony. In *Process safety and environmental protection* (Vol. 183, pp. 708–725). Institution of Chemical Engineers. https://doi.org/10.1016/j.psep.2024.01.055

Wang, J., Faraji, F., Ramsay, J., & Ghahreman, A. (2021). A review of biocyanidation as a sustainable route for gold recovery from primary and secondary low-grade resources. *Journal of Cleaner Production, 296*. https://doi.org/10.1016/j.jclepro.2021.126457. Elsevier Ltd.

Withanage, S. V., & Habib, K. (2021). Life cycle assessment and material flow analysis: Two under-utilized tools for informing e-waste management. *Sustainability (Switzerland), 13*(14). https://doi.org/10.3390/su13147939. MDPI AG.

Wu, W., Liu, X., Zhang, X., Zhu, M., & Tan, W. (2018). Bioleaching of copper from waste printed circuit boards by bacteria-free cultural supernatant of iron–sulfur-oxidizing bacteria. *Bioresources and Bioprocessing, 5*(1). https://doi.org/10.1186/s40643-018-0196-6

Yadav, A. N., Suyal, D. C., Kour, D., Rajput, V. D., Rastegari, A. A., & Singh, J. (2022). Bioremediation and waste management for environmental sustainability. *Journal of Applied Biology and Biotechnology, 10*, 1–5. Open Science Publishers LLP Inc. https://doi.org/10.7324/JABB.2022.10s201

Yang, X., Zhao, Z., Zhang, G., Hirayama, S., van Nguyen, B., Lei, Z., Shimizu, K., & Zhang, Z. (2021). Insight into Cr(VI) biosorption onto algal-bacterial granular sludge: Cr(VI) bioreduction and its intracellular accumulation in addition to the effects of environmental factors. *Journal of Hazardous Materials, 414*. https://doi.org/10.1016/j.jhazmat.2021.125479

Zabochnicka-Świątek, M., & Krzywonos, M. (2014). Introduction potentials of biosorption and bioaccumulation processes for heavy metal removal. *Polish Journal of Environmental Studies, 23*(2).

Zhang, X. D., Wu, H. Y., Wu, D., Wang, Y. Y., Chang, J. H., Zhai, Z. B., Meng, A. M., Liu, P. X., Zhang, L. A., & Fan, F. Y. (2010). Toxicologic effects of gold nanoparticles in vivo by different administration routes. *International Journal of Nanomedicine, 5*(1), 771–781. https://doi.org/10.2147/IJN.S8428

Zhao, X., Hao, F., Lu, D., Liu, W., Zhou, Q., & Jiang, G. (2015). Influence of the surface functional group density on the carbon-nanotube-induced α-chymotrypsin structure and activity alterations. *ACS Applied Materials & Interfaces, 7*(33), 18880–18890. https://doi.org/10.1021/acsami.5b05895

Zhu, N., Zhang, B., & Yu, Q. (2020). Genetic engineering-facilitated coassembly of synthetic bacterial cells and magnetic nanoparticles for efficient heavy metal removal. *ACS Applied Materials and Interfaces, 12*(20), 22948–22957. https://doi.org/10.1021/acsami.0c04512

Chapter 11
Impact of Microbial Nanobiotechnology on the Economy of Developing Countries

Sarita Aryal, Priyanka Kumari, Chandan Kumar Gautam, and Rishikesh Kumar Gupta

Abstract Sustainable economic growth around the globe essentially involves an amalgamation of several approaches which include but are not limited to effective policy-making, scientific innovations, and technological interventions to achieve a robust economy. Among these approaches, the recent advancements in microbial nanobiotechnology hold strong potential to impact healthcare, industrial technology, agriculture, and environmental issues that are directly associated with the overall growth of a country's economy. Microbial nanobiotechnology merges microbiology, microbial technology, and nanotechnology has emerged as an important tool with the potential to significantly transform the socio-economic and environmental growth of developing economies. This chapter focuses on how microbial nanobiotechnology can reshape the economy of developing countries by overviewing various aspects of the economy. After the comprehensive analysis of case studies and empirical data, we have revealed that this field is commending sustainable growth. Here we overview the challenges in healthcare, environmental sustainability, technology, and innovation and reveal the role of microbial nanobiotechnology in solving them. Additionally, we address the underlying challenges and ethical considerations such as genetic modification, regulation, and safety issues. In conclusion, a nation's economic progress is intricately linked to advancements in science and technology. Integrating microbial nanobiotechnology into innovative

S. Aryal (✉)
Medical Oncology, Dana Farber Cancer Institute, Harvard Medical School, Boston, MA, USA

P. Kumari
Department of Forensic Science, Sam Higginbottom University of Agriculture, Technology and Sciences, Prayagraj, Uttar Pradesh, India

Regional Forensic Science Laboratory, Muzaffarpur, Bihar, India

C. K. Gautam
Department of Microbiological Sciences, North Dakota State University, Fargo, ND, USA

R. K. Gupta
Amity Institute of Neuropsychology and Neurosciences, Amity University Uttar Pradesh, Noida, India

Y. K. Mohanta et al. (eds.), *Nano-microbiology for Sustainable Development*,
https://doi.org/10.1007/978-3-031-78845-1_11

ecosystems fosters entrepreneurship, research opportunities, and technology transfer. This not only attracts foreign investments but also lays the foundation for a technology-driven economy in developing countries. These developments have the potential to significantly enhance the overall quality of life and economic prospects in these nations.

Keywords Nanobiotechnology · Healthcare · Sustainability · Innovation · Economic growth

11.1 Introduction

11.1.1 Introduction to Microbial Nanobiotechnology

Microbes, including viruses, fungi, and bacteria, along with nanotechnology form the fundamental building blocks of microbial nanobiotechnology (Grasso et al., 2019). At the nanoscale, their distinct biological characteristics and abilities are exploited and controlled for various purposes (Ball et al., 2019; Grasso et al., 2019). By merging their inherent biological functions with the unique attributes of nanotechnology, a world of possibilities emerges for addressing urgent problems across industries, environmental management, agriculture, and healthcare. The ability to manipulate matter at the nanoscale through nanotechnology gives materials and structures distinctive qualities (Belkin et al., 2015; Drexler, 1992). Through the integration of the concepts from nanotechnology with the intricate biology of microbes, researchers can delve into a domain where even the tiniest constituents can have profound effects (Bayda et al., 2019; Lateef et al., 2021a). When several disciplines come together, macroscopic challenges can be tackled on a nanoscale platform. This convergence enables harnessing the extraordinary nanoscale capabilities of microbes, creating opportunities for transformative applications in a variety of industries (Buzea et al., 2007). As we set out on this expedition, we investigate the microscopic state to unveil the major influence and inventive possibilities awaiting Microbial Nanobiotechnology. A plethora of applications in microbial nanobiotechnology holds the potential to transform industries and improve people's quality of life (Daniel & Astruc, 2004; Grasso et al., 2019; Li et al., 2011) (Fig. 11.1). The potential of this discipline is as broad as the microbial world itself, spanning from environmental remediation and sustainable agriculture to improved food and materials and precision medicine (Li et al., 2011). The synergy of microbial entities and nanotechnology fosters innovations that hold the key to addressing global challenges.

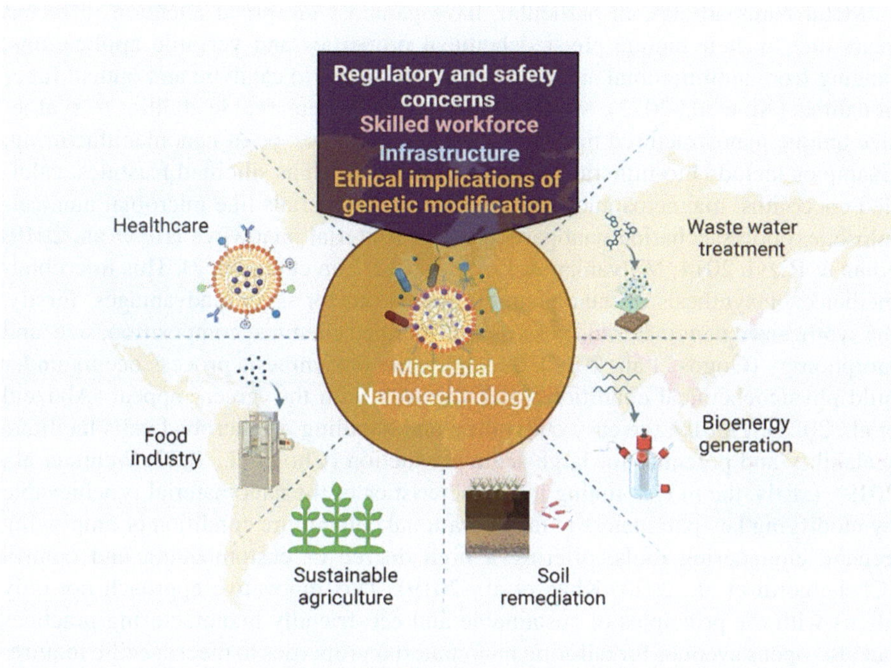

Fig. 11.1 An overview of microbial nanotechnology and its applications in various fields, as well as its associated challenges. Key application areas include healthcare, the food industry, sustainable agriculture, soil remediation, bioenergy generation, and wastewater treatment, depicted around the central concept. Highlighted challenges encompass regulatory and safety concerns, the need for a skilled workforce, infrastructure requirements, and ethical implications of genetic modification. The background world map underscores the global significance of microbial nanotechnology. (Figure created with BioRender)

11.1.2 Significance of Microbial Nanobiotechnology in Developing Countries

Microbial nanobiotechnology offers hitherto underexplored opportunities for innovation and improved resource efficiency, particularly important to developing nations. Research has shown that the integration of microbiology and nanotechnology generates novel solutions, resulting in agricultural improvements (Dwivedi et al., 2016). Nanoscale capabilities enable tailored microbes to improve crop yields and nutrient absorption, addressing food security concerns (Shang et al., 2019). The widespread utilization of nanomaterials has witnessed a remarkable surge across various industries, spanning from sensor technologies to biomedicine, exerting a significant impact on scientific and technological advancements. A noteworthy development in this arena is the biosynthesis of environment-friendly nanomaterials by microorganisms, gaining traction as an alternative to conventional chemical and physical manufacturing methods (Kröger & Poulsen, 2008).

Metal nanoparticles, in particular, have gained widespread attention in recent years due to their unique physicochemical properties and versatile applications, ranging from antimicrobial and anti-cancer properties to catalytic and optical functionalities (Xu et al., 2022). Microorganisms exhibit remarkable abilities to synthesize unique nanostructured materials, broadening the scope of nanomanufacturing. Examples include bio-mineralized nanostructures such as silicified frustules, calcified coccoliths, magnetosomes, and organic nanomaterials like microbial nanocellulose, exopolysaccharide nanoparticles, and bacterial nanowires (He et al., 2019; Khan & Rizvi, 2014; Malvankar & Lovley, 2012; Yan et al., 2017). This microbial-mediated biosynthesis of nanomaterials stands out for several advantages. Firstly, the synthesized nanomaterials possess well-defined chemical composition, size, and morphology (Gogos et al., 2012). Secondly, the biosynthesis process occurs under mild physicochemical conditions, aligning well with the 'green' appeal (Abuzeid et al., 2023). Thirdly, the easy cultivation and handling of microbial cells facilitate scalability and potential for large-scale production (Cho et al., 2022; Wehrs et al., 2019). Lastly, the *in vivo* tuning in characteristics of the nanomaterial is achievable by modifying key parameters in the operational cell culture condition or employing genetic engineering tools, offering a high degree of customization and control (Chehelgerdi et al., 2023; Khan et al., 2019). This innovative approach not only aligns with the principles of sustainable and eco-friendly manufacturing practices but also opens avenues for tailoring nanomaterial properties to meet specific requirements. As the exploration of microbial-mediated biosynthesis continues, it holds immense promise for revolutionizing nanomanufacturing practices and driving advancements across various fields.

11.2 Economic Challenges in Developing Countries

Why does Microbial Nanobiotechnology matter, particularly in the context of developing economies? This question invites us to consider the unique challenges faced frequently by these nations; challenges often characterized by limited resources, urgent healthcare concerns, and the imperative for environmentally sustainable practices. Microbial nanobiotechnology, with its capacity for accuracy, scalability, and adaptability, emerges as a promising prospect, offering innovative solutions that could potentially surpass conventional approaches and unlock new avenues for inclusive economic development (Fig. 11.1).

11.2.1 Healthcare Development

The persistent shortage of funds and resources in the healthcare sector of developing nations is one of the biggest economic obstacles. This shortage stems from competing priorities for expenditure and limited government funds, resulting in underfunded health systems (Basu et al., 2012). Consequently, this hinders the provision of high-quality healthcare services delivery by causing shortages of essential medical supplies, infrastructure, and healthcare personnel are often insufficient (Palmer, 2000). Disparities in wealth are also a contributing factor in unequal access to medical treatment, especially for marginalized populations and rural areas, due to transportation infrastructure gaps and physical barriers (Dassah et al., 2018; Syed et al., 2013). Health disparities are exacerbated by this since individuals with low incomes find it difficult to access timely and quality medical care (Garcia Popa-Lisseanu et al., 2005; Kasthuri, 2018). The healthcare systems of developing nations are severely strained by the financial cost of common diseases, including both communicable and non-communicable disorders (Dassah et al., 2018). The financial burden of common diseases overly stretched healthcare systems in developing countries when it comes to treating and preventing diseases, diverting funds from other essential services (Filip et al., 2022). Furthermore, the advent of epidemics, as observed in the case of infectious diseases, results in imposing unforeseen financial strains and demands prompt and resource-intensive responses (Filip et al., 2022; Haldane et al., 2021; Hui, 2006). Insufficient investment in healthcare infrastructure is a pervasive economic challenge. Inadequate hospitals, clinics, and diagnostic facilities hinder the capacity to deliver comprehensive healthcare services (Mosadeghrad, 2014). The lack of sufficient sanitation and clean water sources contributes further to the spread of diseases, compounding the economic burden on the health sector.

The field of microbial nanobiotechnology offers promising solutions to manufacture affordable diagnostic and therapeutic instruments that will not only be user-friendly but portable as well (Kaittanis et al., 2010). By developing affordable diagnostic and therapeutic tools such as engineered nanoparticles and nanoscale materials for targeted drug delivery, microbial nanobiotechnology lad to achieve the effective and efficient medical treatments (Patra et al., 2018; Grumezescu, 2017; Hassan et al., 2017). Facilitated by microbial nanobiotechnology, it is possible to create nanoscale miniature diagnostic instruments that function at the point of care. This is particularly beneficial in underserved distant locations with limited access to conventional healthcare services (Bobo et al., 2016). Furthermore, nanomaterials and engineered microbes can be used for effective water filtration to reduce water-borne infections and improve public health in underdeveloped nations (Amin et al., 2014; Ojha, 2020; Yaqoob et al., 2020).

Microbial nanobiotechnology enables sustainable and environmentally friendly pharmaceutical biosynthesis. Therapeutic agents and pharmaceuticals can be produced using engineered microbes, which could lessen reliance on expensive imports and increase accessibility to essential medications (Newman & Cragg, 2020; Salam et al., 2023). Initiatives about microbial nanobiotechnology may also include training programs and capacity building for regional healthcare professionals. This not only improves the workforce's skill set but also fosters knowledge transfer and sustainability in improving cutting-edge technological applications for better healthcare.

11.2.2 Agricultural Advancement and Environmental Sustainability

The economic obstacles to agricultural advancement and environmental sustainability are substantial and have an effect on community livelihoods and ecosystem health (Pata, 2021). One major issue facing farmers in developing nations is their limited access to financial resources (Cai et al., 2017). The adoption of contemporary agricultural technologies, investments in environment-friendly farming methods, and the purchase of high-quality inputs are all hampered by financial constraints, which ultimately lowers overall agricultural output (Barrows et al., 2014). Moreover, farmers may also be exposed to financial risk due to trade imbalances and fluctuating global commodity prices, which makes it challenging for them to invest in resilient agricultural systems and formulate long-term sustainability plans (Ahmad et al., 2021; Basso & Antle, 2020). Economic issues often lead to the abuse of chemical pesticides and fertilizers because of their apparent immediate benefits, further jeopardizing agricultural sustainability (Tudi et al., 2021). Although these inputs have the potential to increase yields in the short term, the long-term environmental consequences, such as soil erosion and water contamination, entail significant financial costs (Kopittke et al., 2019; Tudi et al., 2021). Farmers with limited financial resources may find it difficult to make the upfront investments in alternative technologies and necessary training for transitioning to sustainable agriculture methods (Giller et al., 2009; Piñeiro et al., 2020). To overcome these economic obstacles, a comprehensive strategy incorporating financial assistance, education, legislative incentives, and integration of emerging technologies including but not limited to microbial nanobiotechnology is required, to support sustainable farming techniques that enhance agricultural productivity and environmental health.

11.2.3 Industry and Technology with a Sustainable Economy

In the landscape of industry and technology in developing nations, economic challenges are fundamentally reshaping manufacturing processes (Yang et al., 2021). The ever-changing market's demands and narrow profit margins are compelling manufacturing organizations to reassess their functional process structures (Handoyo et al., 2023). Recognizing the need to align economic considerations with environmental and societal aspects, business managers actively seek new technologies to enhance their entire supply chains and achieve sustainability (Ahmed et al., 2023; Rehman et al., 2021; Shi et al., 2022; Yadav et al., 2020). Considering the competitive environment of the global market, the development of sustainable products has become imperative (Hayes & Wheelwright, 1979). Some research has emphasized on adoption of green design and disposal systems as critical drivers for manufacturing sustainability, highlighting the integration of economic, environmental, and societal factors as essential for a nation's economic development (Pezzoli, 1997; Shahzad et al., 2022).

The urgency for sustainable development has gained prominence on a global scale, driven by concerns over climate change, resource scarcity, and energy security (Kumar & Majid, 2020). Introduced by the United Nations in 1972, sustainable development seeks to balance economic growth, environmental conservation, and social well-being. The urgency to transition to low-carbon energy systems is emphasized by the Intergovernmental Panel on Climate Change (IPCC), emphasizing the significant contribution of renewable energy by 2050. Achieving such a transition requires significant investments, and the recent global pandemic has revealed the interconnectedness of energy systems with economic development. Policymakers now face the challenge of designing recovery plans that not only address immediate economic concerns but also shape energy infrastructure for a sustainable future (Fallah Shayan et al., 2022; Salvador et al., 2017).

Developing nations, with abundant renewable potential and growing energy demand, play a crucial role in this global energy transition, offering an opportunity to implement radical transformations more swiftly and manageably than their developed counterparts. To navigate economic challenges in the industry and technology sectors, developing nations must focus on drivers for sustainable development (Akbari & Hopkins, 2022). Sustainability in manufacturing organizations, including the adoption of sustainable energy resource systems, effective sustainability performance metrics, industrial ecology initiatives, and heightened customer awareness for sustainability, are crucial drivers (Alayón et al., 2022; Hariyani & Mishra, 2022).

The economic and climate crises necessitate a reevaluation of energy systems, urging a significant increase in investments in renewable energy technologies (Sokołowski & Heffron, 2022). The current global challenges present a potential opportunity for developing nations to lead in the energy transition, leveraging their renewable potential and influencing energy infrastructure for a sustainable and resilient future (Hariram et al., 2023). Policies that prioritize sustainability, focusing on advancements in technological integration and awareness, can pave the way for economic growth that aligns with environmental and societal well-being (Malik et al., 2022; Nasrollahi et al., 2020).

11.3 Role of Microbial Nanobiotechnology in Developing Countries

Microbial nanobiotechnology holds immense promise in addressing several challenges faced by developing nations (Fig. 11.1). Applications span in multi-facet range, including but not limited to sanitation, agriculture, healthcare, environmental cleanup, energy provision, bioremediation, disease monitoring and management, waste treatment, capacity building, and affordable technology. This interdisciplinary field harnesses the potential of microorganisms, including bacteria, fungi, and viruses, conjugated with nanostructures and nanoparticles (Chhipa, 2019). This section aims to delve into the extensive potential of microbial nanobiotechnology across diverse domains.

11.3.1 Scope of Microbial Nanobiotechnology

Microbial nanotechnology encompasses a broad scope of applications across diverse fields. Microorganisms such as actinomycetes, algae, bacteria, fungi, viruses, and yeast have been demonstrated to synthesize nanoscale materials like carbon dots, metals, and their oxides (Altammar, 2023; Kousheh et al., 2020; Li et al., 2011). While nanotechnology as a whole has been in existence for decades, the specific domain of microbial nanotechnology, which represents a smaller domain within it, remains underexplored in terms of its full potential. They have potential applications in key priority sectors of a developing economy including medicine, materials science, agriculture, industries, and environmental remediation (Gkantzou et al., 2021; Mandeep & Shukla, 2020; Pandit et al., 2022; Parveen et al., 2012; Zain et al., 2023). The ongoing exploration of these applications highlights the dynamic and expanding scope of microbial nanotechnology (Verma et al., 2023).

11.3.1.1 Application in Agriculture

Agriculture accounts for a huge portion of economic activities in developing nations. When microbially synthesized or associated nanomaterials are integrated with modern farming practices, offer promising solutions to mitigate the environmental dispersion of agro-chemicals such as pesticides, herbicides, and fertilizers (Mandeep & Shukla, 2020; Mishra et al., 2022). Highlighting their dual benefit not only lowers overall cropping input costs but also reduces pollution demonstrating their significant effect. Further, nanomaterials incorporated into microbial formulations enhance protection against pathogens and pests by enhancing their adherence to plant surfaces and seed coatings, thereby fostering optimal conditions for germination (Mishra et al., 2022). Furthermore, they contribute to water and nutrient management by designing nanoscale delivery systems for gradual nutrient release and facilitating water absorption in the soil (Elemike et al., 2019). However, ongoing research is crucial to comprehend the long-term impacts and address ethical considerations associated with these transformative technologies. The recent introduction of nanomaterial-based fertilizers and agrochemicals signifies the importance of investigating interactions between nanoparticles and microbes (Kumar et al., 2019; Yadav et al., 2023). The study of the long-term effects of nanoparticles on microbial ecosystems necessitates prolonged studies and interdisciplinary collaboration which can empower regional farmers to build capacity and promote the adoption of sustainable agricultural practices (Javaid et al., 2022; Westmeier et al., 2018).

11.3.1.2 Healthcare Advancements

Individuals in developing nations often experience limited access to healthcare compared to those in developed countries (Javaid et al., 2022; Kruk et al., 2018). While still underutilized, microbial nanotechnology holds promise for improving the healthcare sector. Infectious diseases emerge from diverse sources such as food-borne illnesses, water contamination, and hospital-acquired infections and further pandemics persist as a major global challenge (Singh et al., 2017). Despite advancements in pathogen identification, established diagnostic methods encounter challenges in labor-intensive sample preparation, bulky instrumentation, and slow data readout (Kaittanis et al., 2018). Microbial nanobiotechnology has emerged as a transformative force in healthcare, particularly in diagnostics, therapeutics, and drug delivery systems (Malik et al., 2023; Yuwen et al., 2023). Owing to their low cytotoxicity, production cost, and sustainability, microbial nanomaterials have garnered significant attention (Ansari & Rehman, 2021). Engineered microbes have demonstrated the potential to produce nanoparticles with antimicrobial properties, offering applications in wound healing, and controlling bacterial as well as fungal infection (Durán et al., 2016). They also have shown effectiveness against drug-resistant pathogens and have the potential to disrupt microbial biofilm signaling cascade (Buszewski et al., 2017; Lahiri et al., 2021; Sunkar & Nachiyar, 2012). Furthermore, the interaction between microbes and nanoparticles holds promising implications for cancer therapy. Bacteria-associated engineered nanoparticles have been investigated for their efficacy in precisely delivering and improving the effectiveness of anticancer drugs (Ansari & Rehman, 2021). This underscores the substantial potential of microbial nanobiotechnology in advancing the field of personalized medicine. However, it is important to note that the potential of microbial nanobiotechnology has only been slightly realized so far, leaving a vast potential still to be achieved in the healthcare sector.

11.3.1.3 Industrial and Environmental Impact

Microbial nanobiotechnology has triggered a revolutionary transformation across various industrial sectors, including fermentation, biofuel production, and biosensor development (Malik et al., 2023; Mandeep & Shukla, 2020). While the potential of nanobiotechnology is immense, developing nations have yet to fully explore its capabilities. In the context of biofuel production, nanomaterials and nanocomposites play a crucial role in enhancing the catalytic efficiency of microorganisms (Arya et al., 2021; Wang et al., 2016), exemplifies the utilization of nanoscale metal catalysts, resulting in a significant increase in ethanol production by bacteria are actively being used for biofuel production. Many developing nations are actively pushing to enhance bioethanol production, intending to use it as a blend of fuels to further reduce carbon emissions (Malode et al., 2021). This strategic initiative not only lessens dependence on expensive petrochemicals but also indirectly supports environmental sustainability, economic growth, and overall development (Popp et al., 2014).

Managing industrial and domestic waste presents a significant challenge in developing nations striving for rapid growth, often hindered by a lack of state-of-the-art technology for effective implementation and monitoring. The fusion of microbes and nanoparticles facilitates the creation of biosensors for detecting various analytes which can be used in monitoring the water and soil quality. For instance, bioengineering of bacteria with gold nanoparticles, results in a highly sensitive biosensor for detecting heavy metal ions in industrial wastewater (Malekzad et al., 2017). A recent study has shown a successful application of microbial-nanoparticle systems for the bioremediation of heavy metal-contaminated soils was documented (Jiang et al., 2018). The researchers employed bacteria adorned with iron nanoparticles, demonstrating their capability to efficiently eliminate toxic metals such as lead and cadmium. Microbial nanobiotechnology also finds application in the treatment of wastewater and pollutant removal. Microbes often form a biofilm which makes it resilient to removal; microbe-derived nanoparticles can easily disrupt the biofilm signaling cascade and potentially reduce the microbial load (Gautam et al., 2013; Lahiri et al., 2021). Moreover, engineered microbial systems can be utilized to break down antibiotic residues in wastewater, offering sustainable solutions for environmental remediation and wastewater treatment (Bala et al., 2022; Li et al., 2022).

11.3.2 Microbial Nanobiotechnology in Sustainable Growth

As highlighted above, microbial nanobiotechnology can play a critical role in addressing the challenges faced by agricultural, environmental, healthcare, and industrial sectors which are pivotal in promoting sustainable growth in developing nations. Furthermore, the application of nanobiotechnology in renewable energy facilitates bioenergy production, offering sustainable and clean energy sources for developing nations (Singh et al., 2023). To ensure the success of these applications, capacity-building initiatives, and educational programs are crucial, empowering regional research scientists and technicians (Armstrong, 2010). Affordability and accessibility remain key considerations to ensure that nanobiotechnological solutions are accessible to the populations in developing nations (Tiwari & Tiwari, 2022). Ethical practices are also essential, emphasizing responsible research and development for the safe and ethical use of these technologies in promoting economic and social development (Bruce, 2006).

Microbial Nanobiotechnology stands as a transformative force driving sustainable growth across diverse sectors. By leveraging the unique capabilities of microorganisms at the nanoscale, this cutting-edge field offers innovative solutions for fostering environmentally conscious practices, optimizing industrial processes, and addressing global challenges. Microbial nanobiotechnology is applicable at the forefront in advancing sustainable growth, promising a future where technology and nature collaborate for the benefit of society and the planet (Verma et al., 2023).

11.4 Challenges and Ethical Considerations

11.4.1 Ethical Implications of Genetic Modification

Microbial nanobiotechnologies represent a burgeoning field anchored in the complex interactions within microbial systems. This area of study is distinguished by ongoing efforts aimed to enhance the effectiveness and breadth of these technologies through strategic manipulation of microbial agents via genetic engineering. Despite the growing prevalence of genetically modified organisms (GMOs) worldwide, particularly within microbial and plant realms, persistent concerns regarding their safety endure (Bawa & Anilakumar, 2013).

Unlike genetically modified plants and animals, microbial systems exhibit a heightened propensity for exchanging genetic material across species boundaries, thereby introducing an element of unpredictability into the outcomes of genetic modifications. Engineered microbes often carry selection markers conferring resistance to specific antibiotics or chemicals. Mishandling of these genetically modified microorganisms poses risks of water and soil contamination, which could also lead to disturbance in microbial equilibrium in nature. The introduction of foreign DNA by engineered microbes, facilitated by horizontal gene transfer, paves the way for the emergence of antibiotic resistance in prevalent microbial pathogens (Aminov, 2011; Michaelis & Grohmann, 2023). Although the use of auxotrophic mutants can overcome the antibiotic resistance concerns associated with public health (Amrofell et al., 2023), currently it is being mostly used for yeasts and very few bacterial genetic modifications. Additionally, manipulation involving the addition, deletion, or mutation of certain genes may yield potent toxins or undesired byproducts, potentially leading to the creation of invasive species or novel human pathogens. Nevertheless, the use of precise tools like clustered regularly interspaced short palindromic repeats (CRISPR) has the potential to address the issue of untargeted modification (Redman et al., 2016).

These implications underscore ethical considerations, particularly in developing nations where constraints such as lax regulations, limited skilled personnel, and awareness gaps pose significant challenges. Moreover, our comprehensive understanding of the ecological ramifications and human health implications of this emerging technology is hindered by significant knowledge gaps and limitations in our ability to predict or regulate outcomes. While these issues also resonate in the application of such technologies in developed nations, robust regulatory frameworks and proficient human resources enable the mitigation of associated risks.

11.4.2 Regulatory and Safety Concerns

11.4.2.1 Challenges and Ethical Considerations

The adoption of microbial nanobiotechnology in developing economies poses specific challenges and ethical concerns demanding concerted action from governmental bodies, researchers, international organizations, and public participation (Lateef et al., 2021b; Roco & Bainbridge, 2005). These challenges can be broadly classified into four categories: economic, social, technological, and environmental (Lateef et al., 2021b). Economic challenges stemming from financial constraints and inadequate infrastructure hinder comprehensive research and development (Bürgi & Thalappil, 2006). Social challenges, including the need for social acceptability and robust regulatory frameworks, necessitate effective communication and investment in education and training programs (Bürgi & Thalappil, 2006). Technological hurdles require knowledge exchange, skill development, and the adaptation of legal frameworks to facilitate technology transfer (Rajput et al., 2021). Environmental challenges entail careful consideration of potential impacts on ecosystems, human health, and biodiversity, demanding collaborative efforts to establish regulatory and biosafety guidelines.

The advancement of microbial nanobiotechnology in developing countries upfront ethical considerations related to potential risks and benefits. Emphasizing responsible use and transparent utilization of microbial nanobiotechnologies, addressing concerns related to informed consent, and recognizing the sensitivity of biological information are paramount in this context, there is a crucial need to prioritize privacy and data security, with social justice and equity concerns underscoring the importance of ensuring the equitable sharing of benefits and burdens resulting from nanobiotechnological progress (Ekins & Zenghelis, 2021; Shahcheraghi et al., 2022). To promote responsible use and informed consent, prioritizing privacy, data security, and mitigating environmental impact. Efforts should be directed toward minimizing any potential environmental impact resulting from the implementation of these technologies, including strategies to reduce the environmental footprint associated with microbial nanobiotechnology (Shahcheraghi et al., 2022). Given the potential dual-use nature of microbial nanobiotechnologies, necessitating robust ethical guidelines to prevent misuse (Rahmati et al., 2020), collaborative efforts involving governments, research institutions, ethical committees, and international organizations are crucial to establish and enforce ethical standards in deploying microbial nanobiotechnology in developing nations (Have, 2007). Moreover, creating regulatory bodies and institutions to bridge the gap between policymakers, innovators, and investors, becomes indispensable to addressing these challenges and ethical considerations.

11.4.2.2 Addressing Regulatory and Safety Concerns

Microbial nanobiotechnology encounters regulatory and safety obstacles in developing nations due to the lack of tailored regulations, leading to uncertainty regarding ethical guidelines and safety standards (Chakrabarti & Sharma, 2019). Existing regulations have gaps that require addressing to ensure comprehensive coverage of unique aspects specific to microbial nanobiotechnology, given the rapid advancements in the field (Amenta et al., 2015) (Fig. 11.1). International harmonization of regulatory frameworks is essential to facilitate technology transfer and collaborative research efforts (Tarhan, 2020). Robust ethical review processes are necessary to evaluate ethical implications. Addressing safety concerns entails evaluating potential risks from engineered nanomaterial exposure, including toxicity and long-term effects, and implementing comprehensive risk assessments and mitigation measures (Xuan et al., 2023). Understanding nano-bio interactions is crucial for safe use, along with implementing protocols for safe handling, transportation, and disposal of nanomaterials to minimize occupational hazards and environmental contamination.

11.5 Case Studies

11.5.1 Implications and Success Story of Microbial Nanobiotechnology in Developing Countries

Microbial nanobiotechnology holds promise to bring significant advancements in various sectors, especially in developing countries where innovative and cost-effective solutions are crucial, by addressing various challenges and promoting sustainable development. Here we have categorized (Table 11.1) some of the most successful microbial nano-biotechnological approaches and a success story of microbial nanobiotechnology in developing countries into the following major fields: agriculture and food security, healthcare and disease control, water and soil remediation, renewable energy, waste management, drug delivery, and enzymatic processes to illustrate the potential impact.

These case studies illustrate the potential impact and success of microbial nanobiotechnology in addressing various challenges and promoting sustainable development and eco-friendly solutions across various sectors in developing countries. In India, nano-fertilizers were introduced- nanoparticles carrying essential nutrients and growth-promoting substances with microbial components lead to increased crop yields and have shown improved nutrient uptake efficiency increasing soil fertility, reduced environmental impact, and crop productivity (Kumar et al., 2020). Nanoparticles with antimicrobial properties were also developed which helped to combat crop diseases, reducing the reliance on chemical pesticides . In some African countries, nanocarriers designed with microbial components have been explored for

Table 11.1 Different domains of technological domains where microbial nanobiotechnology was implicated

Implicated domain	Country name	Technological development	Success story	References
Agriculture and food security	India	Nano-fertilizers	Nano-fertilizers developed in India have demonstrated improved nutrient uptake efficiency, reduced environmental impact, and increased food production.	Kumar et al. (2020) and Upadhyay et al. (2023)
Healthcare and disease control	Africa	Affordable diagnostic kits	Nanoparticle-based drug delivery systems have shown enhanced drug stability, improved bioavailability, and targeted delivery, contributing to more effective treatments	Nasiruddin et al. (2017) and Salie and Saidi (2023)
Water and soil remediation	Southeast Asia	Water purification systems in rural areas	Applications of nanomaterials in microbial remediation have resulted in the removal of contaminants, such as heavy metals, from water sources, contributing to environmental sustainability	Koch et al. (2021)
Renewable energy	Latin America	Microbial fuel cells and biofuel initiatives	Successful integration of microbial fuel cells with nanomaterials has shown promise in generating clean energy from organic waste, offering a sustainable solution for off-grid communities.	Chukwuma Chris et al. (2020) and Khan et al. (2021)
Waste management	Africa	Efficient waste treatment and recycling	Utilized nanomaterials with embedded microbial agents to remediate contaminated water sources. This approach effectively reduced the levels of pollutants, providing clean water for communities.	Apollon (2023), Bagchi & Behera (2019), and Uddin et al. (2021)
Drug delivery	Asia	Drug delivery for cancer therapy	Nanoparticle-based drug delivery systems have shown enhanced therapeutic efficacy, reduced side effects, and improved patient outcomes in cancer treatment.	Malvankar and Lovley (2014), Nath et al. (2023), and Rahmati et al. (2020)
Enzymatic processes	South America	Enzyme immobilization	The incorporation of nanomaterials in enzymatic processes has led to more efficient industrial processes, including biofuel production and biocatalysis for pharmaceutical synthesis.	Gkantzou et al. (2021) and Hegedús & Nagy (2014)

targeted drug delivery to combat diseases like malaria and tuberculosis (Salie & Saidi, 2023). These nanocarriers showed significant enhancement in the drug stability, improved bioavailability and efficacy of medications, especially for poorly soluble drugs, and contributed to more effective treatments along with a significant reduction in side effects. Since several African countries possess limited medical resources, microbial nanosensor-based affordable diagnostic kits were developed for rapid and cost-effective disease detection to improve disease management, reduce the burden on the healthcare system, and improve overall public health. In Brazil, nanotechnology has been employed for the targeted delivery of antimicrobial agents to treat tropical diseases, improving patient compliance and treatment outcomes (Nasiruddin et al., 2017). Microbial nanobiotechnology was implemented for bioremediation in Southeast Asia countries facing pollution issues (Koch et al., 2021). Engineered microbes with nanomaterials were used for water to degrade pollutants and improve water quality (Khan et al., 2021). Applications of these nanomaterials in microbial remediation have resulted in the removal of contaminants, such as heavy metals and microbial contaminants, providing access to clean water in rural areas. This ultimately contributed to environmental sustainability by reducing waterborne diseases and improving public health. In Nigeria, researchers have developed nano-enabled microbial systems to remediate oil-contaminated soil, contributing to environmental restoration in oil-producing regions (Chukwuma Chris et al., 2020; Khan et al., 2021). Latin American countries developed microbial fuel cells using nanomaterial-coated electrodes for sustainable electricity generation (Apollon, 2023; Bagchi & Behera, 2019; Uddin et al., 2021). This was successfully able to generate clean energy from organic waste, offering a sustainable solution for off-grid communities and reducing further reliance on traditional fuels. Some other developing countries (e.g. Bangladesh) are focused on using microbial nano biohybrids for bioenergy production to enhance the efficiency of biofuel production from organic waste, contributing to sustainable energy solutions in rural areas (Nath et al., 2023). Microbial nanobiotechnology has successfully been utilized as a sustainable waste management system in African nations (Malvankar & Lovley, 2014; Rahmati et al., 2020). Engineered microbes with nanomaterials have been employed to break down and digest organic waste efficiently which eventually helped in reducing environmental pollution, generation of valuable by-products, and reduced recycling leads on traditional waste management. In South American industries microbial nano bioconjugates have been used to immobilize enzymes, improving their stability and catalytic activity which has led to more efficient industrial processes, including textile processing, biofuel production, and biocatalysis for pharmaceutical synthesis resulting in increased efficiency and reduced environmental impact (Gkantzou et al., 2021; Hegedús & Nagy, 2014).

11.5.2 New Initiatives to Promote Microbial Nanobiotechnology in Developing Countries

In developing countries, a strategic approach is required to promote microbial nanobiotechnology, that involves the consideration of the unique challenges and opportunities within these regions. By leveraging existing resources, policymakers, researchers, and industry leaders are aiming to contribute efficiently. Expert researchers are emphasizing establishing training programs/workshops and international collaborations to train more scientists, researchers, and technicians in microbial nanobiotechnology. In that direction, regional policymakers and researchers have established dedicated research grants and funding opportunities for projects in microbial nanobiotechnology to encourage collaboration between research institutions, industry partners, and government agencies to support innovative research and promote open-access platforms for sharing research findings, data, and protocols. To foster interest and raise awareness about the potential applications of microbial nanobiotechnology among students pursuing careers in this field, government agencies have implemented educational programs at schools and universities. Pushing the boundaries, regional policymakers are providing regulatory support to facilitate the development and commercialization of new technologies. Although the above-mentioned initiatives remain limited to a few developing nations and that too at a nascent stage, by implementing a combination of these initiatives, policymakers, researchers, and industry stakeholders can contribute collectively to the growth and adoption of microbial nanobiotechnology and fostering sustainable development in developing countries.

11.6 Future Directions of Microbial Nanobiotechnology

There exists a palpable potential for this innovative field to positively impact the economies of developing countries. Early implementations have shown promising outcomes, hinting at the transformative power of microbial nanobiotechnology to enhance economic landscapes. These initial strides have begun to attract foreign investment, indicating a pathway toward long-term, inclusive development. However, it's essential to approach these technologies with caution, acknowledging the ethical, social, and environmental implications that accompany their integration. Notably, while progress has been made, significant challenges such as regulatory frameworks, ethical considerations, and technology transfer issues remain to be addressed for the widespread and sustainable implementation of microbial nanobiotechnology in developing countries. Through continued research, collaboration, and conscientious decision-making, the full potential of microbial nanobiotechnology to drive economic growth and societal progress in these regions can be realized.

11.7 Conclusion

In conclusion, the impact of microbial nanobiotechnology on the economy of developing countries is profound and multifaceted. By offering practical and scalable solutions across healthcare, agriculture, and industry, microbial nanobiotechnology has the potential to revolutionize economic trajectories in these nations. Its ability to provide affordable healthcare solutions, promote agricultural sustainability, and foster environmentally conscious industrial processes underscores its pivotal role in addressing complex economic challenges. As developing nations navigate towards a sustainable and inclusive economic future, nurturing local expertise and ensuring equitable access to the benefits of microbial nanobiotechnology will be imperative. This chapter has illuminated the transformative potential of microbial nanobiotechnology in shaping the economies of developing countries and underscores the importance of continued research and investment in this burgeoning field.

Acknowledgments We thank the university and institute admins there for administrative support. We also that our colleagues whose work we could not cite due to the limited scope of this book chapter.

Author Contributions Conceptualization; formal analysis; resources, S.A.; writing- original draft preparation, S.A., P.K., and R.K.G.; writing- review and editing, S.A., P.K., C.K.G. and R.K.G.; supervision, S.A. All authors read and agreed to the published version of the manuscript.

References

Abuzeid, H. M., Julien, C. M., Zhu, L., & Hashem, A. M. (2023). Green synthesis of nanoparticles and their energy storage, environmental, and biomedical applications. *Crystals, 13*, 1576.

Ahmad, M., Muslija, A., & Satrovic, E. (2021). Does economic prosperity lead to environmental sustainability in developing economies? Environmental Kuznets curve theory. *Environmental Science and Pollution Research, 28*, 22588–22601.

Ahmed, N., Hamid, Z., Rehman, K. U., Senkus, P., Khan, N. A., Wysokińska-Senkus, A., & Hadryjańska, B. (2023). Environmental regulation, fiscal decentralization, and agricultural carbon intensity: A challenge to ecological sustainability policies in the United States. *Sustainability, 15*, 5145.

Akbari, M., & Hopkins, J. L. (2022). Digital technologies as enablers of supply chain sustainability in an emerging economy. *Operations Management Research, 15*, 689–710.

Alayón, C. L., Säfsten, K., & Johansson, G. (2022). Barriers and enablers for the adoption of sustainable manufacturing by manufacturing SMEs. *Sustainability, 14*, 2364.

Altammar, K. A. (2023). A review on nanoparticles: Characteristics, synthesis, applications, and challenges. *Frontiers in Microbiology, 14*, 1155622.

Amenta, V., Aschberger, K., Arena, M., Bouwmeester, H., Botelho Moniz, F., Brandhoff, P., Gottardo, S., Marvin, H. J. P., Mech, A., Quiros Pesudo, L., Rauscher, H., Schoonjans, R., Vettori, M. V., Weigel, S., & Peters, R. J. (2015). Regulatory aspects of nanotechnology in the agri/feed/food sector in EU and non-EU countries. *Regulatory Toxicology and Pharmacology, 73*, 463–476.

Amin, M. T., Alazba, A. A., & Manzoor, U. (2014). A review of removal of pollutants from water/wastewater using different types of nanomaterials. *Advances in Materials Science and Engineering, 2014*, 825910.
Aminov, R. I. (2011). Horizontal gene exchange in environmental microbiota. *Frontiers in Microbiology, 2*.
Amrofell, M., Rengarajan, S., Vo, S., Tovar, E., LoBello, L., Dantas, G., & Moon, T. S. (2023). Engineering E. coli strains using antibiotic-resistance-gene-free plasmids. *Cell Reports Methods, 3*, 100669.
Ansari, M., & Rehman, S. (2021). *Microbial nanotechnology: Green synthesis and applications*. Springer.
Apollon, W. (2023). An overview of microbial fuel cell technology for sustainable electricity production. *Membranes, 13*(11), 884.
Armstrong, R. (2010). Systems architecture: A new model for sustainability and the built environment using nanotechnology, biotechnology, information technology, and cognitive science with living technology. *Artificial Life, 16*, 73–87.
Arya, I., Poona, A., Dikshit, P. K., Pandit, S., Kumar, J., Singh, H. N., Jha, N. K., Rudayni, H. A., Chaudhary, A. A., & Kumar, S. (2021). Current trends and future prospects of nanotechnology in biofuel production. *Catalysts, 11*, 1308.
Bagchi, S., & Behera, M. (2019). Microbial fuel cells. In *Bioelectrochemical interface engineering* (pp. 91–116).
Bala, S., Garg, D., Thirumalesh, B. V., Sharma, M., Sridhar, K., Inbaraj, B. S., & Tripathi, M. (2022). Recent strategies for bioremediation of emerging pollutants: A review for a green and sustainable environment. *Toxics, 10*, 484.
Ball, A. S., Patil, S., & Soni, S. (2019). Introduction into nanotechnology and microbiology. *Methods in Microbiology, 46*, 1–18. https://doi.org/10.1016/bs.mim.2019.04.003. Epub 2019 May 10.
Barrows, G., Sexton, S., & Zilberman, D. (2014). Agricultural biotechnology: The promise and prospects of genetically modified crops. *Journal of Economic Perspectives, 28*, 99–120.
Basso, B., & Antle, J. (2020). Digital agriculture to design sustainable agricultural systems. *Nature Sustainability, 3*, 254–256.
Basu, S., Andrews, J., Kishore, S., Panjabi, R., & Stuckler, D. (2012). Comparative performance of private and public healthcare systems in low- and middle-income countries: A systematic review. *PLoS Medicine, 9*, e1001244.
Bawa, A. S., & Anilakumar, K. R. (2013). Genetically modified foods: Safety, risks, and public concerns-A review. *Journal of Food Science and Technology, 50*, 1035–1046.
Bayda, S., Adeel, M., Tuccinardi, T., Cordani, M., & Rizzolio, F. (2019). The history of nanoscience and nanotechnology: From chemical-physical applications to nanomedicine. *Molecules (Basel, Switzerland), 25*.
Belkin, A., Hubler, A., & Bezryadin, A. (2015). Self-assembled wiggling nano-structures and the principle of maximum entropy production. *Scientific Reports, 5*, 8323.
Bobo, D., Robinson, K. J., Islam, J., Thurecht, K. J., & Corrie, S. R. (2016). Nanoparticle-based medicines: A review of FDA-approved materials and clinical trials to date. *Pharmaceutical Research, 33*, 2373–2387.
Bruce, D. (2006). Ethical and social issues in nanobiotechnologies: Nano2Life provides a European ethical 'think tank' for research in biology at the nanoscale. *EMBO Reports, 7*, 754–758.
Bürgi, B., & Thalappil, P. (2006). Societal implications of nanoscience and nanotechnology in developing countries. *Current Science, 90*.
Buszewski, B., Railean-Plugaru, V., Pomastowski, P., Rafinska, K., Szultka-Mlynska, M., & Kowalkowski, T. (2017). Antimicrobial effectiveness of bioactive silver nanoparticles synthesized by actinomycetes HGG16n strain. *Current Pharmaceutical Biotechnology, 18*, 168–176.
Buzea, C., Pacheco, I. I., & Robbie, K. (2007). Nanomaterials and nanoparticles: Sources and toxicity. *Biointerphases, 2*, Mr17-71.
Cai, Y., Golub, A. A., & Hertel, T. W. (2017). Agricultural research spending must increase in light of future uncertainties. *Food Policy, 70*, 71–83.

Chakrabarti, K. K., & Sharma, S. K. (2019). Microbial nanobiotechnology: Challenges and opportunities in developing countries. *Journal of Nanobiotechnology, 16*, 15–28.

Chehelgerdi, M., Chehelgerdi, M., Allela, O. Q. B., Pecho, R. D. C., Jayasankar, N., Rao, D. P., Thamaraikani, T., Vasanthan, M., Viktor, P., Lakshmaiya, N., Saadh, M. J., Amajd, A., Abo-Zaid, M. A., Castillo-Acobo, R. Y., Ismail, A. H., Amin, A. H., & Akhavan-Sigari, R. (2023). Progressing nanotechnology to improve targeted cancer treatment: Overcoming hurdles in its clinical implementation. *Molecular Cancer, 22*, 169.

Chhipa, H. (2019). Chapter 6—Applications of nanotechnology in agriculture. In V. Gurtler, A. S. Ball, & S. Soni (Eds.), *Methods in microbiology* (pp. 115–142). Academic.

Cho, J. S., Kim, G. B., Eun, H., Moon, C. W., & Lee, S. Y. (2022). Designing microbial cell factories for the production of chemicals. *JACS Au, 2*, 1781–1799.

Chukwuma Chris, O., Francis, E., & Louis Chukwuemeka, O. (2020). Environmental nanoremediation in Nigeria: A review of its potentials. *Algerian Journal of Engineering and Technology, 3*.

Daniel, M.-C., & Astruc, D. (2004). Gold nanoparticles: assembly, supramolecular chemistry, quantum-size-related properties, and applications toward biology, catalysis, and nanotechnology. *Chemical Reviews, 104*, 293–346.

Dassah, E., Aldersey, H., McColl, M. A., & Davison, C. (2018). Factors affecting access to primary health care services for persons with disabilities in rural areas: A "best-fit" framework synthesis. *Global Health Research and Policy, 3*, 36.

Drexler, K. E. (1992). *Nanosystems: Molecular machinery, manufacturing, and computation*. Wiley.

Durán, N., Nakazato, G., & Seabra, A. B. (2016). Antimicrobial activity of biogenic silver nanoparticles, and silver chloride nanoparticles: An overview and comments. *Applied Microbiology and Biotechnology, 100*, 6555–6570.

Dwivedi, S., Saquib, Q., Al-Khedhairy, A., & Musarrat, J. (2016). Understanding the role of nanomaterials in agriculture. In *Microbial inoculants in sustainable agricultural productivity 2016*. Springer.

Ekins, P., & Zenghelis, D. (2021). The costs and benefits of environmental sustainability. *Sustainability Science, 16*, 949–965.

Elemike, E. E., Uzoh, I. M., Onwudiwe, D. C., & Babalola, O. O. (2019). The role of nanotechnology in the fortification of plant nutrients and improvement of crop production. *Applied Sciences, 9*, 499.

Fallah Shayan, N., Mohabbati-Kalejahi, N., Alavi, S., & Zahed, M. A. (2022). Sustainable Development Goals (SDGs) as a framework for Corporate Social Responsibility (CSR). *Sustainability, 14*, 1222.

Filip, R., Gheorghita Puscaselu, R., Anchidin-Norocel, L., Dimian, M., & Savage, W. K. (2022). Global challenges to public health care systems during the COVID-19 pandemic: A review of pandemic measures and problems. *Journal of Personalized Medicine, 12*.

Garcia Popa-Lisseanu, M. G., Greisinger, A., Richardson, M., O'Malley, K. J., Janssen, N. M., Marcus, D. M., Tagore, J., & Suarez-Almazor, M. E. (2005). Determinants of treatment adherence in ethnically diverse, economically disadvantaged patients with rheumatic disease. *The Journal of Rheumatology, 32*, 913–919.

Gautam, C., Madhav, M., Bind, S., Srivastav, A., & Veerappapillai, S. (2013). An insight into biofilm ecology and its applied aspects. *International Journal of Pharmacy and Pharmaceutical Sciences, 5*, 69–73.

Giller, K. E., Witter, E., Corbeels, M., & Tittonell, P. (2009). Conservation agriculture and smallholder farming in Africa: The heretics' view. *Field Crops Research, 114*, 23–34.

Gkantzou, E., Chatzikonstantinou, A. V., Fotiadou, R., Giannakopoulou, A., Patila, M., & Stamatis, H. (2021). Trends in the development of innovative nanobiocatalysts and their application in biocatalytic transformations. *Biotechnology Advances, 51*, 107738.

Gogos, A., Knauer, K., & Bucheli, T. D. (2012). Nanomaterials in plant protection and fertilization: Current state, foreseen applications, and research priorities. *Journal of Agricultural and Food Chemistry, 60*, 9781–9792.

Grasso, G., Zane, D., & Dragone, R. (2019). Microbial nanotechnology: Challenges and prospects for green biocatalytic synthesis of nanoscale materials for sensoristic and biomedical applications. *Nanomaterials (Basel), 10.*

Grumezescu, A. M. (2017). *Nanoscale fabrication, optimization, scale-up and biological aspects of pharmaceutical nanotechnology*. William Andrew.

Haldane, V., De Foo, C., Abdalla, S. M., Jung, A.-S., Tan, M., Wu, S., Chua, A., Verma, M., Shrestha, P., Singh, S., Perez, T., Tan, S. M., Bartos, M., Mabuchi, S., Bonk, M., McNab, C., Werner, G. K., Panjabi, R., Nordström, A., & Legido-Quigley, H. (2021). Health systems resilience in managing the COVID-19 pandemic: lessons from 28 countries. *Nature Medicine, 27*, 964–980.

Handoyo, S., Suharman, H., Ghani, E. K., & Soedarsono, S. (2023). A business strategy, operational efficiency, ownership structure, and manufacturing performance: The moderating role of market uncertainty and competition intensity and its implication on open innovation. *Journal of Open Innovation: Technology, Market, and Complexity, 9*, 100039.

Hariram, N. P., Mekha, K. B., Suganthan, V., & Sudhakar, K. (2023). Sustainalism: An integrated socio-economic-environmental model to address sustainable development and sustainability. *Sustainability, 15*, 10682.

Hariyani, D., & Mishra, S. (2022). Organizational enablers for sustainable manufacturing and industrial ecology. *Cleaner Engineering and Technology, 6*, 100375.

Hassan, S., Prakash, G., Ozturk, A., Saghazadeh, S., Sohail, M. F., Seo, J., Dockmeci, M., Zhang, Y. S., & Khademhosseini, A. (2017). Evolution and clinical translation of drug delivery nanomaterials. *Nano Today, 15*, 91–106.

Have, H. A. M. J. (2007). *Nanotechnologies, ethics and politics*. UNESCO Publishing.

Hayes, R. H., & Wheelwright, S. C. (1979). The dynamics of product-process life cycles. *Harvard Business Review, 57*, 127–136.

He, X., Deng, H., & Hwang, H. M. (2019). The current application of nanotechnology in food and agriculture. *Journal of Food and Drug Analysis, 27*, 1–21.

Hegedús, I., & Nagy, E. (2014). Stabilization techniques of single enzymes as nanoparticles or enzyme nano-bioconjugates. *Biotechnology, 10*, 323–360.

Hui, E. K. (2006). Reasons for the increase in emerging and re-emerging viral infectious diseases. *Microbes and Infection, 8*, 905–916.

Javaid, M., Haleem, A., Singh, R. P., & Suman, R. (2022). Enhancing smart farming through the applications of Agriculture 4.0 technologies. *International Journal of Intelligent Networks, 3*, 150–164.

Jiang, D., Zeng, G., Huang, D., Chen, M., Zhang, C., Huang, C., & Wan, J. (2018). Remediation of contaminated soils by enhanced nanoscale zero valent iron. *Environmental Research, 163*, 217–227.

Kaittanis, C., Santra, S., & Perez, J. M. (2010). Emerging nanotechnology-based strategies for the identification of microbial pathogenesis. *Advanced Drug Delivery Reviews, 62*, 408–423.

Kaittanis, C., Andreou, C., Hieronymus, H., Mao, N., Foss, C. A., Eiber, M., Weirich, G., Panchal, P., Gopalan, A., Zurita, J., Achilefu, S., Chiosis, G., Ponomarev, V., Schwaiger, M., Carver, B. S., Pomper, M. G., & Grimm, J. (2018). Prostate-specific membrane antigen cleavage of vitamin B9 stimulates oncogenic signaling through metabotropic glutamate receptors. *The Journal of Experimental Medicine, 215*, 159–175.

Kasthuri, A. (2018). Challenges to healthcare in India—The five A's. *Indian Journal of Community Medicine: Official Publication of Indian Association of Preventive & Social Medicine, 43*, 141–143.

Khan, M. R., & Rizvi, T. F. (2014). Nanotechnology: Scope and application in plant disease management. *Plant Pathology Journal, 13*, 214–231.

Khan, I., Saeed, K., & Khan, I. (2019). Nanoparticles: Properties, applications and toxicities. *Arabian Journal of Chemistry, 12*, 908–931.

Khan, S., Naushad, M., Al-Gheethi, A., & Iqbal, J. (2021). Engineered nanoparticles for removal of pollutants from wastewater: Current status and future prospects of nanotechnology for remediation strategies. *Journal of Environmental Chemical Engineering, 9*, 106160.

Koch, N., Islam, N. F., Sonowal, S., Prasad, R., & Sarma, H. (2021). Environmental antibiotics and resistance genes as emerging contaminants: Methods of detection and bioremediation. *Current Research in Microbial Sciences, 2*, 100027.

Kopittke, P. M., Menzies, N. W., Wang, P., McKenna, B. A., & Lombi, E. (2019). Soil and the intensification of agriculture for global food security. *Environment International, 132*, 105078.

Kousheh, S. A., Moradi, M., Tajik, H., & Molaei, R. (2020). Preparation of antimicrobial/ultraviolet protective bacterial nanocellulose film with carbon dots synthesized from lactic acid bacteria. *International Journal of Biological Macromolecules, 155*, 216–225.

Kröger, N., & Poulsen, N. (2008). Diatoms-from cell wall biogenesis to nanotechnology. *Annual Review of Genetics, 42*, 83–107.

Kruk, M. E., Gage, A. D., Arsenault, C., Jordan, K., Leslie, H. H., Roder-DeWan, S., Adeyi, O., Barker, P., Daelmans, B., Doubova, S. V., English, M., García-Elorrio, E., Guanais, F., Gureje, O., Hirschhorn, L. R., Jiang, L., Kelley, E., Lemango, E. T., Liljestrand, J., Malata, A., Marchant, T., Matsoso, M. P., Meara, J. G., Mohanan, M., Ndiaye, Y., Norheim, O. F., Reddy, K. S., Rowe, A. K., Salomon, J. A., Thapa, G., Twum-Danso, N. A. Y., & Pate, M. (2018). High-quality health systems in the Sustainable Development Goals era: Time for a revolution. *The Lancet Global Health, 6*, e1196–e1252.

Kumar, J. C. R., & Majid, M. A. (2020). Renewable energy for sustainable development in India: Current status, future prospects, challenges, employment, and investment opportunities. *Energy, Sustainability and Society, 10*, 2.

Kumar, A., Singh, A., & Choudhary, K. K. (2019). *Role of plant growth promoting microorganisms in sustainable agriculture and nanotechnology*.

Kumar, Y., Tiwari, K., Nayak, R., Rai, A., Singh, S., Singh, A., Kumar, Y., Tomar, H., Singh, T., & Raliya, R. (2020). Nanofertilizers for increasing nutrient use efficiency, yield and economic returns in important winter season crops of Uttar Pradesh. *Indian Journal of Fertilizers, 16*, 772–786.

Lahiri, D., Nag, M., Sheikh, H. I., Sarkar, T., Edinur, H. A., Pati, S., & Ray, R. R. (2021). Microbiologically-synthesized nanoparticles and their role in silencing the biofilm signaling cascade. *Frontiers in Microbiology, 12*, 636588.

Lateef, A., Darwesh, O., & Matter, I. (2021a). *Microbial nanobiotechnology: The melting pot of microbiology, microbial technology and nanotechnology* (pp. 1–19).

Lateef, A., Eb, G.-K, Dasgupta, N., & Ranjan, S. (2021b). *Microbial nanobiotechnology: Principles and applications*.

Li, X., Xu, H., Chen, Z.-S., & Chen, G. (2011). Biosynthesis of nanoparticles by microorganisms and their applications. *Journal of Nanomaterials, 2011*, 270974.

Li, S., Show, P. L., Ngo, H. H., & Ho, S. H. (2022). Algae-mediated antibiotic wastewater treatment: A critical review. *Environmental Science and Ecotechnology, 9*, 100145.

Malekzad, H., Zangabad, P. S., Mirshekari, H., Karimi, M., & Hamblin, M. R. (2017). Noble metal nanoparticles in biosensors: Recent studies and applications. *Nanotechnology Reviews, 6*, 301–329.

Malik, S., Dhasmana, A., Preetam, S., Mishra, Y. K., Chaudhary, V., Bera, S. P., Ranjan, A., Bora, J., Kaushik, A., Minkina, T., Jatav, H. S., Singh, R. K., & Rajput, V. D. (2022). Exploring microbial-based green nanobiotechnology for wastewater remediation: A sustainable strategy. *Nanomaterials (Basel), 12*.

Malik, S., Muhammad, K., & Waheed, Y. (2023). Emerging applications of nanotechnology in healthcare and medicine. *Molecules (Basel, Switzerland), 28*.

Malode, S. J., Prabhu, K. K., Mascarenhas, R. J., Shetti, N. P., & Aminabhavi, T. M. (2021). Recent advances and viability in biofuel production. *Energy Conversion and Management: X, 10*, 100070.

Malvankar, N. S., & Lovley, D. R. (2012). Microbial nanowires: A new paradigm for biological electron transfer and bioelectronics. *ChemSusChem, 5*, 1039–1046.

Malvankar, N. S., & Lovley, D. R. (2014). Microbial nanowires for bioenergy applications. *Current Opinion in Biotechnology, 27*, 88–95.

Mandeep, & Shukla, P. (2020). Microbial nanotechnology for bioremediation of industrial wastewater. *Frontiers in Microbiology, 11*, 590631.
Michaelis, C., & Grohmann, E. (2023). Horizontal gene transfer of antibiotic resistance genes in biofilms. *Antibiotics (Basel, Switzerland), 12.*
Mishra, P., Muralidaran, Y., Vaishnavi, G., Dhoni, S., Jyotsana, S., Singh, M., & Kumar, V. (2022). *Microbial nanotechnology and its indispensable role in sustainable agriculture* (pp. 53–72).
Mosadeghrad, A. M. (2014). Factors affecting medical service quality. *Iranian Journal of Public Health, 43*, 210–220.
Nasiruddin, M., Neyaz, M. K., & Das, S. (2017). Nanotechnology-based approach in tuberculosis treatment. *Tuberculosis Research and Treatment, 2017*, 4920209.
Nasrollahi, Z., Hashemi, M.-s., Bameri, S., & Mohamad Taghvaee, V. (2020). Environmental pollution, economic growth, population, industrialization, and technology in weak and strong sustainability: Using STIRPAT model. *Environment, Development and Sustainability, 22*, 1105–1122.
Nath, P. C., Ojha, A., Debnath, S., Sharma, M., Sridhar, K., Nayak, P. K., & Inbaraj, B. S. (2023). Biogeneration of valuable nanomaterials from agro-wastes: A comprehensive review. *Agronomy, 13*, 561.
Newman, D. J., & Cragg, G. M. (2020). Natural products as sources of new drugs over the nearly four decades from 01/1981 to 09/2019. *Journal of Natural Products, 83*, 770–803.
Ojha, A. (2020). Chapter 19—Nanomaterials for removal of waterborne pathogens: Opportunities and challenges. In M. N. Vara Prasad & A. Grobelak (Eds.), *Waterborne pathogens* (pp. 385–432). Butterworth-Heinemann.
Palmer, N. (2000). The use of private-sector contracts for primary health care: Theory, evidence and lessons for low-income and middle-income countries. *Bulletin of the World Health Organization, 78*, 821–829.
Pandit, C., Roy, A., Ghotekar, S., Khusro, A., Islam, M. N., Emran, T. B., Lam, S. E., Khandaker, M. U., & Bradley, D. A. (2022). Biological agents for synthesis of nanoparticles and their applications. *Journal of King Saud University, Science, 34*, 101869.
Parveen, S., Misra, R., & Sahoo, S. K. (2012). Nanoparticles: A boon to drug delivery, therapeutics, diagnostics and imaging. *Nanomedicine: Nanotechnology, Biology and Medicine, 8*, 147–166.
Pata, U. K. (2021). Linking renewable energy, globalization, agriculture, CO_2 emissions and ecological footprint in BRIC countries: A sustainability perspective. *Renewable Energy, 173*, 197–208.
Patra, J. K., Das, G., Fraceto, L. F., Campos, E. V. R., Rodriguez-Torres, M. P., Acosta-Torres, L. S., Diaz-Torres, L. A., Grillo, R., Swamy, M. K., Sharma, S., Habtemariam, S., & Shin, H.-S. (2018). Nano based drug delivery systems: Recent developments and future prospects. *Journal of Nanobiotechnology, 16*, 71.
Pezzoli, K. (1997). Sustainable development: A transdisciplinary overview of the literature. *Journal of Environmental Planning and Management, 40*, 549–574.
Piñeiro, V., Arias, J., Dürr, J., Elverdin, P., Ibáñez, A. M., Kinengyere, A., Opazo, C. M., Owoo, N., Page, J. R., Prager, S. D., & Torero, M. (2020). A scoping review on incentives for adoption of sustainable agricultural practices and their outcomes. *Nature Sustainability, 3*, 809–820.
Popp, J., Lakner, Z., Harangi-Rákos, M., & Fári, M. (2014). The effect of bioenergy expansion: Food, energy, and environment. *Renewable and Sustainable Energy Reviews, 32*, 559–578.
Rahmati, F., Hosseini, S. S., Mahuti Safai, S., Asgari Lajayer, B., & Hatami, M. (2020). New insights into the role of nanotechnology in microbial food safety. *3 Biotech, 10*, 425.
Rajput, V. D., Singh, A., Minkina, T., Rawat, S., Mandzhieva, S., Sushkova, S., Shuvaeva, V., Nazarenko, O., Rajput, P., Komariah, Verma, K. K., Singh, A. K., Rao, M., & Upadhyay, S. K. (2021). Nano-enabled products: Challenges and opportunities for sustainable agriculture. *Plants (Basel, Switzerland), 10.*
Redman, M., King, A., Watson, C., & King, D. (2016). What is CRISPR/Cas9? *Archives of Disease in Childhood. Education and Practice Edition, 101*, 213–215.

Rehman, A., Ma, H., Ahmad, M., Irfan, M., Traore, O., & Chandio, A. A. (2021). Towards environmental Sustainability: Devolving the influence of carbon dioxide emission to population growth, climate change, Forestry, livestock and crops production in Pakistan. *Ecological Indicators, 125*, 107460.

Roco, M., & Bainbridge, W. S. (2005). Societal implications of nanoscience and nanotechnology: Maximizing human benefit. *Journal of Nanoparticle Research, 7*, 1–13.

Salam, M. A., Al-Amin, M. Y., Salam, M. T., Pawar, J. S., Akhter, N., Rabaan, A. A., & Alqumber, M. A. A. (2023). Antimicrobial resistance: A growing serious threat for global public health. *Healthcare (Basel, Switzerland), 11*.

Salie, F., & Saidi, T. (2023). Nanomedicine drug delivery in South Africa: A retrospective study on research, funding and collaboration. *Frontiers in Pharmacology, 14*, 1317137.

Salvador, R., Piekarski, C. M., & Francisco, A. C. (2017). Approach of the two-way influence between lean and green manufacturing and its connection to related organisational areas. *International Journal of Production Management and Engineering, 5*, 73–83.

Shahcheraghi, N., Golchin, H., Sadri, Z., Tabari, Y., Borhanifar, F., & Makani, S. (2022). Nanobiotechnology, an applicable approach for sustainable future. *3 Biotech, 12*, 65.

Shahzad, M., Qu, Y., Rehman, S. U., & Zafar, A. U. (2022). Adoption of green innovation technology to accelerate sustainable development among manufacturing industry. *Journal of Innovation & Knowledge, 7*, 100231.

Shang, Y., Hasan, M. K., Ahammed, G. J., Li, M., Yin, H., & Zhou, J. (2019). Applications of nanotechnology in plant growth and crop protection: A review. *Molecules (Basel, Switzerland), 24*.

Shi, R., Irfan, M., Liu, G., Yang, X., & Su, X. (2022). Analysis of the impact of livestock structure on carbon emissions of animal husbandry: A sustainable way to improving public health and green environment. *Frontiers in Public Health, 10*, 835210.

Singh, L., Kruger, H. G., Maguire, G. E. M., Govender, T., & Parboosing, R. (2017). The role of nanotechnology in the treatment of viral infections. *Therapeutic Advances in Infectious Disease, 4*, 105–131.

Singh, B., Sharma, K., Tyagi, S., Gautam, D., Chaudhary, M., Kumar, A., Vikal, S., & Gautam, Y. (2023). *Green and sustainable technology for clean energy production: Applications* (pp. 563–585).

Sokołowski, M. M., & Heffron, R. J. (2022). Defining and conceptualising energy policy failure: The when, where, why, and how. *Energy Policy, 161*, 112745.

Sunkar, S., & Nachiyar, C. V. (2012). Biogenesis of antibacterial silver nanoparticles using the endophytic bacterium Bacillus cereus isolated from Garcinia xanthochymus. *Asian Pacific Journal of Tropical Biomedicine, 2*, 953–959.

Syed, S. T., Gerber, B. S., & Sharp, L. K. (2013). Traveling towards disease: Transportation barriers to health care access. *Journal of Community Health, 38*, 976–993.

Tarhan, Ö. (2020). 16—Safety and regulatory issues of nanomaterials in foods. In S. M. Jafari (Ed.), *Handbook of food nanotechnology* (pp. 655–703). Academic.

Tiwari, R., & Tiwari, S. (2022). *Leveraging bionanotechnology for sustainable and inclusive growth: Prospects for Indo-German collaboration in ensuring affordable healthcare*.

Tudi, M., Daniel Ruan, H., Wang, L., Lyu, J., Sadler, R., Connell, D., Chu, C., & Phung, D. T. (2021). Agriculture development, pesticide application and its impact on the environment. *International Journal of Environmental Research and Public Health, 18*, 1112.

Uddin, M. J., Jeong, Y.-K., & Lee, W. (2021). Microbial fuel cells for bioelectricity generation through reduction of hexavalent chromium in wastewater: A review. *International Journal of Hydrogen Energy, 46*, 11458–11481.

Upadhyay, P. K., Singh, V. K., Rajanna, G. A., Dwivedi, B. S., Dey, A., Singh, R. K., Rathore, S. S., Shekhawat, K., Babu, S., Singh, T., Kumar, Y., Singh, C., Rangot, M., Kumar, A., Sarkar, S., Dash, S., & Rawat, S. (2023). Unveiling the combined effect of nano fertilizers and conventional fertilizers on crop productivity, profitability, and soil well-being. *Frontiers in Sustainable Food Systems, 7*.

Verma, M. L., Joseph, P. J., & Sen Karaman, D. (2023). Editorial: Microbial nanotechnology: A new frontier in microbiology. *Frontiers in Microbiology, 14*, 1296573.

Wang, C., Ehrhardt, C. J., & Yadavalli, V. K. (2016). Nanoscale imaging and hydrophobicity mapping of the antimicrobial effect of copper on bacterial surfaces. *Micron, 88*, 16–23.

Wehrs, M., Tanjore, D., Eng, T., Lievense, J., Pray, T. R., & Mukhopadhyay, A. (2019). Engineering robust production microbes for large-scale cultivation. *Trends in Microbiology, 27*, 524–537.

Westmeier, D., Hahlbrock, A., Reinhardt, C., Fröhlich-Nowoisky, J., Wessler, S., Vallet, C., Pöschl, U., Knauer, S. K., & Stauber, R. H. (2018). Nanomaterial–microbe cross-talk: Physicochemical principles and (patho)biological consequences. *Chemical Society Reviews, 47*, 5312–5337.

Xu, J. J., Zhang, W. C., Guo, Y. W., Chen, X. Y., & Zhang, Y. N. (2022). Metal nanoparticles as a promising technology in targeted cancer treatment. *Drug Delivery, 29*, 664–678.

Xuan, L., Ju, Z., Skonieczna, M., Zhou, P.-K., & Huang, R. (2023). Nanoparticles-induced potential toxicity on human health: Applications, toxicity mechanisms, and evaluation models. *MedComm, 4*, e327.

Yadav, G., Kumar, A., Luthra, S., Garza-Reyes, J. A., Kumar, V., & Batista, L. (2020). A framework to achieve sustainability in manufacturing organisations of developing economies using industry 4.0 technologies' enablers. *Computers in Industry, 122*, 103280.

Yadav, A., Yadav, K., & Abd-Elsalam, K. A. (2023). Exploring the potential of nanofertilizers for a sustainable agriculture. *Plant Nano Biology, 5*, 100044.

Yan, Y. L., Desvignes, T., Bremiller, R., Wilson, C., Dillon, D., High, S., Draper, B., Buck, C. L., & Postlethwait, J. (2017). Gonadal soma controls ovarian follicle proliferation through Gsdf in zebrafish. *Developmental Dynamics: An Official Publication of the American Association of the Anatomists, 246*, 925–945.

Yang, T., Yi, X., Lu, S., Johansson, K. H., & Chai, T. (2021). Intelligent manufacturing for the process industry driven by industrial artificial intelligence. *Engineering, 7*, 1224–1230.

Yaqoob, A. A., Parveen, T., Umar, K., & Mohamad Ibrahim, M. N. (2020). Role of nanomaterials in the treatment of wastewater: A review. *Water, 12*, 495.

Yuwen, L., Zhang, S., & Chao, J. (2023). Recent advances in DNA nanotechnology-enabled biosensors for virus detection. *Biosensors, 13*, 822.

Zain, M., Ma, H., Nuruzzaman, M., Chaudhary, S., Nadeem, M., Shakoor, N., Azeem, I., Duan, A., Sun, C., & Ahamad, T. (2023). Nanotechnology based precision agriculture for alleviating biotic and abiotic stress in plants. *Plant Stress, 10*, 100239.

Chapter 12
Nanotechnology and Microbes: Revolutionizing Water Management

Vanga Sri Varsha, Tanmayi Boreda, Sanjana Reddy Pailla, Yashasvi Kambhampati, T. Gourav, Rajasri Yadavalli, G. Vijaya Laxmi, Sphoorthy Nadimpalli, and C. Nagendranatha Reddy

Abstract Driven by a growing global demand for safe drinking water, coupled with escalating threats of water scarcity and rising pollution, innovative solutions are paramount for long-term water resource management. The recent investigations on nanotechnology's multidimensional role in water treatment focus on its applications in adsorption, membrane filtration, immobilization, catalysis, disinfection, sensing, and multifunctional device creation. The use of nano-assisted bioremediation technologies, demonstrating how bacteria, fungus, yeast, microalgae, and enzymes may eliminate inorganic and organic pollutants is researched. In addition, the incorporation of microbial-assisted nanotechnology is being studied for its potential uses in microbial fuel cells (MFCs), bio-nanocomposites, and biologically enhanced nanoparticle (NP) production, all of which contribute to sustainable water management techniques. Using these modern methods, also addresses the removal of various impurities such as heavy metals and dyes. This chapter explores the advancements, challenges, and economic feasibility of microbial-assisted nanotechnologies for water treatment, addressing environmental, safety, regulatory, and ethical considerations. It also outlines future trends and contributes to the knowledge base on sustainable water resource management.

Keywords Bio-nanocomposites · Bioremediation · Catalysis · Microbial fuel cells · Multifunctional devices · Nanotechnology

---

V. S. Varsha · T. Boreda · S. R. Pailla · Y. Kambhampati · T. Gourav · R. Yadavalli · G. Vijaya Laxmi · C. Nagendranatha Reddy (✉)
Department of Biotechnology, Chaitanya Bharathi Institute of Technology, Hyderabad, Telangana, India
e-mail: nagendranath_biotech@cbit.ac.in

S. Nadimpalli
Department of Biotechnology, Chaitanya Bharathi Institute of Technology, Hyderabad, Telangana, India

Department of Chemistry, Graduate School of Science, Kyushu University, Fukuoka, Japan

Y. K. Mohanta et al. (eds.), *Nano-microbiology for Sustainable Development*,
https://doi.org/10.1007/978-3-031-78845-1_12

12.1 Introduction

Many cities throughout the world are wrestling with the problem of guaranteeing sustainable freshwater supplies in the growing urbanization and rising water demands. It now is estimated that 780 million people lack access to safe drinking water, and 2.5 billion people lack access to basic sanitation. Consequently, factors associated with water, sanitation, and hygiene claim the lives of around 3.4 million people annually (UNICEF, 2024). Over the previous century, conversion and development have destroyed more than half of the world's wetlands. Because they serve as filters and flood buffers, wetlands are crucial to the well-being of both people and natural systems. The world's water supply is being threatened by water contamination by various pollutions (from human developments) and water-related diseases, drought and aridification, mismanagement of groundwater, destruction and degradation of freshwater ecosystems, saltwater intrusion, pollution, land degradation, over-extraction, climate change, etc. (UNEA, 2022; Gleick et al., 2001) (Fig. 12.1).

The issue is exacerbated by microbes, salts, and pollution from industry and agriculture. Global warming is anticipated to have significant implications for the world's freshwater resources (Alikhani et al., 2021). Rising urbanization, along with sustained economic expansion in populous nations such as India and China, has resulted in increased industrial outputs across several sectors, putting significant strain on water resources owing to untreated industrial waste. These difficulties highlight the critical importance of technical innovation, resource stewardship, and collaborative efforts to provide long-term solutions. Water's interconnectedness with food, climate, health, and other facets of life underscores the importance of coordinated action. Notably, the dramatic reduction in freshwater species over the last 50 years emphasizes the relevance of water as a multidisciplinary study issue (Mandeep & Shukla, 2020). The different types, but not limited to, of water pollutants, include organic and inorganic pollutants, pathogens, nutrients, agricultural

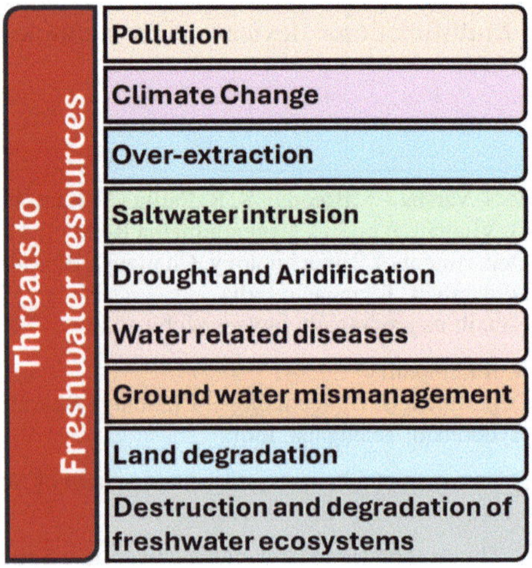

Fig. 12.1 Classification of various threats to freshwater resources

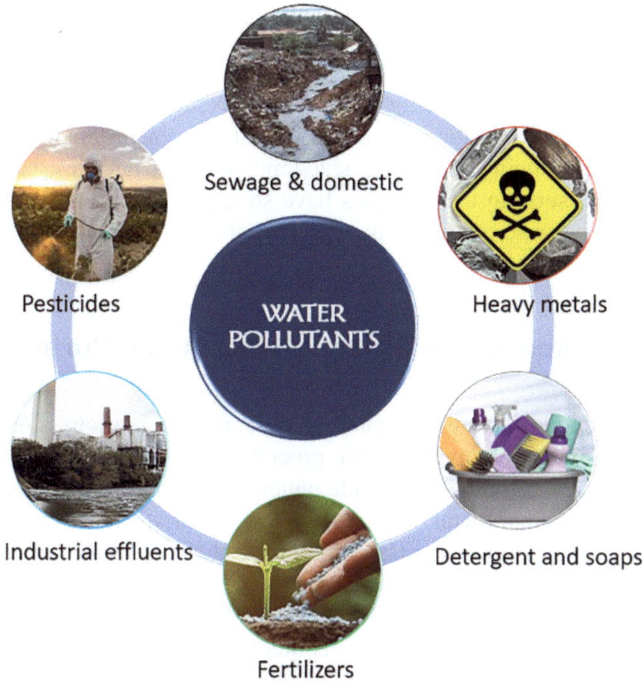

Fig 12.2 Types of various pollutants present in waterbodies

runoff, suspended solids and sediments, and thermal and radioactive components (Wasewar et al., 2020) (Fig. 12.2).

Primarily the water resource management relied on large-scale infrastructure projects and established chemical treatment methods. This included building dams and reservoirs for water storage and distribution, alongside traditional filtration systems utilizing sand and gravel. Throughout history and into the present, carbon has been used to manufacture water filtration and purification components. As early as 3750 BC, Egyptians employed wood charcoal (carbon) for water purification. Chemical disinfection with chlorine or chloramines was a mainstay for eliminating pathogens, while sedimentation and coagulation processes aimed to remove larger particles. However, these conventional methods faced limitations in addressing emerging contaminants like microplastics and complex organic pollutants. Additionally, energy consumption and potential environmental side effects from chemical usage presented ongoing challenges (Shah & Shah, 2020). Despite these limitations, new advances in nanotechnology provide intriguing answers to water scarcity and pollution concerns. Nanotechnology has shown promise in the improvisation of water purification systems. Nanoscale decrease in size not only improves pollutant removal capability but also makes these technologies more cost-effective and scalable. The World Health Organization's declining restrictions on toxins like arsenic, as well as the unique features of nanomaterials, highlight their critical role in developing long-term water treatment solutions (Sharma et al., 2023).

Along with these, activated carbon and its derivatives can be used for treatment as well as sensing and monitoring applications. Nanoparticles (metal and metal oxide) have also been used for adsorption, oxidation, sensing, photocatalysis, and disinfection. Despite being a debated academic issue for decades, nanomaterials-based therapy and sensing technologies have struggled to gain a substantial commercial foothold because of sensitivity unpredictability (Mishra et al., 2023).

12.2 Nanotechnology in Water Treatment and Purification

Because of the unique features of nanomaterials, nanotechnology aids in modernizing water treatment and purification procedures. Nanomaterials, which have dimensions less than 100 nm and include nanotubes or carbon-based nanocomposites as well as metals and metal oxide-based nanomaterials, have special chemical, biological, and physical characteristics that make them ideal for wastewater treatment. By using these features, nanotechnology strives to improve the efficacy of existing treatment techniques and develop novel ones. Higher surface area for adsorption, antimicrobial property for disinfection, (photo) catalysis, and unique electronic and optical properties useful in innovative therapeutic procedures and water quality monitoring sensing devices are all desirable nanomaterial properties for wastewater applications (Qu et al., 2013a). Some of the significant and efficient mechanisms that aid in wastewater treatment and purification include adsorption using different NPs, filtration using different membranes, immobilization, disinfection and microbial control, photo-catalysis, sensing and monitoring, multifunctional devices, etc.

12.2.1 Adsorption

Nanomaterials, particularly carbon-based ones like nanocomposites or nanotubes, offer a great surface area conducive to adsorption. This property is harnessed in wastewater management practices to remove effluents through efficient adsorption processes (Vázquez-Núñez et al., 2020).

12.2.1.1 Carbon-Based Nano-adsorbents (CNTs)

Carbon nano-adsorbents have outperformed activated carbon in the adsorption of several organic complexes. The vast surface area and various CNTs pollutants interaction contribute to their high adsorption capability. Peptide interactions with the organic molecules containing benzene rings or C=C bonds, such as polar aromatic chemicals and polycyclic aromatic hydrocarbons, are possible because of the electron-rich CNT surface. At a proper pH, electrostatic forces enhance the adsorption of the positively charged organic compounds like a few antibiotics (Qu et al., 2013b). Oxidized CNTs have a high metal ion adsorption capability with quick kinetics. CNT surface functional groups (e.g., hydroxyl, phenol, and carboxyl) are the primary metal ion adsorption sites, mostly via chemical bonding and electrostatic attraction. These CNTs are superior adsorbents for heavy metals than activated carbon. These are cost-effectiveness of adsorbents is heavily influenced by regeneration and reuse. Metal ions on nanotubes may be easily reversed by reducing the pH of the solution, and their adsorption capacity following regeneration stays rather constant (Kolya & Kang, 2023).

12.2.1.2 Metal Based Nano-adsorbents

Effective and economical, metal oxides viz., titanium dioxide, alumina, and iron oxide serve as potent adsorbents for radionuclides heavy metals in water treatment. The absorption mechanism hinges on the intricate interplay between dissolved oxygen and metals in its oxides, involving swift adsorption of few metal ions at the outer surfaces following that intraparticle diffusion alongside micropore walls, which becomes rate-limiting step. Nano-sized counterparts of these used metal oxides exhibit fast kinetics and heightened adsorption ability, attributed to their larger surface area, reduced intraparticle diffusion rate, and increased at reaction sites (Dar et al., 2020). A notable example involves nano-magnetite, where decreasing particle size ranges in 300 nm to 11 nm exponentially increase in adsorption of arsenic, primarily due to augmented surface area. However, as the particle size diminishes, and the specific surface area-normalized adsorption capacity increases noticeably imbibing 3 times higher arsenic. Metal oxide nanocrystals exhibit adaptability, with the ability to undergo compression into porous pellets with no significant loss of surface area under medium pressure. This flexibility allows their application as fine powders or porous pellets, catering to diverse industrial needs. In arsenic removal, metal-based nanomaterials showcase remarkable potential, surpassing the efficacy of activated carbon (You et al., 2022). Metal-resistant bacteria and fungi, when combined with engineered nanomaterials, offer a powerful tool for removing the heavy metals from water. NPs, like nano-zero valent iron, interact with heavy metal ions, facilitating their precipitation or adsorption onto microbial surfaces, leading to efficient removal (Kishore et al., 2022).

12.2.1.3 Polymeric Nano-adsorbents

Due to its cation exchange characteristics, the phosphate of a metal [$M(HPO_4)_2$] has employed as an effective adsorbent for heavy metal-ions. However, ultrafine particles are not ideal for flow-through systems or fixed bed because of concerns related to washing away and significant pressure loss. Polymer nanocomposite adsorbents were created to overcome these technological problems. When the polymer cation exchanger was used as additional material for amorphous zirconium phosphate (ZrP), the negatively charged functional group that was immobilized on the porous polymer matrix increases heavy metal cation penetration owing to the Donnan membrane effect (Ali & Gupta, 2006). Amorphous zirconium phosphate was impregnated into a highly acidic cation exchange resin (D001), resulting in the nanocomposite adsorbent (ZrP-001). This nanocomposite exhibited good adsorption property towards lead ion over others and ubiquitous cations like group 1 and 2 metal cations. Inorganic nanoparticles with a specific interaction towards heavy metal ions, such as hydrated ferric oxides, $M(HPO_4)_2$, hydrous zirconium oxide, $Zr(HPO_3S)_2$, and hydrated manganese oxide, can be introduced into the porous polymer cation exchangers to get nanocomposite adsorbents that are specific for the heavy metals (Kolya & Kang, 2023) (Table 12.1).

12.2.2 *Membrane Filtration*

Water treatment is the process of eliminating unwanted impurities from the water using membranes as a physical barrier. Membranes provide high automation, need less land and chemicals, and have a modular design for flexibility. The intrinsic trading between permeability and membrane selectivity, on the other hand, it is a significant challenge. Pressure-driven membrane processes are complicated by high energy consumption and membrane fouling. Incorporating functional nanomaterials into membranes can improve fouling resistance, penetrability, thermal and mechanical stability, and pollutant degradation, and self-cleaning activities are provided (Jatoi et al., 2021).

12.2.2.1 Nanofiber Membranes

The method of electrospinning is a low-cost, straightforward process of manufacturing ultrafine fibers from a range of materials (including ceramics, polymers, and even metals). Their nanofibers have a high specific porosity and surface area, resulting in nanofibers with complicated pore architectures. The size, shape, composition, primary and secondary structure, and alignment (spatial) of electro-spun nanofibers may all be easily modified for specific applications. As a result, they've been suggested as a pretreatment before ultrafiltration or reverse osmosis (RO) (Kolya & Kang, 2023). Functional NPs may easily be doped into spinning solutions or generated *in situ* to produce NP-impregnated nanofibers. The excellent properties and adaptability make them a standard foundation for fabricating membrane filters or media which are multifunctional, or by adding functional components to the nanofibers. This may be explained by designing an affinity nanofiber membrane by combining ceramic NPs or adhesives on the nanofiber scaffold to remove heavy metals and organic pollutants during filtering (Kuppusamy et al., 2015).

Table 12.1 Application of various nanomaterials using different technologies for sustainable water resource management

Application	Nanomaterial used	Available technologies	References
Adsorption	Carbon based nanotubes	Adsorption of recalcitrant contaminants, contaminant detection	Yu et al. (2014)
	Nanoscale metal-oxide	Slurry reactors, adsorptive media filters	Aziz et al. (2015)
	Nanofibers with core-shell structure	Reactive nano-adsorbents	Egodawatte et al. (2016)
Membrane filtration	Nano-zeolites	Highly permeable thin film nanocomposite (TFN) membranes	Lee et al. (2020)
	Nano-Ag	Anti-biofouling membranes	Yin et al. (2013)
	CNTs	Anti-biofouling membranes	Alizadeh et al. (2019)
	Nano-TiO_2	High-performance TNF membranes, reactive membranes	Zheng et al. (2013)
	Nano-magnetite	Forward osmosis	Irena et al. (2021)
Immobilization	Glycidyl Methacrylate	Cryogel immobilization	Dogan et al. (2015)
	Polyhydroxy Butyrate beads	Bead immobilization	Binhayeeding et al. (2020)
	Manganese peroxidase	Nanocomposite formation	Siddeeg et al. (2020)
Photocatalysis	Nano-TiO_2	Photocatalytic reactors	Riaz and Park (2020)
	Fullerene derivatives	Solar disinfection systems	Pan et al. (2020)
Disinfection and microbial control	Nano-Ag	Water disinfection	Bharadwaj et al. (2021)
	CNTs	Anti-biofouling surface	Qi et al. (2018)
	Nano-TiO_2	Disinfection and decontamination	Petti and Messano (2016)
Sensing and monitoring	Quantum dots (QDs)	Optical detection	Lou et al. (2014)
	Dye-doped silica	Optical and electrochemical detection	Bochkova et al. (2020)
	CNTs	Electrochemical detection	Cho et al. (2022)
	Magnetic NPs	Sample preconcentration and purification	Yildiz (2016)

12.2.2.2 Nanocomposite Membranes

By incorporating NPs into inorganic or polymeric membranes, membrane nanotechnology strives to produce multifunction or synergy. Hydrophilic metal oxide NPs (ex: zeolite, Al_2O_3, and TiO_2), antimicrobial NPs, and photocatalytic nanomaterials (e.g., TiO_2 and bi-metallic particles) are examples of these nanomaterials. The major purpose of introducing hydrophilic metal-oxide NPs is to prevent fouling by improving membrane's hydrophilicity. Metal-oxide NPs like silica, alumina, zeolite, and TiO_2 has demonstrated to improve membrane water permeability, surface hydrophilicity, and fouling resistance. Additionally, inorganic NPs improve the thermal and mechanical resilience of polymeric membranes, minimizing a detrimental impact of heat and compaction on membrane permeability. Antimicrobial nanomaterials like CNTs and Nano-Ag may help to minimize membrane biofouling. However, the biofouling of membranes for their prolonged effectiveness is not reported yet. Long-term filtering studies are required to assess the fouling of antibacterial activity of CNTs (Dar et al., 2020).

12.2.2.3 Thin-Film Nanocomposite(TFN) Membranes

The integration of NPs on the layer of TFN membranes is important to the development of TFN membranes. Nanomaterials include Nano-Ag, nano-zeolites, Nano-TiO_2, and carbon nanotubes. The effect of NPs on membrane selectivity and permeability is determined by their size, kind, and quantity. The most often utilized dopants in TFN are nano-zeolites, which improve membrane permeability and result in a more permeable, thicker polyamide active layer, and negatively charged. They act as transporters for the antimicrobial compounds like Ag+, giving the membrane anti fouling capabilities (Giri & Kumari, 2023). The addition of nano-TiO_2 (up to 5% weight) to the TFC active layer enhances membrane rejection while retaining permeability. When the concentration exceeds 5% wt, the water flow rises, indicating the creation of defects in the active layer. Because of their antibacterial properties, CNTs have also found use in TFN membranes. Single wall CNTs were covalently attached to a TFN membrane surface, resulting in mild anti-bacterial characteristics, and potentially lowering or delaying membrane biofouling (Roy et al., 2021).

12.2.2.4 Biological Membranes

Aquaporins regulate the flow of water across cell membranes through highly permeable and selective protein channels. They were integrated into triblock-polymer amphiphilic vesicles, that has water permeability that was at lowest order of the magnitude greater than original vesicles. One potential idea is to encapsulate commercial nanofiltration membranes with aquaporin-incorporated lipid bilayers, however this has had minimal success. Because of their atomic smoothness and water molecules in one-dimensional single-file ordering, aligned CNTs have been proven to produce substantially quicker water penetration than the Hagen Poiseuille equation predicts (Kishore et al., 2022). The scaling up of membrane fabrication and nanomaterial manufacturing is a major challenge for both aligned and aquaporin carbon nanotube membranes. The most frequent method of producing aligned nanotubes is chemical vapor deposition (CVD). Nanocomposite and TFN membranes are scalable because they can be manufactured using existing industrial production techniques. Enhanced water permeability has the potential to decrease the required applied membrane area or pressure, consequently reducing costs and potentially improving energy efficiency in treatments with low osmotic pressure (Ajith et al., 2021).

12.2.2.5 Forward Osmosis

The process of forward osmosis (FO) harnesses the osmotic gradient to pull water from a solution with lower pressure to one with higher osmotic pressure. This process requires a draw solute with high energy-intensive and osmolality to recover water. Magnetic NPs have been investigated as a novel form of draw solute for simple reuse and separation. Hydrophilic coating aids dissolution and increases osmotic pressure. A study showed that 0.065 M poly(ethylene glycol) diacid-coated magnetic nano particles achieved forward osmosis permeate flux greater than 10 L m^2 h^{-1} when deionized water is utilized as feed solution (Phian et al., 2022).

12.2.3 Immobilization

On the basis of iron oxide reaction, the magnetic NPs, or n-Fe_3O_4, with surface coating as phthalic acid and protective membrane, a biosorbent was constructed. *Staphylococcus aureus* was then added to the resultant product to cause the development of the n-Fe_3O_4- Phth-*S. aureus*. Sorption processes of Ni(II), Cu(II), and Pb(II) is significantly influenced by the functional groups of the core iron oxide NPs and *S. aureus*. Cu(II) was extracted from aqueous solutions using *S. cerevisiae* biomass adsorbed on the surface of chitosan coated magnetic NPs. Conglomerations are not visible, and analysis using TEM demonstrated that biomass has effectively immobilized on the surface of chitosan-coated magnetic NPs. The spinal-shaped, pure Fe_3O_4 NPs were verified by the XRD images (Benjamin et al., 2019). Manganese peroxidase that has immobilized on the chitosan beads via glutaraldehyde cross-linking enhances dye decolorization in effluents of textiles by up to 97%. The mutagenicity and cytotoxicity tests were dramatically decreased, even by up to 70%. The function of this has support in application of dye removal and enzyme immobilization. They created a nanocomposite of chitosan and Fe_3O_4 NPs coated with N-(3-dimethylaminopropyl)-N′-ethyl carbodiimide hydrochloride, which has large surface area that is ideal for manganese peroxidase attachment (Yang et al., 2022).

12.2.4 Photo-Catalysis

Photocatalytic oxidation is a sophisticated technology that is used to remove trace pollutants and microbial pathogens as well as to treat refractory organic molecules. It can be used to improve the biodegradability of hazardous and non-biodegradable pollutants. However, due to restricted photocatalytic activity and light fluence, its extensive applicability confronts limitations such as delayed kinetics. Nano-TiO_2 is vastly used as a semiconductor photocatalyst in wastewater treatment due to its chemical stability, low cost, low toxicity, and availability as a raw material. When it absorbs a UV photon, it produces an electron/hole (e^-/h^+) pair that either migrates to the surface and thereby creates reactive oxygen species (ROS) or experiences unwanted recombination(Punia et al., 2023; Nagendranatha Reddy et al., 2021). TiO_2 nanotubes are observed to be working efficiently than the conventional TiO_2 and because of smaller carrier-diffusion routes and fast transfer of reactants mass towards surface of nanotube, NPs are also used in the breakdown of organic molecules. Metal oxide nanomaterials like CeO_2, TiO_2, and CNTs, are investigated as an effective catalyst in the heterogeneous catalytic ozonation procedures thereby it rapidly allows and eliminates the organic pollutant matter. Nanomaterials have a high catalytic activity due to their huge surface area and conveniently accessible surface (Tarkeshwar et al., 2023).

12.2.5 Disinfection and Microbial Control

Traditional disinfection practices have increased public health, but a challenge of providing efficient disinfection without forming harmful disinfection byproducts (DBP) requires innovations in technology. Nanomaterials with active antimicrobial properties, such as, nano-TiO_2, nano-Ag, nano-ZnO, CNTs, nano-Ce_2O_4, and fullerenes, can inactivate the microorganisms by producing toxic metal ions and also generating few ROS. Nano-Ag is commonly used for usage wastewater treatment devices of its wide-spectrum, strong antimicrobial activity and lower toxicity to human beings. Release of silver ions, that attaches to functional groups with the higher affinity of Ag^+, is crucial for its bioavailability and toxicity (Khan, 2021; Phian et al., 2022). CNTs, with their antibacterial activity, fibrous structure, and conductivity, enables their usage in the antimicrobial filters. Antibacterial properties of carbon nanotubes and some carbon-based nanomaterials involves electronic structure-dependent oxidation stress and membrane perturbation. Metallic, short, and dispersed CNTs with small diameters have more toxicity, but they can also be used in electrochemical processes to inactivate physically trapped microorganisms through oxidation. TiO_2 photocatalysts effectively produce ROS under inactivating pathogens, UV-A irradiation, and degrading organic contaminants with less amount of DBP formation. Nano-TiO_2 has higher photocatalytic property because of its large specific surface area, lower volume e^-/h^+ recombination, and faster interfacial charge transfer. However, when particle size becomes smaller, the photocatalytic activity reduces due to surface recombination of e^-/h^+ complex (Al-Issai et al., 2019).

12.2.6 Sensing and Monitoring

Monitoring of water quality is difficult because of low micropollutant concentrations, enormous complexity in wastewater matrices, also the unavailability of low-cost, quick pathogen detecting technologies. Rapid microbe identification is critical for biofilm or diagnosis-based disinfection management, that rely on improved sensors for responsive, high-efficiency, and targeted treatment (Sharma et al., 2023). The combination of nanomaterials with recognition agents might result in microbial detection sensors that are rapid, sensitive, and selective. Using their unique optical, magnetic or electrochemical capabilities, nanomaterials can increase speed, sensor sensitivity, and also multiplex target detection. Magnetic nanoparticles and carbon nanotubes are being researched for sample concentration and purification. QDs, noble metal NPs, dye-doped NPs, and CNTs are commonly used in nano sensing research. QDs feature broad absorption bands but limited fluorescence emission spectra, allowing them to detect many targets with a single excitation source. Noble metal NPs increase surface enhanced Raman spectroscopy, whereas dye-doped polymeric and silica NPs display higher luminous intensity. CNTs are ideal for field-effect transistors and electrodes because they promote electron transport and analyte-sensor interactions (Ali & Gupta, 2006; Modestra et al., 2020).

12.2.7 Multifunctional Devices

Advancements in functional nanomaterials and their integration with traditional technology provide prospects for building multifunctional processes capable of performing numerous functions in one reactor, such as water disinfection, decontamination, and separation. Nanomaterials are ideally suited to multifunctional systems due to their ease of assembly, even on small carriers such as nanofibers. Membranes have been extensively researched as a platform for multifunctional devices, and electro spun nanofiber-based multifunctional devices for low-cost water treatment have been proposed. Electrospinning is a simple and low-cost technology for producing ultrafine nonwoven fibres that can be utilized for purposes other than filtering. Electro spun nanofibers are an excellent substrate for multifunctional filters, and it is predicted that nanofibers with diverse functions can be stacked in layers or cartridges for independent regeneration or optimization of each functionality (Abdelbasir & Shalan, 2019).

12.3 Nanotechnology Assisted Bioremediation (NAB) for Wastewater Treatment

Nanotechnology-assisted bioremediation (NAB) is an advanced approach to address water pollution that blends the power of NPs with biological processes. This has several advantages, including greater efficiency, cost-effectiveness, and reduced environmental impact. However, possible dangers linked with the usage of NPs, such as their cumulative impact on ecosystems and human health, must be considered (Singh et al., 2020). Microorganisms viz., bacteria, microalgae, fungi, yeast, etc. along with enzymes play a major and significant role in bioremediating different components of wastewater by utilizing the additional features or advantages available with nanotechnology. Combining microorganisms with nanotechnology has become a best method for effective water filtration, addressing various pollutants in water systems. Heavy metals like mercury, cadmium, and lead can be eliminated from water through microbe-assisted nano remediation, where microorganisms bioaccumulate and change these heavy metals. Nanomaterials like iron, titanium, or carbon NPs can increase microbial activity and create a synergistic interaction that speeds up removal of heavy metals from water. Organic pollutants, like pesticides, medications, and industrial chemicals, can be broken down by microbes into safe byproducts. Nanotechnology facilitates this biodegradation process, creating an environment favourable for microbial activity and growth. Nanomaterials can help optimize and control microbial activities, such as nutrient cycling and bio flocculation, improvising the overall quality of water (Megharaj et al., 2011).

12.3.1 Removal of Organic and Inorganic Pollutants by Bacteria

Microbial-assisted nanotechnology can target organic pollutants using nanomaterials with catalytic properties. For instance, nano catalysts supported by microbes can enhance the degradation of organic compounds through advanced oxidation processes, breaking down pollutants into harmless byproducts (Vázquez-Núñez et al., 2020). Bacterial cells may concentrate metals, which is used in wastewater treatment. Bioremediation methods that use indigenous microorganisms have proven to be environmentally beneficial, effective, and self-sustaining in removing contaminants from their surroundings. Nanoparticles (NPs) can be used to remove organic contaminants directly or through adsorption, assisting bioremediation by boosting microorganism growth or trapping contaminants using microbial enzymes such as alcohol dehydrogenase, manganese peroxidase and catalase. NPs are known to increase biosurfactant synthesis in microorganisms, hence enhancing hydrocarbon solubility and pollutant degradation. Immobilizing microbial cells is known to boost microbe activity, ensuring their survival in difficult environments. When immobilized cells are grown with oil as a carbon source, *Pseudomonas* sp. and *Micrococcus luteus* has been shown to improve crude oil sorption and consuming capabilities. The addition of microbial enzymes to NP matrices improves their performance under a variety of circumstances, including pH, temperature, and degradation capability. The environmentally friendly synthesis of NPs from plant extracts, and bacterial, and fungal enzymes provides a sustainable alternative (Saxena et al., 2020).

12.3.1.1 Petrochemical Degradation

Petroleum hydrocarbons are further classified as aromatics, asphaltenes, resins, and saturates. Microbes degrade them in a distinct ways, based on the type and quantity of hydrocarbons available. Higher molecular weight polycyclic hydrocarbons may not be decomposed. *Acinetobacter sp.* degrades n-alkanes in chain lengths ranging from C10 to C40, with carbon serving as the only carbon source. Microbe-assisted nanotechnology is pivotal in addressing both oil pollution and hydrocarbon contamination in water (Benjamin et al., 2019). This collaborative strategy synergizes the metabolic capabilities of oil-degrading microbes with the adsorption properties of nanomaterials, resulting in highly effective oil remediation. These nanomaterials enhance microbial attachment to hydrocarbons, accelerating the degradation process, particularly in addressing hydrocarbon spills and pollution in aquatic environments. The integration of microbes and nanotechnology thus offers a versatile and efficient solution for remediating water contaminated with oil and hydrocarbons, highlighting its potential for sustainable environmental protection and water quality management (Vats & Pandey, 2021).

Mycobacterium can degrade petroleum products, *Flavobacteria, Modococci, Nocardia, Moraxella, Chrobacteria, Cyanobacteria, Arthrobacter, Streptomyces, Bacilli, Burkholderia, Brevibacterium, Gordonia, Dietzia, Aeromicrobium, Pseudomonas, Aeromonas*, and other bacteria. *Sphingomonas* could decompose polyaromatic hydrocarbons (Tarkeshwar et al., 2023). Soil bacteria and saltwater bacteria have different biodegradation rates. In aerobic conditions, microorganisms use specialized enzyme systems (peroxidase, oxygenase, and hydroxylase) to degrade petroleum hydrocarbons, which begin to attach to bacterial cells for the raw materials which further proceeds to biosurfactants generation. Biosurfactants are engaged in increasing solubility and removing the pollutant. They increase the amount of oil and surface area accessible for bacteria to use and lower surface tension to aid in the formation of micelles. The generation of biosurfactants ensures the utilization of substrates (Wang et al., 2020).

Petrochemical wastewater degradation has been a significant concern in recent years, with various investigations on the usage of photocatalytic methods to treat this wastewater. Possible microbial community resistance development is another concern, as persistent exposure to nanomaterials may cause some microorganisms to develop resistance mechanisms, decreasing their efficacy in treating water. Ethical concerns arise from the deliberate discharge of synthetic microorganisms or NPs into the environment, and a comprehensive ethical assessment is needed to balance the advantages of wastewater treatment against potential harm to the environment and microbial communities (Jatoi et al., 2021; Aruna et al., 2022).

12.3.1.2 Pesticides

Pesticides are chemicals used to eradicate pests. They are classified by their use (algicides, herbicides, bactericides, fungicides, rodenticides, nematicides, and insecticides). The generated conventional activated sludge was enhanced with 2,4-dichlorophenol (2,4-DCP) to digest bacteria isolated from 2,4-DCP-contaminated soil. The bio-augmentation of the sludge increased the sludge's removal effectiveness to 88–89% when the initial concentration was between 10 and 100 mg/L. Activated sludge from an oil refinery's wastewater was acclimated and used to remove fluoxil from the water. The herbicide clearance rate reached 80% after 70 h of exposure (Khan, 2021; Phian et al., 2022). Pressurized activated sludge is a very efficient treatment technology that does not require a huge land area, is less expensive than other methods, and is more environmentally benign than chlorination. It does, however, require a sludge disposal location as well as skilled workers for maintenance and operation. The employment of integrated methodologies is the best approach to designing solutions for removing and degrading environmental and public health pollutants such as pesticides. This review included the treatment of a variety of pesticides from water, including 25 herbicides, 27 insecticides, 20 fungicides, one bactericide, one algaecide, one molluscicide, and one nematicide (Al-Issai et al., 2019).

The steps in establishing a wastewater treatment plant should include a thorough influent water analysis to identify contaminants, detailed literature review research to identify effective treatment techniques for the contaminants encountered, and an overall feasibility study to determine the cost of each effective method. For example, the most effective strategies for removing atrazine were nano-ZnO, biological treatment using a specific microbial consortium, adsorption with wood charcoal and biochar in addition to ozonation/H_2O_2, and nanofiltration. Adsorption appears to be the most successful and cost-effective way for removing methyl parathion, according to research. Chestnuts were the most effective absorbent, removing efficiently with the highest starting rate. The optimum treatment procedure for 2,4-D was oxidation (peroxymonosulfate), followed by ozonation in the presence of peroxide, and adsorption with wood charcoal (Sharma et al., 2023). Because of the unstainable removal of both physical and chemical pesticides, microbes can be used as a bio-weapons to combat hazardous chemicals arising from agricultural purposes (Saravanan et al., 2021).

12.3.1.3 Dyes Degradation

Regarding color diversity, rapidity of absorption, solubility and intensity of coloration, and water solubility, artificial dyes have several benefits over natural dyes thereby resulting in over utilization of dyes in various applications. As dye demand and production increase, so does concern about the impact of textile releases on general health of aquatic ecosystem. Additives and other compounds belonging to organic and inorganic classes, as well as color quantities ranging from 10 to 200 mg/L, are found in textile industry wastewaters. Azo dyes that are widely utilized in the textile industry due to their ease in economic and applications (Sreelatha et al., 2015; Kumar et al., 2015). The treatment of water contaminated with dyes and colorants benefits significantly from the integration of microbes and nanotechnology. Nanomaterials with high surface areas, like nanostructured metal oxides or carbon-based nanomaterials, facilitate the adsorption of dyes. Microbes, in turn, contribute enzymatic capabilities for the degradation of complex dye molecules. The synergistic effect enhances the overall efficiency of colorant removal from water systems (Fajardo et al., 2022).

Aeromonas hydrophila, *Bacillus cereus,* and *Bacillus subtilis* have been thoroughly examined and show promise for bioremediation of azo dyes. *Pseudomonas luteola*, *Pseudomonas sp.*, and *Proteus mirabilis* can destroy azo dyes under anoxic circumstances. For color breakdown, these bacteria use oxidoreductive enzymes. To dissolve the azo bonds, aerobic bacteria use oxygen-catalyzed to give azoreductase. Some microbial strains break down dyes in aerobic conditions and oxidize aromatic ring organic compounds with mono dioxygenase. Anaerobic bacteria break down dyes via the enzyme azoreductase. These enzymes are also involved in desulfonation, hydroxylation, and deamination. Under aerobic conditions, *Pseudomonas aeruginosa* could decolorize a variety of azo dyes (Mondal et al., 2019; Nagendranatha Reddy et al., 2022).

The reactive dye binding capacity and pH of maximum adsorption of reactive dyes by various ion-exchange resins. However, the pH of dyehouse reactive dyeing effluent is usually highly alkaline, which requires strong acids for the removal of dyes and neutralization after decolorization treatment (Saxena et al., 2020). Ion-exchange resins are difficult to dispose of since they are not biodegradable. Reactive dye removal has had limited success due to poor hydrodynamic properties, limited recycling and reusability, unpredictable adsorption behavior, and difficulty regeneration when compared to activated carbon adsorbents. Some adsorbents, including lignocellulosic adsorbents, have low dye binding capability due to their anionic nature with reactive dyes. Cationic chitosan, on the other hand, binds reactive dyes well (Kumar et al., 2023) (Fig. 12.3).

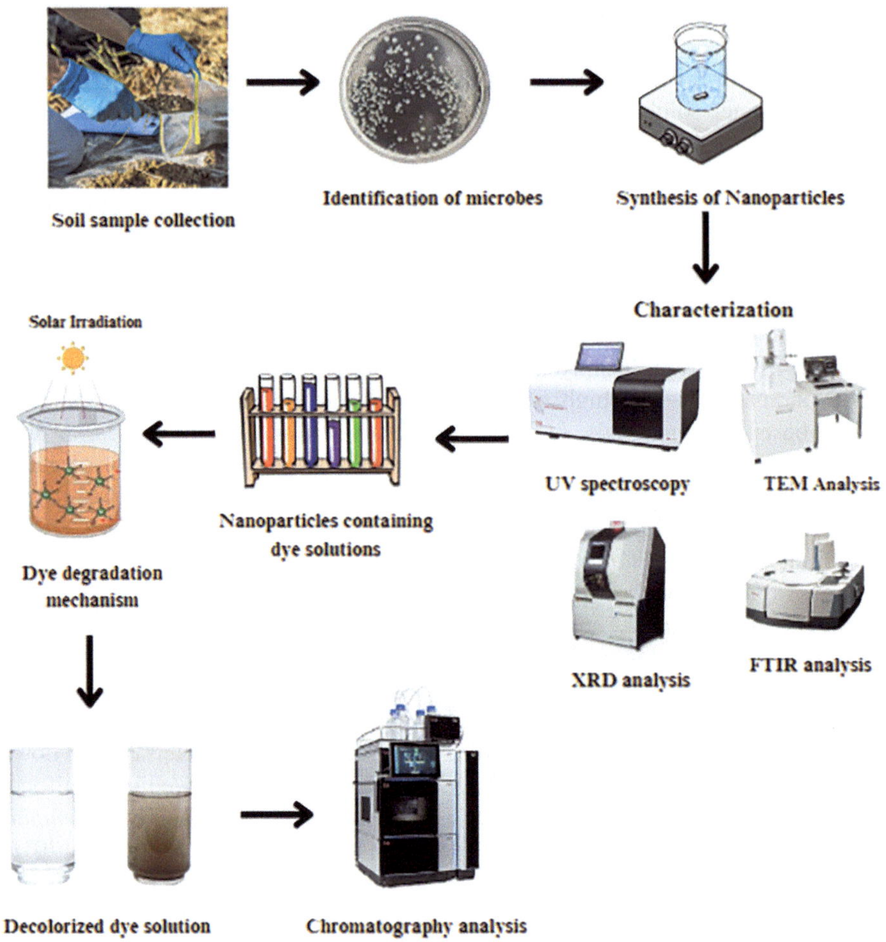

Fig. 12.3 Schematic representation of dye degradation using NPs

12.3.1.4 Pharmaceutical Products and Personal Care (PPCPs)

Pharmaceuticals and personal care products (PPCPs) are new sources of persistent pollution. These have progressively increased globally, particularly since COVID-19 epidemic. This has led to pollution of water by pharmaceutical products, which is a serious health hazard. Upon long-term exposure to humans the pharmaceutical pollutants induce severe genotoxic responses, with south Asia, sub-Saharan Africa, and South America being the most contaminated areas. Increasing access to medicines in middle income countries has led to a higher volume of pharmaceuticals in wastewater compared to low-income countries due to poor wastewater management. Water pollution by pharmaceutical products is a serious health concern, with pharmaceutical pollutants causing severe genotoxic response upon long-term exposure to humans. The top seven microorganisms accounted for four to seven hundred thousand deaths in 2019 due to antibiotic resistance (Kabir et al., 2018).

Membrane-based technologies are the most efficient and dependable way to filter water for drinking, but they are expensive and waste more water than they save. To improve the performance of photocatalysis and adsorption processes, detailed studies of nanocomposites adorned with various types of nanomaterials were undertaken. Consumers introduce PPCPs to marine environment using home cleaners, sanitizing agents, detergents, shampoo, and medications. PPCPs have persistent and complicated compounds that infiltrate the water cycle, causing an increase in antibiotic resistance, reproductive problems, and tumor growth. They convert certain contaminants into recalcitrant and hazardous compounds that get distributed quickly in the water levels. Microorganisms have a challenging time breaking down PPCPs since medications are poisonous to bacteria. Microbes transform complicated structures into a benign or less hazardous version (Shahi Khalaf Ansar et al., 2023).

12.3.1.5 Degradation of Heavy Metals

The existence of a single poisonous metallic species in a contaminated solution is uncommon, but the presence of many metal ions frequently results in an interaction effect. Recently, there has been an increase in interest in using free-living and immobilized microbes to remove multiple metals from metal solutions simultaneously. The presence of other metal ions in the solution affects the overall ability of metal-binding due to competition. E. coli biomass was found to have selective binding behavior towards Pd(II) in a bimetal solution containing Pt(IV) and Pd(II). In a single metal solution containing Pt(IV) and Pd(II), the bacterium absorbed Pt(IV) somewhat more efficiently. However, in binary metal solutions, the absorption rate of Pd(II) increased (Dixit et al., 2015).

The biosorption order of metal ions in binary and ternary metallic solutions was found to increase with the ionic radius of the metal present in the solution. The presence of one metal ion slows the biosorption rate of competing metal ions. In the ternary system, the other two metals in the solution impeded the sorption rate of the third metal, and the absorption rate of all the metals was lower than in their lone metal ion solutions (Verma & Kuila, 2019). The biosorption of the heavy metal ions in wastewater is reacted by various factors, including pH, temperature, biosorbent dosage, and competing ions. pH influences the solution chemistry, chemical nature of metal ions, and ionization state of functional groups on the biosorbent surface. An increase in pH promotes electrochemical attraction and metal ion adsorption, while temperature increases the rate of adsorbate diffusion and decreases liquid viscosity. The amount of biomass also affects the biosorption process, with an increase in biosorption occurring with an increase in biomass concentration. Competing ions can reduce the specificity of the biosorbent. Agitation is crucial for the biosorption mechanism, as it minimizes mass transfer resistance and accelerates metal uptake. The nature of the binding sites, biosorbent, growth rate, and contact time are crucial factors for effective biosorption. Small-sized cells and the age of the culture also affect removal efficiency. The contact time between the biosorbent and adsorbate is essential for maximum biosorption. The reaction medium can be maintained at optimum conditions, such as pH, agitation, temperature, and contact time, to achieve highest amount of sorption (Dixit et al., 2015).

12.3.2 Removal of Pollutants by Fungi and Yeast

The yeast cell wall, composed of organic components and polymers, is a common biosorbent for metal removal. *Saccharomyces cerevisiae*, in immobilized or free form, is a widely studied biosorbent for metal removal due to its ease of cultivation, safety, and potential for identifying the action of biosorption in removal of metal ion. The biosorption of Pb(II) ions from aqueous solutions has been investigated using ethylenediamine-modified yeast biomass coated with magnetic chitosan microparticles (EYMC). The immobilized waste biomass *S. cerevisiae* in Ca-alginate, PVA-Ca-alginate-graphene oxide in $CaCl_2$-boric acid solution, and Ca-alginate with graphene oxide (GO) has shown the best biosorption properties and mechanical stability. The biosorbent of immobilized *S. cerevisiae* biomass in magnetic chitosan nanospheres is utilized for the removal of Co(II), Sr(II), and Cs(I) from one- and two-component solutions. Initial pH values have a great impact on the removal of these ions (Yin et al., 2020).

Fungi plays a significant role in the bioremediation of toxic compounds, mainly in the myconanotechnology or mycosynthesis of metal NPs. These fungi are more metal-tolerant, have a high metal absorption potential, and are ideal for metal NP synthesis due to their direct involvement in the bio reduction and stabilization processes. Some proficient fungal candidates used for metal NP synthesis include *Penicillium, Fusarium, Aspergillus,* and *Verticillium.* Metals such as cadmium, gold–silver alloys, zirconium, silver, platinum, gold, palladium, titanium, selenium, tellurium, magnetite, silica, and uraninite are used for the green synthesis of NPs. Recent research has demonstrated the successful use of these unique nanostructures in bioremediation, with primary mechanisms comprising heavy metal adsorption and the transformation of harmful chemicals into harmless or less toxic forms. White-rot fungus (WRF) and NPs are extremely important for their bioremediation capabilities, with magnetic NPs exhibiting higher stability and elimination of pollutants from wastewater (Mohiuddin et al., 2024).

12.3.3 Removal of Pollutants by Microalgae

Microalgae plays an important role in nano-assisted bioremediation, particularly in water treatment, as they effectively absorb and collect toxins like heavy metals, minerals, and organic pollutants.

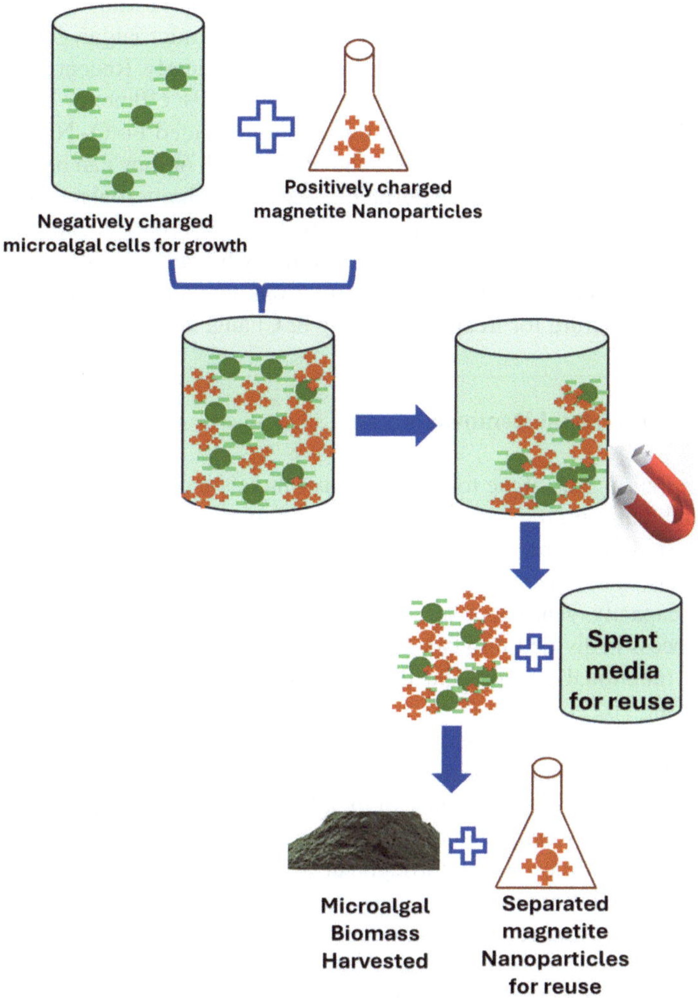

Fig. 12.4 Schematic representation of harvesting microalgal biomass using magnetite NPs

When combined with NPs like nanostructured materials or nanocomposites, their remediation capacity is significantly increased. NPs have a vast surface area for adsorption, allowing contaminants to attach and concentrate. Microalgae's large surface area, rapid growth rates, and adaptability make them an effective choice for water treatment. This long-term and efficient solution to environmental degradation is a promising choice for future bioremediation efforts (Abdelfattah et al., 2023; Nagendranatha Reddy et al., 2019) (Fig. 12.4).

Microalgae harvesting is a high-cost operation, accounting for 20–30% of overall expenditures. Biological approaches such as fungi, organic molecules, exopolymer compounds, and organic polymers have been investigated to lower harvesting costs. NPs enable advanced vertices for microalgae harvesting that do not use mechanical or electrical forces. Magnetic NPs bind to microalgae cells at precise pH levels, allowing them to flocculate and separate efficiently. Recent studies reveal that NPs can harvest biomass in less time, with over 95% efficiency for iron oxide NPs and 96.6% efficiency for cetrimonium bromide coated Fe_3O_4 NPs. Following harvest, magnetic NPs can be recovered and repurposed, or they can be employed in additional biomass processing. The efficiency with which microalgae are harvested is also influenced by agitation speed and culture pH. Microalgae can grow in both autotrophic and heterotrophic modes to consume hydrocarbons. Due to inappropriate textile effluent composition, successive dilutions are required for use. Microalgae are a significant culture for treatment (Leong & Chang, 2020).

12.3.3.1 Heavy Metal Removal

Phycoremediation uses algae to remove heavy metals including arsenic, cadmium, mercury and, chromium, is a less expensive and more efficient way to treat wastewater than traditional methods (evaporation, precipitation, electrochemical treatment, osmosis, and, ion exchange). Algae are especially useful because of their high absorption capacity, year-round availability, and lack of harmful metabolite synthesis. The process is divided into two stages: first, metals are adsorbed on the surface of cell *via* functional groups, and then they are bioaccumulated within the cells (Tripathi et al., 2022).

Cadmium poisoning affects algal growth and photosynthetic efficiency differently depending on strain. Diatoms have higher tolerance and remediation efficiency (up to 99%) than green algae because of their higher adsorption capacity. Freshwater algae such as *Scenedesmus acutus* and *Chlorella pyrenoidosa* remove considerable amounts of cadmium, primarily through surface sorption. Arsenic, which is highly hazardous in its inorganic form, is detoxified by algae by converting it to organic forms via chelators and functional groups. Green algae and marine species such as *Chlorella salina* and *Dunaliella tertiolecta* have been successful, with *Scenedesmus almeriensis* removing 40.7% of arsenic in 3 h. Genetic changes, like those found in *C. reinhardtii*, dramatically improve arsenic abatement. The removal effectiveness varies with temperature and initial arsenic concentration but is unaffected by salinity (Megharaj et al., 2011). Algae such as *Chlorella vulgaris, Spirogyra,* and *Spirulina* may remove both trivalent and hexavalent chromium. *Chlorella* may remove up to 97% of chromium by biosorption and accumulation. Advanced approaches, such as pre-treatment and immobilization, improve chromium removal efficiency. Algae's surface functional groups allow them to effectively absorb mercury, which is highly poisonous. *Chlorella vulgaris* adsorbs mercury via carboxylic groups, but *Scenedesmus obtusus* achieves a high sorption capacity after phosphate treatment. Another method of removal involves the enzymatic conversion of mercury (Hg^{2+}) to elemental mercury (Hg), which is subsequently volatilized. Bioaccumulation induces phytochelators, which aid in mercury detoxification (Zeraatkar et al., 2016).

12.3.3.2 Nutrient Reduction

Membrane photobioreactors (MPBRs) have shown promise in the water treatment and nutrient elimination by utilizing microalgae's capabilities. The performance of MPBRs is influenced by key parameters including hydraulic retention time and solids retention time. Microbial-assisted nanotechnology can be harnessed for nutrient removal. Types: Nitrate, phosphate. Engineered NPs, when used in conjunction with specific microbial strains, enable the efficient uptake and removal of nitrogen and phosphorus from water bodies, addressing concerns related to nutrient pollution (Ajith et al., 2021). Nanocomposites with specific binding sites are used to capture and remove excess nutrients. Microbes aid in transformation of these nutrients into less harmful forms through processes like denitrification.

A mixed culture of microalgae, such as *Chaetophora sp.* and *Navicula sp.*, achieved optimal nutrient removal rates at a hydraulic retention time (HRT) of 1 day and a sludge retention time (SRT) of 3 days. Outdoor pilot-scale MPBRs achieved phosphate and ammonium eliminating rates of 1.17 mg/L/day and 7.68 mg/L/day, respectively. Optimized MPBR configurations, such as those operated showed higher nutrient removal efficiency at low nitrogen and phosphorus contents, suggesting suitability for the treatment with low nutrient levels. Microfiltration and FO in MPBRs and *Chlorella vulgaris* systems exhibited biomass accumulation, but osmotic MPBRs achieved higher nutrient removal efficiencies. The submerged osmotic MPBR achieved higher nutrient removal efficiencies but was prone to flux loss and foulant deposition. Overall, MPBRs show promise for efficient nutrient removal and biomass production, with optimization of operating conditions and configurations enhancing their performance. (Li et al., 2019).

12.3.4 Removal of Pollutants by Enzymes

Enzymatic bioconversion technologies are gaining popularity in wastewater treatment because to their high activity, specificity, and less harmful biodegradation byproducts. One potential approach is to use crude enzyme extracts, which must be purified in order to increase their apparent activity and catalytic efficiency. The functionalized TiO_2 nanoparticles may be a suitable candidate for enzyme immobilization. It remains to be seen if biocatalytic NPs can be employed to address new micropollutants of concern. The crude enzyme was covalently immobilized on TiO_2 NPs using a previously developed sequential approach. The crude enzyme content was evaluated using the Lowry technique, and the activity of both free and immobilized enzymes were assessed using oxidation. To determine the protein components, the culture supernatant of the fungus P. ostreatus was electrophoretically examined with SDS-PAGE. The crude enzyme extract from P. ostreatus was demonstrated to be homogenous with a single band, coinciding with the molecular mass of the purified commercial laccases released from the same white-rot fungus. (Sharma et al., 2018). By diluting the original crude enzyme extracts, the undesirable contact between non-protein components and the immobilization support surface could be reduced. The resulting NPs performed similarly to the purified commercial enzymes. Two sample micropollutants were used to demonstrate the possibility of utilizing biocatalytic NPs for wastewater treatment. Overall, the biocatalytic NPs in this study indicate a significant potential to encapsulate crude enzyme extracts for cost-effective practical applications (Peixoto et al., 2011).

12.4 Microbial-Assisted Nanotechnology for Sustainable Water Management and Its Applications

Microbial-assisted nanotechnology is a revolutionary approach to sustainable water management, combining the unique characteristics of microbes with nanomaterials to develop environmentally friendly water treatment methods. This method enhances the effectiveness of microbial activities by increasing their reliability, viability, and efficiency. It is particularly useful in cleanup of heavy metal-polluted water, where NPs laden with metal-resistant bacteria or fungus target and sequester heavy metals, protecting water quality. Additionally, nanomaterials serve as carriers for microbial enzymes, enhancing the catalytic efficiency of hydrocarbon, pesticide, and pharmaceutical residue degradation in water systems, reducing pollutants' ecological impact (Malik et al., 2022).

12.4.1 Microbial Fuel Cells

Microbial fuel cells (MFCs) are a prominent solution to generate electricity by treating wastewater. However, the design of electrodes remains a challenge due to the need for cost-effective and compatible materials. By altering material properties, MFC technology has made substantial advances in electron generation and transmission. Carbon-based materials are frequently employed because to their high porosity, huge surface area, and excellent electrical conductivity. To improve electrode performance, several nanocomposites with improved characteristics have been proposed and created (Verma et al., 2021). The primary goal of MFC research is to improve the interaction of microorganisms and anode materials in wastewater treatment and identify their compatibility. Researchers are particularly concerned with cost-effective materials because MFCs must be both sustainable and environmentally beneficial. For example, stainless steel improves the power density and coulombic efficiency of MFCs. Incorporating stainless steel into different anode materials has resulted in tremendous progress. Graphene, with its enormous surface area and high electrical conductivity, is commonly utilized as an anode material in MFCs. The power density of graphene-modified stainless steel mesh anodes in MFC is 2668 mW/m^2. N-doping TiO_2 nanosheets can improve the electronic characteristics of carbon paper anodes, resulting in higher energy generation while minimizing electron loss (Umar et al., 2021).

Nanomaterial electrode materials in MFCs provide a promising tool for high hydrogen production, as MFCs have the potential to generate electricity by treating wastewater. Although the yield and purity of hydrogen remain a difficult problem, there are still many opportunities in the development of electrode materials to produce hydrogen. (MFCs), should maintain chemical stability, have a large surface area, and be easy to upgrade. Carbon electrodes are commonly used due to their cost-efficiency, ease of production, chemical stability, biocompatibility, and electrical conductivity. However, carbonaceous electrodes have lower electrical conductivity and higher chemical charges compared to metallic counterparts. Nanomaterials can help address these limitations by modifying electrodes and their potential impact on power transfer processes, mass transfer, and cathodic electron transfer. Nanomaterials can have a favorable impact on critical limits in electrode and surface materials, catalysts in oxygen reduction processes, and media that improve bacterial electron transfer. Overall, nanoparticles represent a viable option for large-scale electrode materials in fuel cells (Bose et al., 2018) (Table 12.2).

Table 12.2 Purpose of various microbial-assisted nanotechnology in wastewater treatment

Technology	Description	Purpose in water treatment	References
MFCs	MFCs are bioelectrochemical systems that utilize microorganisms to convert organic matter into electricity. The electricity generated can be used to power various devices, including water treatment systems. MFCs have the potential to provide a cost-effective and sustainable method for water treatment	Biodegradation of organic contaminants	Du et al. (2007)
Bio nanocomposites	Bio-nanocomposites are materials that combine the properties of biological materials (e.g., bacteria, enzymes) with the properties of nanomaterials (e.g., NPs, nanotubes). They can be used to develop new materials for water treatment, such as filters, adsorbents, and catalysts	Elimination of contaminants, including heavy metals, pharmaceuticals, and pesticides	Annu et al. (2024)
Biologically Enhanced NP Synthesis (BENS)	BENS is a method for synthesizing NPs using biological materials, like bacteria, and fungi. BENS offers several advantages including reduced environmental impact and improved control over NP size and shape. NPs synthesized using BENS can be utilized for wastewater treatment applications, like disinfection and catalysis	Disinfection of waterborne pathogens	Ojha (2020)

12.4.2 Bio-nanocomposites

Bio-nanocomposites are innovative materials used in sustainable water management, particularly in water treatment and purification. These composites combine biodegradable polymers with NPs, resulting in efficient and environmentally friendly solutions. The applications include the elimination of contaminants from water sources, with NPs such as graphene oxide or nanocellulose enhancing adsorption capacity. These materials trap pollutants such as organic compounds, heavy metals, and microbes, aiding in wastewater purification. They also improve the durability and permeability of biodegradable polymers, resulting in more effective filtration membranes for water treatment plants. Additionally, bio-nanocomposites regulate the release of water treatment chemicals, ensuring sustained and focused water quality without regular interventions (Dar et al., 2020).

12.4.2.1 Bacteria-Mediated Bio Composites

Bacteria perform bioaccumulation, bioleaching, and biomineralization to solubilize metal ions, leading to research on using them as nanofactories for nanocomposites. Biogenic magnetic NPs (BMNPs) can be fabricated through controlled mineralization and biologically induced processes. This simple process makes it suitable for intracellular and extracellular biosynthesis of nanocomposites. Although it is time-consuming than physical and chemical methods, linking with ultrasonication and microwave can decrease the time (He et al., 2017).

12.4.2.2 Alginate Beads with Silver NPs

Alginate, a natural polymer derived from seaweed, is mixed with silver NPs to form antibacterial bio-nanocomposites. These beads can be used to remove bacteria and other germs from water, providing a more environmentally friendly alternative to typical disinfection procedures (Vats & Pandey, 2021).

12.4.2.3 Algal-Based Bio Composites

Algae, a microbe species, have been used to create nanocomposites (NPs) for removing pollutants from sewage. These NPs are enriched with bioactive molecules with main functional groups, such as carboxyl, amine, and hydroxyl. Phyconanotechnology, a type of algal-based NP synthesis, where algae-mediated nanoparticles have been used for metal and dye removal from aqueous solutions. Macromolecules from various algal species have also been used as reducing, capping, and stabilizing agents for the synthesis of nanocomposites (Kumar et al., 2023).

12.4.3 Biologically Enhanced NP Synthesis

Biologically accelerated NP synthesis is a green and sustainable method for water management, utilizing microbial extracts as natural sources. These agents function as reducing agents, stabilizers, and shape-directing agents, allowing for the creation of NPs of specific size and morphology. This approach reduces the need for harmful substances and high-energy procedures, contributing to a healthier environment. Biologically produced NPs can be used for water disinfection, preventing water-borne infections, and removing contaminants like dyes, heavy metals, and organic contaminants. This approach aligns with circular economy concepts and is more environmentally friendly, with higher biocompatibility and stability, making them suitable for sustainable wastewater treatment solutions (Al-Saidi et al., 2021).

12.4.3.1 Silver NPs for Antibacterial Water Treatment

Silver NPs produced biologically using microbial agents have powerful antibacterial capabilities. These NPs can be used for water disinfection, efficiently eradicating hazardous bacteria, and assuring safe drinking water production without the use of typical chemical disinfectants. Nanostructures, including silver NPs, possess antimicrobial properties. When integrated with microbial systems, these NPs can effectively inhibit the growth and proliferation of pathogenic bacteria and microorganisms in water, contributing to the disinfection process (Khan, 2021).

12.4.3.2 Iron NPs for Groundwater Remediation

The microbial-assisted production of iron NPs is used for heavy metal cleanup in groundwater. These NPs, which are generated from iron-reducing bacteria, may effectively encapsulate, and remove toxins such as arsenic and chromium, providing a long-term option for cleaning up groundwater resources (Rajput et al., 2024).

12.4.3.3 Gold NPs for Catalytic Degradation

Catalytic properties are demonstrated by biologically manufactured gold NPs, which are frequently produced with the assistance of microbial extracts. These NPs can be used to catalyze decomposition due to organic pollutants in wastewater, supporting a more effective and sustainable way of water treatment (Teimouri et al., 2018).

12.4.3.4 Titanium Dioxide NPs for Photocatalysis

Titanium dioxide NPs generated biologically from microbial sources can be employed in photocatalytic processes. When exposed to sunlight, these NPs demonstrate photocatalytic activity and can aid in the breakdown of organic contaminants, providing a green and sustainable method of water treatment.

12.4.3.5 Zinc Oxide NPs for UV Filtration

The creation of zinc oxide NPs can be used in water filtration systems. The UV-blocking capabilities of these NPs improve the performance of UV water treatment technologies by limiting the passage of damaging UV rays, making water disinfection more energy-efficient and sustainable (Spoială et al., 2021).

12.5 Challenges and Economic Feasibility

Since there are less clean and useable water resources available to fulfill the increased demand for water, sustainable water resource management is a crucial global concern. A viable approach to overcoming this difficulty is microbe-assisted nanotechnology, which offers innovative techniques for water treatment, purification, and conservation. However, the implementation of such technology is not without its challenges and economic considerations. The possible ecological impact is a significant obstacle to the management of water resources using microbe-assisted nanotechnology. Although the aim of these technologies is to enhance the quality of the water, the introduction of nanomaterials and engineered microorganisms into aquatic habitats may have unexpected effects on the local fauna and flora. To make sure that the introduction of these technologies doesn't disturb the delicate balance of aquatic ecosystems, significant environmental risk assessments are required. The creation of reliable and durable microbial nanotechnologies presents another difficulty. The effectiveness of these technologies depends on microbes' capacity to carry out tasks, such as improving water treatment procedures or decomposing contaminants. A lot of R and D is needed to ensure that modified microorganisms remain stable and functional in various environmental conditions (Vishwakarma et al., 2020).

Economic viability towards applying microbe-assisted nanotechnology in water resource management is a crucial factor to consider, in addition to environmental considerations. These technologies can be expensive to develop and implement initially, requiring large investments in infrastructure, staff training, and research. These technologies' economic sustainability is dependent upon several aspects, including scalability, cost-effectiveness, and long-term maintenance. When evaluating if microbial nanotechnologies are economically feasible, scalability is an important consideration. These technologies have to be scalable to satisfy the demands of

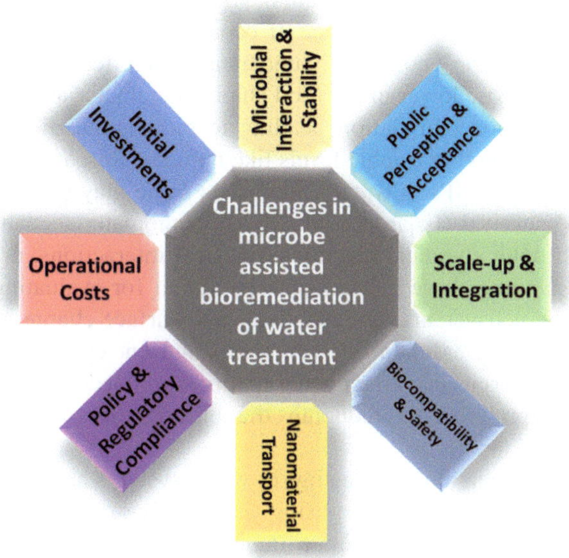

Fig. 12.5 Various challenges in microbe-assisted bioremediation of water treatment

large-scale water treatment plants and distribution systems in order to have a greater impact on the management of water resources. Scalability in the face of efficiency and cost-effectiveness demands careful planning and investment, making it a challenging engineering problem (Mishra et al., 2021).

Despite these difficulties, microbes-assisted nanotechnology holds great promise for managing water resources. These innovations could lower pollution, raise the general sustainability of the water resources, and increase the quality of water. To effectively manage the intricate interactions between technology, ecology, and economy, engineers, scientists, stakeholders and policymakers must work together to address the problems and ensure economic feasibility. Through the resolution of these issues, microbial-aided nanotechnology can be crucial in accomplishing worldwide sustainable water resource management (Kabir et al., 2018) (Fig. 12.5).

12.6 Environmental and Safety concerns

12.6.1 Environmental Impact of Nanomaterials in Water Systems

One of the key concerns about using nanomaterials in water systems is their potential to disturb the environment. Because of their special qualities, nanomaterials may interact differently with aquatic ecosystems than conventional contaminants do. Despite their large surface area and small size, they have the potential to be more reactive, which can change the balance of ecosystems and affect the behavior of living things. We still don't completely understand the effects on aquatic flora and wildlife, especially those lower in the food chain. Nanomaterials accumulate in living things through bioaccumulation, leading to biomagnification and potentially posing health risks to higher trophic levels like humans. Research is ongoing on the toxicological effects of NPs in water systems, as some may be harmful to aquatic life and species' survival (Fajardo et al., 2022). Nanomaterials can cause harmful consequences through chemical reactions, physical interactions, or a combination of both. Understanding these mechanisms is crucial for evaluating and reducing damage to aquatic ecosystems. The reactivity and surface charge of NPs can alter water quality, affecting essential elements and nutrient cycles. Additionally, the presence of nanomaterials can affect water treatment procedures, potentially releasing hazardous byproducts or reducing the elimination of contaminants (Wu et al., 2022).

Nanomaterials dispersion and destiny in water systems are influenced by sedimentation and transport processes, posing risks to far-off ecosystems. Long-term effects of nanomaterial deposition in sediments could affect creatures and benthic populations. Antibacterial resistance (AMR) in aquatic environments may arise due to the usage of antimicrobial NPs in wastewater treatment. The selective pressure these materials apply could facilitate the survival and spread of resistant strains, endangering humans and aquatic health. Also, the persistence of nanomaterials in environment raises the potential for ongoing exposure to aquatic creatures and the cumulative effects on ecosystems (Wadgaonkar et al., 2018).

The creation of thorough laws for the release and monitoring of NPs in the environment is still behind the rapid growth of nanotechnology. It is difficult to adequately evaluate and manage any environmental risks when there are regulatory gaps and uncertainties. Furthermore, it is technically difficult to monitor the behavior and presence of nanomaterials in water systems, requiring the development of advanced detection techniques (Saeed et al., 2022) (Fig. 12.6).

12.6.2 Safety Considerations for Microbial Applications

In sustainable water resource management, ensuring the biological containment of microbial applications is a basic safety precaution. The microorganisms used to treat water have to be chosen carefully or genetically modified to have limited growth and survival outside of the regulated treatment environment. By doing this, unexpected outcomes like the introduction of genetically modified or non-native creatures into natural ecosystems which may displace native species and disturb ecological balance are avoided. It is essential to preserve the genetic stability of modified microorganisms to stop accidental genetic transmission and escape. Mechanisms to reduce horizontal gene transfer should be included in the design of genes added to microbial strains for the goal of increased water treatment. This

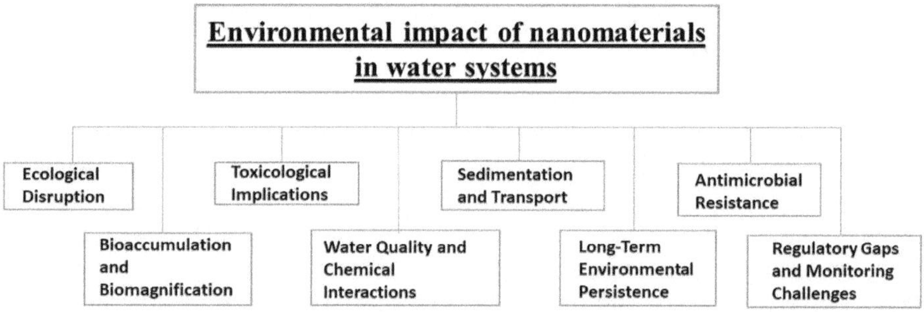

Fig. 12.6 The different environmental impacts of nanomaterials in water system

lowers the possibility that genetic material would inadvertently transfer undesirable features to other microbes or give rise to new, potentially harmful organisms (Zhu et al., 2019).

Human health and environmental health are at risk due to the potential risks of ingesting or inhaling NPs released from the treatment process. These risks include toxicity, bioaccumulation, and bio persistence. Unintended consequences may arise from the interaction of engineered microbes with other pollutants or organisms in wastewater. Microbial safety is crucial, as genetically modified microbes must be followed to prevent their escape into the environment and potential harm to ecosystems. Pathogen concerns exist, as engineered microbes could become pathogenic or carry virulence genes. Waste management is also crucial, as the microbials and NPs must be disposed of safely. Knowledge gaps regarding the long-term environmental impacts of releasing engineered microbes and NPs are not fully understood. To address these risks, new regulations or adaptations of existing ones may be needed. Additional safety measures include strict risk analysis before any large-scale implementation and open communication with the public. By considering these safety considerations and conducting further research, microbe-assisted nanotechnology for wastewater bioremediation can be developed and implemented responsibly and sustainably (El-Kady et al., 2023).

Safety factors necessitate detailed risk assessments and scenario preparation. Proactive management of future problems is possible by identifying potential risks and creating mitigation methods ahead of time. This involves planning for various scenarios, from unintentional releases to unexpected ecological interactions, and developing plans to reduce or mitigate any adverse impact (Ojemaye et al., 2020).

12.6.3 *Regulatory and Ethical Issues in Sustainable Water Management*

The changing nature of nanomaterial regulations poses a significant regulatory obstacle to the implementation of microbe-assisted nanotechnology in sustainable water management. Current regulatory frameworks might not fully cover the special qualities and possible hazards related to nanomaterials. To evaluate and regulate the environmental discharge, usage, and disposal of nanomaterials in water treatment processes, governments, and regulatory organizations need to modify and create new criteria (Giri & Kumari, 2023). The implementation of nanotechnology in water management is guaranteed for compliance with safety and environmental regulations by the establishment of specific regulatory frameworks. Conducting comprehensive risk evaluations for microbe-assisted nanotechnology in water management is a challenging undertaking for regulatory organizations. This entails assessment of the long-term effects of NPs along with any possible dangers to the environment and human health. The implementation of uniform risk assessment procedures promotes uniform assessments among various applications and guarantees the efficient identification and mitigation of any hazards. Regulatory agencies must stay up with new developments in nanotechnology and the growing body of knowledge regarding the behavior of nanomaterials in water systems by regularly updating risk assessment techniques (Roy et al., 2021).

The uses of microbes-assisted nanotechnology in wastewater management require strict reporting and monitoring to track environmental fate and microbial applications. Global standardization of rules promotes international cooperation and knowledge exchange, reducing environmental and socioeconomic inequalities. Ethical considerations are crucial in research and development, ensuring responsible experimentation, transparent reporting, and consideration of potential impacts on society. Ethical issues include environmental justice, fair access to water treatment technology, and equitable allocation of advantages and drawbacks. Principles of informed consent and public participation are essential for ethical water management, promoting understanding, trust, and cooperation (Ezezika & Singer, 2010).

12.7 Conclusions and Future Trends

Microbe-assisted nanotechnology is a promising approach for sustainable water resource management, utilizing the unique properties of microorganisms and engineered nanomaterials. This innovative approach aims to optimize contaminant removal and water treatment procedures by modifying nanomaterial characteristics to establish beneficial interactions with specific microbial strains. The integration of machine learning and artificial intelligence (AI) in this technology is expected to automate water treatment procedures, predict microbial behavior, and optimize system performance. AI algorithms can analyze large-scale datasets, allowing for precise adjustments, adaptive control, and real-time monitoring to maximize contaminant removal and system efficiency. The development of the nanomaterials towards improved properties like stability, reactivity, and surface area, will lead to the creation of next-generation nanocomposites, paving the way for environmentally friendly and economically viable water treatment methods. The expansion of microbe-assisted nanotechnology from lab conditions to real-world applications is a major trend, with field research and pilot projects crucial for evaluating its long-term environmental effects, scalability, and performance.

Future directions in microbe-assisted nanotechnology for sustainable water resource management include improved contaminant removal systems, optimization through artificial intelligence, and developments in nanomaterial design. Public awareness and acceptance of this emerging field are crucial for the successful adoption of microbe-assisted nanotechnology in water management. A novel paradigm for attaining sustainability in the face of growing global water challenges is presented by a convergence of microbial processes and nanotechnology in water resource management. By employing this creative approach, we can unlock game-changing solutions that support resource efficiency and environmental stewardship, as well as improve water quality.

In conclusion, a potent and promising tactic for managing water resources sustainably is the incorporation of nanotechnology aided by microbes. This creative solution offers a way forward for robust and effective water systems as we deal with the growing demands on water resources and the growing threats to water quality. To fully realize the potential of microbes-assisted nanotechnology and usher in a sustainable era of water resource management for future generations, we need to keep researching, work together, and remain committed to ethical practices.

Acknowledgments The authors thank the Principal and Management of CBIT, Hyderabad for their constant support and encouragement in carrying out this work.

References

Abdelbasir, S. M., & Shalan, A. E. (2019). An overview of nanomaterials for industrial wastewater treatment. *Korean Journal of Chemical Engineering, 36*(8), 1209–1225. https://doi.org/10.1007/s11814-019-0306-y

Abdelfattah, A., Ali, S. S., Ramadan, H., El-Aswar, E. I., Eltawab, R., Ho, S.-H., Elsamahy, T., Li, S., El-Sheekh, M. M., Schagerl, M., Kornaros, M., & Sun, J. (2023). Microalgae-based wastewater treatment: Mechanisms, challenges, recent advances, and future prospects. *Environmental Science and Ecotechnology, 13*, 100205. https://doi.org/10.1016/j.ese.2022.100205

Ajith, M. P., Aswathi, M., Priyadarshini, E., & Rajamani, P. (2021). Recent innovations of nanotechnology in water treatment: A comprehensive review. *Bioresource Technology, 342*, 126000. https://doi.org/10.1016/j.biortech.2021.126000

Ali, I., & Gupta, V. K. (2006). Advances in water treatment by adsorption technology. *Nature Protocols, 1*(6), 2661–2667. https://doi.org/10.1038/nprot.2006.370

Alikhani, S., Nummi, P., & Ojala, A. (2021). Urban wetlands: A review on ecological and cultural values. *Water, 13*, 3301. https://doi.org/10.3390/w13223301

Al-Issai, L., Elshorbagy, W., Maraqa, M., Hamouda, M., & Soliman, A. (2019). Use of NPs for the disinfection of desalinated water. *Water, 11*(3), 559. https://doi.org/10.3390/w11030559

Alizadeh, A., Razmjou, A., Ghaedi, M., & Jannesar, R. (2019). Nanoporous solid-state membranes modified with multi-wall carbon nanotubes with anti-biofouling property. *International Journal of Nanomedicine, 14*, 1669–1685. https://doi.org/10.2147/IJN.S189728

Al-Saidi, M., Das, P., & Saadaoui, I. (2021). Circular economy in basic supply: Framing the approach for the water and food sectors of the Gulf cooperation council countries. *Sustainable Production and Consumption, 27*, 1273–1285. https://doi.org/10.1016/j.spc.2021.03.004

Annu, M. M., Tripathi, S., & Shin, D. K. (2024). Biopolymeric nanocomposites for wastewater remediation: An overview on recent progress and challenges. *Polymers (Basel), 16*(2), 294. https://doi.org/10.3390/polym16020294. PMID: 38276702; PMCID: PMC10818902.

Aruna, V., Chada, N., Tejaswini Reddy, M., Geethikalal, V., Dornala, K., & Nagendranatha Reddy, C. (2022). Plant synthesized NPs for dye degradation. In M. P. Shah, & A. Roy (Eds.), *Phytonanotechnology*. Springer. https://doi.org/10.1007/978-981-19-4811-4_1. Print ISBN 978-981-19-4810-7; Online ISBN 978-981-19-4811-4.

Aziz, W. J., Mezher, S. J., & Hassan, M. A. H. (2015). Gas sensor of Au/AgNPs/PSi/c-Si using pulsed laser ablation using a YAG: Nd laser. *Optik, 126*(20), 2584–2587. https://doi.org/10.1016/j.ijleo.2015.06.033

Benjamin, S. R., De Lima, F., Florean, E. O. P. T., & Guedes, M. I. F. (2019). Current trends in nanotechnology for bioremediation. *International Journal of Environment and Pollution, 66*(1/2/3), 19. https://doi.org/10.1504/IJEP.2019.104526

Bhardwaj, A. K., Sundaram, S., Yadav, K. K., & Srivastav, A. L. (2021). An overview of silver nano-particles as promising materials for water disinfection. *Environmental Technology and Innovation, 23*, 101721. https://doi.org/10.1016/j.eti.2021.101721

Binhayeeding, N., Yunu, T., Pichid, N., Klomklao, S., & Sangkharak, K. (2020). Immobilisation of Candida rugosa lipase on polyhydroxybutyrate via a combination of adsorption and cross-linking agents to enhance acylglycerol production. *Process Biochemistry, 95*, 174–185. https://doi.org/10.1016/j.procbio.2020.02.007

Bochkova, O., Khrizanforov, M., Gubaidullin, A., Gerasimova, T., Nizameev, I., Kholin, K., Laskin, A., Budnikova, Y., Sinyashin, O., & Mustafina, A. (2020). Synthetic tuning of CoII-doped silica nanoarchitecture towards electrochemical sensing ability. *Nanomaterials, 10*, 1338. https://doi.org/10.3390/nano10071338

Bose, D., Kandpal, V., Dhawan, H., Vijay, P., & Gopinath, M. (2018). *Energy recovery with microbial fuel cells: Bioremediation and bioelectricity* (pp. 7–33). https://doi.org/10.1007/978-981-10-7413-4_2

Cho, G., Azzouzi, S., Zucchi, G., & Lebental, B. (2022). Electrical and electrochemical sensors based on carbon nanotubes for the monitoring of chemicals in water—A review. *Sensors, 22*, 218. https://doi.org/10.3390/s22010218

Dar, O. A., Malik, M. A., Talukdar, M. I. A., & Hashmi, A. A. (2020). Bionanocomposites in water treatment. In *Bionanocomposites* (pp. 505–518). Elsevier. https://doi.org/10.1016/B978-0-12-816751-9.00019-2

Dixit, R., Wasiullah, Malaviya, D., Pandiyan, K., Singh, U., Sahu, A., Shukla, R., Singh, B., Rai, J., Sharma, P., Lade, H., & Paul, D. (2015). Bioremediation of heavy metals from soil and aquatic environment: An overview of principles and criteria of fundamental processes. *Sustainability, 7*(2), 2189–2212. https://doi.org/10.3390/su7022189

Dogan, T., Bayram, E., Uzun, L., Şenel, S., & Denizli, A. (2015). Trametes versicolor laccase immobilized poly(glycidyl methacrylate) based cryogels for phenol degradation from aqueous media. *Journal of Applied Polymer Science, 132*, 20.

Du, Z., Li, H., & Gu, T. (2007). A state of the art review on microbial fuel cells: A promising technology for wastewater treatment and bioenergy. *Biotechnology Advances, 25*(2007), 464–482.

Egodawatte, S., Greenstein, K. E., Vance, I., Rivera, E., Myung, N. V., Parkin, G. F., Cwiertnyb, D. M., & Larsen, S. C. (2016). Electrospun hematite nanofiber/mesoporous silica core/

shell nanomaterials as an efficient adsorbent for heavy metals. *RSC Advances, 2016*(6), 90516–90525.

El-Kady, M. M., Ansari, I., Arora, C., Rai, N., Soni, S., Verma, D. K., Singh, P., & Mahmoud, A. E. D. (2023). Nanomaterials: A comprehensive review of applications, toxicity, impact, and fate to environment. *Journal of Molecular Liquids, 370*, 121046. https://doi.org/10.1016/j.molliq.2022.121046

Ezezika, O. C., & Singer, P. A. (2010). Genetically engineered oil-eating microbes for bioremediation: Prospects and regulatory challenges. *Technology in Society, 32*(4), 331–335. https://doi.org/10.1016/j.techsoc.2010.10.010

Fajardo, C., Martinez-Rodriguez, G., Blasco, J., Mancera, J. M., Thomas, B., & De Donato, M. (2022). Nanotechnology in aquaculture: Applications, perspectives and regulatory challenges. *Aquaculture and Fisheries, 7*(2), 185–200. https://doi.org/10.1016/j.aaf.2021.12.006

Giri, V. P., & Kumari, M. (2023). Microbial approaches in fabrication of nanoscale materials effectively enhance the antimicrobial and crop protection potential—A review. *Plant Nano Biology, 3*, 100027. https://doi.org/10.1016/j.plana.2023.100027

Gleick, P. H., Singh, A., & Shi, H. (2001). *Threats to the world's freshwater resources. emerging threats to the world's freshwater resources*. A report of the Pacific Institute for Studies in Development, Environment, and Security, Oakland, California. ISBN Number: 1-893790-06-1.

He, S., Zhong, L., Duan, J., Feng, Y., Yang, B., & Yang, L. (2017). Bioremediation of wastewater by iron oxide-biochar nanocomposites loaded with photosynthetic bacteria. *Frontiers in Microbiology, 8*. https://doi.org/10.3389/fmicb.2017.00823

Irena, P., Stergar, J., Bukšek, H., Drofenik, M., Gyergyek, S., Hélix-Nielsen, C., & Ban, I. (2021). Superparamagnetic Fe_3O_4@CA NPs and their potential as draw solution agents in forward osmosis. *Nanomaterials, 11*, 2965. https://doi.org/10.3390/nano11112965

Jatoi, A. S., Akhter, F., Mazari, S. A., Sabzoi, N., Aziz, S., Soomro, S. A., Mubarak, N. M., Baloch, H., Memon, A. Q., & Ahmed, S. (2021). Advanced microbial fuel cell for waste water treatment—A review. *Environmental Science and Pollution Research, 28*(5), 5005–5019. https://doi.org/10.1007/s11356-020-11691-2

Kabir, E., Kumar, V., Kim, K.-H., Yip, A. C. K., & Sohn, J. R. (2018). Environmental impacts of nanomaterials. *Journal of Environmental Management, 225*, 261–271. https://doi.org/10.1016/j.jenvman.2018.07.087

Khan, A. G. (2021). Potential of coupling heavy metal (HM) phytoremediation by bioenergy plants and their associated HM-adapted rhizosphere microbiota (Arbuscular Mycorrhizal fungi and plant growth promoting microbes) for bioenergy production. *Journal of Energy and Power Technology, 3*(4), 1–1. https://doi.org/10.21926/jept.2104049

Kishore, S., Malik, S., Shah, M. P., Bora, J., Chaudhary, V., Kumar, L., Sayyed, R. Z., & Ranjan, A. (2022). A comprehensive review on removal of pollutants from wastewater through microbial nanobiotechnology -based solutions. *Biotechnology and Genetic Engineering Reviews*, 1–26. https://doi.org/10.1080/02648725.2022.2106014

Kolya, H., & Kang, C.-W. (2023). Next-generation water treatment: Exploring the potential of biopolymer-based nanocomposites in adsorption and membrane filtration. *Polymers, 15*(16), 3421. https://doi.org/10.3390/polym15163421

Kumar, A. N., Nagendranatha Reddy, C., & Venkata Mohan, S. (2015). Biomineralization of azo dye bearing wastewater in periodic discontinuous batch reactor: Effect of microaerophilic conditions on treatment efficiency. *Bioresource Technology, 188*, 56–64. https://doi.org/10.1016/j.biortech.2015.01.098

Kumar, A., Nighojkar, A., Varma, P., Prakash, N. J., Kandasubramanian, B., Zimmermann, K., & Dixit, F. (2023). Algal mediated intervention for the retrieval of emerging pollutants from aqueous media. *Journal of Hazardous Materials, 455*, 131568. https://doi.org/10.1016/j.jhazmat.2023.131568

Kuppusamy, S., Thavamani, P., Megharaj, M., & Naidu, R. (2015). Bioremediation potential of natural polyphenol rich green wastes: A review of current research and recommendations for future directions. *Environmental Technology and Innovation, 4*, 17–28. https://doi.org/10.1016/j.eti.2015.04.001

Lee, T. H., Roh, J. S., Yoo, S. Y., Roh, J. M., Choi, T. H., & Park, H. B. (2020). High-performance polyamide thin-film nanocomposite membranes containing ZIF-8/CNT hybrid nanofillers for reverse osmosis desalination. *Industrial and Engineering Chemistry Research, 59*(12), 5324–5332.

Leong, Y. K., & Chang, J.-S. (2020). Bioremediation of heavy metals using microalgae: Recent advances and mechanisms. *Bioresource Technology, 303*, 122886. https://doi.org/10.1016/j.biortech.2020.122886

Li, K., Liu, Q., Fang, F., Luo, R., Lu, Q., Zhou, W., Huo, S., Cheng, P., Liu, J., Addy, M., Chen, P., Chen, D., & Ruan, R. (2019). Microalgae-based wastewater treatment for nutrients recovery: A review. *Bioresource Technology, 291*, 121934. https://doi.org/10.1016/j.biortech.2019.121934

Lou, Y., Zhao, Y., Chen, J., & Zhu, J. J. (2014). Metal ions optical sensing by semiconductor quantum dots. *Journal of Materials Chemistry C, 2*, 595–613.

Malik, S., Dhasmana, A., Preetam, S., Mishra, Y. K., Chaudhary, V., Bera, S. P., Ranjan, A., Bora, J., Kaushik, A., Minkina, T., Jatav, H. S., Singh, R. K., & Rajput, V. D. (2022). Exploring microbial-based green nanobiotechnology for wastewater remediation: A sustainable strategy. *Nanomaterials, 12*(23), 4187. https://doi.org/10.3390/nano12234187

Mandeep, & Shukla, P. (2020). Microbial nanotechnology for bioremediation of industrial wastewater. *Frontiers in Microbiology, 11*. https://doi.org/10.3389/fmicb.2020.590631

Megharaj, M., Ramakrishnan, B., Venkateswarlu, K., Sethunathan, N., & Naidu, R. (2011). Bioremediation approaches for organic pollutants: A critical perspective. *Environment International, 37*(8), 1362–1375. https://doi.org/10.1016/j.envint.2011.06.003

Mishra, B., Rajasri Yadavalli, Y. V., & Nagendranatha Reddy, C. (2021). Recent advancements and challenges of nanomaterials application in biofuel production. In R. Praveen Kumar et al. (Eds.), *Nanomaterials application in biofuels and bioenergy production systems*. Academic. ISBN 978-0-12-822401-4.

Mishra, S., Sahoo, S., Kumar Sahoo, P., & Kumar Sahoo, N. (2023). Recent advancements in the plant and microbial assisted green synthesis of nanomaterials. *Materials Today: Proceedings*. https://doi.org/10.1016/j.matpr.2023.10.144

Modestra, J. A., Nagendranatha Reddy, C., Krishna, K. V., Min, B., & Venkata Mohan, S. (2020). Regulated surface potential impacts bioelectrogenic activity, interfacial electron transfer and microbial dynamics in microbial fuel cell. *Renewable Energy, 149*, 424–434. https://doi.org/10.1016/j.renene.2019.12.018

Mohiuddin, O., Harvey, A., Teresa Orta Ledesma, M., & Velasquez-Orta, S. (2024). Bioremediation of waste by yeast strains. *Electronic Journal of Biotechnology*. https://doi.org/10.1016/j.ejbt.2024.01.005

Mondal, M., Halder, G., Oinam, G., Indrama, T., & Tiwari, O. N. (2019). Bioremediation of organic and inorganic pollutants using microalgae. In *New and future developments in microbial biotechnology and bioengineering* (pp. 223–235). Elsevier. https://doi.org/10.1016/B978-0-444-63504-4.00017-7

Nagendranatha Reddy, C., Nguyen, H. T. H., Noori, M. T., & Min, B. (2019). Potential applications of algae in the cathode of microbial fuel cells for enhanced electricity generation with simultaneous nutrient removal and algae biorefinery: Current status and future perspectives. *Bioresource Technology, 292*, 122010. https://doi.org/10.1016/j.biortech.2019.122010

Nagendranatha Reddy, C., Kondaveeti, S., & Min, B. (2021). Influence of trace metals concentration on methane generation using microbial electrochemical systems. *Process Biochemistry, 102*, 213–219. https://doi.org/10.1016/j.procbio.2020.12.021

Nagendranatha Reddy, C., Kondaveeti, S., Krishna, G. M., & Min, B. (2022). Application of bioelectrochemical systems to regulate and accelerate the anaerobic digestion processes. *Chemosphere, 287*, 132299. https://doi.org/10.1016/j.chemosphere.2021.132299

Ojemaye, M. O., Adefisoye, M. A., & Okoh, A. I. (2020). Nanotechnology as a viable alternative for the removal of antimicrobial resistance determinants from discharged municipal effluents and associated watersheds: A review. *Journal of Environmental Management, 275*, 111234. https://doi.org/10.1016/j.jenvman.2020.111234

Ojha, A. (2020). Nanomaterials for removal of waterborne pathogens: Opportunities and challenges. *Waterborne Pathogens*, 385–432. https://doi.org/10.1016/B978-0-12-818783-8.00019-0. Epub 2020 Feb 14. PMCID: PMC7153326.

Pan, Y., Liu, X., Zhang, W., Liu, Z., Zeng, G., Shao, B., Liang, Q., He, Q., Yuan, X., Huang, D., & Chen, M. (2020). Advances in photocatalysis based on fullerene C60 and its derivatives: Properties, mechanism, synthesis, and applications. *Applied Catalysis B: Environmental, 265*, 118579. https://doi.org/10.1016/j.apcatb.2019.118579

Peixoto, R. S., Vermelho, A. B., & Rosado, A. S. (2011). Petroleum-degrading enzymes: Bioremediation and new prospects. *Enzyme Research, 2011*, 1–7. https://doi.org/10.4061/2011/475193

Petti, S., & Messano, G. A. (2016). Nano-TiO_2-based photocatalytic disinfection of environmental surfaces contaminated by meticillin-resistant Staphylococcus aureus. *Journal of Hospital Infection, 93*(1), 78–82. https://doi.org/10.1016/j.jhin.2016.01.020

Phian, S., Nagar, S., Kaur, J., & Rawat, C. D. (2022). Emerging issues and challenges for microbes-assisted remediation. In *Microbes and microbial biotechnology for green remediation* (pp. 47–89). Elsevier. https://doi.org/10.1016/B978-0-323-90452-0.00026-8

Punia, A., Pratap Singh, R., Singh, V., & Chauhan, N. S. (2023). Environment sustainability with microbial nanotechnology. In *Environmental applications of microbial nanotechnology* (pp. 289–314). Elsevier. https://doi.org/10.1016/B978-0-323-91744-5.00012-6

Qi, L., Hu, Y., Liu, Z., An, X., & Bar-Zeev, E. (2018). Improved anti-biofouling performance of thin -film composite forward-osmosis membranes containing passive and active moieties. *Environmental Science & Technology, 52*(17), 9684–9693.

Qu, X., Alvarez, P. J. J., & Li, Q. (2013a). Applications of nanotechnology in water and wastewater treatment. *Water Research, 47*(12), 3931–3946. https://doi.org/10.1016/j.watres.2012.09.058

Qu, X., Brame, J., Li, Q., & Alvarez, P. J. J. (2013b). Nanotechnology for a safe and sustainable water supply: Enabling integrated water treatment and reuse. *Accounts of Chemical Research, 46*(3), 834–843. https://doi.org/10.1021/ar300029v

Rajput, P., Singh, A., Agrawal, S., et al. (2024). Effects of environmental metal and metalloid pollutants on plants and human health: Exploring nano-remediation approach. *Stress Biology, 4*, 27. https://doi.org/10.1007/s44154-024-00156-y

Riaz, S., & Park, S.-J. (2020). An overview of TiO_2-based photocatalytic membrane reactors for water and wastewater treatments. *Journal of Industrial and Engineering Chemistry, 84*, 23–41. https://doi.org/10.1016/j.jiec.2019.12.021

Roy, A., Sharma, A., Yadav, S., Jule, L. T., & Krishnaraj, R. (2021). Nanomaterials for remediation of environmental pollutants. *Bioinorganic Chemistry and Applications, 2021*, 1–16. https://doi.org/10.1155/2021/1764647

Saeed, M. U., Hussain, N., Sumrin, A., Shahbaz, A., Noor, S., Bilal, M., Aleya, L., & Iqbal, H. M. N. (2022). Microbial bioremediation strategies with wastewater treatment potentialities—A review. *Science of the Total Environment, 818*, 151754. https://doi.org/10.1016/j.scitotenv.2021.151754

Saravanan, A., Kumar, P. S., Varjani, S., Jeevanantham, S., Yaashikaa, P. R., Thamarai, P., Abirami, B., & George, C. S. (2021). A review on algal-bacterial symbiotic system for effective treatment of wastewater. *Chemosphere, 271*, 129540. https://doi.org/10.1016/j.chemosphere.2021.129540

Saxena, G., Goutam, S. P., Mishra, A., Mulla, S. I., & Bharagava, R. N. (2020). Emerging and ecofriendly technologies for the removal of organic and inorganic pollutants from industrial wastewaters. In *Bioremediation of industrial waste for environmental safety* (pp. 113–126). Springer. https://doi.org/10.1007/978-981-13-3426-9_5

Shah, A., & Shah, M. (2020). Characterisation and bioremediation of wastewater: A review exploring bioremediation as a sustainable technique for pharmaceutical wastewater. *Groundwater for Sustainable Development, 11*, 100383. https://doi.org/10.1016/j.gsd.2020.100383

Shahi Khalaf Ansar, B., Kavusi, E., Dehghanian, Z., et al. (2023). Removal of organic and inorganic contaminants from the air, soil, and water by algae. *Environmental Science and Pollution Research, 30*, 116538–116566. https://doi.org/10.1007/s11356-022-21283-x

Sharma, B., Dangi, A. K., & Shukla, P. (2018). Contemporary enzyme based technologies for bioremediation: A review. *Journal of Environmental Management, 210*, 10–22. https://doi.org/10.1016/j.jenvman.2017.12.075

Sharma, M., Agarwal, S., Agarwal Malik, R., Kumar, G., Pal, D. B., Mandal, M., Sarkar, A., Bantun, F., Haque, S., Singh, P., Srivastava, N., & Gupta, V. K. (2023). Recent advances in microbial engineering approaches for wastewater treatment: A review. *Bioengineered, 14*(1). https://doi.org/10.1080/21655979.2023.2184518

Siddeeg, S. M., Tahoon, M. A., Mnif, W., & Ben Rebah, F. (2020). Iron oxide/Chitosan magnetic nanocomposite immobilized manganese peroxidase for decolorization of textile wastewater. *Processes, 8*, 5. https://doi.org/10.3390/pr8010005

Singh, R., Behera, M., & Kumar, S. (2020). Nano-bioremediation: An innovative remediation technology for treatment and management of contaminated sites. In *Bioremediation of industrial waste for environmental safety* (pp. 165–182). Springer. https://doi.org/10.1007/978-981-13-3426-9_7

Spoială, A., Ilie, C. I., Trușcă, R. D., Oprea, O. C., Surdu, V. A., Vasile, B. Ș., Ficai, A., Ficai, D., Andronescu, E., & Dițu, L. M. (2021). Zinc oxide NPs for water purification. *Materials (Basel), 14*(16), 4747. https://doi.org/10.3390/ma14164747. PMID: 34443269; PMCID: PMC8397993.

Sreelatha, S., Nagendranatha Reddy, C., Velvizhi, G., & Venkata Mohan, S. (2015). Reductive behaviour of acid azo dye based wastewater: Biocatalyst activity in conjunction with enzymatic and bio-electro catalytic evaluation. *Bioresource Technology, 188*, 2–8. https://doi.org/10.1016/j.biortech.2015.02.053

Tarkeshwar, Pandit, M. A., & Kapinder. (2023). Application of microbial nanobiotechnology for combating water pollution. In *Environmental applications of microbial nanotechnology* (pp. 365–380). Elsevier. https://doi.org/10.1016/B978-0-323-91744-5.00006-0

Teimouri, M., Khosravi-Nejad, F., Attar, F., Saboury, A. A., Kostova, I., Benelli, G., & Falahati, M. (2018). Gold NPs fabrication by plant extracts: Synthesis, characterization, degradation of 4-nitrophenol from industrial wastewater, and insecticidal activity—A review. *Journal of Cleaner Production, 184*, 740–753. https://doi.org/10.1016/j.jclepro.2018.02.268

Tripathi, S., Sanjeevi, R., Anuradha, J., Chauhan, D. S., & Rathoure, A. K. (2022). Nano-bioremediation. In *Research anthology on emerging techniques in environmental remediation* (pp. 135–149). IGI Global. https://doi.org/10.4018/978-1-6684-3714-8.ch007

Umar, M. F., Rafatullah, M., Abbas, S. Z., Mohamad Ibrahim, M. N., & Ismail, N. (2021). Advancement in benthic microbial fuel cells toward sustainable bioremediation and renewable energy production. *International Journal of Environmental Research and Public Health, 18*(7), 3811. https://doi.org/10.3390/ijerph18073811

UNEA. (2022). https://www.unep.org/news-and-stories/story/five-threats-water-sustains-our-farms#:~:text=The%20problem%20can%20be%20exacerbated,to%20farming%20and%20food%20security

UNICEF. (2024). *Water, Sanitation and Hygiene (WASH) Safe water, toilets and good hygiene keep children alive and healthy*. https://www.unicef.org/wash

Vats, T., & Pandey, P. (2021). Utility of metal oxide-based bionanocomposites for wastewater treatment. In *Bionanomaterials for environmental and agricultural applications*. IOP Publishing. https://doi.org/10.1088/978-0-7503-3863-9ch7

Vázquez-Núñez, E., Molina-Guerrero, C. E., Peña-Castro, J. M., Fernández-Luqueño, F., & de la Rosa-Álvarez, M. G. (2020). Use of nanotechnology for the bioremediation of contaminants: A review. *Processes, 8*(7), 826. https://doi.org/10.3390/pr8070826

Verma, S., & Kuila, A. (2019). Bioremediation of heavy metals by microbial process. *Environmental Technology and Innovation, 14*, 100369. https://doi.org/10.1016/j.eti.2019.100369

Verma, J., Kumar, D., Singh, N., Katti, S. S., & Shah, Y. T. (2021). Electricigens and microbial fuel cells for bioremediation and bioenergy production: A review. *Environmental Chemistry Letters, 19*(3), 2091–2126. https://doi.org/10.1007/s10311-021-01199-7

Vishwakarma, G. S., Bhattacharjee, G., Gohil, N., & Singh, V. (2020). Current status, challenges and future of bioremediation. In *Bioremediation of pollutants* (pp. 403–415). Elsevier. https://doi.org/10.1016/B978-0-12-819025-8.00020-X

Wadgaonkar, S. L., Nancharaiah, Y. V., Esposito, G., & Lens, P. N. L. (2018). Environmental impact and bioremediation of seleniferous soils and sediments. *Critical Reviews in Biotechnology, 38*(6), 941–956. https://doi.org/10.1080/07388551.2017.1420623

Wang, W., Zhang, Y., Li, M., Wei, X., Wang, Y., Liu, L., Wang, H., & Shen, S. (2020). Operation mechanism of constructed wetland-microbial fuel cells for wastewater treatment and electricity generation: A review. *Bioresource Technology, 314*, 123808. https://doi.org/10.1016/j.biortech.2020.123808

Wasewar, K. L., Singh, S., & Kansal, S. K., (2020). Process intensification of treatment of inorganic water pollutants. In P. Devi, P. Singh, & S. K. Kansal (Eds.), *Inorganic pollutants in water* (pp. 245–271). Elsevier. ISBN 9780128189658. https://doi.org/10.1016/B978-0-12-818965-8.00013-5

Wu, Z., Chan, B., Low, J., Chu, J. J. H., Hey, H. W. D., & Tay, A. (2022). Microbial resistance to nanotechnologies: An important but understudied consideration using antimicrobial nanotechnologies in orthopaedic implants. *Bioactive Materials, 16*, 249–270. https://doi.org/10.1016/j.bioactmat.2022.02.014

Yang, Y., Waterhouse, G. I. N., Chen, Y., Sun-Waterhouse, D., & Li, D. (2022). Microbial-enabled green biosynthesis of nanomaterials: Current status and future prospects. *Biotechnology Advances, 55*, 107914. https://doi.org/10.1016/j.biotechadv.2022.107914

Yildiz, I. (2016). Applications of magnetic NPs in biomedical separation and purification. *Nanotechnology Reviews, 5*(3), 331–340. https://doi.org/10.1515/ntrev-2015-0012

Yin, J., Yang, Y., Hu, Z., & Deng, B. (2013). Attachment of silver NPs (AgNPs) onto thin-film composite (TFC) membranes through covalent bonding to reduce membrane biofouling. *Journal of Membrane Science, 441*, 73–82. https://doi.org/10.1016/j.memsci.2013.03.060

Yin, Z., Zhu, L., Li, S., Hu, T., Chu, R., Mo, F., Hu, D., Liu, C., & Li, B. (2020). A comprehensive review on cultivation and harvesting of microalgae for biodiesel production: Environmental pollution control and future directions. *Bioresource Technology, 301*, 122804. https://doi.org/10.1016/j.biortech.2020.122804

You, X., Yang, L., Zhou, X., & Zhang, Y. (2022). Sustainability and carbon neutrality trends for microalgae-based wastewater treatment: A review. *Environmental Research, 209*, 112860. https://doi.org/10.1016/j.envres.2022.112860

Yu, J.-G., Zhao, X.-H., Yang, H., Chen, X.-H., Yang, Q., Yu, L.-Y., Jiang, J.-H., & Chen, X.-Q. (2014). *Aqueous adsorption and removal of organic contaminants by carbon nanotubes*. Web. https://doi.org/10.1016/J.SCITOTENV.2014.02.129

Zeraatkar, A. K., Ahmadzadeh, H., Talebi, A. F., Moheimani, N. R., & McHenry, M. P. (2016). Potential use of algae for heavy metal bioremediation, a critical review. *Journal of Environmental Management, 181*, 817–831. https://doi.org/10.1016/j.jenvman.2016.06.059

Zheng, X., Chen, D., Wang, Z., Lei, Y., & Cheng, R. (2013). Nano-TiO_2 membrane adsorption reactor (MAR) for virus removal in drinking water. *Chemical Engineering Journal, 230*, 180–187. https://doi.org/10.1016/j.cej.2013.06.069

Zhu, Y., Liu, X., Hu, Y., Wang, R., Chen, M., Wu, J., Wang, Y., Kang, S., Sun, Y., & Zhu, M. (2019). Behavior, remediation effect and toxicity of nanomaterials in water environments. *Environmental Research, 174*, 54–60. https://doi.org/10.1016/j.envres.2019.04.014

Chapter 13
Bioengineered Nano-sorbents: Microbial Allies in Pollutant Sequestration

Sowjanya Sivakumar, Mohamad Zuber Nurgat, Laxmikarthika V. Srinivasan, Akashi Bora, Mrunal S. Wagh, Neelam Amit Kungwani, and Mahesh Narayan

Abstract The global issue of toxic contaminants removal from wastewater is a significant concern, exacerbated by rapid urbanization and industrialization, which generate heavy metal-laden wastewater affecting the environment and human health. Traditional wastewater treatment methods are expensive and energy-intensive due to the complex as well as large volume of associated pollutants. Microbes, immune to metal toxicity, play a crucial role in the ecosystem's resource cycle and energy flow, offering promising benefits in wastewater detoxification. This has led to the development of processes like biomineralization, oxidation-reduction, bio-precipitation, bioaccumulation, biosurfactant technology, bioleach-

S. Sivakumar
School of Biosciences and Technology, Vellore Institute of Technology, Vellore, Tamil Nadu, India

VIT School of Agricultural Innovations and Advanced Learning, Vellore Institute of Technology, Vellore, Tamil Nadu, India

M. Z. Nurgat
Department of Biotechnology, St Xavier's College (Empowered Autonomous Institution), Mumbai, Maharashtra, India

L. V. Srinivasan
School of Biosciences and Technology, Vellore Institute of Technology, Vellore, Tamil Nadu, India

A. Bora
Nano-biotechnology and Translational Knowledge Laboratory, Department of Applied Biology, School of Biological Sciences, University of Science and Technology Meghalaya, Ri-Bhoi, Meghalaya, India

M. S. Wagh (✉)
Research and Development, Organica Biotech Private Limited, Mumbai, Maharashtra, India

N. A. Kungwani
Marine Biotechnology, Gujarat Biotechnology University, Gandhinagar, Gujarat, India

M. Narayan
Department of Chemistry and Biochemistry, The University of Texas at El Paso (UTEP), El Paso, TX, USA

Y. K. Mohanta et al. (eds.), *Nano-microbiology for Sustainable Development*,
https://doi.org/10.1007/978-3-031-78845-1_13

ing, bio-volatilization, and biosorption. Combining bioremediation with nanotechnology provides a cost-effective solution for pollutant removal. Nanomaterials exhibit exceptional characteristics such as improved catalysis, adsorption properties, and high reactivity. Bacterial, fungal, and algal cultures are used as catalysts for sustainable nanomaterial production, facilitating bioremediation by immobilizing or inducing microbial enzymes. Understanding the interaction between microorganisms, contaminants, and nanoparticles is crucial for developing innovative remediation strategies. Bio-engineered nano-sorbents, combined with microorganisms, offer an effective solution for pollutant sequestration due to their high surface areas and specific functionalization. Microbes possess natural or engineered abilities to degrade or transform contaminants, enhancing the remediation process. This chapter explores the interaction between nano-sorbents and microbes, current bio-engineering technologies, and practical applications in water treatment and soil remediation. It covers different types of microbe-derived nanomaterials used in pollutant remediation, their characteristics, and synthesis methods. The adsorption mechanisms and kinetics of bio-engineered nanomaterials are briefly discussed, along with their applications in wastewater treatment and soil remediation.

Keywords Biogenic nanoparticles · Microbes · Adsorption kinetics · Heavy metals · Wastewater

13.1 Introduction

Numerous contaminants, including potentially dangerous ones that endanger both human health and the environment, can be found in wastewater (Liu et al., 2019). Aquatic ecosystems, animals, and general water quality may all be negatively impacted by these dangerous contaminants, which can come from a variety of sources including industrial processes, agricultural runoff, and home sewage (Rathi et al., 2021). For efficient wastewater management and the preservation of water resources, it is essential to comprehend the many kinds of harmful chemicals found in wastewater as well as their possible environmental impacts. Our ability to manage wastewater sustainably and create plans to lessen the impact of these contaminants will increase as we gain expertise in them (Saravanan et al., 2021a).

Hazardous contaminants identified in wastewater frequently consist of heavy metals (like lead, mercury, and cadmium), organic compounds (like pesticides, medicines, and industrial chemicals), nutrients (like nitrogen and phosphorus), and pathogens (like bacteria and viruses) (Koul et al., 2021). These contaminants have a long environmental half-life, bioaccumulate in living things, and upset natural equilibrium (Elbehiry et al., 2021). Risky contaminants in wastewater have a variety of effects on the ecosystem. Aquatic ecosystems may become unhealthy due to the buildup of heavy metals in sediments and aquatic life, which can be harmful.

Organic molecules have the potential to contaminate water sources, impacting the quality of drinking water and presenting health concerns to individuals (Yang et al., 2019a, b). Over-abundance of nutrients can result in eutrophication, which deteriorates aquatic environments and causes algal blooms and hypoxia. Discharged human and wildlife pathogens can transmit water-borne diseases (Filote et al., 2021).

Robust monitoring, treatment, and management techniques are needed to address the dangerous contaminants found in wastewater (Toy et al., 2022). Mitigating the environmental impact of hazardous pollutants and safeguarding water resources for future generations can be achieved by implementing sophisticated wastewater treatment technology, reducing pollutant discharges at the source, and promoting sustainable behaviors (Rathi et al., 2021). To manage wastewater and safeguard the ecosystem sustainably, we must first understand the many kinds of harmful chemicals found in wastewater and the effects they have on the environment (Kishor et al., 2021). The use of nanomaterial-based adsorption as a practical and sustainable way to remove contaminants from a variety of environmental media has grown in popularity (Ahmed et al., 2022). The customizable surface chemistry, high surface area, and reactivity of nanomaterials, along with their distinctive characteristics at the nanoscale, make them a possible solution to the problems associated with pollutant removal (Rani et al., 2021). By investigating the potential of nanoparticles as adsorbents, we can gain a deeper comprehension of their function in the elimination of pollutants and their significance in environmentally sustainable management (Bhatt et al., 2022). By employing nanomaterial-based adsorption, we may create creative and effective plans to reduce pollution, safeguard ecosystems, and encourage a more hygienic and salubrious environment for generations to come (Gupta & Bharadwaj, 2020).

The unique features and prospective applications of microbe-derived nanomaterials (MDNs) in numerous domains have made them an intriguing research subject in recent times (Mathew et al., 2022). These nanoparticles differ from traditionally created nanomaterials in that they are produced by microbes using natural processes, giving rise to structures with unique properties (Dhanker et al., 2021). By investigating the characteristics and synthesis mechanisms of MDNs, we can learn more about their possible applications in healthcare, environmental remediation, and biotechnology. Knowing the special qualities of MDNs can help develop novel and environmentally friendly ways to use microorganism's power to produce sophisticated nanomaterials with customized qualities for certain uses (Annamalai et al., 2021).

This summary will look at the several dangerous contaminants that are frequently present in wastewater, where they come from, and how they might affect ecosystems and public health. Overviewing the fundamentals and uses of nanomaterial-based adsorption in environmental remediation is the goal of this introduction. In this, the remarkable qualities of MDNs-like their stability, biocompatibility, and variety of functions—will be briefly discussed.

13.2 Types of Microbe-Derived Nanomaterials

Nanoparticles could be fabricated by employing either of the three methods that are physical, chemical, and biological. Microbe-derived nanomaterials fall into the biological approach of synthesis. Biological synthesis could be further subdivided based on the type of microbe (Bacteria, Fungi, Yeast, Viruses and Algae) employed for synthesis, type and whether the synthesis is occurring intracellularly or extracellularly. Biological systems overpower the other two methods as they are eco-friendly, feasible and rapid. Easy maintenance, fast growth and ability to grow in various habitats are the added advantages of employing microbes for the synthesis of nano-materials (Koul et al., 2021).

13.2.1 Biogenic Nanoparticles

Bacteria are a good candidate for nanoparticle synthesis due to their characteristics such as the ability to adapt to different environments, rapid growth and easy isolation. Physical conditions such as oxygenation and temperature could be easily managed and manipulated which in turn helps in the fabrication of various sizes of nanoparticles for varied applications (Shah et al., 2021). Metal nanoparticles such as silver, gold, zinc, etc. could be synthesized using bacteria, examples of which are listed in Table 13.1. Nanoparticles synthesized from organic materials using bacteria have also been reported such as cellulose nanocomposites and bacterial cellulose. Additionally, they are capable of producing oxide metals, magnetosomes and nano mineral crystals (Mughal et al., 2021).

Fungi have an edge over bacteria in terms of economic and ecological feasibility along with easy scale-up and downstream processes. Many studies show the production of silver nanoparticles employing fungi (Shah et al., 2021). However, the synthesis of gold nanoparticles using fungi is not strongly supported by various studies. Metal nanoparticles such as magnetite, have application in wastewater treatment as it aids in removing heavy metals and radioactive materials (Martins et al., 2017). Additionally, magnetite has superparamagnetic behavior which helps in easy separation after wastewater treatment. Nuclear waste management is done using bio-nanocomposites fabricated using fungus (Mughal et al., 2021).

Algae is been considered as a model organism for the synthesis of nanoparticles as they have a high ability to chelate and transform metals into malleable states (Martins et al., 2017). Additionally, they produce a variety of active compounds such as chlorophyll, phycobilins (phycocyanin and phycoerythrin), pigments like carotenoids (carotene and xanthophyll), vitamins, oil, antioxidants (polyphenols and tocopherols) and polyunsaturated fatty acids. Theoretically, these active compounds are thought to contribute significantly to the synthesis of nanoparticles. Various studies show the production of silver, gold, and iron oxide nanoparticles employing algae (Mughal et al., 2021; Mukherjee et al., 2021).

Table 13.1 List of nanoparticles synthesized employing different microbes

Metal nanoparticle	Microbe	Species	Use	Site of synthesis	References
Platinum and Palladium	Bacteria	*Desulfovibrio vulgaris*	Catalysts	Extracellularly	Martins et al. (2017)
Silver	Bacteria	*Cupriavidus* sp. and *Bacillus subtilis*	Antibacterial	Extracellularly	Markus et al. (2016) andPatil et al. (2019)
Nanoselenium	Bacteria	*Bacillus* sp.	Biomedical	Extracellularly	Bharathi et al. (2020)
Gold	Bacteria	*Lactobacillus kimchicus* ,*Staphylococcus epidermidis* and *Paracoccus haeundaensis*	Drug delivery, cancer diagnostic, Catalysts and Antioxidants	Both	Mathivanan et al. (2019), Pantidos, (2014), and Srinath and Ravishankar Rai (2015)
Zinc Oxide	Bacteria	*Bacillus subtillis* and *Staphylococcus aureus*	Synthetic dyes and Antibacterial	Both	Dhandapani et al. (2020) and Rauf et al. (2017)
Selenium	Fungi	*Mariannaea* sp. HJ	Medicinal and electronics	Both	Vijayanandan and Balakrishnan (2018)
Iron Oxide	Fungi	*Aspergillus niger*	Wastewater treatment	Extracellularly	Kumari et al. (2023)
Cobalt Oxide	Fungi	*Aspergillus nidulans*	Energy storage	Extracellularly	Joshi et al. (2017)
Copper	Fungi	*Aspergillus niger*	Antidiabetic and Antibacterial	Extracellularly	Zhang et al. (2019)
Silver	Algae	*Portieria hornemannii* and *Botryococcus braunii*	Antibacterial and catalysis	–	Arya et al., (2019) and Manikandakrishnan et al. (2019)
Gold	Algae	*Caulerpa racemosa, Egregia sp. and Cystoseira baccata*	Biomedical and cancer	–	Salem et al. (2019), González-Ballesteros et al. (2017), and Mourdikoudis et al. (2018)
Iron Oxide	Algae	*Colpomenia sinuosa* and *Pterocladia capillacea*	Antibacterial	–	Colin et al. (2018)

13.2.2 Extracellular Polymeric Substances

Microbes can break down inorganic materials into nanoparticles through their extracellular or intracellular routes. The approach for induced synthesis would be different for both routes. For instance, in the extracellular pathway, the supernatant containing the enzymes is obtained by centrifugation of broth by separating the microbes as pellets (Jain et al., 2015). The microbe-free supernatant containing enzymes is used for the synthesis of nanoparticles in a cell-free system. Whereas, in intracellular synthesis, the microbial cellular mechanism is employed for the synthesis of nanoparticles. The pellet obtained after centrifugation is used and exposed to the substrate, this leads to the entrapment of positively charged substrate on the negatively charged cell wall of microbial cells (Fadiji et al., 2022). Subsequently, bio-reduction initiates as a result of enzyme activity to yield nano-clusters which eventually leads to the formation and diffusion of nanoparticles (Koul et al., 2021).

Irrespective of whether the microbe is synthesizing nanoparticles through an extracellular or intracellular route the extracellular polymeric substances (EPSs) have a significant impact on it. EPSs are heavy polymers synthesized by microbes (typically bacteria), these EPSs could be soluble in the environment or bound to the cells either loosely or tightly (Jain et al., 2015). This impact is due to physical, chemical, or biological mechanisms. The physical mechanisms include several interactions such as Van der Waals forces, electrostatic double-layer force, steric interactions, polymer bridging interaction, hydrophobic interactions and magnetic attractive forces. The chemical mechanisms encompass redox reactions, photocatalytic reactions, re-precipitation, complexation, cation bridging, ligand exchange, hydrogen bonding and π bonding. Lastly, biological mechanisms contain endocytosis, mitochondrial effects and biodegradation (Huangfu et al., 2019).

Employing the above-mentioned mechanism the EPSs can lead to stabilization, chemical reduction, determination of charge on nanoparticles, resistance to nanoparticle toxicity, and controlling of the activity of nanoparticles (Kroll et al., 2014; Kang et al., 2014; Jain et al., 2015; Joshi et al., 2012). One such example is the growth of elemental selenium to biogenic selenium nanoparticles depicted in Fig. 13.1.

13.2.3 Microbial Cell Surfaces

The cell outer covering of the cells could be effectively used for modifying the nanoparticle synthesized to make it more effective. This approach was initially concentrated around the use of mammalian cells, however, multiple studies are coming up with the use of microbes as a source of the membrane (Purohit et al., 2019). Molecules, subunits and bound macro-molecules derived from membranes are employed to increase the functionality of nanoparticle synthesized (Annamalai et al., 2021). On the other hand, in cell membrane-based nanostructure whole

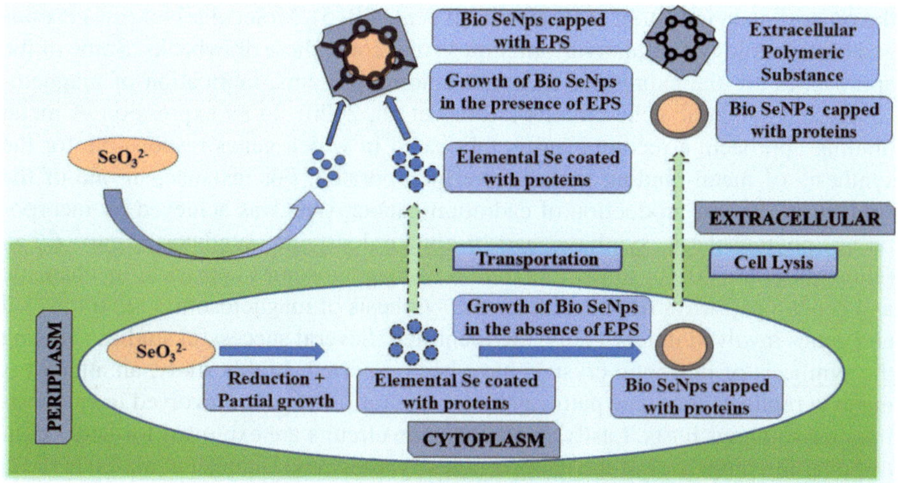

Fig. 13.1 Schematic representation of the conversion of elemental selenium to biogenic selenium nanoparticles. (Jain et al., 2015)

membrane is used rather than just using solitary components. Cell membrane-based nanostructure resolves the complication of conjugating, identifying, and synthesizing discrete components (Fang et al., 2017, 2018).

Cell membrane-based nanostructure includes vesicular structures, membrane-coated nanoparticles and hybrid cell membrane coating. Vesicular structures are natural parallels of liposomes, for example, exosomes which could be synthesized from large membranes or could be naturally occurring (Qu et al., 2017). Outer membrane vesicles (OMVs) are analogous structures exhibited by bacteria. In one of the studies, gold nanoparticles were coated with OMVs which became more stable than independent gold nanoparticles in a physiological environment. Membrane-coated nanoparticles are based on a core-shell system, which consists of the nanoparticle core and shell of the cell membrane (Fang et al., 2018). Various studies employing various combinations of different cell membranes and nanoparticles are reported. A hybrid cell membrane is a modification of membrane-coated nanoparticles in which multiple cell membrane is employed instead of single cell type cell membranes (Li et al., 2015). One of the studies provided proof of this idea wherein they synthesized nanoparticles coated with the hybrid membrane of erythrocytes and platelets (Fang et al., 2017, 2018).

13.2.4 Engineered Microbial Cells

Naturally occurring metal-resistant microorganisms are employed for the synthesis of nanoparticles. However, this approach has some drawbacks such as non-homogenous dimensions of nanoparticles synthesized and difficulty in scaling up the process at an industrial scale (Dwivedi et al., 2015). Molecular biology, genetics and biotechnology cumulatively attempt to overcome these drawbacks. Some of the approaches are the expression of metal-binding proteins, fabrication of magnetosomes and synthetic gene circuits (Abbas et al., 2020). In an expression of metal binding approach, a recombinant is fabricated in which genes responsible for the synthesis of metal-binding proteins are incorporated. For instance, in one of the studies, successful production of cadmium nanocrystals was achieved by incorporating phytochelatin synthase and γ-glutamylcysteine synthetase into *E.coli* (Huangfu et al., 2019). In the fabrication of magnetosomes approach, mechanistic and genetic pathways responsible for the synthesis of magnetosomes are unraveled and genes involved in fabrication are identified. Several successful studies showing the synthesis of magnetic crystals have been reported. In this study, an attempt is made to replicate the same pathway by incorporation of genes involved in synthesis into heterologous hosts. Lastly, synthetic gene circuits are exploited for better control over the properties of the nanoparticles synthesized (Dhanker et al., 2021).

13.3 Characteristics and Synthesis of Microbe-Derived NMs

Microbes can build a wide variety of nanostructures. As a result, scientists are now eager to employ these microbes to create nanostructures for a variety of applications. Both bacteria and fungi are capable of producing inorganic chemicals through biologically mediated and induced synthesis. Controlling the biological synthesis can produce nanostructures with the necessary composition and geometries. The biological sources for the synthesis of nanoparticles are depicted in Fig. 13.2.

13.3.1 Synthesis of Nanomaterials Using Fungi

More research is being done on fungi as an environmentally acceptable and long-term substitute for synthesizing nanoparticles. By employing fungal biomass or fungal extracts, metal ions are reduced to create nanoparticles (Saravanan et al., 2021a). Silver nanoparticles (AgNPs) have been synthesized by fungi, including *Trichoderma spp., Aspergillus niger*, and *Fusarium oxysporum*. These nanoparticles have antibacterial qualities and could be used in agriculture and health, among other areas (Kulkarni et al., 2023). Gold nanoparticles (AuNPs) have been synthesized using fungi such as *Penicillium spp.* and *Trichoderma harzianum*. Due to their special optical characteristics, these nanoparticles are employed in medication delivery, sensing, and imaging applications (Anjum et al., 2023). Copper nanoparticles (CuNPs) have been synthesized by *A. flavus* and *P. brevicompactum* among other fungi. These nanoparticles have demonstrated encouraging catalytic and antibacterial qualities (Fouda et al., 2022). Zinc oxide nanoparticles (ZnO NPs) have been synthesized using fungi such as *Aspergillus fumigatus* and *Penicillium spp.*

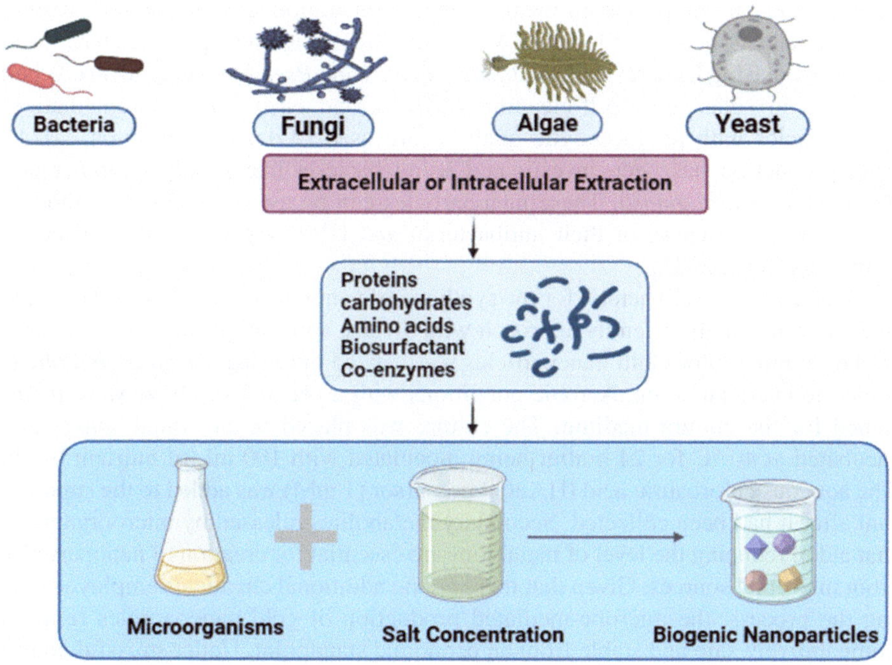

Fig. 13.2 Overview of biogenic nanoparticles and their sources for synthesis

Because of their antibacterial and UV-blocking qualities, these nanoparticles are used in electronics, cosmetics, and sunscreen (Mani et al., 2022). In general, using fungus to synthesize nanoparticles is a cost- and sustainability-effective method with potential applications in many different industries.

13.3.2 Synthesis of Nanomaterials Using Bacteria

Due to their capacity to reduce metal ions and generate nanoparticles, bacteria have also been investigated for the synthesis of nanomaterials. Bacterial biomass or bacterial extracts are used in the procedure to speed up the creation of nanoparticles (Saravanan et al., 2021b). AuNPs have been synthesized using bacteria including *Pseudomonas aeruginosa, Bacillus subtilis*, and *Escherichia coli*. These nanoparticles have been used in medication delivery, sensing, and imaging due to their distinctive optical characteristics (Kulkarni et al., 2023). The synthesis of AgNPs has been carried out using bacteria such as *Bacillus licheniformis, Bacillus cereus*, and *Lactobacillus spp*. These antibacterial nanoparticles are used in medical devices, water purification, and wound healing (Sharma et al., 2022). Iron oxide nanoparticles (Fe_3O_4) have been synthesized using bacteria like *Shewanella oneidensis* and *Geobacter sulfurreducens*. Due to their magnetic characteristics, these nanoparticles are employed in medication administration, environmental cleanup, and biomedical imaging (El-Gendy & Nassar, 2021). TiO_2 nanoparticles have been synthesized using bacteria such as *Bacillus cereus* and *Pseudomonas stutzeri*. When it comes to solar cells, self-cleaning surfaces, and environmental remediation, nanoparticles with photocatalytic qualities are used (Azizi & Daneshjou, 2024). ZnO production has been investigated using bacteria like *Bacillus subtilis* and *Pseudomonas aeruginosa*. These nanoparticles can be used in sunscreen, fabrics, and cosmetics because of their antibacterial and UV-blocking qualities (Raha & Ahmaruzzaman, 2022).

Overall, the use of bacteria for the synthesis of nanomaterials offers a sustainable and environmentally friendly approach with diverse applications in various fields.

The synthesis for Gold nanoparticles was carried out using *Vibrio alginolyticus* a marine bacterial strain. Aerobic conditions with a pH of 7 or above were maintained for the culture medium. The culture was placed in an orbital shaker and incubated at 40 °C for 24 h after being inoculated with 100 mL of nutrient broth. The aqueous chloroauric acid (HAuCl) precursor (1 mM) was added to the supernatant after it had been collected. Secondary metabolites released by microorganisms that aid in reducing the level of metal ions are essential for creation of nanoparticles from microbial sources. Given that there are no additional chemicals employed during the process, the microbe-mediated production of gold nanoparticles remains comparatively safe and stable from an economic standpoint. Under stressful conditions, the bacteria begin to create proteins or enzymes that aid in the reduction of hazardous chemicals like gold chloride in gold nanoparticles. A reddish-brown solution with a maximal absorbance peak at 530 nm was obtained. SEM electron microscopy demonstrated the production of amorphous gold nanoparticles with a size range of 100–150 nm as shown in Fig. 13.3.

13.3.3 Synthesis of Nanomaterials Using Yeast

A versatile microbe well-known for its fermentation skills, yeast has also shown promise as a technique for the production of nanomaterials (Yang et al., 2022). Yeast is a desirable option for the environmentally friendly and sustainable manufacture of nanoparticles due to its capacity to decrease metal ions and generate proteins and enzymes. To demonstrate the wide range of uses for yeast-mediated nanotechnology, the recent study explored the synthesis of nanomaterials using yeast (Rana et al., 2020). Yeast biomass or extracts were used in the reduction of metal ions to create nanoparticles. The method uses proteins and enzymes, which are natural reducing agents found in yeast cells, to catalyze the bioreduction of metal ions. Because it doesn't require strong chemicals or high temperatures, this biogenic technique is both economical and environmentally beneficial (Kumari et al., 2023). It has been extensively explored how to synthesize different nanoparticles, such as silver, gold, and selenium nanoparticles, using yeast species like *Saccharomyces cerevisiae* and *Candida utilis* (Golhani et al., 2020). In another

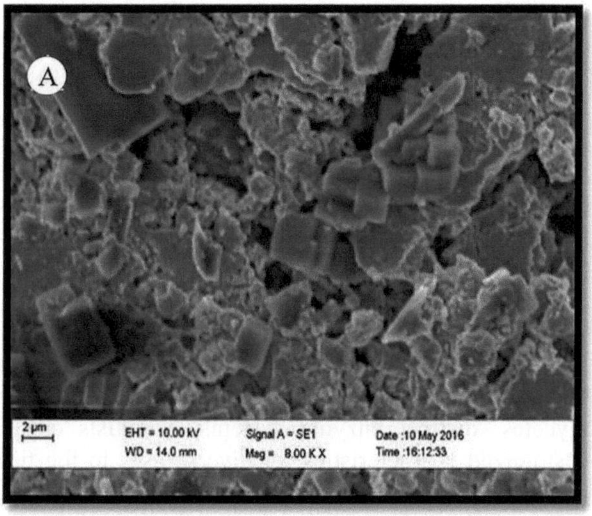

Fig. 13.3 SEM image of microbe-derived gold nanoparticles synthesized form *Vibrio alginolyticus*

study, *Saccharomyces cerevisiae* was used to synthesize silver nanoparticles. Strong antimicrobial qualities make AgNPs made from yeast perfect for application in antibacterial coatings, wound dressings, and water purification systems (Roychoudhury, 2020). *Candida utilis* was used to synthesize gold nanoparticles., and because of their distinct optical characteristics and biocompatibility, yeast-produced gold nanoparticles were used in drug delivery, biosensing, and catalysis (Khan & Lee, 2020). *Saccharomyces cerevisiae* can be used to synthesize selenium nanoparticles (SeNPs), and because of their antioxidant qualities, yeast-produced nanoparticles used in nutritional supplements, medical devices, and environmental cleanup (Bapte et al., 2022). As nanotechnology develops further, yeast-based synthesis methods will likely contribute significantly to the advancement of nanoscience and technology (Singhal et al., 2023).

13.3.4 Synthesis of Nanomaterials Using Actinomycetes

A class of filamentous bacteria called *actinomycetes* is well-known for their wide range of metabolic activities, and these bacteria have shown great promise in the production of nanomaterials (Venkateswaran et al. 2022). These microbes have special enzymatic systems that allow them to use bioreduction techniques to decrease metal ions and create nanoparticles (Yaradoddi et al., 2022). *Actinomycetes* are desirable candidates for the synthesis of nanoparticles due to the benefits of this green synthesis technique, which include sustainability, biocompatibility, and scalability. The ability to synthesize nanoparticles has been investigated in *actinomycetes* genera such as *Micromonospora, Actinomadura*, and *Streptomyces* (Golińska,

2020). Few examples of nanomaterials synthesized using *actinomycetes* are discussed below.

Streptomyces sp was used to reduce silver ions to form AgNPs that shown to have antimicrobial properties that make them useful in antibacterial coatings, wound dressings, and water treatment applications (Venkateswaran et al., 2022). *Actinomadura sp.* produces gold nanoparticles (AuNPs), that have special optical properties and are useful in drug delivery systems, biosensors, and catalysis (Bahrami et al., 2022). *Micromonospora sp.* synthesizes iron oxide nanoparticles (Fe_3O_4 NPs), that has potential in using in magnetic resonance imaging, medicine delivery, and environmental remediation (Rajesh et al., 2024).

Utilizing actinomycetes for the manufacture of nanomaterials is a biogenic and sustainable method that makes use of these bacteria's metabolic capabilities to produce nanoparticles (Venkateswaran et al., 2022). Through the utilization of actinomycetes' distinct enzyme systems, scientists can fabricate nanoparticles with customized characteristics for diverse uses in the fields of biomedicine, environmental science, and materials science (Fadiji et al., 2022). The synthesis of nanoparticles mediated by actinomycetes is a sustainable method that presents a great opportunity for the creation of sophisticated materials with a wide range of applications. Actinomycetes-based synthesis methods are expected to be important in promoting innovation and tackling global concerns as nanotechnology advances (Yaradoddi et al., 2022).

13.3.5 Synthesis of Nanomaterials Using Algae

Algae's considerable capacity for metal reduction and accumulation gives it an advantage in the creation of nanoparticles. The proteins in the algae extract, which are bioactive substances, serve as a reductant and stabilize the metallic nanoparticle. Algal biomass, both living and dead, is helpful for the biogenic production of nanoparticles. Due to its high sequestration of carbon dioxide and quick growth rate, algae are the preferred candidate for the synthesis of nanoparticles. When algae are utilized to make nanoparticles, the least amount of energy is used. The ability of algae to produce harmless byproducts during the biogenic material synthesis process for metallic nanoparticles is one of their main advantages. In bioreduction, a variety of pigments and enzymes derived from algae are crucial. A polysaccharide known as fucoidans is secreted by certain brown algae and is helpful in the synthesis of gold nanoparticles as a substitute for chemical processes. Numerous algal strains, including *Acanthophora spicifera, Laminaria japonica, Tetraselmis kochinensis*, and *Turbinaria conoides*, may find use in the production of gold nanoparticles. Antibiotic-resistant *Staphylococcus aureus* has been demonstrated to be susceptible to bactericidal activity by the biosynthesis of nanocomposites derived from *Chlorella vulgaris* (Kulkarni et al., 2023).

The synthesis of gold nanoparticles using algal species *S. plantensis* was carried out and these contain bioactive compounds that can prevent illness. The protein extract of *S. plantensis* was mixed with 10mM of Hydrogen tetrachloroaurate in equal volumes and stirred continuously. Subsequently, the solution was mixed with 1 N NaOH. The color of the solution followed a shift from green to a yellow-greenish shade. The solution became ruby red after a further 48 h of room temperature incubation, indicating the production of AuNPs. Arya et al. (2019) developed the standard procedure for algal-based CuNP synthesis using a UV-visible spectrometer. They started with 5 ml of *Botryococcus braunii* extract, which they added gradually to 50 ml of 0.001 M copper acetate ($Cu(CO_2CH_3)_2$) solution while stirring continuously at 100 °C for 24 h. This allowed them to monitor the formation of NPs based on the change in color of the mixture solution containing copper acetate solution and algal extract. In a recent study, 100 ml of distilled water and 15 g of biomass of *Arthrospira platensis* were combined, and the mixture was left for 48 h at room temperature. The biomass-free filtrate was then collected by centrifugation. Then, to get the concentration of 0.002 M, 98 ml of the biomass filtrate was added after 0.44 g of zinc acetate ($Zn(CH_3CO_2)_2$) was heated to melt in 2 ml of distilled water. Subsequently, the solution combination was kept in a shaker at 150 rpm for 24 h at 30 °C. ZnONPs were obtained in powder form by collecting the resulting white precipitate and oven-drying it for 24 h at 200 °C. *Colpomenia sinuosa* or *Pterocladia capillacea* seaweed extracts were added in a 1:1 ratio to a 100 mM $FeCl_3$ solution and stirred for 1 h at 800 rpm to create Fe_3O_4 NPs. After that, the mixture was incubated for 30 min at room temperature. The development of Fe_3O_4 NPs was indicated by the color changing to dark brown (Salem et al., 2019).

13.4 Biogenic Synthesis Routes and Mechanisms

The processes by which biological organisms, including bacteria, fungi, plants, and algae, are employed to synthesize nanoparticles are known as "biogenic synthesis routes". By using biological systems' innate ability to reduce, these biogenic pathways transform metal ions into nanoscale particles. Depending on the organism and the particular nanoparticle being generated, different pathways are involved in biogenic synthesis (Mughal et al., 2021). Figure 13.4 shows the synthesis of biogenic nanoparticles. Mechanisms of biogenic synthesis involves the following essential factors,

1. **Reduction:** The primary mechanism involved in biogenic synthesis is the reduction of metal ions to form nanoparticles. Biological systems produce enzymes, proteins, and other biomolecules that act as reducing agents, facilitating the conversion of metal ions into nanoscale particles (Shah et al., 2021).
2. **Stabilization:** In addition to reducing metal ions, biological systems also play a role in stabilizing the resulting nanoparticles. Biomolecules present in the biological extracts help prevent agglomeration and maintain the stability of the nanoparticles (Soni et al., 2021).

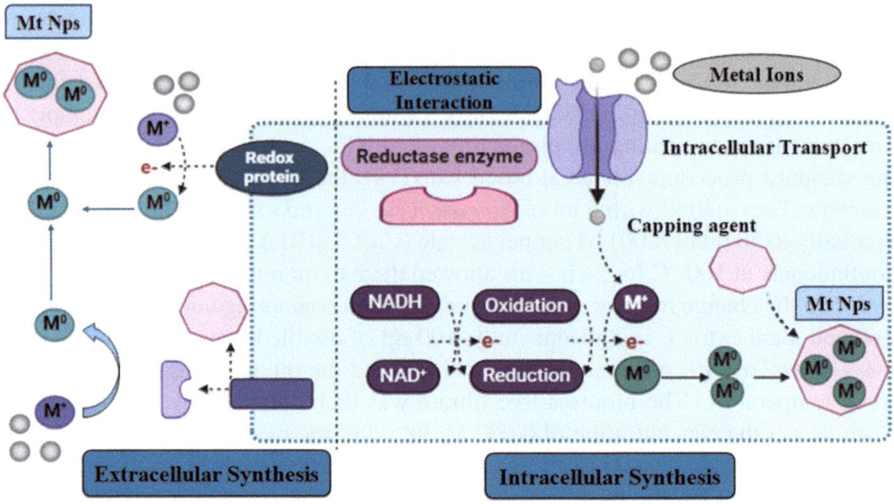

Fig. 13.4 Mechanism-based visualization for the synthesis of micro-derived nanoparticles. (Kulkarni et al., 2023)

3. **Nucleation and Growth:** The nucleation and growth of nanoparticles occurs through a series of chemical reactions and interactions between the reducing agents and metal ions. The size, shape, and properties of the nanoparticles are influenced by the reaction conditions and the biological components involved (Zhang et al., 2023).

13.5 Properties and Characteristics of Microbe-Derived Nanomaterials

Microbe-derived nanomaterials, synthesized using biological organisms such as bacteria, fungi, and algae, possess unique properties and characteristics that make them attractive for various applications. These properties are influenced by the biological synthesis route, the type of microorganism used, and the specific nanoparticle being produced (Mughal et al., 2021).

Microbe-derived nanomaterials typically exhibit nanoscale dimensions, with sizes ranging from a few nanometers to hundreds of nanometers. The shape of the nanoparticles can vary, including spherical, rod-shaped, triangular, and more complex structures, depending on the synthesis conditions and the biological system used (Adetunji et al., 2023). The composition of microbe-derived nanomaterials depends on the metal ions or precursors used in the synthesis process. Common nanomaterials synthesized by microbes include metal nanoparticles (e.g., silver, gold, copper), metal oxide nanoparticles (e.g., zinc oxide, iron oxide), and semiconductor nanoparticles (e.g., quantum dots) (Yang et al., 2022). The surface of microbe-derived nanomaterials is often coated with biomolecules, such as proteins, enzymes, and polysaccharides, which can influence their stability, reactivity, and interactions with other molecules. These surface functional groups can also be modified for specific applications (Adetunji et al., 2023).

These nanomaterials are often biocompatible and non-toxic, making them suitable for biomedical applications such as drug delivery, imaging, and tissue engineering. The presence of biological molecules on the nanoparticle surface can enhance their biocompatibility (Yang et al., 2022).

Also, nanomaterials synthesized by microbes, particularly silver nanoparticles, exhibit strong antimicrobial activity against a wide range of pathogens. These nanoparticles can disrupt bacterial cell membranes, inhibit microbial growth, and have potential applications in wound healing, water treatment, and antimicrobial coatings (Khan & Lee, 2020). Some microbe-derived nanomaterials possess catalytic properties due to their high surface area and unique surface chemistry. These nanoparticles can act as catalysts for various chemical reactions, including degradation of pollutants, hydrogen production, and organic synthesis (Fouda et al., 2022).

Microbe-derived nanomaterials can exhibit unique optical and magnetic properties, depending on their composition and structure. For example, gold nanoparticles show plasmonic properties, while iron oxide nanoparticles exhibit magnetic behavior, making them useful for imaging and sensing applications (Anjum et al., 2023).

Microbe-derived nanomaterials have shown promise in environmental remediation, such as the removal of heavy metals, organic pollutants, and dyes from contaminated water and soil. These nanoparticles can efficiently adsorb, degrade, or immobilize pollutants, contributing to sustainable environmental solutions (Rathi et al., 2021).

Overall, microbe-derived nanomaterials offer a sustainable and eco-friendly approach to nanoparticle synthesis, with properties and characteristics that make them versatile for a wide range of applications in healthcare, environmental science, catalysis, and materials science. Further research and development in this field holds great potential for advancing nanotechnology and addressing global challenges.

13.6 Adsorption Mechanisms of Microbe-Derived NMs

Adsorption embodies the process of accumulation of atoms, ions, or molecules on a solid surface. The solids employed for absorbing dissolved substances are referred to as adsorbents, while the term adsorbate pertains to the molecules that undergo adsorption. The formation of an adsorbate layer occurs, which significantly builds up on the surface of the adsorbent. The phenomenon of adsorption is observed in various physical, chemical, and biological systems. Its utilization is evident in industries such as water treatment, where activated carbon or synthetic resins are utilized for purification purposes (Naser et al., 2021). The capacity to regulate the synthesis process is predominantly influenced by factors such as the pH of the

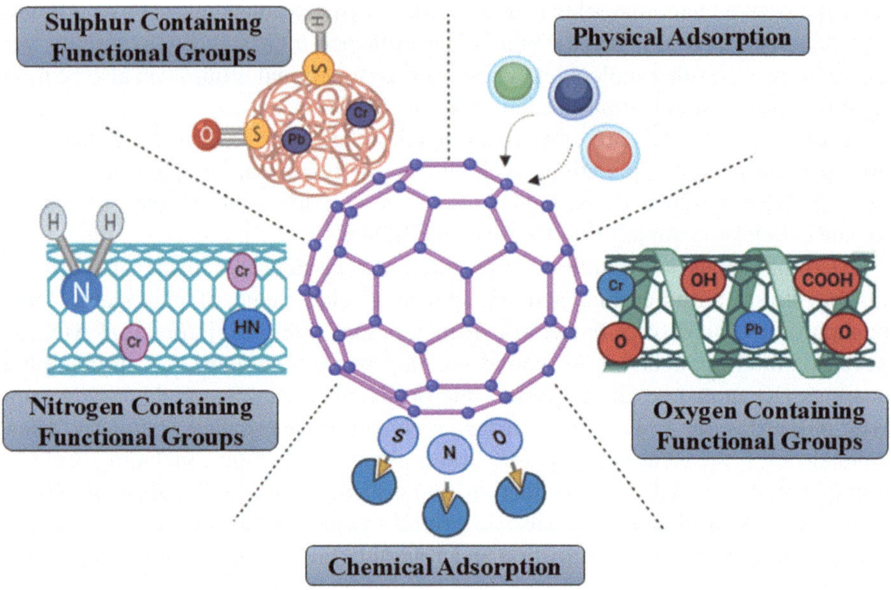

Fig. 13.5 Schematics of different adsorption mechanisms of microbe-derived nanomaterials

system, the speed of stirring, the rate of titration, the duration of the reaction, and the temperature. These five factors explicitly or implicitly establish the adsorption capability, a quality that is vital in terms of the suitability of nanoparticles as adsorbents (Manyangadze et al., 2020). Functional structures and heavy metals have complicated intermolecular interactions that are influenced by the adsorbate's nature, the aqueous solution's ionic surroundings, and the varied composition and chemical reactions of the carbon surface. Several possible processes lead to heavy metals being adsorbed onto carbon adsorbents, including the exchange of ions, surface complexation, physical adsorption, electrostatic contact, and precipitation. Each mechanism's specific function for heavy metal adsorption differs greatly depending on the target metal ion, the solution's ionic environment, and the carbon adsorbent (Alsharari et al., 2018). The differentadsorption mechanisms are depicted in Fig. 13.5.

13.6.1 Physical Adsorption

A deficient process called physical adsorption causes heavy metals to diffuse into carbon-adsorbent pores and deposit within without synthesizing chemical bonds. The surface area and pore size distribution of the adsorbents have a significant impact on physical adsorption inevitably. The interactions between van der Waals and electrostatic forces control physisorption. All systems have Vander Waals forces, but only systems with persistent charges may experience electrostatic interactions (Petrovic et al., 2022). Adsorption predominantly takes place within pore walls or at certain locations inside particles in extremely porous carbons. Variations in a molecule's weight, shape, or polarity can cause separation and increase the affinity for that molecule by strengthening the molecule's interaction with the adsorbent surface. Adsorption works based on exploiting the variations in contact strengths (Petrovic et al., 2022). The polarizability of the absorbent is greatly affected by the incorporation of heteroatoms that are in the format of Surface functional groups, and this has the power to raise the reported thermal conductivity of adsorption to levels exceeding those usually associated with the physisorption of carbon dioxide (Youk et al., 2020). The primary adsorption method for heavy metals is physical adsorption, which is prevalent but improbable. Yet, a few studies have highlighted the significance of physical adsorption in the process of aqueous heavy metal adsorption on carbon-based materials. For instance, Nayak et al. (2017) found that pore diffusion and physical adsorption processes play a major role in the removal of cadmium II and nickel (II) from aqueous solutions employing two different forms of activated carbon. According to Liu et al. (2019) ion exchange is the predominant mechanism for the absorption of copper (II) onto activated carbon produced from walnut shells, though physical force additionally plays an essential role.

13.6.2 Chemical Adsorption

New bond formation that has a binding energy larger than around 0.5 eV per molecule is the phenomenon that commonly characterizes chemisorption. In certain cases, chemisorption is referred to as selective adsorption, wherein the adsorbate (CO_2) interacts with a designated location (Webley & Danaci, 2019). The chemical diversity of a carbon's surface determines its chemical properties, which in turn influences the viability and magnitude of chemisorption. The existence of heteroatoms like O, N, and S is necessary for surface chemical heterogeneity. The manner of modification and/or the substance from which the carbon is sourced determines the particular form and corresponding amount of heteroatoms (Chiang & Juang, 2017). The oxidative treatments that generate oxygen-containing functionalities provide the adsorbents an acidic character, which in turn enables electron-acceptor characteristics. On the other hand, capabilities that incorporate N usually have a fundamental nature. As an outcome, these functions acquire electron-donor qualities, making them desirable adsorption locations for the CO_2 molecule's electron-deficient C atom (Adelodun et al., 2015).

13.6.3 Oxygen-Containing Functional Groups

Presently, functional groups containing oxygen have been found to have the greatest effect on the surface reactions, surface behavior, hydrophilicity, electrical, and catalytic characteristics of carbons. It has been discovered that the most prevalent functional groups, hydroxy and carboxyl groups, have a significant role concerning the adsorption of heavy metals, including Pb (II), Cr (VI), Hg (II), Cd II), and Ni (II) ions, onto a variety of carbon adsorbents (Lu & Astruc, 2018). The degree of hydrophilicity is determined by the number of polar O-groups present, and oxygen-containing groups regulate the two qualities of C-adsorbents: hydrophobicity/hydrophilicity and acidity/basicity (Saha & Kienbaum, 2019). The significant part that oxygen-containing groups on GO surfaces played in Pb (II) and Cd (II) adsorption, respectively. The process of Cr (VI) adsorption on H_3PO_4 treated biochar involved the formation of an inner sphere complex between the biochar's surface oxygenic functional groups (Liang et al., 2017). Biochars made from different feedstocks have been oxidized using a variety of chemical and physical modification techniques. These modifications have significantly increased the amount of surface oxygen-containing functional groups in biochars, which in turn has increased the rate of adsorption and capacities of the material to different heavy metals, such as lead (II), Cd (II), Cu (II), Ni (II), and others (Inyang et al., 2016).

13.6.4 Nitrogen-Containing Functional Groups

The fundamental characteristics of carbon surfaces can be improved by adding nitrogen groups, which has led to a rising interest in carbon adsorbent research due to the potential benefits of nitrogen-containing functional groups for heavy metal adsorption (Yang et al., 2019a, b). Pyridinic, graphite, pyridine, amine (1,2,3), nitro and nitroso type, and pyrrolic are the N-groups that are formed on the C-adsorbents. Except quaternary all the N-groups occur at the graphitic plane edge site (Saha & Kienbaum, 2019). It is often reported that the N concentration is the primary factor affecting the selectivity of CO_2 adsorption in carbon sorbents (Yang et al., 2019a, b). By pyrolyzing biomass treated with diammonium hydrogen phosphate, nitrogen-fictionalized carbon (biochar/activated carbon) may be produced immediately and has a potent sorption capacity for aqueous Cr (VI) (Guo et al., 2017). To increase the ability of carbon surfaces to adsorb different heavy metals in aqueous solutions, nitrogen has been doped into several other carbon materials, including carbon film, mesoporous carbon, microporous carbon, and magnetic carbon nanoparticles (Yang et al., 2019b).

13.6.5 Sulphur Containing Functional Groups

In contrast to conventional dopants like nitrogen, sulfur is significantly larger. It extends beyond the graphene plane, creating strain and flaws within the framework that may lead to charge localization and facilitate CO_2 adsorption (Saha & Kienbaum, 2019). In general, there are two possible means to create S-doped carbons: through (i) carbonize precursors that contain sulfur (such as organic polymers, aromatic compounds, ionic liquids, or biological materials) and then activate them later to create porosity, or (ii) activate precursors that contain sulfur by activating them (Petrovic et al., 2022). Functional groups containing sulfur, however, can either enhance or reduce specific surface area according to the experimental setup. Furthermore, elemental sulfur has been added to carbon adsorbents by employing sulfuration processes such as chemical reaction deposition and solution infiltration, which can increase the adsorption capacity of the materials to a variety of aqueous heavy metals. (Yang et al., 2019a, b). In addition to having partial oxygen-containing groups and sulfonic acid groups that show good adsorption capacity towards aqueous Cd (II) and a fast adsorption kinetic rate, sulfonated graphene nanosheets also have a relatively complete sp2-hybridized plane along with a high affinity towards aromatic pollutants in water (Shen & Chen, 2015a, b). Dithiocarbamate groups functionalized multi-walled carbon nanotubes made of carbon disulfide and ethylenediamine have also been reported to exhibit extremely high adsorption capabilities for Cd (II), Zn (II), and Cu (II), in aqueous solutions (Li et al., 2015).

13.6.6 Adsorption Isotherms and Kinetics

13.6.6.1 Adsorption Isotherm

To determine the adsorption capacity of the adsorbent materials, adsorption isotherms are crucial. Adsorption isotherms correlate the equilibrium relationship between the adsorbent and the adsorbates at constant temperature. To illustrate the adsorption isotherms, several models, including the Langmuir, Temkin, Freundlich,Dubinin–Radushkevich, and Sips models, have been presented thus far (Yang et al., 2019a, b).

Langmuir Model

The Langmuir model states that adsorption occurs uniformly on the adsorbents' active sites and that adsorption behaviors cease after the adsorbates have utilized all the available adsorptive space. According to the Langmuir model, each adsorption active site can only bind one adsorbate, and all these sites have identical binding energies. The Langmuir model's linear version is given as

$$\frac{C_e}{q_e} = \frac{1}{(bq_m)} + \frac{c_e}{q_m} \tag{13.1}$$

Here, q_e is given as the adsorption capacity of the adsorbent in equilibrium (mg/g), C_e is given as the adsorbate concentration in equilibrium (mg/L), b is the adsorption equilibrium constant, qm is given by monolayer adsorption capacity (mg/g) (Yang et al., 2019a, b).

Temkin Isotherm

Temkin isotherms connect the highest binding energy employed during adsorption to the continuous distribution of bonding energies. It explains how the temperature of adsorbate molecules decreases linearly as the contact between adsorbate and adsorbent molecules increases. The Temkin isotherm equation is provided as

$$q_e = \text{B}\ln A_T + \text{B}\ln C_e$$

Where C_e is the concentration of the adsorbate in equilibrium (mg/L), q_e is the adsorption capacity of the adsorbent in equilibrium (mg/g), A_T is the Temkin constant, B is the heat constant (J/mol) and T is the temperature at 25 °C (Wadhawan et al., 2020).

Freundlich Isotherm Model

Reversible and heterogeneous adsorption systems are described by the Freundlich model. It is not solely restricted to the formation of monolayers.

$$q_e = k_F C_e^n$$

$$\ln q_e = \ln k_F + (\ln C_e / n)$$

Where q_e is the amount of pollutant adsorbed in equilibrium (mg/g), k_F is the equilibrium constant of the Freundlich model concerning adsorption capacity (mg/g) × (L/mg) ^ (1/n), where n is the adsorption intensity (Naser et al., 2021).

Dubinin-Radushkevich Isotherm Model

Investigating how heavy metals adsorb on nanoparticles, particularly carbon nanomaterials is best accomplished with the help of this model. Compared to the Freundlich and Langmuir isotherms, it provides more information and accurate data on the adsorption of molecules on nanoparticles of carbon and nanotubes. Dubinin-Radushkevich (DRK) isotherms include an adsorption process where surface curvature is important. For the adsorption process on a heterogeneous surface, Gaussian energy distribution is used. When the concentration range is moderate and the solute concentration activities are high, this model fits the data best. The equation is given below:

$$q_e = (q_s)\exp + (-K_{ad} \in^2)$$

Where q_e is the adsorption capacity of the adsorbent in equilibrium (mg/g), K_{ad} is the Dubinin-Radushkevich constant (mol^2/KJ^2), and q_s is the theoretical saturation capacity (mg/g) (Wadhawan et al., 2020).

Sips Model

A combination of the Freundlich and Langmuir isotherms is the Sips isotherm. By predicting heterogeneous adsorption systems, the Sips model may circumvent the Freundlich isotherm model's restriction of increasing adsorbate concentration. The Sips isotherm efficiently decreases to the Freundlich isotherm at low adsorbate concentrations. The Langmuir isotherm's monolayer sorption capacity is predicted by the Sips model at high adsorbate concentrations. The Sips model may be written as

$$q_e = \frac{q_m\left(k_s c_e\right)^{nS}}{1+\left(k_s c_e\right)^{nS}}$$

Where q_e is the adsorption capacity of the adsorbent in equilibrium (mg/g), C_e is the concentration of the adsorbate in equilibrium (mg/L), q_m is the maximum adsorption capacity of Sips (mg/g), ks is the Langmuir equilibrium constant (L/mg) and nS is compared to the Freundlich heterogeneity factor nF (Singh et al., 2021).

13.6.6.2 Adsorption Kinetics

Adsorption kinetics research is undoubtedly crucial in demonstrating the underlying processes and pathways of the adsorption process. The calculated kinetic rate is used to determine how best to use the adsorbent. The adsorption system under study presents a challenge in terms of practical application, as indicated by its sluggish adsorption rate. Adsorbed pollutant molecules adhere to the surface of the adsorbent through three stages in the adsorption mechanism: layer diffusion, pores diffusion, and adhesion (Naser et al., 2021)

Pseudo-First Order Kinetics

The pseudo-first-order kinetics determines the adsorption capacity of solid and liquid systems. The linear expression is given as:

$$\frac{dq_t}{dt} = K\left(q_e - q_t\right)$$

On integration,

$$\log\left(q_e - q_t\right) = \log q_e - \frac{k_1}{2.303}t$$

q_e is the adsorption capacity of the adsorbent in equilibrium (mg/g), qt is given as the adsorption capacity (mg/g), the contact time is given as t, k_1 and K are the rate constants for the pseudo-first-order kinetics (Yang et al., 2019a).

Pseudo-Second Order kinetics

The pseudo-second-order kinetics are used to describe the adsorption capacity of solid phase systems. The linear equation for the second order is given as follows:

$$\frac{t}{q_1} = \frac{1}{K_2 q_e^2} + \frac{1}{q_e} t$$

Here, q_e is the adsorption capacity of the adsorbent in equilibrium (mg/g), qt is given as the adsorption capacity at a given time t, and K_2 is the rate constant for pseudo-second-order per minute (Wadhawan et al., 2020).

Double Exponential Model

The adsorption of heavy metal ions is reported in two mechanisms. Ion absorption through both internal and exterior diffusion occurs in the first phase. The uptake of the metal ions till the state of equilibrium is noted in the second phase. The mathematical expression is given as:

$$q_t = q_e - D_1 m_{ads} \, e^{(-k_{D2t})}$$

D_1 is the adsorption in a rapid step (m/mol), D_2 is the adsorption parameter for a slow process (m/mol), q_e and q_t are the amount of ion adsorbed at equilibrium and at the time (min) (Wadhawan et al., 2020).

Intra-particle Diffusion Model

The Morris and Weber hypothesis, which describes the multistep process of adsorption and includes moving molecules into the adsorbent's solid pores, provides the foundation for this model.

$$q_t = k_{id} t^{0.5} + C$$

Where $t^{0.5}$ is the half-time for diffusion (sec), C is the boundary thickness, K_{id} is the intra-particle diffusion constant, q_t is the adsorption capacity at a particular time t (Singh et al., 2021).

13.7 Applications and Relevant Case Studies of Microbe-Derived NMs in Hazardous Pollutant Removal

The massive volume of dangerous chemicals released from industry and the polluted wastes created by the material cycle combined to pollute the environment worldwide. Industrialization is one of the primary causes of metal and metalloid contamination in wastewater (Kumari & Tripathi, 2020). Due to their inability to decompose, metals can continue to exist in the atmosphere for extended periods, raising significant risks to the ecosystem and human health. The heavy metals like lead (Pb), cadmium (cd), chromium (Cr), nickel (Ni), copper (Cu), and iron (Fe) are used for various operational activities in different industries. Metal intoxication modifies sulfhydryl neutrality, destroys DNA, and promotes the production of free radicals (Sharma & Kumar, 2021). The toxic effect of the contaminant can be reduced through microbial cells when they convert from one state to another (Sharma, 2021). Microorganisms contribute to the production of signaling molecules that can aid in the biodegradation processes (Sharma, 2021). Eliminating toxic metals from the environment has garnered a lot of interest and is very important. To date, a wide range of methods have been developed to address this issue, such as electrochemical treatment, membrane filtration, adsorption, chemical precipitation, and ion exchange (Yang et al., 2019a, b).

The usage of nanomaterials satisfies the requirement for a wide surface area in heavy metal removal processes for contamination detection, adsorption, and immobilization (Lu & Astruc, 2018). Adsorbents composed of nanomaterials possess several unique characteristics, such as their nanoscale size, huge surface area, strong mechanical properties, strong solution mobility, porosity, hydrophilicity, dispersibility, and hydrophobicity (Yaqoob et al., 2020). Researchers explored several nanomaterials as effective adsorbents to remove heavy metals from raw wastewater, including biosorbents, graphene, activated carbon, and magnetic nanoparticles (Huangfu et al., 2019). Since various studies have reported success using microbe-mediated nanoparticles in remediation processes, microbes have been used to improve the use of nanoparticles for the nano bioremediation process. Microbes have been widely accepted for this scientific technique because of their biological, physical, optical, and chemical characteristics, including their high surface-to-volume ratio, highly changeable functionality, super-hydrophobic and filtering nature, and sensitive affinity membranes (Torimiro et al., 2021). The applications of biogenic nanoparticles in various fields are illustrated in Fig. 13.6.

13.7.1 Removal of Heavy Metals from Wastewater

Bacterial cells have been used as a bio-nano factory to create a variety of metal nanoparticles (NPs), including Ag, Au, Cu, Se, and Fe, as well as metal oxide NPs, including zinc oxide (ZnO), copper oxide (CuO), and silver oxide (Ag_2O) (Grasso et al., 2020). Many researchers have been interested in metal oxide nano-adsorbents because of their effective removal of different heavy metals, including lead, arsenic, copper, nickel, cadmium, mercury, and chromium. Metal oxide nanoparticles (NPs) comprising titanium, nickel, and aluminum are particularly useful for degrading dyes, cleaning up phosphates, and neutralizing nickel and aluminum. TiO_2 NPs can act as photocatalysts, facilitating the ozonation process which ultimately results in the total solubilization of organic pollutants (Pantidos, 2014).

Silver nanoparticle's extensive variety of applications across several disciplines has led to enormous growth in their synthesis, as reported. As was previously noted,

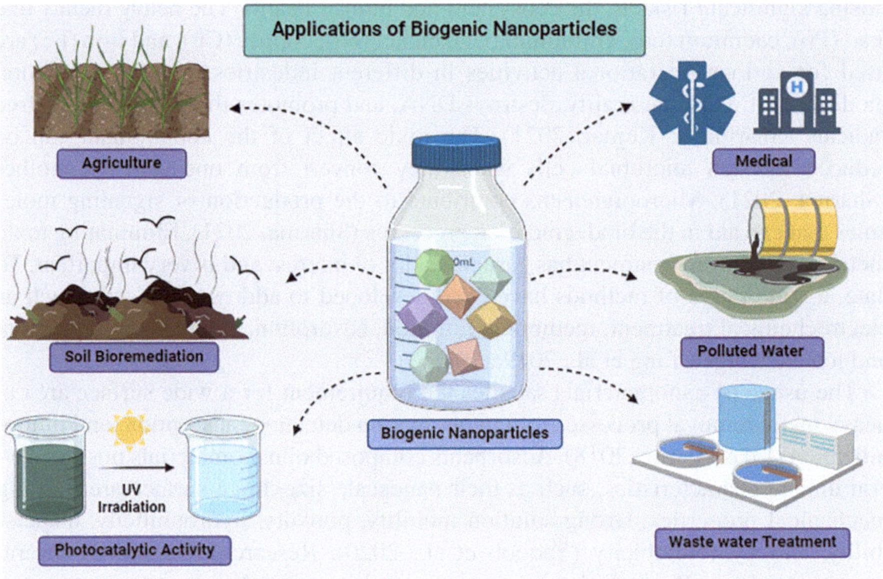

Fig. 13.6 Applications of biogenic nanoparticles in various fields

bacteria are particularly capable of converting heavy metals into nanoparticles and have a great capacity to tolerate abiotic stressors. Some bacteria can grow and live in high quantities of metals, especially heavy metals, such as facultative aerobes *Pseudomonas stutzeri* and facultative anaerobes *P. aeruginosa* (Koul et al., 2021). *Lactococcus lactis* and *Kluyveromyces marxianus* were used in the bioremediation of heavy metals, including Ni and Cd, with removal efficiencies of 75% (Cherni et al., 2020). Iron-sulfur (FeS) magnetic nanoparticles that have undergone biosynthesis can adsorb a variety of heavy metals and a small number of anions. Therefore, they have the potential to be used as an adsorbent for heavy metals in bioremediation processes. It has been noted that complete anaerobic conditions are necessary for the fabrication of magnetite NPs, which is a very slow process. Nonetheless, in fully aerobic circumstances, magnetite nanoparticles using the nonmagnetotactic bacteria *Acinetobacter spp.* (Koul et al., 2021). Analyzing a study on the bioremediation and removal of cadmium-polluted soil, it was found that cadmium was successfully removed by co-treating with nanohydroxyapatite (NHAP) and *Bacillus subtilis* (Liu et al., 2019). Different kinds of heavy metal nanoparticles, such as palladium (Pd), tellurium (Te), and platinum (Pt), are also being developed for certain applications in a variety of industries. For example, Zonaro et al. (2017) showed that poisonous tellurite oxyanions may be converted into beneficial Te NPs by using the bacteria *Ochrobactrum sp.*. *Shewanella loihica PV-4* was utilized in a recent investigation to create Pt and Pd NPs (size 2–4 nm) using electrochemically active biofilms (Ahmed et al., 2018). An efficient way to enhance the bioremediation of Pb and Cd-induced pollution was to employ FeO nanoparticles made by coating polyvinylpyrrolidone (PVP) with a Gram-negative bacteria species called *Halomonas sp*. After 24 hours, this approach eliminated over 100% of the Pb and almost 100% of the Cd, when compared to bacteria or even Nanoparticles alone (Alabresm et al., 2018).

Studies have demonstrated that *Cunninghamella elegans* can produce chitosan nanoparticles (NCt) through the biosynthetic process with the use of chitosan, a bioactive polymer (Alsharari et al., 2018). According to Raj et al., (2016), the generated NCt particles outperformed bulk chitosan in their ability to biosorb and bioremediate lead (II) and Copper (II) ions in a contaminated environment. The ability of *Pseudomonas aeruginosa* to increase cadmium bio reduction accelerated the manufacture of Cadmium Sulphide Nanoparticles, as demonstrated by the effective removal of cadmium from cadmium-contaminated water.

The significance of algae in the photo-removal method of Cr (VI) under UV irradiation was explained by the introduction of *Chlorella vulgaris* as a modified agent in nanoscale bi-metallic (TiO_2/Ag) chitosan nanofiber mats (Wang et al., 2017). It has been found that *Chlorella vulgaris* releases organic compounds (chlorophylls, carboxylate acids) that improve the photocatalytic degradation of chromium VI on TiO_2/Ag chitosan nanofiber mats. This supports the synergistic effect of algae and TiO_2/Ag hybridized nanomaterial. A similar study was done on pharmaceutical effluents that included heavy metals, namely lead (Pb) and chromium (Cr) (Kumari & Tripathi, 2020).

13.7.2 Adsorption of Organic Pollutants and Emerging Contaminants

Particulates, hazardous colors, medicines, organic and inorganic micropollutants, hazardous heavy metals, pigments, poisonous dyes, and pathogenic microorganisms are among the emerging pollutants (EPs) that are being documented in a broad range of research (Xiao et al., 2017). A growing number of residential, commercial, and agricultural activities are causing these pollutants' concentration and volume to increase steadily (Zhou et al., 2016). Over the past few decades, biogenic nanoparticles, or BNPs, have attracted a great deal of study attention since the idea of microbial interactions with metals was developed (Gillan, 2016). BNPs have been used in biological, electrical, and wastewater treatment applications because of their special qualities. It is now widely known that microbial interaction may be used to design the synthesis of BNPs (Ali et al., 2019). Toxic textile dyes, pharmaceuticals, pigments, personal care products (PPCP), heavy metals, halogenated recalcitrant pollutants, pesticides, refractory organic micro-pollutants, and pathogenic microorganisms have all been removed using BNPs derived from various bacterial species (Martins et al., 2017). The toxic and emerging pollutants that are treated using microbe mediated nanoparticles are given in Table 13.2.

Table 13.2 Toxic and emerging pollutants and their removal mechanisms through biogenic nanoparticles

Pollutants	Biogenic nanoparticles (BNPs)	Microbes/Strain	Reaction conditions	Removal mechanism	References
2- Chlorophenol, 2,4-dichlorophenol	MnOx	*Pseudomonas sp.G7*	Equilibrium time for 12 h, 20 °C Temperature with a pH of 7	Oxidation, Co-precipitation, and electrostatic interaction	Tu et al. (2015)
Rhodamine B (RhB)	ZnS	*Shewanella oneidensis MR-1*	Elqulibrium time at 3 h	Oxidation and reduction	Xiao et al. (2015)
Congo red	Bio-Pd, Pd-AGS	Anaerobic granules (UASB)	Equilibrium time from 4 to 6 h and at a temperature of 35 °C	Combination of microbial metabolism of oxidation and reduction	Quan et al. (2015)
2-nitrophenol (2-Np), 3-nitrophenol (3-Np), 4-nitrophenol (4-Np), o-nitrophenyl amine (o-NPA),m-nitrophenyl amine(m-NPA)	AuNPs	*Trichosporon monyevideense* WIN		Combination of microbial metabolic and biomass sorption	Shen et al. (2016)
Methyl Orange	SeNps	*Bacillus sp.*		Oxidation and reduction	Ahluwalia et al. (2016)
Methylene Blue	Pbs nanocrystals	SRB C*lostridiaceae sp*	Equilibrium time for 20 h	Oxidation and reduction	Yue et al. (2016)
Acid Brilliant Scarlet green, Acid Red B, Acid Orange G, Acid Black1, Reactive red X-3B, Reactive Black, Reactive Red, Cation Red	Bio-AuNps	Trichoderma sp. WL-Go (ASS)	Equilibrium time of 100–180 min at 30 °C	Oxidation and reduction	Qu et al. (2017)

2-nitrophenols, 3-nitrophenol, 4-nitrophenol,o-nitrophenol, m-nitroaniline, Cationic Red X-GRL, Acid Orange II, Acid Scarlet Green, Catoin Red, Acid yellow 11, Reactive Red, Acid Red B, Reactive Red X-3B, Acid Orange G, Acid Black 10B, Reactive green KE-4B, Reactive Black	AuNps +NaBH4	*Aspergillus sp. WL-Au*	Reaction time ranging from 120 to 540 s	Oxidation and reduction	Qu et al. (2017)
Nitrite (NO2-N)	(Hydro) oxides biogenic Fe (II) magnetite	*Shewanella oneidensis*	Equilibrium time of 20 h at 30 °C	Oxidation and reduction	Lu et al. (2017)
Ciprofloxacin, Sulfamethoxazole, Ibuprofen, 17b-estradiol	Bio-Pt + H_2	*Desulfovibrio vulgaris*	Equilibrium time of 24 hours at 25°C	Combination of adsorption,co-metabolism of microorganisms, and oxidation-reduction	Martins et al. (2017)
Ciprofloxacin, Sulfamethoxazole, Ibuprofen,17β-oestradiol	Bio-Pd + H_2				

13.7.3 Utilization of Microbe-Derived Nanomaterials in Soil Remediation

Nanoremediation, a new application of nanotechnology, addresses environmental pollution by treating hazardous waste. Gillham's research on zero-valent iron (ZVI) as a permeability barrier was the first to utilize nanotechnology in decontamination (Aliyari Rad et al., 2023). Studies have demonstrated the ability of zero-valent iron nanoparticles (NPs) contained in silica microspheres to break down polybrominated diphenyl ethers, a class of environmental contaminant that quickly builds up in the soil (Ray et al., 2023). Aqueous solution of decabromodiphenyl ether was treated with tetrahydrofuran (THF) to break it down. When subjected to environmental and temperature stress, the researchers found that it was effective in a THF and water solution. Furthermore, the (Xie et al., 2016) research revealed that this degrading process produced a 78% removal efficiency, or elimination efficiency, for decabromodiphenyl ether in soil. It had greater importance than the biomass of plants exposed to NPs and they are shown in Fig. 13.7.

Investigations into the phytotoxicity of Cr (VI) were conducted, and tests using iron nanoparticles supported by bio-carbon on cabbage mustard revealed reduced levels of Cr (VI) and enhanced growth. The immobilization effectiveness for chromium (VI) and total chromium (Cr) in remediation studies was 100% as well as 91.94%, respectively, with the infusion of 8 g per kg of soil (Su et al., 2016). TiO_2 nanoparticles (NPs) and other pollutants can be dangerous to humans if they are discovered in organic material, according to research done (Fan et al., 2016) on the effects of TiO_2 on Daphnia magna's cytotoxicity. When exposed to humic acid,

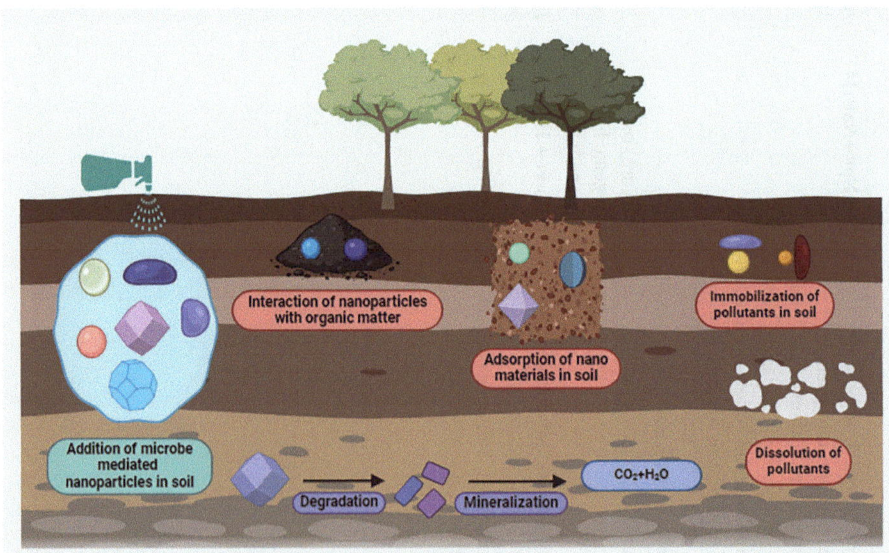

Fig. 13.7 Microbe mediated nanomaterials in soil remediation

Nano-TiO_2 significantly reduced the toxicity of Cu in D. magna, indicating that the presence of humus in the soil may reduce the effectiveness of these NPs as a treatment when they encounter certain pollutants. The degree of adsorption that nano-TiO_2 may attain determines its capacity to collect heavy metals. The capacity of nano-TiO_2 to collect Cu is dependent on both the degree of metal absorption and the amount of humic substances in the solution (Fan et al., 2016). *Avicennia officinalis* rhizosphere biofabricated iron oxide nanoparticles (NPs) including Aspergillus tubingensis (STSP 25) were discovered by Mahanty in the Sundarbans. The synthesized NPs showed a five-cycle regeneration potential and eliminated or removed nearly ninety percent of the heavy metals (HMs) in wastewater, including Ni (II), Cu (II), Pb (II), and Zn (II). An endothermic process chemically attached the metal ions to the NPs' surface (Mahanty et al., 2020).

It's been contended that producing NPs using bacteria is a beneficial and environmentally friendly process. Copper NPs were synthesized using *SINT7*, an *Escherichia* species that is resistant to copper. Biogenic nanoparticles have been shown to break down textile effluents and azo dyes. Consequently, reaction black-5, malachite green, Congo red, and direct blue-1 were dropped by 83.6%, 90.6%, 97.1%, and 88.4% at a lower concentration of 25 mg/L, respectively. Upon reaching a 100 mg/L concentration, the corresponding reductions were 76.84%, 31.1%, 83.90%, and 62.32%. In addition, treated industrial sewage samples had lower levels of suspended particles as well as phosphate and chloride ions. These biogenic nanoparticles' performance might improve long-term and affordable industrial manufacturing (Noman et al., 2020).

13.8 Environmental Implications and Safety Considerations

Microbe-derived nanomaterial as discussed in the previous section has a significant potential in contributing to bioremediation. However, it is crucial to consider some of the key factors such as the fate and transport of microbe-derived nanomaterial in the environment, associated risks, and potential strategies to evade or minimize the pernicious effect of employing this technology.

13.8.1 Fate and Transport of Microbe-Derivnanomaterialial in Aquatic and Terrestrial Environments

Nanomaterials can enter into the aquatic and terrestrial environments via various routes. For instance, routes into an aquatic environment include runoff from adulterated soil, direct discharge of effluents, aerial precipitation, emissions from sewage treatment plants, and drizzle drift load of agricultural products. However, the routes into the terrestrial environment encompass of use of agro-products, implementation of remediation procedures, aerial accumulation, and the employment of sewage sludge as a fertilizer (Batley et al., 2013). In terms of the life cycle of a nano-based product, the above-mentioned routes may fall into either of the following phases that are either, release during production, use, or after disposal. From various studies, it has been evident that most of the nanoparticle deposition occurs during use and after disposal phase. Moreover, studies show that wastewater treatment is the largest source of deposition of nanoparticles into the environment. Additionally, estimates from research suggest that deposition in terrestrial environment is highest, followed by aquatic environment and air (Bundschuh et al., 2018).

After the transport of nanomaterials into the environment, it can have various fates. It is driven by chemical transformation, aggregation and disaggregation processes. The chemical transformation includes amendments in the surface characteristics of nanomaterials, dissolution, degradation and speciation. Change in the surface characteristics of a nanomaterial is majorly the result of adsorption, desorption, or precipitation. However, the dissolution of nanoparticles is a cumulative result of intrinsic properties (particle chemistry, shape, size, surface coating and state of aggregation) of nanomaterial and environmental parameters such as temperature, dissolved organic compound, and pH. Endocytosis of nanomaterial could occur if they are made available to microbes by dissolution and release of toxic ions or directly in nano form (Dwivedi et al., 2015). Additionally, it is reported that under a distinct environmental setup, the nanomaterial could be made inert by the process called passivation. This passivation is the result of sulfidation and makes the surface of nanomaterial inert, for instance, sulfidation of silver nanoparticles leads to the fabrication of hollow Ag_2S NP or core-shell Ag_0–Ag_2S structures (Bundschuh et al., 2018).

Aggregation is also one of the fates that a nanomaterial can undergo when released into the environment. The aggregation is affected by interaction with organic, inorganic, particulate, or dissolved compounds. Aggregation could be of two types of homo-aggregation or hetero-aggregation. Homo aggregation refers to the interaction between the same nanoparticles. Derjaguin-Landau-Verwey-Overbeek (DLVO) theory is used to analyze homo-aggregation. Additionally, in terms of efficiency multivalent cations are a step ahead of monovalent cations, within each the efficiency is dependent on critical coagulation concentrations (CCCs). Lastly, the isoelectric point affects the interaction with the organism and natural organic matter in the environment contributing to either stabilizing or destabilizing of nanoparticles. On the other hand, heterogeneous aggregation is the interaction between nanoparticle and natural colloids such as montmorillonite, maghemite, and kaolinite, also microorganisms, algae, and proteins. Heterologous interaction resembles more to an actual scenario in the environment compared to homologous interaction because homologous interaction occurs rarely due to low concentration of identical nanoparticles. Hetero aggregation occurs rapidly if the interacting component is of the opposite charge. In conclusion, the fate of a nanoparticle is determined by the concentration and surface characteristics of the nanoparticle and environmental factors such as pH, type, and concentration of natural organic matter, and ionic strength, type, and charge of ions (Bundschuh et al., 2018).

13.8.2 Potential Risks Associated with the Release and Accumulation of Nanomaterials

There are several risks associated with the entry of nanoparticles into the environment via various routes and interaction of nanoparticles with the environment resulting in various fates of the nanomaterials. This fate nanomaterials may have deleterious effects such as climate change, and nanotoxicity to microbial community flora and fauna (Abbas et al., 2020). Carbon nanotubes and graphene are examples of carbon-based nanomaterials contributing to global climate change, they do so by fusing with water vapors and gases. Moreover, living organisms are negatively majorly due to excess production of reactive oxygen species. Physical contact, dissolution, and release of toxic ions are other mechanisms to cause toxicity. Microbial communities (bacteria, fungi, protozoa and algae) interaction with engineered nanomaterials changes microbial community functioning and layout. Oxidative stress and altering the bio-availability of nutrients are considered to be direct impacts by the nanomaterials, whereas, toxicity caused after interaction with other components in the environment such as toxic pollutants and naturally occurring organic compounds falls into indirect impacts (Abbas et al., 2020).

In plants, studies suggest that phytotoxicity due to nanomaterial is concentration dependent i.e., at low concentration it's beneficial, whereas high concentration causes toxicity. Deleterious impacts of phytotoxicity caused by nanomaterials include structural and physiological changes such as amended rates of transpiration and photosynthesis, declined growth, and impaired seed germination (Abbas et al., 2020). Lastly, other multicellular organisms experience toxicity as a result of biomagnification. For instance, humans are the ultimate recipient in the food chain as a result they experience various harmful effects due to exposure to high concentrations of nanomaterial. Using animals as a model, various studies have reported several concerning effects such as genotoxicity, cell proliferation, imbalance of various systems, cytotoxicity, and hepatic and neurological damage (Rajput et al., 2020).

13.8.3 Strategies for Minimizing Environmental Impact and Ensuring Safe Use

As discussed in the earlier section, nanomaterials have various potential risks to the environment, microbes, plants, and animals. Strategies and methods for reducing this effect differ from conventionally used methods for waste management. Classification of nano waste and accessing its exposure capability aids in achieving efficient nano waste handling and disposal. Methods of nano waste treatment include physical, chemical, and biological approaches (Abbas et al., 2020).

The use of the high absorption property of clay shows significant potential for the evasion of nanoparticles. Their high absorption property is due to high surface area, enlargement capacity, and cation interchange potential. The mode of action of clay in reducing nanotoxicity is by hetero aggregation and sedimentation. As a result, the surface area to volume of nanomaterial decreases which affects its reactivity. For instance, the teratogenic effect on zebrafish embryos by titanium dioxide-engineered nanoparticles was drastically reduced due to hetero-aggregation caused by montmorillonite clay particles of nanomaterials. Employing clay has various advantages such as being cheap, easily available, and nontoxic (Abbas et al., 2020).

Biochar is synthesized during the pyrolysis process and is made from various waste biomass. It can be effectively employed for remediating environmental contaminants. For instance, soil organisms experienced less uptake, bioavailability, and accumulation of silver and cadmium dioxide nanomaterials due to the use of biochar in the soil system. Additionally, it also shows the potential to evade phytotoxicity. Biochar interacts and adsorb nanomaterials to form irreversible hetero aggregates via different approaches such as the interaction of nanomaterials with functional groups of biochar to avoid repulsive forces, charge transfer, hydrogen bonding, and van der Waals forces. Moreover, it can alter the pH of a medium containing nanomaterials adversely affecting the dissolution of the nanomaterials (Abbas et al., 2020).

The bioremediation approach employing plants and microbes is considered economical and safe for managing nanomaterial-rich effluents. Microbes can degrade the surface or core of the nanomaterial. Extracellular polymeric substances and extracellular enzymes synthesized by microbe's aid in the degradation of nanomaterials such as carbon nanotubes. Additionally, in the rhizosphere interaction of microbes with plants aids in phytoremediation (Abbas et al., 2020).

Phytoremediation involves the cultivation of plants that accumulate nanomaterials within their tissues by removing them from soil, water, and air. However, biomass rich in hyper-accumulated nanomaterial needs to be managed according to the guidelines of a regulatory organization such as the U.S Environmental Protection Agency. Landfilling, combustion and composting are current conventional methods for the management of biomass concentrated with nano waste, however environment is polluted by secondary pollutants. To overcome this issue a conceptual model known as cradle-to-cradle system could be employed in which this biomass is used for energy generation, thus making it an environmentally friendly, economical, and clean management approach (Abbas et al., 2020).

Green synthesis of non-toxic nanomaterials is an emerging sustainable approach. Additionally, the green synthesis approach overpowers energy-intensive chemical and physical approaches by reducing the use of harmful solvents and chemicals. Green synthesis aims to reduce harmful effects during the production phase of the life cycle of nanoparticle-containing products. Apart from the above-mentioned approaches, the adoption of adequate legislative measures is very crucial for regulating the production, use, and disposal of nanomaterial-based products (Purohit et al., 2019). Lastly, it could be concluded that rather than a standalone approach a multifarious approach should be opted, including a blend of regular evaluation of nano-based products, the latest research, adequate legislation, and public awareness would ensure fruitful benefits and help manage deleterious outcomes (Abbas et al., 2020).

13.9 Conclusions and Future Directions

An environmentally friendly method for bioremediating industrial effluents has been made possible by the integration of nanotechnology and microorganisms. The production of nanomaterials using microbes, as described, offers better options for sustainable and affordable heavy metal degradation treatment. Over the past few decades, significant attempts have been made to solve the issue of heavy metal pollution by creating innovative solutions. However, using bioremediation techniques still stands as the most advantageous course of action because of their affordability, environmental safety effects, and applicability. Bioremediation mechanisms are also highly variable, making them cost-effective and accessible. The article includes a summary of some of the mechanisms underlying the various bioremediation techniques, as well as information on the microorganisms involved, potential influencing variables, and ways that recently developed technologies can enhance existing bioremediation methods to increase their efficacy. If properly employed, bioresources like bacteria and microbial enzymes can aid in the production of nanoparticles, which is showing promise as a revolutionary tactic. Because they are highly degradable and less hazardous, microbial assistance in the synthesis of metal nanoparticles holds great promise. Even though stable nanoparticles can be produced by choosing the right microbes and maximizing circumstances, it has also been found that the mechanism underlying biologically synthesized nanoparticles is not well known. Therefore, for the large-scale, sustainable synthesis of nanoparticles, it is necessary to select appropriate bacteria or microbial consortia. High temperatures, high pressures, high energies, stabilizers, and hazardous compounds are not necessary for the microbial synthesis of nanoparticles. To remove many contaminants from wastewater while maintaining selectivity, it is necessary to manipulate microbial enzymes to create nanoparticles with a broad variety of organic functional groups. Numerous benefits come with synthesizing nanoparticles using microbes like yeast, actinomycetes, bacteria, fungi, and algae, including easy production, low cost, high efficiency, safety, and environmental friendliness. Microbial nanoparticles have the potential to treat contaminants at contaminated places. Biocompatible and readily separable by filtration or precipitation techniques are the residues remaining after pollutants are broken down by microbial nanoparticles. There won't be any waste in the end because biochar may be used to manufacture value-added items like building materials from leftover waste. Thus, a more environmentally friendly method of synthesizing nanoparticles creates new opportunities for numerous biotechnological uses.

Metallic nanoparticles can be biologically synthesized at the laboratory size, but bulk manufacturing of these particles requires escalation to the industrial scale. Though just 1% of nanotechnology materials have been commercialized to date, the application of effective microbes-assisted nanotechnology has the potential to significantly increase the industrial economy. Making this technique economically viable and sustainable by industry needs requires the efficient manufacture of microbial nanoparticles. Since there is currently no relevant cost-related data, a cost-benefit analysis should also be performed for its commercial utilization. When it comes to chemical synthesis, the use of chemicals, salts, reducing and stabilizing agents, and media is costly; yet, in microbial synthesis, metal salts, and microbial growth media are equally costly. To reduce costs, recyclable waste biomass might be used as an alternative for producing nanoparticles. Research on genetic engineering and the metabolic pathways of microorganisms may pave the way for the innovative development of these promising nano factories, which might then be scaled up and used commercially as effective, long-term bioremediation solutions. To fully utilize the omics-derived data and get a deeper comprehension of microbial processes, sophisticated computational techniques are necessary. Therefore, the successful manufacturing of nanoparticles by microorganisms using green chemistry will mark a significant advancement towards the usage of green nanotechnology.

References

Abbas, Q., Yousaf, B., Ullah, H., Ali, M. U., Ok, Y. S., & Rinklebe, J. (2020). Environmental transformation and nano-toxicity of engineered nano-particles (ENPs) in aquatic and terrestrial organisms. *Critical Reviews in Environmental Science and Technology, 50*(23), 2523–2581. https://doi.org/10.1080/10643389.2019.1705721

Adelodun, A. A., Kim, K. H., Ngila, J. C., & Szulejko, J. (2015). A review on the effect of amination pretreatment for the selective separation of CO_2. *Applied Energy, 158*, 631–642.

Adetunji, C. O., Egbuna, C., Ajiboye, A. A., Ajayi, O. O., Dauda, W. P., Ghazanfar, S., Ogundolie, F. A., Mathew, J. T., Adetunji, J. B., Inobeme, A., & Titilayo, O. (2023). Biogenic nanomaterials with diverse biological activities in the food and biomedical industries. In *Evaporative coolers for the postharvest management of fruits and vegetables* (pp. 395–420). Academic.

Ahluwalia, S., Prakash, N. T., Prakash, R., & Pal, B. (2016). Improved degradation of methyl orange dye using bio-co-catalyst Se nanoparticles impregnated ZnS photocatalyst under UV irradiation. *Chemical Engineering Journal, 306*, 1041–1048. https://doi.org/10.1016/j.cej.2016.08.028

Ahmed, E., Kalathil, S., Shi, L., Alharbi, O., & Wang, P. (2018). Synthesis of ultra-small platinum, palladium and gold nanoparticles by Shewanella loihica PV-4 electrochemically active biofilms and their enhanced catalytic activities. *Journal of Saudi Chemical Society, 22*(8), 919–929. https://doi.org/10.1016/j.jscs.2018.02.002

Ahmed, S. F., Mofijur, M., Ahmed, B., Mehnaz, T., Mehejabin, F., Maliat, D., Hoang, A. T., & Shafiullah, G. M. (2022). Nanomaterials as a sustainable choice for treating wastewater. *Environmental Research, 214*, 113807.

Alabresm, A., Chen, Y. P., Decho, A. W., & Lead, J. (2018). A novel method for the synergistic remediation of oil-water mixtures using nanoparticles and oil-degrading bacteria. *Science of the Total Environment, 630*, 1292–1297. https://doi.org/10.1016/J.SCITOTENV.2018.02.277

Ali, I., Peng, C., Khan, Z. M., Naz, I., Sultan, M., Ali, M., Abbasi, I. A., Islam, T., & Ye, T. (2019). Overview of microbes based fabricated biogenic nanoparticles for water and wastewater treatment. *Journal of Environmental Management, 230*, 128–150. https://doi.org/10.1016/j.jenvman.2018.09.073. Academic.

Aliyari Rad, S., Nobaharan, K., Pashapoor, N., Pandey, J., Dehghanian, Z., Senapathi, V., Minkina, T., Ren, W., Rajput, V. D., & Asgari Lajayer, B. (2023). Nano-microbial remediation of polluted soil: A brief insight. *Sustainability (Switzerland), 15*(1). https://doi.org/10.3390/su15010876. MDPI.

Alsharari, S. F., Tayel, A. A., & Moussa, S. H. (2018). Soil emendation with nano-fungal chitosan for heavy metals biosorption. *International Journal of Biological Macromolecules, 118*, 2265–2268. https://doi.org/10.1016/j.ijbiomac.2018.07.103

Anjum, S., Vyas, A., & Sofi, T. (2023). Fungi-mediated synthesis of nanoparticles: Characterization process and agricultural applications. *Journal of the Science of Food and Agriculture, 103*(10), 4727–4741. https://doi.org/10.1002/JSFA.12496

Annamalai, J., Ummalyma, S. B., Pandey, A., & Bhaskar, T. (2021). Recent trends in microbial nanoparticle synthesis and potential application in environmental technology: A comprehensive review. *Environmental Science and Pollution Research, 28*(36), 49362–49382. https://doi.org/10.1007/S11356-021-15680-X

Arya, A., Mishra, V., & Chundawat, T. S. (2019). Green synthesis of silver nanoparticles from green algae (Botryococcus braunii) and its catalytic behavior for the synthesis of benzimidazoles. *Chemical Data Collections, 20*. https://doi.org/10.1016/j.cdc.2019.100190

Azizi, Z. L., & Daneshjou, S. (2024). Bacterial nano-factories as a tool for the biosynthesis of TiO_2 nanoparticles: characterization and potential application in wastewater treatment. *Applied Biochemistry and Biotechnology*. https://doi.org/10.1007/S12010-023-04839-6

Bahrami, Y., Bouk, S., Kakaei, E., & Taheri, M. (2022). Natural products from Actinobacteria as a potential source of new therapies against colorectal cancer: A review. *Frontiers in Pharmacology, 13*, 929161.

Bapte, P. S., Pansambal, S. S., & Borgave, S. (2022). Biosynthesized selenium nanoparticles: an excellent bait for antioxidant therapy. *Journal of Nanostructures, 12*(1), 178–193.

Batley, G. E., Kirby, J. K., & McLaughlin, M. J. (2013). Fate and risks of nanomaterials in aquatic and terrestrial environments. *Accounts of Chemical Research, 46*(3), 854–862. https://doi.org/10.1021/ar2003368

Bharathi, S., Kumaran, S., Suresh, G., Ramesh, M., Thangamani, V., Pugazhvendan, S. R., & Sathiyamurthy, K. (2020). Extracellular synthesis of nanoselenium from fresh water bacteria Bacillus sp., and its validation of antibacterial and cytotoxic potential. *Biocatalysis and Agricultural Biotechnology, 27*. https://doi.org/10.1016/j.bcab.2020.101655

Bhatt, P., Pandey, S., Joshi, S., et al. (2022). *Nanobioremediation: A sustainable approach for the removal of toxic pollutants from the environment*. Elsevier. Retrieved June 18, 2024, from https://www.sciencedirect.com/science/article/pii/S0304389421030028

Bundschuh, M., Filser, J., Lüderwald, S., McKee, M. S., Metreveli, G., Schaumann, G. E., Schulz, R., & Wagner, S. (2018). Nanoparticles in the environment: where do we come from, where do we go to? *Environmental Sciences Europe, 30*(1). https://doi.org/10.1186/s12302-018-0132-6. Springer.

Cherni, Y., Botta, C., Kasmi, M., Franciosa, I., Cocolin, L., Chatti, A., Trabelsi, I., & Elleuch, L. (2020). Mixed culture of Lactococcus lactis and Kluyveromyces marxianus isolated from kefir grains for pollutants load removal from Jebel Chakir leachate. *Water Environment Research, 92*(12), 2041–2048. https://doi.org/10.1002/WER.1363

Chiang, Y. C., & Juang, R. S. (2017). Surface modifications of carbonaceous materials for carbon dioxide adsorption: A review. *Journal of the Taiwan Institute of Chemical Engineers, 71*, 214–234. https://doi.org/10.1016/j.jtice.2016.12.014

Colin, J. A., Pech-Pech, I. E., Oviedo, M., Águila, S. A., Romo-Herrera, J. M., & Contreras, O. E. (2018). Gold nanoparticles synthesis assisted by marine algae extract: Biomolecules shells from a green chemistry approach. *Chemical Physics Letters, 708*, 210–215. https://doi.org/10.1016/j.cplett.2018.08.022

Dhandapani, P., Prakash, A. A., AlSalhi, M. S., Maruthamuthu, S., Devanesan, S., & Rajasekar, A. (2020). Ureolytic bacteria mediated synthesis of hairy ZnO nanostructure as photocatalyst for decolorization of dyes. *Materials Chemistry and Physics, 243*. https://doi.org/10.1016/j.matchemphys.2020.122619

Dhanker, R., Hussain, T., Tyagi, P., Singh, K. J., & Kamble, S. S. (2021). The emerging trend of bio-engineering approaches for microbial nanomaterial synthesis and its applications. *Frontiers in Microbiology, 12*. https://doi.org/10.3389/fmicb.2021.638003. Frontiers Media S.A.

Dwivedi, A. D., Dubey, S. P., Sillanpää, M., Kwon, Y. N., Lee, C., & Varma, R. S. (2015). Fate of engineered nanoparticles: Implications in the environment. *Coordination Chemistry Reviews, 287*, 64–78. https://doi.org/10.1016/j.ccr.2014.12.014. Elsevier B.V.

Elbehiry, F., Alshaal, T., Elhawat, N., & Elbasiouny, H. (2021). Environmental-friendly and cost-effective agricultural wastes for heavy metals and toxicants removal from wastewater. *Handbook of Environmental Chemistry, 117*, 107–127. https://doi.org/10.1007/698_2021_786

El-Gendy, N., & Nassar, H. N. (2021). Biosynthesized magnetite nanoparticles as an environmental opulence and sustainable wastewater treatment. *Science of the Total Environment*. Elsevier. Retrieved June 20, 2024, from https://www.sciencedirect.com/science/article/pii/S0048969721006781

Fadiji, A. E., Mortimer, P. E., Xu, J., Ebenso, E. E., Babalola, O. O., Fadiji, A. E., Mortimer, P. E., Xu, J., Ebenso, E. E., & Babalola, O. O. (2022). Microbe-mediated biosynthesis of nanoparticles: Applications and future prospects. *Mdpi.Com*. https://doi.org/10.3390/su141710839

Fan, W., Peng, R., Li, X., Ren, J., Liu, T., & Wang, X. (2016). Effect of titanium dioxide nanoparticles on copper toxicity to Daphnia magna in water: Role of organic matter. *Water Research, 105*, 129–137. https://doi.org/10.1016/J.WATRES.2016.08.060

Fang, R. H., Jiang, Y., Fang, J. C., & Zhang, L. (2017). Cell membrane-derived nanomaterials for biomedical applications. *Biomaterials, 128*, 69–83. https://doi.org/10.1016/j.biomaterials.2017.02.041. Elsevier Ltd.

Fang, R. H., Kroll, A. V., Gao, W., & Zhang, L. (2018). Cell membrane coating nanotechnology. *Advanced Materials, 30*(23). https://doi.org/10.1002/adma.201706759. Wiley-VCH Verlag.

Filote, C., Roşca, M., Hlihor, R. M., Cozma, P., Simion, I. M., Apostol, M., & Gavrilescu, M. (2021). Sustainable application of biosorption and bioaccumulation of persistent pollutants in wastewater treatment: Current practice. *Processes, 9*(10), 1696. https://doi.org/10.3390/PR9101696

Fouda, A., Hassan, S. E. D., Eid, A. M., Awad, M. A., Althumayri, K., Badr, N. F., & Hamza, M. F. (2022). Endophytic bacterial strain, Brevibacillus brevis-mediated green synthesis of copper oxide nanoparticles, characterization, antifungal, in vitro cytotoxicity, and larvicidal activity. *Green Processing and Synthesis, 11*(1), 931–950. https://doi.org/10.1515/GPS-2022-0080/HTML

Gillan, D. C. (2016). Metal resistance systems in cultivated bacteria: Are they found in complex communities? *Current Opinion in Biotechnology, 38*, 123–130. https://doi.org/10.1016/J.COPBIO.2016.01.012

Golhani, D. K., Khare, A., Burra, G. K., Jain, V. K., & Rao Mokka, J. (2020). Microbes induced biofabrication of nanoparticles: A review. *Inorganic and Nano-Metal Chemistry, 50*(10), 983–999. https://doi.org/10.1080/24701556.2020.1731539

Golińska, P. (2020). Synthesis of nanoparticles by actinomycetes: Mechanism and applications. In *Microbial nanotechnology* (pp. 1–19). CRC Press.

González-Ballesteros, N., Prado-López, S., Rodríguez-González, J. B., Lastra, M., & Rodríguez-Argüelles, M. C. (2017). Green synthesis of gold nanoparticles using brown algae Cystoseira baccata: Its activity in colon cancer cells. *Colloids and Surfaces B: Biointerfaces, 153*, 190–198. https://doi.org/10.1016/j.colsurfb.2017.02.020

Grasso, G., Zane, D., & Dragone, R. (2020). Microbial nanotechnology: Challenges and prospects for green biocatalytic synthesis of nanoscale materials for sensoristic and biomedical applications. *Nanomaterials, 10*(1). https://doi.org/10.3390/nano10010011. MDPI AG.

Guo, Z., Zhang, J., Liu, H., & Kang, Y. (2017). Development of a nitrogen-functionalized carbon adsorbent derived from biomass waste by diammonium hydrogen phosphate activation for Cr(VI) removal. *Powder Technology, 318*, 459–464. https://doi.org/10.1016/j.powtec.2017.06.024

Gupta, C., & Bharadwaj, A. (2020). *Summary and future perspectives of nanomaterials and technologies: Special emphasis on energy and environment*. Elsevier. Retrieved June 18, 2024, from https://www.sciencedirect.com/science/article/pii/B978012819355600011X

Huangfu, X., Xu, Y., Liu, C., He, Q., Ma, J., Ma, C., & Huang, R. (2019). A review on the interactions between engineered nanoparticles with extracellular and intracellular polymeric substances from wastewater treatment aggregates. *Chemosphere, 219*, 766–783. https://doi.org/10.1016/j.chemosphere.2018.12.044. Elsevier Ltd.

Inyang, M. I., Gao, B., Yao, Y., Xue, Y., Zimmerman, A., Mosa, A., Pullammanappallil, P., Ok, Y. S., & Cao, X. (2016). A review of biochar as a low-cost adsorbent for aqueous heavy metal removal. *Critical Reviews in Environmental Science and Technology, 46*(4), 406–433. https://doi.org/10.1080/10643389.2015.1096880

Jain, R., Jordan, N., Weiss, S., Foerstendorf, H., Heim, K., Kacker, R., Hübner, R., Kramer, H., Van Hullebusch, E. D., Farges, F., & Lens, P. N. L. (2015). Extracellular polymeric substances govern the surface charge of biogenic elemental selenium nanoparticles. *Environmental Science and Technology, 49*(3), 1713–1720. https://doi.org/10.1021/es5043063

Joshi, N., Ngwenya, B. T., & French, C. E. (2012). Enhanced resistance to nanoparticle toxicity is conferred by overproduction of extracellular polymeric substances. *Journal of Hazardous Materials, 241–242*, 363–370. https://doi.org/10.1016/j.jhazmat.2012.09.057

Joshi, C. G., Danagoudar, A., Poyya, J., Kudva, A. K., & Dhananjaya, D. (2017). Biogenic synthesis of gold nanoparticles by marine endophytic fungus-Cladosporium cladosporioides isolated from seaweed and evaluation of their antioxidant and antimicrobial properties. *Process Biochemistry, 63*, 137–144. https://doi.org/10.1016/j.procbio.2017.09.008

Kang, F., Alvarez, P. J., & Zhu, D. (2014). Microbial extracellular polymeric substances reduce Ag+ to silver nanoparticles and antagonize bactericidal activity. *Environmental Science and Technology, 48*(1), 316–322. https://doi.org/10.1021/es403796x

Khan, S. A., & Lee, C. S. (2020). Green biological synthesis of nanoparticles and their biomedical applications. *Nanotechnology in the Life Sciences*, 247–280. https://doi.org/10.1007/978-3-030-44176-0_10

Kishor, R., Purchase, D., Dattatraya Saratale, G., Ganesh Saratale, R., Fernando, L., Ferreira, R., Bilal, M., Chandra, R., & Bharagava, R. N. (2021). *Ecotoxicological and health concerns of persistent coloring pollutants of textile industry wastewater and treatment approaches for environmental safety*. Elsevier. https://doi.org/10.1016/j.jece.2020.105012

Koul, B., Poonia, A., & Yadav, D. (2021). Microbe-mediated biosynthesis of nanoparticles: Applications and future prospects. *Mdpi.Com*. https://doi.org/10.3390/biom11060886

Kroll, A., Behra, R., Kaegi, R., & Sigg, L. (2014). Extracellular polymeric substances (EPS) of freshwater biofilms stabilize and modify CeO_2 and Ag nanoparticles. *PLoS ONE, 9*(10). https://doi.org/10.1371/journal.pone.0110709

Kulkarni, D., Sherkar, R., Shirsathe, C., Sonwane, R., Varpe, N., Shelke, S., More, M. P., Pardeshi, S. R., Dhaneshwar, G., Junnuthula, V., & Dyawanapelly, S. (2023). Biofabrication of nanoparticles: sources, synthesis, and biomedical applications. *Frontiers in Bioengineering and Biotechnology, 11*. https://doi.org/10.3389/fbioe.2023.1159193. Frontiers Media S.A.

Kumari, V., & Tripathi, A. K. (2020). Remediation of heavy metals in pharmaceutical effluent with the help of Bacillus cereus-based green-synthesized silver nanoparticles supported on alumina. *Applied Nanoscience (Switzerland), 10*(6), 1709–1719. https://doi.org/10.1007/s13204-020-01351-9

Kumari, K., Mangatayaru, K., & Reddy, G. B. (2023). *Fungal and yeast-mediated biosynthesis of metal nanoparticles: Characterization and bio applications*. Elsevier. Retrieved June 20, 2024, from https://www.sciencedirect.com/science/article/pii/B9780323999229000180

Li, Q., Yu, J., Zhou, F., & Jiang, X. (2015). Synthesis and characterization of dithiocarbamate carbon nanotubes for the removal of heavy metal ions from aqueous solutions. *Colloids and Surfaces A: Physicochemical and Engineering Aspects, 482*, 306–314. https://doi.org/10.1016/j.colsurfa.2015.06.034

Liang, J., Li, X., Yu, Z., Zeng, G., Luo, Y., Jiang, L., Yang, Z., Qian, Y., & Wu, H. (2017). Amorphous MnO_2 modified biochar derived from aerobically composted swine manure for adsorption of Pb(II) and Cd(II). *ACS Sustainable Chemistry and Engineering, 5*(6), 5049–5058. https://doi.org/10.1021/ACSSUSCHEMENG.7B00434

Liu, X., Tian, J., Li, Y., Sun, N., Mi, S., Xie, Y., & Chen, Z. (2019). Enhanced dyes adsorption from wastewater via Fe_3O_4 nanoparticles functionalized activated carbon. *Journal of Hazardous Materials, 373*, 397–407. https://doi.org/10.1016/j.jhazmat.2019.03.103

Lu, F., & Astruc, D. (2018). Nanomaterials for removal of toxic elements from water. *Coordination Chemistry Reviews, 356*, 147–164. https://doi.org/10.1016/j.ccr.2017.11.003. Elsevier B.V.

Lu, Y., Zhang, Z., Li, Y., & Liao, W. (2017). Extraction and recovery of cerium(IV) and thorium(IV) from sulphate medium by an α-aminophosphonate extractant. *Journal of Rare Earths, 35*(1), 34–40. https://doi.org/10.1016/S1002-0721(16)60170-9

Mahanty, S., Chatterjee, S., Ghosh, S., Tudu, P., Gaine, T., Bakshi, M., Das, S., Das, P., Bhattacharyya, S., Bandyopadhyay, S., & Chaudhuri, P. (2020). Synergistic approach towards the sustainable management of heavy metals in wastewater using mycosynthesized iron oxide nanoparticles: Biofabrication, adsorptive dynamics and chemometric modeling study. *Journal of Water Process Engineering, 37*. https://doi.org/10.1016/j.jwpe.2020.101426

Mani, V., Nivetha, S., Sabarathinam, S., et al. (2022). *Multifunctionalities of mycosynthesized zinc oxide nanoparticles (ZnONPs) from Cladosporium tenuissimum FCBGr: Antimicrobial additives for paints coating*. Elsevier. Retrieved June 18, 2024, from https://www.sciencedirect.com/science/article/pii/S030094402100521X

Manikandakrishnan, M., Palanisamy, S., Vinosha, M., Kalanjiaraja, B., Mohandoss, S., Manikandan, R., Tabarsa, M., You, S. G., & Prabhu, N. M. (2019). Facile green route synthesis of gold nanoparticles using Caulerpa racemosa for biomedical applications. *Journal of Drug Delivery Science and Technology, 54*. https://doi.org/10.1016/j.jddst.2019.101345

Manyangadze, M., Chikuruwo, N. H. M., Narsaiah, T. B., Chakra, C. S., Radhakumari, M., & Danha, G. (2020). Enhancing adsorption capacity of nano-adsorbents via surface modification: A review. *South African Journal of Chemical Engineering, 31*, 25–32. https://doi.org/10.1016/j.sajce.2019.11.003. Elsevier B.V.

Markus, J., Mathiyalagan, R., Kim, Y. J., Abbai, R., Singh, P., Ahn, S., Perez, Z. E. J., Hurh, J., & Yang, D. C. (2016). Intracellular synthesis of gold nanoparticles with antioxidant activity by probiotic Lactobacillus kimchicus DCY51T isolated from Korean kimchi. *Enzyme and Microbial Technology, 95*, 85–93. https://doi.org/10.1016/j.enzmictec.2016.08.018

Martins, M., Mourato, C., Sanches, S., Noronha, J. P., Crespo, M. T. B., & Pereira, I. A. C. (2017). Biogenic platinum and palladium nanoparticles as new catalysts for the removal of pharmaceutical compounds. *Water Research, 108*, 160–168. https://doi.org/10.1016/j.watres.2016.10.071

Mathew, J., Jose, M., Radhakrishnan, E., & Kumar, A. (2022). *Recent advances in aquaculture microbial technology*. https://books.google.com/books?hl=enandlr=andid=3udhEAAAQBAJandoi=fndandpg=PP1andots=35WrMIPOWNandsig=1mbhz5xQz9MfDALxU1TehMUP-Do

Mathivanan, K., Selva, R., Chandirika, J. U., Govindarajan, R. K., Srinivasan, R., Annadurai, G., & Duc, P. A. (2019). Biologically synthesized silver nanoparticles against pathogenic bacteria: Synthesis, calcination and characterization. *Biocatalysis and Agricultural Biotechnology, 22*. https://doi.org/10.1016/j.bcab.2019.101373

Mourdikoudis, S., Pallares, R. M., & Thanh, N. T. K. (2018). Characterization techniques for nanoparticles: Comparison and complementarity upon studying nanoparticle properties. *Nanoscale, 10*(27), 12871–12934. https://doi.org/10.1039/c8nr02278j. Royal Society of Chemistry.

Mughal, B., Zohaib, S., Zaidi, J., Zhang, X., & Hassan, S. U. (2021). Green synthesis, characterization, and evaluation of the antimicrobial activity of Camellia sinensis silver nanoparticles. *Wiley Online Library*. https://doi.org/10.3390/app11062598

Mukherjee, A., Sarkar, D., & Sasmal, S. (2021). A review of green synthesis of metal nanoparticles using algae. *Frontiers in Microbiology, 12*. https://doi.org/10.3389/fmicb.2021.693899. Frontiers Media S.A.

Naser, J. A., Ahmed, Z. W., & Ali, E. H. (2021). Nanomaterials usage as adsorbents for the pollutants removal from wastewater: A review. *Materials Today Proceedings, 42*, 2590–2595. https://doi.org/10.1016/j.matpr.2020.12.584

Nayak, A., Bhushan, B., Gupta, V., & Sharma, P. (2017). Chemically activated carbon from lignocellulosic wastes for heavy metal wastewater remediation: Effect of activation conditions. *Journal of Colloid and Interface Science, 493*, 228–240. https://doi.org/10.1016/j.jcis.2017.01.031

Noman, M., Shahid, M., Ahmed, T., Niazi, M. B. K., Hussain, S., Song, F., & Manzoor, I. (2020). Use of biogenic copper nanoparticles synthesized from a native Escherichia sp. as photocatalysts for azo dye degradation and treatment of textile effluents. *Environmental Pollution, 257*, 113514. https://doi.org/10.1016/J.ENVPOL.2019.113514

Pantidos, N. (2014). Biological synthesis of metallic nanoparticles by bacteria, fungi and plants. *Journal of Nanomedicine and Nanotechnology, 5*(5). https://doi.org/10.4172/2157-7439.1000233

Patil, H. V., Nithin, K., Sachhidananda, S., Chandrashekara, K., & Kumar, S. (2019). Mycofabrication of bioactive silver nanoparticle: Photo catalysed synthesis and characterization to attest its augmented bio-efficacy. *Arabian Journal of Chemistry, 12*, 4596–4611.

Petrovic, B., Gorbounov, M., & Masoudi Soltani, S. (2022). Impact of surface functional groups and their introduction methods on the mechanisms of CO_2 adsorption on porous carbonaceous adsorbents. *Carbon Capture Science and Technology, 3*. https://doi.org/10.1016/j.ccst.2022.100045. Elsevier Ltd.

Purohit, J., Chattopadhyay, A., & Singh, N. K. (2019). Green synthesis of microbial nanoparticle: Approaches to application. *Nanotechnology in the Life Sciences*, 35–60. https://doi.org/10.1007/978-3-030-16534-5_3. Springer Science and Business Media B.V.

Qu, Y., Pei, X., Shen, W., Zhang, X., Wang, J., Zhang, Z., Li, S., You, S., Ma, F., & Zhou, J. (2017). Biosynthesis of gold nanoparticles by Aspergillum sp. WL-Au for degradation of aromatic pollutants. *Physica E: Low-dimensional Systems and Nanostructures, 88*, 133–141. https://doi.org/10.1016/j.physe.2017.01.010

Quan, X., Zhang, X., & Xu, H. (2015). In-situ formation and immobilization of biogenic nanopalladium into anaerobic granular sludge enhances azo dyes degradation. *Water Research, 78*, 74–83. https://doi.org/10.1016/J.WATRES.2015.03.024

Raha, S., & Ahmaruzzaman, M. (2022). ZnO nanostructured materials and their potential applications: Progress, challenges and perspectives. *Nanoscale Advances*. https://doi.org/10.1039/d1na00880c

Raj, R., Dalei, K., Chakraborty, J., & Das, S. (2016). Extracellular polymeric substances of a marine bacterium mediated synthesis of CdS nanoparticles for removal of cadmium from aqueous solution. *Journal of Colloid and Interface Science, 462*, 166–175. https://doi.org/10.1016/j.jcis.2015.10.004

Rajesh, K., Pitchiah, S., Kannan, K., & Suresh, V. (2024). Biosynthesis of silver nanoparticles from marine actinobacterium micromonospora sp. and their bioactive potential. *Cureus, 16*.

Rajput, V., Minkina, T., Mazarji, M., Shende, S., Sushkova, S., Mandzhieva, S., Burachevskaya, M., Chaplygin, V., Singh, A., & Jatav, H. (2020). Accumulation of nanoparticles in the soil-plant systems and their effects on human health. *Annals of Agricultural Sciences, 65*(2), 137–143. https://doi.org/10.1016/j.aoas.2020.08.001. Faculty of Agriculture, Ain-Shams University.

Rana, A., Yadav, K., & Jagadevan, S. (2020). *A comprehensive review on green synthesis of nature-inspired metal nanoparticles: Mechanism, application and toxicity*. Elsevier. Retrieved June 20, 2024, from https://www.sciencedirect.com/science/article/pii/S0959652620329255

Rani, M., Yadav, J., & Chaudhary, S. (2021). *An updated review on synthetic approaches of green nanomaterials and their application for removal of water pollutants: Current challenges, assessment and future*. Elsevier. Retrieved June 18, 2024, from https://www.sciencedirect.com/science/article/pii/S2213343721017401

Rathi, B. S., Kumar, P. S., & Vo, D. V. N. (2021). Critical review on hazardous pollutants in water environment: Occurrence, monitoring, fate, removal technologies and risk assessment. *Science of the Total Environment, 797*, 149134. https://doi.org/10.1016/J.SCITOTENV.2021.149134

Rauf, M. A., Owais, M., Rajpoot, R., Ahmad, F., Khan, N., & Zubair, S. (2017). Biomimetically synthesized ZnO nanoparticles attain potent antibacterial activity against less susceptible: S. aureus skin infection in experimental animals. *RSC Advances, 7*(58), 36361–36373. https://doi.org/10.1039/c7ra05040b

Ray, M. K., Panda, J., Panda, B. P., Mohanta, T. K., & Mohanta, Y. K. (2023). Mycoremediation of heavy metals and/or metalloids in soil BT. In H. Sarma & S. Joshi (Eds.), *Land remediation and management: Bioengineering strategies* (pp. 161–190). Springer Nature. https://doi.org/10.1007/978-981-99-4221-3_8

Roychoudhury, A. (2020). Yeast-mediated green synthesis of nanoparticles for biological applications. *Indian Journal of Pharmaceutical and Biological Research, 8*, 26–31.

Saha, D., & Kienbaum, M. J. (2019). Role of oxygen, nitrogen and sulfur functionalities on the surface of nanoporous carbons in CO_2 adsorption: A critical review. *Microporous and Mesoporous Materials, 287*, 29–55. https://doi.org/10.1016/j.micromeso.2019.05.051. Elsevier B.V.

Salem, D. M. S. A., Ismail, M. M., & Aly-Eldeen, M. A. (2019). Biogenic synthesis and antimicrobial potency of iron oxide (Fe_3O_4) nanoparticles using algae harvested from the Mediterranean Sea, Egypt. *Egyptian Journal of Aquatic Research, 45*(3), 197–204. https://doi.org/10.1016/j.ejar.2019.07.002

Saravanan, A., Kumar, P. S., Jeevanantham, S., Karishma, S., Tajsabreen, B., Yaashikaa, P., & Reshma, B. (2021a). Effective water/wastewater treatment methodologies for toxic pollutants removal: Processes and applications towards sustainable development. *Chemosphere, 280*, 130595.

Saravanan, A., Kumar, P. S., Karishma, S., Vo, D.-V. N., Jeevanantham, S., Yaashikaa, P., & George, C. S. (2021b). A review on biosynthesis of metal nanoparticles and its environmental applications. *Chemosphere, 264*, 128580.

Shah, A., Khalil, A., Ahmad, K., Iqbal, J., et al. (2021). *Biogenic nanoparticles: Synthesis, mechanism, characterization and applications*. Elsevier. Retrieved June 18, 2024, from https://www.sciencedirect.com/science/article/pii/B9780128214671000100

Sharma, P. (2021). Efficiency of bacteria and bacterial assisted phytoremediation of heavy metals: An update. *Bioresource Technology, 328*. https://doi.org/10.1016/j.biortech.2021.124835. Elsevier Ltd.

Sharma, P., & Kumar, S. (2021). Bioremediation of heavy metals from industrial effluents by endophytes and their metabolic activity: Recent advances. *Bioresource Technology, 339*. https://doi.org/10.1016/j.biortech.2021.125589. Elsevier Ltd.

Sharma, S., Sharma, N., & Kaushal, N. (2022). Comparative account of biogenic synthesis of silver nanoparticles using probiotics and their antimicrobial activity against challenging pathogens. BioNanoScience, *1*(3), 833–840. https://doi.org/10.1007/s12668-022-01004-x

Shen, Y., & Chen, B. (2015a). Sulfonated graphene nanosheets as a superb adsorbent for various environmental pollutants in water. *Environmental Science and Technology, 49*(12), 7364–7372. https://doi.org/10.1021/acs.est.5b01057

Shen, Y., & Chen, B. (2015b). Sulfonated graphene nanosheets as a superb adsorbent for various environmental pollutants in water. *Environmental Science and Technology*. ACS Publications. Retrieved June 29, 2024, from https://pubs.acs.org/doi/abs/10.1021/acs.est.5b01057

Shen, W., Qu, Y., Pei, X., Zhang, X., Ma, Q., Zhang, Z., & Zhou, J. (2016). Green synthesis of gold nanoparticles by a newly isolated strain Trichosporon montevideense for catalytic hydrogenation of nitroaromatics. *Biotechnology letters, 38*, 1503–1508.

Singh, S., Kapoor, D., Khasnabis, S., Singh, J., & Ramamurthy, P. C. (2021). Mechanism and kinetics of adsorption and removal of heavy metals from wastewater using nanomaterials. *Environmental Chemistry Letters, 19*(3), 2351–2381. https://doi.org/10.1007/s10311-021-01196-w. Springer Science and Business Media Deutschland GmbH.

Singhal, A., Baweja, P., Gupta, S., et al. (2023). Biogenesis of nanoparticles using microorganisms: A review. *Horizonepublishing.Com*. https://doi.org/10.14719/pst.2373

Soni, V., Raizada, P., Singh, P., et al. (2021). *Sustainable and green trends in using plant extracts for the synthesis of biogenic metal nanoparticles toward environmental and pharmaceutical advances: A review*. Elsevier. Retrieved June 18, 2024, from https://www.sciencedirect.com/science/article/pii/S0013935121009166

Srinath, B. S., & Ravishankar Rai, V. (2015). Rapid biosynthesis of gold nanoparticles by Staphylococcus epidermidis: Its characterisation and catalytic activity. *Materials Letters, 146*, 23–25. https://doi.org/10.1016/j.matlet.2015.01.151

Su, H., Fang, Z., Tsang, P. E., Zheng, L., Cheng, W., Fang, J., & Zhao, D. (2016). Remediation of hexavalent chromium contaminated soil by biochar-supported zero-valent iron nanoparticles. *Journal of Hazardous Materials, 318*, 533–540. https://doi.org/10.1016/j.jhazmat.2016.07.039

Torimiro, N., Oshibanjo, O., & Otuyelu, F. (2021). *Ecorestoration of heavy metals and toxic chemicals in polluted environment using microbe-mediated nanomaterials*. https://doi.org/10.12691/ijebb-9-1-2

Toy, J. Y. H., Lu, Y., Huang, D., Matsumura, K., & Liu, S. Q. (2022). Enzymatic treatment, unfermented and fermented fruit-based products: Current state of knowledge. *Critical Reviews in Food Science and Nutrition, 62*(7), 1890–1911. https://doi.org/10.1080/10408398.2020.1848788. Taylor and Francis Ltd.

Tu, J., Yang, Z., & Hu, C. (2015). Efficient catalytic aerobic oxidation of chlorinated phenols with mixed-valent manganese oxide nanoparticles. *Journal of Chemical Technology and Biotechnology, 90*(1), 80–86. https://doi.org/10.1002/JCTB.4289

Venkateswaran, S. P., Palaniswamy, V. K., Vishvanand, R., & Periakaruppan, R. (2022). Actinomycetes-assisted nanoparticles: Synthesis and applications. In *Agri-waste and microbes for production of sustainable nanomaterials* (pp. 375–395). Elsevier.

Vijayanandan, A. S., & Balakrishnan, R. M. (2018). Biosynthesis of cobalt oxide nanoparticles using endophytic fungus Aspergillus nidulans. *Journal of Environmental Management, 218*, 442–450. https://doi.org/10.1016/j.jenvman.2018.04.032

Wadhawan, S., Jain, A., Nayyar, J., and Mehta, S. K. (2020). Role of nanomaterials as adsorbents in heavy metal ion removal from waste water: A review. *Journal of Water Process Engineering, 33*. https://doi.org/10.1016/j.jwpe.2019.101038. Elsevier Ltd.

Wang, L., Zhang, C., Gao, F., Mailhot, G., & Pan, G. (2017). Algae decorated TiO_2/Ag hybrid nanofiber membrane with enhanced photocatalytic activity for Cr(VI) removal under visible light. *Chemical Engineering Journal, 314*, 622–630. https://doi.org/10.1016/j.cej.2016.12.020

Webley, P., & Danaci, D. (2019). *CO_2 capture by adsorption processes*. https://books.rsc.org/books/edited-volume/754/chapter/473893

Xiao, X., Ma, X. B., Yuan, H., Liu, P. C., Lei, Y. B., Xu, H., Du, D. L., Sun, J. F., & Feng, Y. J. (2015). Photocatalytic properties of zinc sulfide nanocrystals biofabricated by metal-reducing bacterium Shewanella oneidensis MR-1. *Journal of Hazardous Materials, 288*, 134–139. https://doi.org/10.1016/j.jhazmat.2015.02.009

Xiao, X., Liu, Q. Y., Lu, X. R., Li, T. T., Feng, X. L., Li, Q., Liu, Z. Y., & Feng, Y. J. (2017). Self-assembly of complex hollow CuS nano/micro shell by an electrochemically active bacterium Shewanella oneidensis MR-1. *International Biodeterioration and Biodegradation, 116*, 10–16. https://doi.org/10.1016/j.ibiod.2016.09.021

Xie, Y., Cheng, W., Tsang, P. E., & Fang, Z. (2016). Remediation and phytotoxicity of decabromodiphenyl ether contaminated soil by zero valent iron nanoparticles immobilized in mesoporous silica microspheres. *Journal of Environmental Management, 166*, 478–483. https://doi.org/10.1016/J.JENVMAN.2015.10.042

Yang, J., Hou, B., Wang, J., Tian, B., Bi, J., Wang, N., Li, X., & Huang, X. (2019a). Nanomaterials for the removal of heavy metals from wastewater. *Nanomaterials, 9*(3). https://doi.org/10.3390/nano9030424. MDPI AG.

Yang, X., Wan, Y., Zheng, Y., He, F., Yu, Z., Huang, J., Wang, H., Ok, Y. S., Jiang, Y., & Gao, B. (2019b). Surface functional groups of carbon-based adsorbents and their roles in the removal of heavy metals from aqueous solutions: A critical review. *Chemical Engineering Journal, 366*, 608–621. https://doi.org/10.1016/j.cej.2019.02.119. Elsevier B.V.

Yang, Y., Waterhouse, G., Chen, Y., Sun-Waterhouse, D., & Li, D. (2022). Microbial-enabled green biosynthesis of nanomaterials: Current status and future prospects. *Biotechnology Advances, Elsevier*. Retrieved June 20, 2024, from https://www.sciencedirect.com/science/article/pii/S0734975022000106

Yaqoob, A. A., Parveen, T., Umar, K., & Ibrahim, M. N. M. (2020). Role of nanomaterials in the treatment of wastewater: A review. *Water (Switzerland), 12*(2). https://doi.org/10.3390/w12020495. MDPI AG.

Yaradoddi, J., Banapurmath, N., et al. (2022). *Recent trends of actinomycetes in nanotechnology*. Springer. Retrieved June 18, 2024, from https://link.springer.com/chapter/10.1007/978-981-16-3353-9_11

Youk, S., Hofmann, J. P., Badamdorj, B., Olkel, A. V., Antonietti, M., & Oschatz, M. (2020). Controlling pore size and pore functionality in sp 2-conjugated microporous materials by precursor chemistry and salt templating. *Journal of Materials Chemistry A, 2020 pubs.Rsc.Org*. https://doi.org/10.1039/d0ta05856d

Yue, L., Wang, J., Zhang, Y., Qi, S., & Xin, B. (2016). Controllable biosynthesis of high-purity lead-sulfide (PbS) nanocrystals by regulating the concentration of polyethylene glycol in microbial system. *Bioprocess and Biosystems Engineering, 39*(12), 1839–1846. https://doi.org/10.1007/S00449-016-1658-X/METRICS

Zhang, H., Zhou, H., Bai, J., Li, Y., Yang, J., Ma, Q., & Qu, Y. (2019). Biosynthesis of selenium nanoparticles mediated by fungus Mariannaea sp. HJ and their characterization. *Colloids and Surfaces A: Physicochemical and Engineering Aspects, 571*, 9–16. https://doi.org/10.1016/j.colsurfa.2019.02.070

Zhang, C., Dong, H., Szczyglewska, P., Feliczak-Guzik, A., & Nowak, I. (2023). Nanotechnology–general aspects: A chemical reduction approach to the synthesis of nanoparticles. *Mdpi.Com*. https://doi.org/10.3390/molecules28134932

Zhou, H., Pan, H., Xu, J., Xu, W., & Liu, L. (2016). Acclimation of a marine microbial consortium for efficient Mn(II) oxidation and manganese containing particle production. *Journal of Hazardous Materials, 304*, 434–440. https://doi.org/10.1016/j.jhazmat.2015.11.019

Zonaro, E., Piacenza, E., Presentato, A., Monti, F., Dell'Anna, R., Lampis, S., & Vallini, G. (2017). Ochrobactrum sp. MPV1 from a dump of roasted pyrites can be exploited as bacterial catalyst for the biogenesis of selenium and tellurium nanoparticles. *Microbial Cell Factories, 16*(1). https://doi.org/10.1186/s12934-017-0826-2

Chapter 14
Bioremediation of Various Industrial Effluents Utilizing Microorganism-Assisted Nanotechnology Within a Circular Economy

Shareefraza J. Ukkund, Bhavna Alke, and Usman T. Syed

Abstract Industrialization plays a crucial role in the economic development of a country; however, the effluents produced as byproducts generally contain toxic substances that are detrimental to living organisms. In this regard, remediation technologies are grabbing the attention of scientists and technologists worldwide to treat toxic industrial effluents before exposing them to the natural environment. Microbial nanotechnology, an amalgamation of the utilization of biotechnology at the micro-level and technological advancements at the nano-level has provided a platform for greener and cleaner remediation methodologies towards sustainable development. This chapter concisely compiles the various aspects of microbial technology that has positively influenced the nuances of nanotechnology for accomplishing remediation of various industrial effluents by removal, immobilization, and detoxification strategies. Subsections include the basics, importance, accomplishments, and challenges of the bioremediation capacity of microorganisms such as bacteria, fungi, algae, and genetically engineered microbes. Another important aspect that is highlighted in this book chapter is the comprehensive understanding of major detoxification pathways and bioremediation technologies. The chapter concludes with an emphasis on the paradigm shift in exploiting nanobiotechnology towards a circular economy wherein the waste can be further processed into nanocomposites amongst others thereby, boosting the prospects of integrating Small and Medium Enterprises (SMEs) into the circular economy.

The original version of the chapter has been revised. A correction to this chapter can be found at https://doi.org/10.1007/978-3-031-78845-1_16

S. J. Ukkund (✉)
Department of Biotechnology, P. A. College of Engineering (Affiliated to Visvesvaraya Technological University, Belagavi), Mangalore, Karnataka, India

B. Alke
LAQV/Requimte, Department of Chemistry, NOVA School of Science and Technology, FCT NOVA, NOVA University Lisbon, Caparica, Portugal

U. T. Syed
Division of Biological and Environmental Science and Engineering, King Abdullah University of Science and Technology (KAUST), Thuwal, Saudi Arabia

Y. K. Mohanta et al. (eds.), *Nano-microbiology for Sustainable Development*,
https://doi.org/10.1007/978-3-031-78845-1_14

Keywords Bioremediation · Microorganisms · Circular economy · Waste management · Nanotechnology

14.1 Introduction

Bioremediation stands as an environmentally friendly and cutting-edge technique that harnesses natural biological processes to entirely eradicate harmful contaminants. It encompasses any method utilizing fungi, microorganisms, plants, or their enzymes to restore the environment which has been altered by pollutants, to its natural state. Bioremediation technologies are broadly categorized as either *in situ* or *ex-situ*. In situ bioremediation involves treating the contaminated material at the actual site, while ex-situ entails removing the contaminated material for treatment elsewhere. Various bioremediation technologies include bioventing, land farming, bioreactors, composting, bioaugmentation, rhizofiltration, and bio-stimulation. Microorganisms undertaking the bioremediation function are referred to as bioremediators (or bioaugmentation). However, bioremediation using microorganisms is not an ideal method to treat all kinds of contaminants. Cadmium and lead (heavy metals) cannot be absorbed or captured by microorganisms easily. Moreover, the incorporation of metals such as mercury into the food chain may exacerbate the situation.

Bioremediation emerges as an economically viable and environmentally friendly strategy for minimizing sludge production. This method, capable of achieving complete dye mineralization, stands out for its reduced reliance on chemicals, energy efficiency, and cost-effectiveness. The core principle involves the biological conversion of biodegradable wastes into less complex and non-toxic entities. The adaptability of diverse microbial species in breaking down various chemicals provides significant advantages to this approach. Remediation in this process follows two pathways: (a) Adsorption on microorganisms and (b) Dye biodegradation by microbial cells. Furthermore, these remediation procedures occur under two conditions: aerobic and anaerobic processes (Fig. 14.1) (Singh et al., 2022).

Addressing wastewater solutions through these methods is an eco-friendly approach that does not necessitate substantial investments or expenses. Diverse microorganisms, such as fungi, algae, and bacteria, and their enzymes, exhibit the ability to degrade a broad spectrum of dyes and facilitate the removal of heavy metals.

14.2 Industrial Effluents

Industrial effluents refer to any waste discharge from industries- treated or untreated – that pollute or contaminate the environment into which it is discharged. Industrial effluents often contain various pollutants, including heavy metals, organic compounds, chemicals, and contaminants that can have adverse effects on the

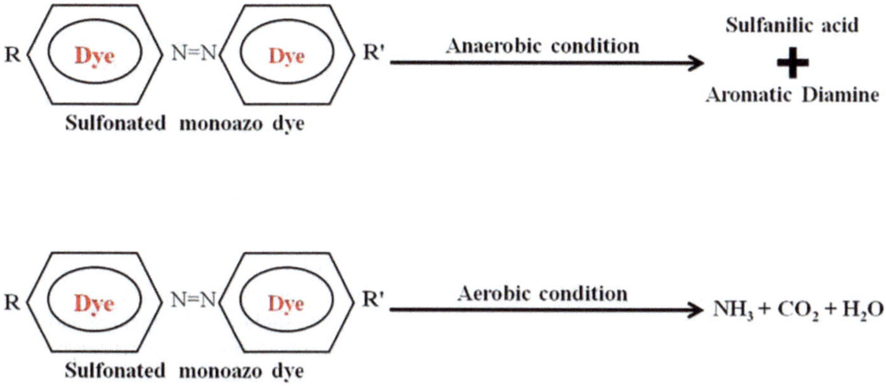

Fig. 14.1 The schematic of mechanism for the removal of dye by microorganisms in anaerobic and aerobic conditions. (Singh et al., 2022)

ecosystems and human health if released into the environment untreated (Rai et al., 2018). Remediation technologies for industrial effluents refer to a set of processes and techniques used to treat and mitigate the environmental impact of wastewater and other byproducts generated during industrial activities.

Polluting industries, as the name suggests, are those that generate significant environmental pollution and can have adverse effects on the air, water, and soil. These industries may release various pollutants into the environment, leading to health hazards and ecological damage (Rai et al., 2018). Table 14.1 depicts various types of polluting industries along with their major pollutants. In this chapter we will be focusing on two of the main industries that are known to introduce an alarming number of pollutants through their industrial effluents.

14.2.1 Textile Industry

Textile manufacturing processes are significant water consumers, with approximately 200 liters of water needed to process just 1 kilogram of fiber. In these textile facilities, a wide array of chemical products is employed for cleaning and dyeing fabrics (Sulthana et al., 2022). Unfortunately, the effluents discharged from these textile plants contain substantial quantities of harmful pollutants and heavy metals. Furthermore, the use of aromatic solvents can lead to the generation of noxious aromatic amines (Jin et al., 2007; Jamil et al., 2024).

Textile industry wastewater comprises aqueous waste containing both organic dyes and a mix of pollutants that may be organic or inorganic. Upon discharge into surface water bodies such as canals, ponds, and rivers, the wastewater imparts a noticeable coloration to the water due to the presence of these organic dyes. This coloration not only diminishes the water's aesthetic appeal but also obstructs or

Table 14.1 Types of pollutants from various industries

Industry	Air pollutants	Water pollutants	Soil pollutants
Textile industry	Particulate matter Volatile Organic Compounds (VOCs) from dyeing and finishing processes	Dyes and chemicals High BOD and COD (Biochemical and Chemical Oxygen Demand)	Disposal of hazardous chemicals and wastes
Polymer industry	VOCs, Particulate Matter, Greenhouse Gases	Chemical Additives, Dyes and Pigments, Resin and Polymer Particles	Spills and leaks, Waste disposal
Agriculture industry	NH_3, CH_4, N_2O, Pesticide drift	Nitrogen and phosphorus from fertilizers, Pesticides and herbicides, Soil erosion and sedimentation, Animal waste runoff	Pesticides, Herbicides, and Salts from irrigation
Chemical industry	VOCs, Ammonia, Chlorinated compounds	Organic chemicals and solvents, Heavy metals, Hazardous waste discharge	Chemical spills and leaks, Contaminated land from chemical production
Manufacturing industry	Particulate matter, Volatile organic compounds, Nitrogen oxides (NOx), SO_2, CO	Heavy metals (e.g., lead, mercury, cadmium), Organic chemicals and solvents, Suspended solids, pH-altering chemicals	Heavy metals, Organic solvents, Contaminated waste products
Energy production (Power plants)	CO_2, SO_2, NOx, Particulate matter, Heavy metals like mercury	Thermal pollution, Ash, and sludge disposal into water bodies	Contaminated land from coal ash disposal
Transportation industry	CO, Nitrogen oxides (NOx), Hydrocarbons, Particulate matter (PM), Lead (from leaded gasoline)	Oil and fuel spills	Heavy metals (from brake and tire wear)
Mining and extractive industries	Dust and particulate matter Emissions from machinery and equipment	Heavy metals (e.g., arsenic, lead, cadmium) Acid mine drainage (sulfuric acid and heavy metals)	Contaminated soil from mining activities
Construction industry	Dust and particulate matter, Emissions from construction equipment	Sediment runoff into water bodies, Construction chemicals, and debris	Soil erosion and compaction
Food processing industry	Organic matter and food odors, Smoke, and emissions from cooking processes	Organic pollutants, High biological oxygen demand (BOD)	Solid waste and byproducts from food processing
Electronics industry	VOCs and ozone-depleting chemicals (in production)	Heavy metals (from electronic waste), Chemicals used in manufacturing processes	Contaminated soil from e-waste disposal

diminishes the penetration of sunlight through the water, subsequently affecting photosynthesis. This, in turn, leads to oxygen depletion, which can have detrimental consequences for aquatic life. Additionally, the various organic and inorganic contaminants present contribute to elevated Biological Oxygen Demand (BOD), rendering the water more toxic and unsuitable for various uses, as highlighted by Pereira and Alves (2012) and Dhanaraj and team (2024).

Presently, there are over 10,000 distinct dyes accessible in the commercial market, with an annual global production exceeding 700,000 tons to cater to the demands of the population. Regrettably, due to inefficient dyeing processes in the textile industry, a substantial portion of the produced dye, ranging from 10% to 25%, is lost during the dyeing procedure, while an additional 2% to 20% of dyes remain unused and are directly discharged as effluents (Tanaya et al., 2024; Naeem et al., 2024). This situation underscores the significance of preventing even minute dye concentrations in water, as concentrations lower than 1 ppm, in some cases, are considered highly undesirable. Synthetic dyes are extensively used in the industry, and the data mentioned above collectively indicate an annual release of approximately 280,000 tons of textile dyes into the environment globally (Jin et al., 2007). This number increased to 300,000 tons in 2022 (Amalina et al., 2022). The quality of groundwater is facing a notable decline, and soil fertility is deteriorating as a consequence of the unregulated release of untreated effluents into various environmental outlets (Carmen & Daniel, 2012; Minhas et al., 2022).

14.2.2 Polymer Industry

Similar to the textile industry, the polymer sector is a significant water consumer in industrial processes. While the water-to-polymer ratio may vary depending on the specific polymers being produced, it's worth noting that considerable amounts of water are still required for various operations. For instance, in some polymer manufacturing processes, an estimated 100–150 L of water are used for every kilogram of polymer produced (Gubanova et al., 2019; Ballús et al., 2023).

Within polymer facilities, a diverse array of chemicals is employed. These chemicals are pivotal in polymer production, contributing to the customization of polymer properties and the enhancement of manufacturing processes. The polymer industry consumes over 2 million tons of various chemicals annually to meet its production demands (Lithner et al., 2011; Thomas et al., 2023)

Unfortunately, the release of effluents from polymer manufacturing can result in environmental challenges. These effluents often contain an array of pollutants, including heavy metals, solvents, and byproducts from polymerization reactions. In certain cases, as observed in studies, up to 15% of these chemicals are released into the environment due to inefficient disposal and treatment methods (Lithner et al., 2011; Mishra et al., 2020).

As in the textile industry, the repercussions of untreated effluent discharge from the polymer industry can be substantial. Groundwater quality can be significantly affected, with studies indicating a decline in the quality of groundwater in regions surrounding polymer facilities. This degradation in groundwater can lead to long-term consequences for local communities, posing health and environmental risks (Bhateria & Jain, 2016; Ukkund et al., 2021a).

Furthermore, soil fertility can diminish due to the unregulated release of polymer effluents on the scale of micro and nanoplastics. It is estimated that in some regions, soil fertility has decreased by 10–15% within a decade of continuous polymer production (Pathan et al., 2020).

Addressing these environmental challenges is not only an ethical imperative but also a regulatory necessity in many regions. Implementing effective wastewater treatment and adopting sustainable industrial practices is critical for the polymer industry to mitigate its environmental impact and protect the quality of local water resources and soil.

14.3 Remediation Technologies

Industrial effluents often contain various pollutants, including heavy metals, organic compounds, chemicals, and contaminants that can be highly toxic to the ecosystem, human health, and the environment if released untreated. Remediation technologies aim to purify or remediate these effluents, making them safe for discharge or reuse.

These technologies can vary widely as physical, chemical, and biological methods. Physical methods include filtration, sedimentation, or adsorption to separate contaminants from the water. Chemical treatments may include processes like coagulation, precipitation, and oxidation to chemically transform or remove pollutants. Biological methods utilize microorganisms to break down or absorb contaminants. The choice of remediation technology depends on the specific characteristics of the effluent and the desired outcome, such as meeting regulatory standards or enabling safe disposal (Rai et al., 2018). In the subsequent sections, we have focused on the removal of dyes, one of the major contaminants of the textile industry, explaining the various kinds of ways in which polluting dyes are treated and their challenges,

Water treatment facilities incorporate a combination of physical, chemical, and biological methods to eliminate contaminants. The effective operation of these treatment plants can address the removal of a wide range of pollutants. However, it's important to note that certain highly toxic substances, such as cyanides and excessive concentrations of heavy metals, may not be completely remediated by these conventional processes, as highlighted in the findings of Singh et al. (2022).

14.3.1 Physical Treatment

Various physical methods are available for the removal of dye. These include adsorption, coagulation and flocculation, granular activated carbon, peat, wood chips, universally available materials, reverse osmosis, and ion exchange methods (Rai et al., 2018). Of this adsorption is the most commonly used physical method.

Activated carbon filters, often referred to as charcoal filters, are employed in the treatment of effluents since they have the remarkable ability to adsorb both organic and inorganic contaminants onto their surfaces. The process of adsorption has garnered significant attention from researchers, primarily because of its efficacy in removing pollutants that are exceptionally stable and resistant to conventional treatment methods. Activated carbon, distinguished by its small and low-volume pores, is specifically engineered for this purpose. These pores significantly augment the material's surface area, resulting in the adsorption of impurities due to its highly porous structure. Notably, activated carbon boasts an extraordinarily high surface area, typically ranging from 500 to 1500 square meters per gram (Kamdi et al., 2012).

In addition to activated carbon, other adsorbents such as corn cobs, silica gel, rice hulls, coal mixtures, and fly ash have demonstrated their utility as high-quality adsorbents. These materials excel in removing contaminants even from dilute solutions, yielding a superior quality of treated effluent without generating any harmful byproducts.

However, it is crucial to consider the limitations of the use of activated carbon. Over time and with repeated use, the adsorbent quality tends to degrade. Furthermore, the initial capital cost of implementing activated carbon filtration systems can be relatively high (Robinson et al., 2001; Periyasamy 2024).

14.3.2 Chemical Methods

Various chemical methods are also available for the dye removal process such as nitrification and denitrification, chlorination, use of ozone, cucurbituril, electrochemical destruction, and photochemical method (Rai et al., 2018). One specific approach within these remediation methods is the photochemical method, which addresses some of the limitations associated with the previous techniques.

The photochemical method represents one of the latest techniques in use, distinguished by its significant advantages. Notably, it stands out due to its complete absence of chemical usage and the absence of sludge formation during the treatment process. Furthermore, the end products resulting from this method are not harmful (Periyasamy 2024). Dye degradation, a pivotal aspect of this method, is regarded as the most effective and practical means of remediation. There are several noteworthy advantages to this approach:

- **Non-Specific Action:** This method is not limited to specific compounds, allowing for the remediation of a wide range of hazardous organic chemicals, fuel hydrocarbons, and dyes.
- **Low Toxicity:** Since the reagents used in the reactions and the final by-products are non-toxic, this process is considered to have very low toxicity.
- **Versatile Application:** This method is suitable for both liquid and gaseous effluents, enhancing its versatility.
- **Solar Energy Utilization:** The primary energy requirement for this process is solar energy, which is readily available, making it an environmentally friendly choice.
- **Economic Viability:** When compared to other remediation processes, this method is often found to be economically viable, offering cost-effective treatment solutions.

Numerous metal oxide semiconductors are at our disposal for harnessing photocatalysis in the degradation of dyes. Some of the notable options in this regard encompass titanium dioxide (TiO_2), zinc oxide (ZnO), and vanadium pentoxide (V_2O_5) (Thennarasu et al., 2024).

14.3.3 Biological Treatment

Various biological methods are also available for the dye removal process such as aerobic and anaerobic treatment and degradation by micro-organism (Rai et al., 2018).

Aerobic and Anaerobic Treatment

Aerobic treatment involves the utilization of bacteria or microorganisms to facilitate the breakdown of organic impurities in the presence of oxygen or air. This biological process transforms these impurities into carbon dioxide, water, and biomass, as illustrated in Fig. 14.1. All these metabolic activities take place within a bioreactor under carefully controlled operating conditions. Conversely, in anaerobic treatment, different strains of bacteria or microorganisms are employed, which do not rely on oxygen or air for the degradation of organic impurities.

Biological Degradation by Microorganisms

In recent decades, microorganisms, including fungi, have gained prominence for their role in eliminating dyes from a wide range of dye-containing effluents. Among these microorganisms, white-rot fungi stand out due to their unique ability to degrade lignin (Torres-Farradá et al., 2024). White-rot fungi employ enzymes such as lignin peroxides to break down dyes. Notably, azo dyes, which are commonly used but challenging to degrade via microbial action, can be effectively broken down by *Phanerochaete chrysosporium.* In addition to its prowess in dye degradation, *Phanerochaete chrysosporium*, a type of white-rot fungi, finds utility in the degradation of xenobiotics, including dioxins and various chloro-organic compounds (Torres-Farradá et al., 2024).

While the potential of white-rot fungi for dye decolorization in liquid fermentation has been demonstrated, there are challenges associated with enzyme production in this context due to the less favorable conditions. A more reliable approach is found in solid-state fermentation (Bibi et al., 2023).

14.4 Microorganism-Assisted Nanotechnology

Microbial technology is a widely researched technology in bioremediation, leveraging the metabolic capabilities of microorganisms to break down or transform pollutants. Bacteria, fungi, and algae have unique abilities to metabolize various pollutants, making them valuable tools in environmental clean-up (El-Sheekh & Mahmoud, 2017).

14.4.1 Mechanisms of Action

The pursuit of an effective, environmentally sustainable, economically feasible, and biological approach to wastewater treatment has led to the development of the biological adsorption method. The biological method for eliminating heavy metals (HMs) and dyes presents a clear advantage in treating large quantities of effluents with low concentrations of biological adsorbent and without production of detrimental secondary chemicals. Another merit is the swift operational timeframe (Singh et al., 2022).

There is a passive uptake of heavy metal (HMs) ions by deceased or living biological materials. This involves a biological absorbent as a solid phase and typically water as a liquid phase. The liquid phase contains the molecules intended for adsorption. The main mechanism involves the attraction of the adsorbent to the adsorbate, which can subsequently be expelled through various methods, such as treatment with organic and mineral acids, or the application of electrochemical and thermal techniques. This process is consistent across these approaches (Patel, 2021).

Several parameters, such as the state of the adsorbate (living or deceased), the biological material used, pH level, and the initial concentrations of heavy metals (HMs) and dyes, play crucial roles in influencing the adsorption process by biological means. This method is an outcome of various metabolism-independent activities primarily occurring within the cell wall. Within the temperature range of 20–35 °C, temperature exhibits minimal influence on the biological adsorption process. Additionally, pH, the presence of other metal ions, and the concentration of biological species strongly impact the biological adsorption process. Notably, pH has a significant effect, influencing the activity of the functional groups in biological species, metal solution chemistry, and the competition among metallic ions and dyes. Owing to their remarkable effectiveness, cost-effectiveness, and abundant availability, a diverse array of biological materials, including algae, bacteria, fungi, and yeasts, have increasingly gained attention for the removal of heavy metals and the decolorization of dyes (Singh et al., 2022).

Below are some of the biological contaminant removal techniques

- Use of bacterial strains to remove pollutants
- Biological route remediation by algae
- Fungi-based biological remediation of pollutants
- Pollutant removal by other biological strains

14.4.2 Detoxification Pathways

Microorganisms employ diverse detoxification pathways to neutralize and eliminate contaminants (Hussain et al., 2023). One common pathway is 'biotransformation', where microbes modify pollutants into less harmful substances (Díaz et al., 2013). For example, bacteria can transform organic pollutants into carbon dioxide and water through processes like oxidation or reduction.

'Biodegradation' is another vital detoxification process, involving the breakdown of complex pollutants into simpler compounds (Lap et al., 2024). Certain bacteria produce enzymes that target specific pollutants, facilitating their degradation. An example is the degradation of hydrocarbons by hydrocarbon-degrading bacteria (Brooijmans et al., 2009).

Bioremediation Technology

Bioremediation technologies harness microbial activities for environmental cleanup. In-situ and ex-situ bioremediation are the two main approaches to this technology (Joshi et al., 2024).

In-situ Bioremediation This method treats contaminants in their original location. Techniques include bioaugmentation (introducing specialized microorganisms), biostimulation (stimulating the growth of existing microorganisms), and phytoremediation (using plants to enhance microbial activity) (Madison et al., 2023).

Ex-situ Bioremediation Contaminated material is removed from its original location for treatment. Common methods include land farming (contaminated soil is spread out and periodically tilled to enhance microbial degradation) and biopiles (similar to land farming but conducted in above-ground piles) (Lin et al., 2010).

Advancement in Microbiological Remediation Techniques

Systematic Biology: Advances in systematic biology, including metagenomics, enable a comprehensive understanding of microbial communities in contaminated environments. Metagenomic techniques analyze the genetic material directly from environmental samples, allowing researchers to identify and characterize diverse microbial species involved in bioremediation (Gunjal et al., 2023).

Gene Editing: Technologies like CRISPR-Cas9 which use gene editing principles offer promising possibilities for tailoring microorganisms to enhance their bioremediation capabilities. By modifying the genetic makeup of bacteria or fungi, researchers can potentially optimize their ability to degrade specific pollutants (Hassan & Ganai, 2023).

OMICS Technologies: Various "OMICS" technologies, including genomics, proteomics, transcriptomics, and metabolomics, provide a detailed view of microbial activities during bioremediation. These approaches allow researchers to monitor gene expression, protein production, and metabolic pathways, providing valuable insights for optimizing bioremediation processes (Hassan & Ganai, 2023).

In conclusion, microbial technology is at the forefront of bioremediation efforts, utilizing the unique capabilities of microorganisms to detoxify and eliminate pollutants. Advances in systematic biology, gene editing, and OMICS technologies further enhance our ability to understand and optimize these microbial processes for effective environmental remediation.

14.5 Nexus of Nanotechnology & Microbial Remediation

Nanotechnology is a crucial field in the remediation of heavy metals from contaminated environments, offering innovative solutions to address environmental pollution (Ukkund et al., 2021a, b). At the heart of this approach is the ability to engineer nanomaterials with unique properties that enhance the efficiency of remediation processes. Nanoparticles, such as zero-valent iron (nZVI), titanium dioxide, and carbon nanotubes, exhibit high surface area and reactivity, enabling them to adsorb, immobilize, or catalytically transform heavy metal ions. These nanoparticles can be tailored to target specific contaminants, providing a versatile and effective means of remediation (Galdames et al., 2020; Ukkund et al., 2021a).

Adsorption or chemical modification by using nanoparticles can be applied to remove organic contaminants. Moreover, nanoparticles can facilitate microbial remediation by immobilization or by inducing the production of microbial enzymes which help in remediation (Benjamin et al., 2019; Ukkund et al., 2022a).

The integration of nanotechnology with microorganisms for bioremediation of industrial effluents emerges as a crucial aspect of a green approach rooted in nanoscience, and biomaterial science. The biosynthesis of nanoparticles generates nanomaterials through microorganisms (Shareef et al., 2017) which has proven to be both eco-friendly and cost-effective in comparison to traditional approaches (Ukkund et al., 2019), providing a sustainable and non-toxic avenue for effluent remediation. The use of bacteria as effective biofactories to reduce and recover heavy metal ions has gained importance, signaling the need for commercial-scale adoption. However, comprehensive optimization of this process is essential.

Anticipated developments in nanoparticle biosynthesis hold the potential to not only enhance environmental remediation but also contribute significantly to the industrial economy through the generation of green energy (Saba et al., 2021). Several approaches have been exploited for Nano-microbial remediation. Below are some of the most researched areas of bioremediation by microbial nanotechnology in industrial effluents.

14.5.1 Bioremediation of Heavy Metals by Microbial Nanotechnology

Toxic metals from industrial waste pose a serious threat due to their accumulation in agricultural soils, subsequently entering the food chain and disrupting natural ecology, impacting human food security (Saba et al., 2021). Microbial nanoparticles are produced employing an eco-friendly nanotechnology method that uses living organisms, including plants and microbes, for the process of biosynthesis (Ukkund et al., 2022b). Toxic metals can be adsorbed by microbial nanoparticles synthesized by microbes effectively. Microbial enzymes play a very important role in the reduction of metal compounds to generate their corresponding nanoparticles. Nanoparticles of this nature offer improved surface reactivity and a more accurate surface area. The redox reaction involved in the biosynthesis of nanoparticles takes place subsequent to introducing the precursor molecule to the extracellular substances present within bacterial cells (Singhal et al., 2023) (Fig. 14.2).

The growth environment of microbial species greatly influences the ability to resist metal ions and facilitate the formation of mineral particles. *Klebsiella aerogenes* is employed in the production of cadmium sulfide particles. Halim et al. (2020) observed a maximum uptake of approximately 20% of the total biomass for *Klebsiella aerogenes* in the presence of 2 mM Cd (NO_3). Yeasts such as *Candida glabrata* and *Schizo saccharomyces* play a role in detoxifying cadmium from the environment through active intracellular metal absorption, coupled with sequestration in small iso-peptides. The resulting cadmium sulfide microcrystallites, with a diameter of approximately 1.8 nm, exhibit characteristics of quantum semiconductor crystallites. These microcrystallites possess size-tunable properties, including optical absorption, photosynthetic capabilities, and electron transfer functions. In the presence of La (NO_3), *Pseudomonas aeruginosa* demonstrates the formation of needle-like deposits around its cells (Hosseini & Sarvi, 2015).

To upscale bioremediation using microbes, it is imperative to conduct a thorough investigation encompassing processes, environmental conditions, and growth requirements, including factors such as temperature, pH, and nutrient availability (Saba et al., 2021).

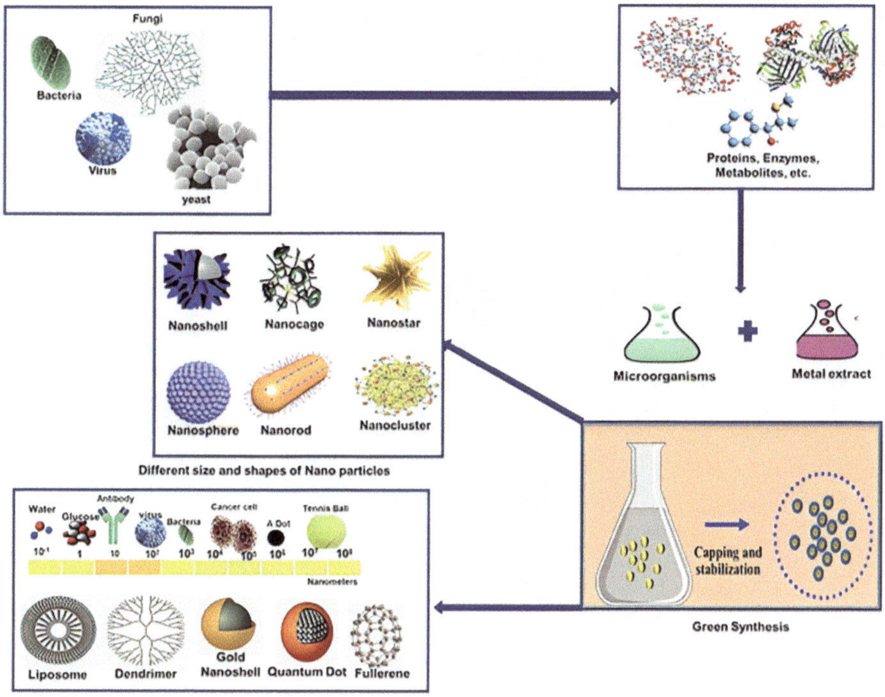

Fig. 14.2 A graphical representation depicting the mechanistic scheme of metal nanoparticle synthesis from microbes. (Saba et al., 2021)

14.5.2 Bioremediation of Industrial Wastewater by Microbial Nanotechnology

Diverse electrochemical processes, advanced oxidation techniques, and valorization methods have been implemented to mitigate the toxicity of wastewater effluents, promoting sustainable use (Gupta & Shukla, 2020). However, the cost-effectiveness of these methods may not be universal across industries. The emergence of nanotechnology and nanoscience has introduced innovative approaches to water pollutant remediation (see Fig. 14.3) (Mandeep & Shukla, 2020). The diminutive size of nanomaterials renders them well-suited for wastewater treatment. Their distinct physical, biological, and chemical properties increase their applicability in various treatment scenarios. A range of nanomaterials, including carbon-based ones like nanocomposites or nanotubes, as well as metal and oxide-based nanomaterials, have been effectively employed for removing wastewater effluents. Wastewater management strategies encompass diverse practices such as filtration using nanoparticles, adsorption, photocatalytic degradation, and the monitoring of various pollutants and contaminants (Palit & Hussain, 2020).

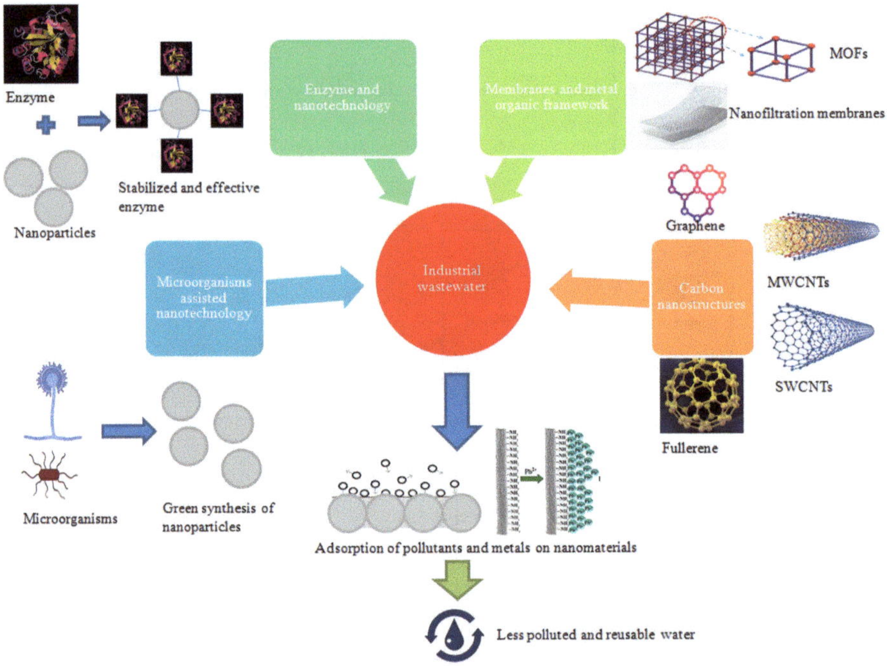

Fig. 14.3 The approaches of the combination of microbial technology and nanotechnology for the bioremediation of wastewater. *MOFs* metal-organic frameworks, *MWCNTs* multi-walled carbon nanotubes, *SWCNTs* single-walled carbon nanotubes. (Mandeep & Shukla, 2020)

Nanotechnological pathways prove more efficient than traditional methods owing to their reduced size, heightened surface area-to-volume ratio, and enhanced chemical properties (Thakur et al., 2024).

The use of extracts of various organisms and microorganisms for the production of green nanomaterials has facilitated environmentally friendly pollutant remediation. Iron nanoparticles, characterized by their green and eco-friendly attributes, are employed in remediation processes because of their non-toxicity, magnetic susceptibility, and their redox potential in water (Bolade et al., 2020).

Utilizing membrane-associated nanomaterials represents an effective approach for the removal of effluents. These nanomaterials contribute to resistance to foul odors, enhanced membrane permeability, improved temperature and mechanical strength, and introduced novel functionalities for the degradation of pollutants. Nano-catalysts play a significant role in augmenting degradation reactions (Corsi et al., 2018). In addition to nano-catalysts and membranes, the utilization of metal-organic frameworks (MOFs) is prevalent in the removal of heavy metals from wastewater. The synthesis of MOFs involves coordinating organic ligands with metal ion precursors. To enhance the effectiveness of MOFs, the coordination of functional groups with metals, rather than the organic ligand, is preferred due to the reduced steric hindrance of metals (Deshpande et al., 2020).

Nano-adsorbents and Nanofiltration Membranes

Nanoparticles have found widespread application as adsorbents for the removal of harmful contaminants from industrial wastewater. Nano-adsorbents demonstrate efficacy in eliminating both organic and inorganic pollutants. These nano-adsorbents are primarily classified into metal, metal oxide-based, and carbon-based nanoparticles. Carbon-based nanoparticles encompass graphene, activated carbon, carbon nanotubes (CNTs), and, to some extent, fullerene (Kumari et al., 2019).

Carbon nanotubes, in particular, serve as effective adsorbents for toxic chemicals present in pharmaceutical or manufacturing wastewater. Kariim et al. (2020) utilized the chemical vapor deposition (CVD) technique to prepare multi-walled carbon nanotubes (MWCNTs) supported on activated carbon with Fe-Ni. The adsorption characteristics revealed values of 2.5961 and 2.1363 for metronidazole and levofloxacin, respectively, falling within the range of good adsorption (2–10). The enthalpy change indicated chemisorption for metronidazole and physisorption for levofloxacin. Additionally, MWCNTs demonstrated the capability to adsorb metals from wastewater, with carboxylated MWCNTs achieving increased adsorption of As(V) and Mn(VII) at 250 and 298 mg/g, respectively. Thermodynamic results suggested that metal removal occurs through chemisorption (Abdulkareem et al., 2020).

The excessive use of water in textile industries has resulted in a significant water pollution issue (Ukkund et al., 2021a, b). Throughout various processes, the industry releases wastewater containing a diverse array of dyes, presenting a formidable challenge for the environment and the overall health of the ecosystem (Ukkund et al., 2021a).

Innovative laboratory-scale experiments have been devised for the remediation of Direct Blue 15 (DB15) dye, utilizing abundantly available and non-toxic halloysite nanotubes (HNT). These experiments extend to the application of DB15 adsorbed-HNT "sludge" in addressing textile industrial effluent (TIE). The resultant sludge, incorporating DB15 adsorbed-HNT, serves as a reinforcing material alongside plastic waste in the fabrication of composites (Ukkund et al., 2021a, b).

14.5.3 Microbial-Based Green Nanobiotechnology for Wastewater Remediation

Among various sustainable methods, the utilization of biogenic Pd NPs, synthesized by *Caldicellulosiruptor saccharolyticus* (a thermophilic, anaerobic cellulolytic bacterium), stands out for the removal of dyes such as diatrizoate and Methyl Orange (MO) from water bodies. These biogenic Pd NPs exhibit a significantly higher degradation efficiency compared to their chemically synthesized counterparts. The rate of dye degradation by biogenic Pd NPs is twice as fast as that observed with chemically synthesized Pd NPs. Additionally, the by-products released during dye degradation are readily absorbed by the biogenic Pd NPs, effectively mitigating and suppressing the toxic effects (Ahmed et al., 2022).

Recent reports indicate that the pharmaceutical industry and medical waste effluents contain numerous toxic drug compounds and xenobiotics. These substances find their way into the aquatic system either directly or through contamination of polluted sites (AlMasoud et al. 2021).

Moreover, further research is essential to optimize the procedures for biofabrication, considering factors such as economic viability, microorganism culture, and handling. The integration of contemporary advanced technologies, including 5G communication, Internet of Things (IoT), cloud computing, and Artificial Intelligence (AI) with bio-nanotechnology-based green nanomaterials represents the upcoming frontier in strategies for the treatment of wastewater. This integration aims to address the global concern of drinking water scarcity (Malik et al., 2022).

14.6 Circular Economy Approach

A circular economy (CE) can be characterized as an economic paradigm focused on optimizing resource utilization by minimizing waste, preserving long-term value, decreasing reliance on primary resources, and establishing closed loops for products, their components, and materials—all within the framework of environmental conservation and socioeconomic advantages. The potential of a circular economy lies in fostering sustainable development by disentangling economic growth from the adverse impacts of resource depletion and environmental degradation (Murray et al., 2017; Babbitt et al., 2018; Hofmann, 2019). The goal of the circular economy is to prolong, to the greatest extent possible, the lifespan of materials, products, and resources within the production system (Reno et al., 2020).

The concept of the circular bioeconomy holds appeal for the reuse, reduction, and recycling of resources. The extraction of nutrients from waste and effluents using microalgae has the potential to substantially address the increasing need for energy and nutraceutical sources in the expanding global population (Taqui et al., 2022). Wastewaters from diverse origins contain valuable nutrients and carbon, presenting an opportunity to recover and employ these resources within the framework of the circular bioeconomy approach (Ummalyma et al., 2021). The economic microalgae biorefinery necessitates careful consideration of three key segments: securing a low-cost nutrient source, implementing efficient harvesting methods, and generating by-products with significant market value as illustrated in Fig. 14.4 (Sarma et al., 2021).

The circular economy (CE) is presently a widely embraced concept advocated by the European Union, various national governments, and several global businesses. CE appears to be a compilation of ambiguous and disparate ideas stemming from various fields and semi-scientific concepts (Korhonen et al., 2018). Notably, the European Commission has recently projected that economic transitions embracing circular economy principles could yield annual economic gains of 600 billion euros for the manufacturing sector within the EU alone (Korhonen et al., 2018).

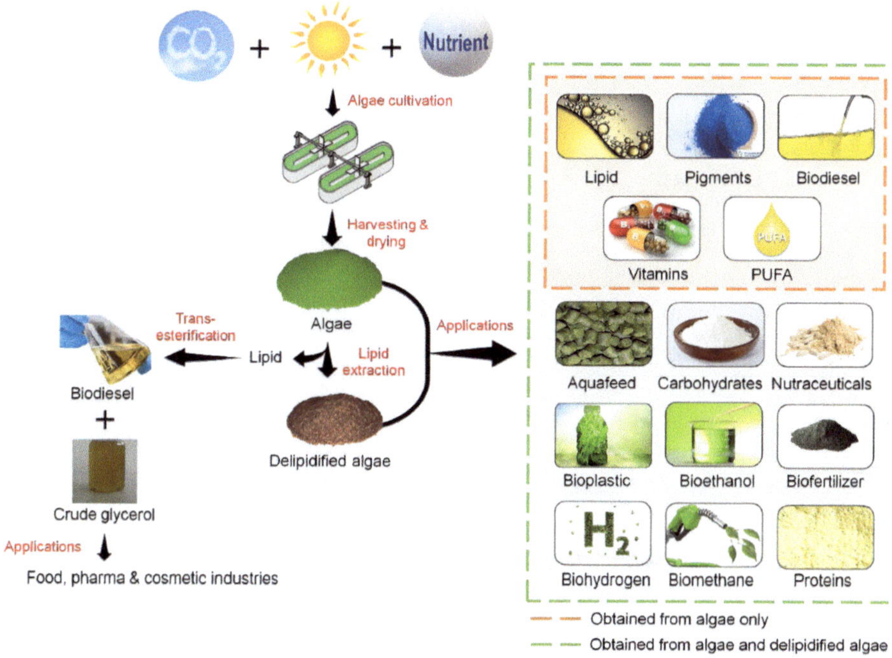

Fig. 14.4 Diagram illustrating the array of products obtained from both raw and delipidified microalgae. (Sarma et al., 2021)

The discharge of wastewater into aquatic environments constitutes a noteworthy environmental challenge, resulting in the contamination of ecosystems and giving rise to a spectrum of issues such as the depletion of marine biodiversity, soil impairment, habitat deterioration, economic ramifications, and health concerns for the surrounding communities (Masoud et al., 2022). On a global scale, an estimated 2.2 million m^2 of wastewater is generated every year across agricultural, industrial, and domestic sectors (Abedi & Mojiri, 2019). The food industry, owing to its substantial freshwater consumption, emerges as a major contributor to this issue (Shrivastava et al., 2022). A notable example of damaging food-processing wastewater is that found in Mexico. These include nejayote, tequila vinasses, and whey, which cause environmental contamination when discharged in large quantities (see Fig. 14.5) (Najar-Almanzor et al., 2023).

Various strategies have been devised for contaminant removal and remediation from aquatic bodies, employing a range of biological and physicochemical processes. Physicochemical approaches such as evaporation, precipitation, osmosis, electrochemical treatment, ion exchange, and membrane filtration are commonly used for the treatment of wastewater. However, many of the methods mentioned are not economically sustainable or environmentally friendly (Alazaiza et al., 2022). Furthermore, the energy consumption associated with wastewater treatment operations constitutes a significant portion, accounting for 2–4% of a nation's total

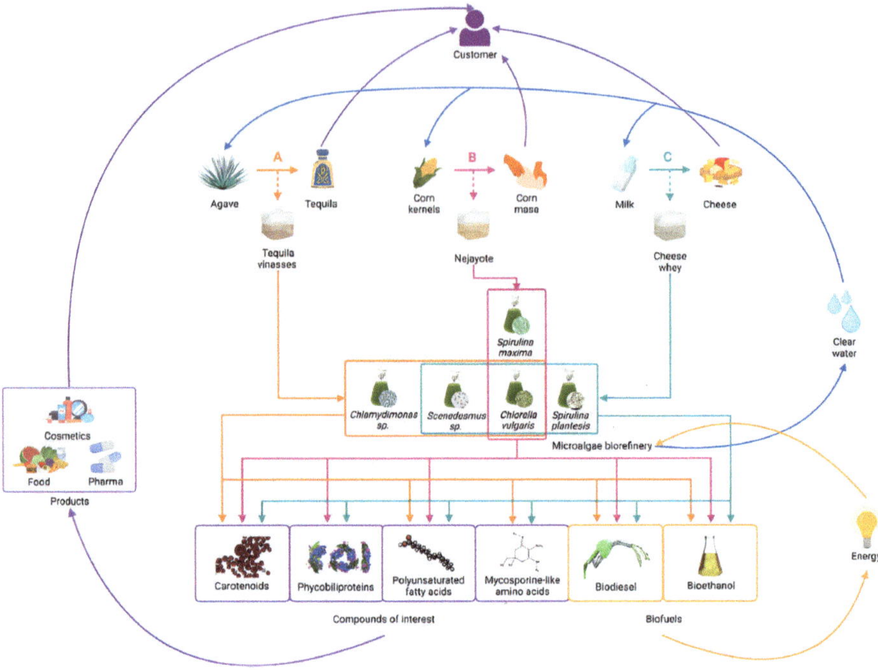

Fig. 14.5 Circular economy approach by microalgae. (Najar-Almanzor et al., 2023)

electricity usage. Also, the management of treatment plants requires specialized labor, and the infrastructure capital costs are substantial (Quintero-Dallos et al., 2019; Sparn & Hunsberger, 2015).

Due to these limitations, researchers are exploring treatment of wastewater and resource recovery systems by microalgae. Phycoremediation, which is a method utilizing algae for the remediation of contaminated water (Phang et al., 2015), offers a promising avenue. In phycoremediation, microalgae exhibit the capability to eliminate or break down organic and inorganic environmental contaminants, alongside assimilating nutrients like nitrogen and phosphorus essential for their growth (Phang et al., 2015). This approach presents a more sustainable and environmentally friendly alternative to conventional wastewater treatment methods (Alazaiza et al., 2022).

Microalgae-based wastewater treatment (MbWT) is a highly promising avenue in the realm of bioremediation technologies for addressing food-processing effluents (López-Sánchez et al., 2022). Microalgae present numerous merits, encompassing swift growth, carbon dioxide fixation, elevated photosynthetic efficiency, resilience in challenging conditions, and proficient utilization of nutrients present in wastewater (Deb et al., 2021; El-Sheekh et al., 2022; Hosseini et al., 2020).

Based on current research discoveries, the utilization of microalgae for alternative biological wastewater remediation appears to be both technically viable and environmentally benign (Alazaiza et al., 2022).

In this context, essential insights are presented to underscore the efficacy of utilizing *Sargassum* sp. biomass for remediation of environmental contaminants, with a specific emphasis on coastal ecosystems (Saldarriaga-Hernandez et al., 2020). The pollutants assimilated by the biomass can be reclaimed and reintroduced into the industrial value chain. Simultaneously, the residual biomass is subjected to refinement processes to extract high-value products, such as bioactive compounds, opening avenues for diverse applications in industries including biofuel, agriculture, cosmetics, nutraceuticals, pharmaceuticals, and more. This approach maximizes the utilization of renewable resources for sustainable practices. *Sargassum* sp. biomass has shown great potential as an environmentally friendly and renewable matrix to treat pollutants in aqueous coastal ecosystems. Its bioremediation capacity depends on the active absorption of pollutants (bioaccumulation) and also on the passive binding of toxic elements to the non-living biomass (biosorption) (Saldarriaga-Hernandez et al., 2020).

Along with circular economy we have one more concept called bioeconomy. The concept of bioeconomy pertains to the innovation, advancement, manufacturing, and utilization of biological products and processes. It holds the promise of environmentally sustainable improvements in global well-being. Bioeconomy specifically emphasizes the application of biotechnology to prevent pollution, recover resources, generate renewable resources, contribute to agriculture and forestry, enhance healthcare, and support industrial processes that involve biodiversity (Prasad, 2024).

Certain species of microalgae have been explored in the treatment of swine wastewater as a means to diminish the organic load in the effluent. Utilizing swine wastewater as a culture medium for microalgae presents notable economic advantages, as this waste can be repurposed into valuable products through the metabolic activities of microalgae (Fig. 14.6). This approach aligns with the principles of a circular economy, where the waste generated from one production process is transformed into the raw material for another, fostering a more sustainable and resource-efficient system (López-Pacheco et al., 2021).

Coupling bioremediation and biorefinery prospects of microalgae for circular economy was illustrated by Chhandama and Rai (2023) (Fig. 14.7). Cultivating microalgae in wastewater holds the potential to integrate environmentally friendly approaches, combining strategies such as biosorption for wastewater treatment and biodiesel production. This dual-purpose application demonstrates a synergistic relationship where microalgae not only contribute to the removal of contaminants through biosorption in wastewater but also serve as a valuable resource for the production of biodiesel. This integrated approach showcases a green and sustainable method that addresses both wastewater management and bioenergy production (Chhandama & Rai, 2023).

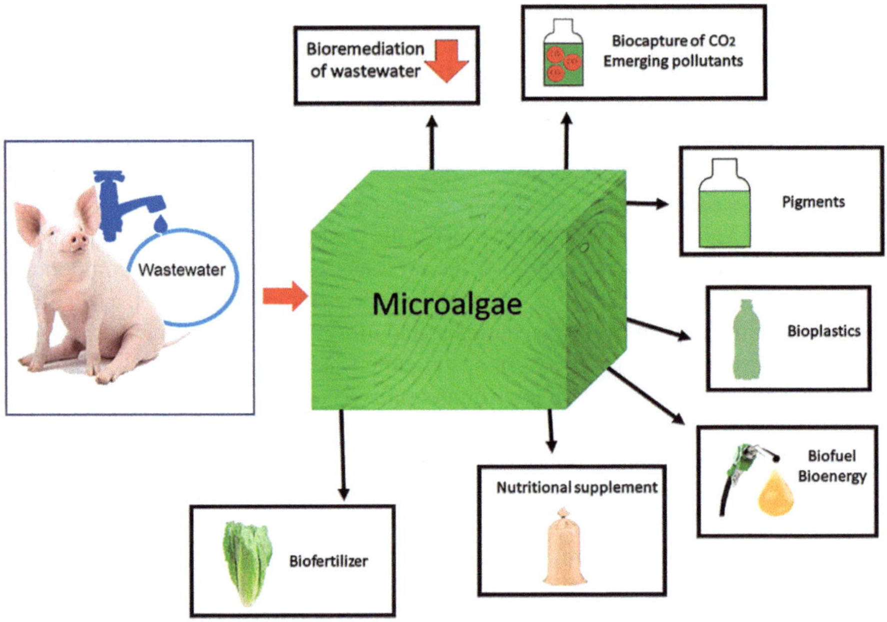

Fig. 14.6 Circular economy by Microalgae for swine wastewater. (López-Pacheco et al., 2021)

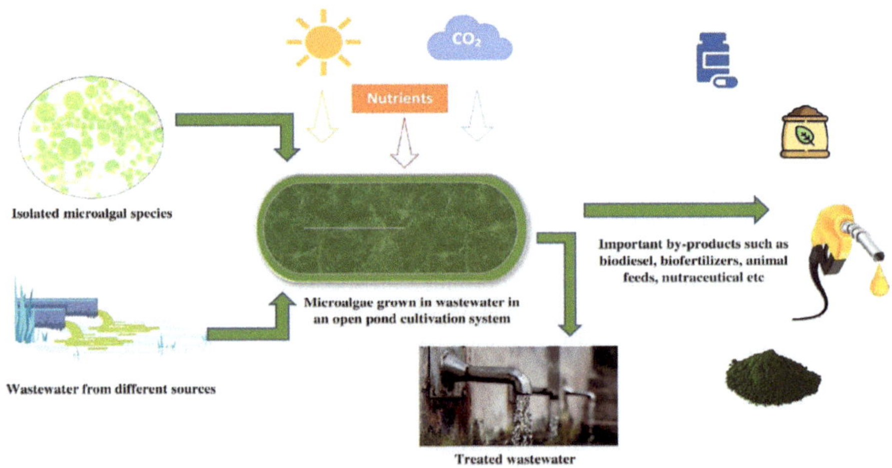

Fig. 14.7 Coupling bioremediation and biorefinery prospects of microalgae for the circular economy. (Chhandama & Rai, 2023)

14.6.1 Valorization of Value-Added Products

In recent years, there has been a growing emphasis on the valorization of value-added products across various industries (Dey et al., 2021). Valorization involves the enhancement of the value and utility of products or by-products through innovative processes, leading to economic, environmental, and societal benefits (Kover et al., 2022). This section explores the diverse facets of value-added product valorization, encompassing its significance, methods, and applications across different sectors (Yadav et al., 2022).

The valorization of value-added products holds paramount significance in the context of sustainable resource management and circular economy principles (Falade, 2021). By maximizing the utility of raw materials and by-products, valorization contributes to reducing waste, minimizing environmental impact, and fostering economic growth (Adesra et al., 2021). This approach aligns with the overarching goal of achieving a more sustainable and resource-efficient global industrial landscape.

14.6.2 Methods and Applications of Valorization in Various Industries

Numerous methods are employed for the valorization of value-added products, each tailored to specific industries and objectives (Langsdorf et al., 2021). Bioconversion processes, such as enzymatic hydrolysis and microbial fermentation, play a crucial role in transforming organic waste into bio-based products with enhanced value (Yadav et al., 2022). Additionally, thermochemical processes, including pyrolysis and gasification, are employed to convert biomass into biofuels and bioenergy (Akhtar et al., 2018). Upcycling and repurposing strategies are also prevalent in valorization efforts, transforming waste materials into high-value products.

The applications of value-added product valorization are widespread, spanning diverse industries such as agriculture (Dey et al., 2021; Kover et al., 2022), food and beverage (Kannah et al., 2020; Kover et al., 2022), pharmaceuticals (Leal et al., 2020), and manufacturing (Ferreira et al., 2020). In agriculture, valorization involves the conversion of crop residues into biofertilizers or biopesticides (Nayak et al., 2019). In the food industry, by-products are repurposed for the creation of functional foods or bioactive ingredients (Kannah et al., 2020). Pharmaceutical valorization focuses on extracting valuable compounds from plant waste or by-products for medicinal purposes (Jiménez-Moreno et al., 2020). The manufacturing sector leverages valorization to recycle and repurpose industrial waste, contributing to a more sustainable production cycle (Ferreira et al., 2020).

14.7 Challenges and Future Perspectives

While value-added product valorization offers promising solutions, several challenges exist. These include technological barriers, economic feasibility, and regulatory frameworks. Future research should address these challenges by developing innovative technologies, enhancing cost-effectiveness, and advocating for supportive policies. Additionally, the integration of emerging technologies, such as artificial intelligence and advanced analytics, can optimize valorization processes and contribute to the efficient utilization of resources.

The valorization of value-added products stands as a pivotal strategy in the pursuit of sustainability and resource efficiency. This section highlights the significance of valorization, the diverse methods employed, and its applications across various industries. As we navigate towards a more sustainable future, ongoing advancements in research and technology will enable the exploration of the full potential of value-added product valorization.

Effluent remediation is a critical aspect of sustainable industrial practices, as it helps industries minimize their environmental footprint, comply with regulations, and protect ecosystems and public health. Advances in remediation technologies continue to play a crucial role in addressing the environmental challenges associated with industrial processes. As the research is witnessing a shift in the paradigm from traditional methods to greener methods in compliance with the United Nations 2030 Sustainable Development Goals, bioremediation of industrial effluents seems to be an obvious choice. Furthermore, the crescent yet concrete approach of amalgamating microorganisms with advanced nanotechnological strategies has gained momentum, specifically with the rise of nano-adsorbents and nano-membranes that offer enhanced effectiveness owing to their flexibility in modification of active surface along with increased surface area. Lastly, to close the loop in terms of environmental sustainability and monetary sustainability, circular economy-based approaches provide a launchpad to boost the role of scientific community in the global development. Consequently, technologies for the bioremediation of the effluents have been steered by the concepts of biorefinery and micro-based nanotechnologies for a win-win scenario resulting in the nowadays accepted terminologies of green chemistry and engineering and 'green economy'. This compilation, thus, aims for a healthy merger of the key entities namely, scientists, environmentalists, and economists.

References

Abdulkareem, A. S., Kovo, A. S., & Afolabi, E. A. (2020). Enhanced adsorption of As (V) and Mn (VII) from industrial wastewater using multi-walled carbon nanotubes and carboxylated multi-walled carbon nanotubes. *Chemosphere, 254*, 126780.

Abedi, T., & Mojiri, A. (2019). Constructed wetland modified by biochar/zeolite addition for enhanced wastewater treatment. *Environmental Technology & Innovation, 16*, 100472.

Adesra, A., Srivastava, V. K., & Varjani, S. (2021). Valorization of dairy wastes: Integrative approaches for value added products. *Indian Journal of Microbiology, 61*(3), 270–278.

Ahmed, S. F., Mofijur, M., Ahmed, B., Mehnaz, T., Mehejabin, F., Maliat, D., et al. (2022). Nanomaterials as a sustainable choice for treating wastewater. *Environmental Research, 214*, 113807.

Akhtar, A., Krepl, V., & Ivanova, T. (2018). A combined overview of combustion, pyrolysis, and gasification of biomass. *Energy & Fuels, 32*(7), 7294–7318.

Alazaiza, M. Y., Albahnasawi, A., Ahmad, Z., Bashir, M. J., Al-Wahaibi, T., Abujazar, M. S. S., et al. (2022). Potential use of algae for the bioremediation of different types of wastewater and contaminants: Production of bioproducts and biofuel for green circular economy. *Journal of Environmental Management, 324*, 116415.

AlMasoud, N., Alhaik, H., Almutairi, M., Houjak, A., Hazazi, K., Alhayek, F., et al. (2021). Green nanotechnology synthesized silver nanoparticles: Characterization and testing its antibacterial activity. *Green Processing and Synthesis, 10*(1), 518–528.

Amalina, F., Abd Razak, A. S., Krishnan, S., Zularisam, A. W., & Nasrullah, M. (2022). Dyes removal from textile wastewater by agricultural waste as an absorbent–A review. *Cleaner Waste Systems, 3*, 100051.

Babbitt, C. W., Gaustad, G., Fisher, A., Chen, W. Q., & Liu, G. (2018). Closing the loop on circular economy research: From theory to practice and back again. *Resources, Conservation and Recycling, 135*, 1–2.

Ballús, O., Guix, M., Baquero, G., & Bacardit, A. (2023). Life cycle environmental impacts of a biobased acrylic polymer for leather production. *Polymers, 15*(5), 1318.

Benjamin, S. R., Lima, F. D., Florean, E. O. P. T., & Guedes, M. I. F. (2019). Current trends in nanotechnology for bioremediation. *International Journal of Environment and Pollution, 66*(1-3), 19–40.

Bhateria, R., & Jain, D. (2016). Water quality assessment of lake water: A review. *Sustainable Water Resources Management, 2*, 161–173.

Bibi, F., Ilyas, N., Saeed, M., Shabir, S., Shati, A. A., Alfaifi, M. Y., Amesho, K. T., Chowdhury, S., & Sayyed, R. Z. (2023). Innovative production of value-added products using agro-industrial wastes via solid-state fermentation. *Environmental Science and Pollution Research, 30*(60), 125197–125213.

Bolade, O. P., Williams, A. B., & Benson, N. U. (2020). Green synthesis of iron-based nanomaterials for environmental remediation: A review. *Environmental Nanotechnology, Monitoring & Management, 13*, 100279.

Brooijmans, R. J., Pastink, M. I., & Siezen, R. J. (2009). Hydrocarbon-degrading bacteria: The oil-spill clean-up crew. *Microbial Biotechnology, 2*(6), 587.

Carmen, Z., & Daniela, S. (2012). *Textile organic dyes-characteristics, polluting effects and separation/elimination procedures from industrial effluents-A critical overview* (Vol. 3, pp. 55–86). IntechOpen.

Chhandama, M. V. L., & Rai, P. K. (2023). Coupling bioremediation and biorefinery prospects of microalgae for circular economy. *Bioresource Technology Reports*, 101479.

Corsi, I., Winther-Nielsen, M., Sethi, R., Punta, C., Della Torre, C., Libralato, G., et al. (2018). Ecofriendly nanotechnologies and nanomaterials for environmental applications: Key issue and consensus recommendations for sustainable and ecosafe nanoremediation. *Ecotoxicology and Environmental Safety, 154*, 237–244.

Deb, D., Mallick, N., & Bhadoria, P. B. S. (2021). Engineering culture medium for enhanced carbohydrate accumulation in Anabaena variabilis to stimulate production of bioethanol and other high-value co-products under cyanobacterial refinery approach. *Renewable Energy, 163*, 1786–1801.

Deshpande, B. D., Agrawal, P. S., Yenkie, M. K. N., & Dhoble, S. J. (2020). Prospective of nanotechnology in degradation of waste water: A new challenges. *Nano-Structures & Nano-Objects, 22*, 100442.

Dey, T., Bhattacharjee, T., Nag, P., Ghati, A., & Kuila, A. (2021). Valorization of agro-waste into value added products for sustainable development. *Bioresource Technology Reports, 16*, 100834.

Dhanaraj, C. J. (2024). Harmful effects of water pollution. In *Handbook of water pollution* (pp. 123–148).

Díaz, E., Jiménez, J. I., & Nogales, J. (2013). Aerobic degradation of aromatic compounds. *Current Opinion in Biotechnology, 24*(3), 431–442.

El-Sheekh, M. M., & Mahmoud, Y. A. (2017). Technological approach of bioremediation using microbial tools: Bacteria, fungi, and algae. In *Handbook of research on inventive bioremediation techniques* (pp. 134–154). IGI Global.

El-Sheekh, M., El-Dalatony, M. M., Thakur, N., Zheng, Y., & Salama, E. S. (2022). Role of microalgae and cyanobacteria in wastewater treatment: Genetic engineering and omics approaches. *International Journal of Environmental Science and Technology*, 1–22.

Falade, A. O. (2021). Valorization of agricultural wastes for production of biocatalysts of environmental significance: Towards a sustainable environment. *Environmental Sustainability, 4*(2), 317–328.

Ferreira, I. A., Godina, R., & Carvalho, H. (2020). Waste valorization through additive manufacturing in an industrial symbiosis setting. *Sustainability, 13*(1), 234.

Galdames, A., Ruiz-Rubio, L., Orueta, M., Sánchez-Arzalluz, M., & Vilas-Vilela, J. L. (2020). Zero-valent iron nanoparticles for soil and groundwater remediation. *International Journal of Environmental Research and Public Health, 17*(16), 5817.

Gubanova, E., Kupinets, L., Deforzh, H., Koval, V., & Gaska, K. (2019). Recycling of polymer waste in the context of developing circular economy. *Architecture, Civil Engineering, Environment, 12*(4), 99–108.

Gunjal, A., Gupta, S., Nweze, J. E., & Nweze, J. A. (2023). Metagenomics in bioremediation: Recent advances, challenges, and perspectives. In *Metagenomics to bioremediation* (pp. 81–102).

Gupta, G. K., & Shukla, P. (2020). Insights into the resources generation from pulp and paper industry wastes: Challenges, perspectives and innovations. *Bioresource Technology, 297*, 122496.

Halim, M. A., Rahman, M. M., Megharaj, M., & Naidu, R. (2020). Cadmium immobilization in the rhizosphere and plant cellular detoxification: Role of plant-growth-promoting rhizobacteria as a sustainable solution. *Journal of Agricultural and Food Chemistry, 68*(47), 13497–13529.

Hassan, S., & Ganai, B. A. (2023). Deciphering the recent trends in pesticide bioremediation using genome editing and multi-omics approaches: A review. *World Journal of Microbiology and Biotechnology, 39*(6), 151.

Hofmann, F. (2019). Circular business models: Business approach as driver or obstructer of sustainability transitions? *Journal of Cleaner Production, 224*, 361–374.

Hosseini, M. R., & Sarvi, M. N. (2015). Recent achievements in the microbial synthesis of semiconductor metal sulfide nanoparticles. *Materials Science in Semiconductor Processing, 40*, 293–301.

Hosseini, A., Jazini, M., Mahdieh, M., & Karimi, K. (2020). Efficient superantioxidant and biofuel production from microalga Haematococcus pluvialis via a biorefinery approach. *Bioresource Technology, 306*, 123100.

Hussain, N., Goswami, L., & Bhattacharya, S. S. (2023). Microbial detoxification of contaminated land. In *Bio-inspired land remediation* (pp. 189–219).

Jamil, P. A. S. M., Aziz, N. A., Bashir, M. J., Aziz, H. A., & Hung, Y. T. (2024). Treatment of textile effluent. In *Industrial waste engineering* (pp. 43–86). Springer.

Jiménez-Moreno, N., Esparza, I., Bimbela, F., Gandía, L. M., & Ancín-Azpilicueta, C. (2020). Valorization of selected fruit and vegetable wastes as bioactive compounds: Opportunities and challenges. *Critical Reviews in Environmental Science and Technology, 50*(20), 2061–2108.

Jin, X. C., Liu, G. Q., Xu, Z. H., & Tao, W. Y. (2007). Decolorization of a dye industry effluent by Aspergillus fumigatus XC6. *Applied Microbiology and Biotechnology, 74*, 239–243.

Joshi, S. S., Jadhav, M. P. K., Chourasia, N. K., Jadhav, M. P. J., & Pathade, K. N. (2024). Bioremediation strategies for soil and water pollution harnessing the power of microorganisms. *Migration Letters, 21*(S8), 1–20.

Kamdi, M. S., Khedikar, I. P., & Shrivastava, R. R. (2012). Physical & chemical parameter of effluent treatment plant for thermal power plant. *International Journal of Engineering Research & Technology (IJERT), 1*(4), 2278-0181.

Kannah, R. Y., Merrylin, J., Devi, T. P., Kavitha, S., Sivashanmugam, P., Kumar, G., & Banu, J. R. (2020). Food waste valorization: Biofuels and value added product recovery. *Bioresource Technology Reports, 11*, 100524.

Kariim, I., Abdulkareem, A. S., & Abubakre, O. K. (2020). Development and characterization of MWCNTs from activated carbon as adsorbent for metronidazole and levofloxacin sorption from pharmaceutical wastewater: Kinetics, isotherms and thermodynamic studies. *Scientific African, 7*, e00242.

Korhonen, J., Honkasalo, A., & Seppälä, J. (2018). Circular economy: The concept and its limitations. *Ecological Economics, 143*, 37–46.

Kover, A., Kraljić, D., Marinaro, R., & Rene, E. R. (2022). Processes for the valorization of food and agricultural wastes to value-added products: Recent practices and perspectives. *Systems Microbiology and Biomanufacturing, 2*, 50–66.

Kumari, P., Alam, M., & Siddiqi, W. A. (2019). Usage of nanoparticles as adsorbents for waste water treatment: An emerging trend. *Sustainable Materials and Technologies, 22*, e00128.

Langsdorf, A., Volkmar, M., Holtmann, D., & Ulber, R. (2021). Material utilization of green waste: A review on potential valorization methods. *Bioresources and Bioprocessing, 8*(1), 1–26.

Lap, B., Debnath, A., Singh, G. K., Chaturvedi, P., Dey, J. K., & Saha, S. (2024). Microbial enzymes in biodegradation of organic pollutants: Mechanisms and applications. In *Microbes based approaches for the management of hazardous contaminants* (pp. 12–19).

Leal, C., Gouvinhas, I., Santos, R. A., Rosa, E., Silva, A. M., Saavedra, M. J., & Barros, A. I. (2020). Potential application of grape (Vitis vinifera L.) stem extracts in the cosmetic and pharmaceutical industries: Valorization of a by-product. *Industrial Crops and Products, 154*, 112675.

Lin, T. C., Pan, P. T., & Cheng, S. S. (2010). Ex situ bioremediation of oil-contaminated soil. *Journal of Hazardous Materials, 176*(1–3), 27–34.

Lithner, D., Larsson, Å., & Dave, G. (2011). Environmental and health hazard ranking and assessment of plastic polymers based on chemical composition. *Science of the Total Environment, 409*(18), 3309–3324.

López-Pacheco, I. Y., Silva-Núñez, A., García-Perez, J. S., Carrillo-Nieves, D., Salinas-Salazar, C., Castillo-Zacarías, C., et al. (2021). Phyco-remediation of swine wastewater as a sustainable model based on circular economy. *Journal of Environmental Management, 278*, 111534.

López-Sánchez, A., Silva-Gálvez, A. L., Aguilar-Juárez, Ó., Senés-Guerrero, C., Orozco-Nunnelly, D. A., Carrillo-Nieves, D., & Gradilla-Hernández, M. S. (2022). Microalgae-based livestock wastewater treatment (MbWT) as a circular bioeconomy approach: Enhancement of biomass productivity, pollutant removal and high-value compound production. *Journal of Environmental Management, 308*, 114612.

Madison, A. S., Sorsby, S. J., Wang, Y., & Key, T. A. (2023). Increasing in situ bioremediation effectiveness through field-scale application of molecular biological tools. *Frontiers in Microbiology, 13*, 1005871.

Malik, S., Dhasmana, A., Preetam, S., Mishra, Y. K., Chaudhary, V., Bera, S. P., et al. (2022). Exploring microbial-based green nanobiotechnology for wastewater remediation: A sustainable strategy. *Nanomaterials, 12*(23), 4187.

Mandeep, & Shukla, P. (2020). Microbial nanotechnology for bioremediation of industrial wastewater. *Frontiers in Microbiology, 11*, 590631.

Masoud, M., El Osta, M., Alqarawy, A., Elsayed, S., & Gad, M. (2022). Evaluation of groundwater quality for agricultural under different conditions using water quality indices, partial least squares regression models, and GIS approaches. *Applied Water Science, 12*(10), 244.

Minhas, P. S., Saha, J. K., Dotaniya, M. L., Sarkar, A., & Saha, M. (2022). Wastewater irrigation in India: Current status, impacts and response options. *Science of the Total Environment, 808*, 152001.

Mishra, B., Varjani, S., Agrawal, D. C., Mandal, S. K., Ngo, H. H., Taherzadeh, M. J., et al. (2020). Engineering biocatalytic material for the remediation of pollutants: A comprehensive review. *Environmental Technology & Innovation, 20*, 101063.

Murray, A., Skene, K., & Haynes, K. (2017). The circular economy: An interdisciplinary exploration of the concept and application in a global context. *Journal of Business Ethics, 140*, 369–380.

Naeem, W., Haq, F., Ullah, H., Khan, A., Kundi, R., Bukhari, I., … & Farid, A. (2024). Sustainable starch strategies: Nano and macro adsorbents for the detoxification of synthetic dyes and heavy metals. In *Sustainable chemical engineering* (pp. 426–449).

Najar-Almanzor, C. E., Velasco-Iglesias, K. D., Nunez-Ramos, R., Uribe-Velázquez, T., Solis-Bañuelos, M., Fuentes-Carrasco, O. J., et al. (2023). Microalgae-assisted green bioremediation of food-processing wastewater: A sustainable approach toward a circular economy concept. *Journal of Environmental Management, 345*, 118774.

Nayak, M., Swain, D. K., & Sen, R. (2019). Strategic valorization of de-oiled microalgal biomass waste as biofertilizer for sustainable and improved agriculture of rice (Oryza sativa L.) crop. *Science of the Total Environment, 682*, 475–484.

Palit, S., & Hussain, C. M. (2020). Functionalization of nanomaterials for industrial applications: recent and future perspectives. In *Handbook of functionalized nanomaterials for industrial applications* (pp. 3–14). Elsevier.

Patel, H. (2021). Review on solvent desorption study from exhausted adsorbent. *Journal of Saudi Chemical Society, 25*(8), 101302.

Pathan, M., Patel, N., Yagnik, H., & Shah, M. (2020). Artificial cognition for applications in smart agriculture: A comprehensive review. *Artificial Intelligence in Agriculture, 4*, 81–95.

Pereira, L., & Alves, M. (2012). Dyes—Environmental impact and remediation. In *Environmental protection strategies for sustainable development* (pp. 111–162).

Periyasamy, A. P. (2024). Recent advances in the remediation of textile-dye-containing wastewater: Prioritizing human health and sustainable wastewater treatment. *Sustainability, 16*(2), 495.

Phang, S. M., Chu, W. L., & Rabiei, R. (2015). Phycoremediation. In *The algae world* (pp. 357–389).

Prasad, M. N. V. (2024). Bioremediation, bioeconomy, circular economy, and circular bioeconomy—Strategies for sustainability. In *Bioremediation and bioeconomy* (pp. 3–32). Elsevier.

Quintero-Dallos, V., García-Martínez, J. B., Contreras-Ropero, J. E., Barajas-Ferrerira, C., Barajas-Solano, A. F., Lavecchia, R., & Zuorro, A. (2019). *Distillery wastewater as a sustainable medium for microalgae production.*

Rai, A., Chauhan, P. S., & Bhattacharya, S. (2018). Remediation of industrial effluents. In *Water remediation* (pp. 171–187).

Reno, U., Regaldo, L., & Gagneten, A. M. (2020). Circular economy and agro-industrial wastewater: Potential of microalgae in bioremediation processes. In *Valorisation of agro-industrial residues–volume I: Biological approaches* (pp. 111–129).

Robinson, T., McMullan, G., Marchant, R., & Nigam, P. (2001). Remediation of dyes in textile effluent: A critical review on current treatment technologies with a proposed alternative. *Bioresource Technology, 77*(3), 247–255.

Saba, I., Wani, K., Syed, A., & Rehman, S. (2021). Role of microbial nanotechnology in bioremediation of heavy metals. In *Microbial nanotechnology: Green synthesis and applications* (pp. 307–319).

Saldarriaga-Hernandez, S., Hernandez-Vargas, G., Iqbal, H. M., Barcelo, D., & Parra-Saldívar, R. (2020). Bioremediation potential of Sargassum sp. biomass to tackle pollution in coastal ecosystems: Circular economy approach. *Science of the Total Environment, 715*, 136978.

Sarma, S., Sharma, S., Rudakiya, D., Upadhyay, J., Rathod, V., Patel, A., & Narra, M. (2021). Valorization of microalgae biomass into bioproducts promoting circular bioeconomy: A holistic approach of bioremediation and biorefinery. *3 Biotech, 11*, 1–29.

Shareef, J. U., Rani, M. N., Anand, S., & Rangappa, D. (2017). Synthesis and characterization of silver nanoparticles from Penicillium sps. *Materials Today Proceedings, 4*(11), 11923–11932.

Shrivastava, V., Ali, I., Marjub, M. M., Rene, E. R., & Soto, A. M. F. (2022). Wastewater in the food industry: Treatment technologies and reuse potential. *Chemosphere, 293*, 133553.

Singh, A., Pal, D. B., Mohammad, A., Alhazmi, A., Haque, S., Yoon, T., et al. (2022). Biological remediation technologies for dyes and heavy metals in wastewater treatment: New insight. *Bioresource Technology, 343*, 126154.

Singhal, A., Baweja, P., Gupta, S., Chopra, H., & Gandhi, P. B. (2023). Biogenesis of nanoparticles using microorganisms: A review. *Plant Science Today, 10*(4), 97–104.

Sparn, B., & Hunsberger, R. (2015). *Opportunities and challenges for water and wastewater industries to provide exchangeable services.*

Sulthana, R., Taqui, S. N., Syed, U. T., Soudagar, M. E. M., Mujtaba, M. A., Mir, R. A., et al. (2022). Biosorption of crystal violet by nutraceutical industrial fennel seed spent equilibrium, kinetics, and thermodynamic studies. *Biocatalysis and Agricultural Biotechnology, 43*, 102402.

Tanaya, K., Kumari, A., Singh, A. K., & Singh, D. (2024). Bioremediation: An economical approach for treatment of textile dye effluents. *Water, Air, & Soil Pollution, 235*(8), 1–36.

Taqui, S. N., Syed, U. T., Syed, R. T., Alqahtani, M. S., Abbas, M., & Syed, A. A. (2022). Bioremediation of textile industrial effluents using nutraceutical industrial spent: Laboratory-scale demonstration of circular economy. *Nanomaterials, 12*(10), 1684.

Thakur, V., Kumar, A., & Singh, J. (2024). Mycobial nanotechnology in bioremediation of wastewater. In *Microbes based approaches for the management of hazardous contaminants* (pp. 1–11).

Thennarasu, G., Rajendran, S., Kalairaj, A., Rathore, H. S., Panda, R. C., & Senthilvelan, T. (2024). A comprehensive review on the application of semiconductor nanometal oxides photocatalyst for the treatment of wastewater. *Clean Technologies and Environmental Policy*, 1–22.

Thomas, J., Patil, R. S., Patil, M., & John, J. (2023). Addressing the sustainability conundrums and challenges within the polymer value chain. *Sustainability, 15*(22), 15758.

Torres-Farradá, G., Thijs, S., Rineau, F., Guerra, G., & Vangronsveld, J. (2024). White rot fungi as tools for the bioremediation of xenobiotics: A review. *Journal of Fungi, 10*(3), 167.

Ukkund, S. J., Ashraf, M., Udupa, A. B., Gangadharan, M., Pattiyeri, A., Marigowda, Y. K., et al. (2019). Synthesis and characterization of silver nanoparticles from Fuzarium Oxysporum and investigation of their antibacterial activity. *Materials Today Proceedings, 9*, 506–514.

Ukkund, S. J., Puthiyillam, P., Alshehri, H. M., Goodarzi, M., Taqui, S. N., Anqi, A. E., et al. (2021a). Adsorption method for the remediation of brilliant green dye using halloysite nanotube: Isotherm, kinetic and modeling studies. *Applied Sciences, 11*(17), 8088.

Ukkund, S. J., Puthiyillam, P., Anqi, A. E., Taqui, S. N., Ali, M. A., Syed, U. T., et al. (2021b). A recent study on remediation of direct blue 15 dye using halloysite nanotubes. *Applied Sciences, 11*(17), 8196.

Ukkund, S. J., Alke, B., Taqui, S. N., & Syed, U. T. (2022a). Dextran nanoparticles: Preparation and applications. In *Polysaccharide nanoparticles* (pp. 1–31). Elsevier.

Ukkund, S. J., Alke, B., Taqui, S. N., & Syed, U. T. (2022b). Role of microbial nanotechnology in energy devices. In *Handbook of microbial nanotechnology* (pp. 517–547). Academic.

Ummalyma, S. B., Sahoo, D., & Pandey, A. (2021). Resource recovery through bioremediation of wastewaters and waste carbon by microalgae: A circular bioeconomy approach. *Environmental Science and Pollution Research, 28*(42), 58837–58856.

Yadav, A., Rene, E. R., Sharma, M., Jatain, I., Mandal, M. K., & Dubey, K. K. (2022). Valorization of wastewater to recover value-added products: A comprehensive insight and perspective on different technologies. *Environmental Research*, 113957.

Chapter 15
Microorganism-Assisted Bionanotechnology for Textile Dye Wastewater Treatment

Kusumlata, Rameshwari Ashish Banjara, and Ashish Kumar

Abstract The last century has witnessed an increase in the global population and rapid industrialization, which has led to environmental pollution and negatively impacted air and water quality, which is essential for life on Earth. It is estimated that 3.7 million liters of water are contaminated daily worldwide by pollutants from the textile industry. The discharge of synthetic dyes from wastewater by the textile industry is one of the leading causes of water pollution. Dye and metal are the principal components of the effluents emitted by the textile industry. Most textile dyes are carcinogenic and cause a variety of skin problems. Currently used physical and chemical treatment methods for textile dyes are more labor intensive, expensive, and ineffective. As an alternative, nanoparticles have emerged as a better dye removal and degradation option due to their superior surface characteristics and chemical reactivity. It is easy to use, environmentally friendly, and quickly meets the growing demand for safe, nontoxic, and biodegradable nanoparticles on a worldwide scale. Microbe-based biological nanoparticle production and nanotechnology have drawn significant interest in industrial wastewater treatment because of their uncomplicated handling procedures, ecologically safe disposal, and superior downstream processing. Thus, this chapter focus on the current status of natural nanoparticles and other recent techniques for the treatment of wastewater from the textile industry for environmental sustainability.

Keywords Textile effluent · Dye degradation · Pollutant · Environmental toxicity · Bionanoparticles

Kusumlata · A. Kumar (✉)
Department of Biotechnology, Guru Ghasidas Vishwavidyalaya (A Central University), Bilaspur, Chhattisgarh, India

R. A. Banjara
Department of Chemistry, Rajeev Gandhi Govt. Postgraduate College, Ambikapur, Chhattisgarh, India

Y. K. Mohanta et al. (eds.), *Nano-microbiology for Sustainable Development*,
https://doi.org/10.1007/978-3-031-78845-1_15

Abbreviations

AMBR	Algal membrane bioreactor
BOD	Biological oxygen demand
COD	Chemical oxygen demand
EPS	Xopolysaccharide-stabilised
HRT	Hydraulic retention time
MBRs	Membrane bioreactors
MF	Microfiltration
MFCs	Microbial fuel cells
MMM	Mixed matrix membranes
MNPs	Metallic NPs
MOFs	Metal-organic frameworks
NaBH4	Sodium borohydride
NPs	Nanoparticles
NZVI	Nanoscale zero-valent iron
PEM	Proton exchange membrane
SBRs	Sequencing batch reactors
TMO	Transition metal oxide
TSS	Total suspension solids
UASB	Upflow anaerobic sludge blanket
UBRs	Upflow blanket bioreactors
UF	Ultra filtration

15.1 Introduction

The environment is seriously threatened by several very hazardous compounds produced by many industries, including dyes, pigments, and other aromatics (Srivastava et al., 2022). The textile, leather tanning, paper, pharmaceutical, cosmetic, and plastic sectors are the only businesses that use more than 10,000 industrial dyes (Mondal et al., 2015). The textile industry in India is the second-largest and one of the fastest-evolving sectors of the economy, which causes a significant amount of waste to be produced throughout the dyeing, printing, and finishing processes. Synthetic dyes are notable pollutants and are generally not degraded biologically because of their complicated structures, which prevents them from dissolving in water (Srivastava et al., 2022). In several industries, especially the textile industry, synthetic dyes are widely used. Azo dyes, which constitute a significant amount of synthetic dyes, have been observed to be potential genotoxic and carcinogenic agents (Sha et al., 2016). In the traditional textile industry, 1 kg of textile compounds is extracted using approximately 100 L of water. Surface and groundwater in the surrounding areas of dyeing enterprises are seriously contaminated by this procedure since the dyes do not entirely bind to the textiles and are lost in some amount in the wastewater, which can range from 2% for simple dyes to 50% for reactive dyes (Srivastava et al., 2022). During the dyeing of fibres, yarns, and fabrics, the number of synthetic dyes used in the textile industry decreases by 10–15%. The presence of unfixed dyes from the aqueous dyeing process is a major factor in the release of coloured water. Thus, the presence of these dyes in water has the potential to reduce or entirely prevent sunlight from entering the stream, which would reduce photosynthesis in aquatic plants and have an impact on aquatic ecology (Thakare et al., 2022).

Simultaneously, a blessing and a curse for human society have been demonstrated by modernisation and inventive industrial innovations, with water resources being particularly negatively impacted. Detoxifying water systems containing these harmful substances is consequently essential (Mondal et al., 2015). Organic dyes constitute a significant amount of the many contaminants that are dumped into water bodies and are produced by the textile, paint, culinary, leather, and printing industries. The presence of these synthetic organic dyes in water bodies increases the chemical oxygen demand, which has an impact on aquatic life even if a large portion of them is nontoxic or less harmful (Gupta et al., 2020). Aquatic bodies are impacted by coloured organic dyes because they reduce dissolved oxygen levels by inhibiting sunlight (Mondal et al., 2015). Numerous non-eco-friendly physico-chemical treatment methods produce large volumes of sludge. Therefore, there is an urgent need for the development of inexpensive and effective methods for removing dangerous chemicals, including textile dyes, from polluted water.

Compared to other traditional technologies, nanotechnology is currently thought to be a promising solution for water purification. It can breakdown longer-chained, more complex chemical molecules—which are usually toxic—into simpler ones (Hamada & Jabal, 2022). This technology may be used in conjunction with other methods for treating wastewater, such as coagulation, adsorption, photocatalysis, and membrane technology. These techniques, which use nanotubes, nanocatalysts, nanoadsorbents, nanostructured catalytic membranes, and micromolecules, can be applied to resolve water quality issues. An overview of the applications of nanobiotechnology, either alone or in conjunction with other technologies, for the treatment of industrial wastewater is provided in this chapter. The potential and difficulties of combining biotechnology and nanomaterials for the treatment of water and wastewater are also discussed (Sahu et al., 2021).

Researchers have altered the green synthesis of nanoparticles to prevent the toxic effects of chemicals because chemical approaches involve methods that are harmful to the environment (Roy et al., 2022). Biosynthesis using new, nontoxic, environmentally benign, and easily obtainable biological materials—fungi, bacteria, biomolecules, and plant extracts—has recently been the subject of extensive research (Moritz & Moritz, 2013). Various microorganisms, viz. Algae, fungi, and bacteria, including *Escherichia coli*, have been used in the synthesis of copper, zinc, iron, silver, and gold nanoparticles. Both algae biomass and algae nanoparticles can remove pollutants from wastewater; however, due to their higher surface-to-volume ratio, algae nanoparticles are far more effective than algae biomass. NPs derived through biological synthesis have all these qualities, making them good substitutes for physically and chemically synthesised NPs. Emerging research is primarily focused on selecting environmentally friendly, commercially viable, harmless, and competent compounds (Sharma et al., 2023). The rapid growth of nanotechnology has created new opportunities for the creation of creative, low-cost, and highly effective wastewater treatment techniques for dye contamination (Al-Buriahi et al., 2022). Therefore, it is necessary to investigate the possible uses of low-cost nanomaterials in the production of potable water and the treatment of wastewater.

15.1.1 Socioeconomic and Environmental Impacts of Textile Dyes

Synthetic textile dyes are responsible for 17–20% of all industrial water pollution, and approximately 40% of synthetic dyes contain identified carcinogens. Chemical dyes may accumulate in rivers, releasing hazardous, hormone-disrupting compounds that result in animal mutations and death in addition to damaging human health (Srivastava & Singh, 2019). The dyeing and finishing processes used in the manufacturing of textiles are thought to be among the main causes of water pollution. The textile industry and its wastewater, including dyes, are major causes of serious pollution-related issues on a global scale (Carmen & Daniela, 2012). However, 20% of lost dyes are present in industrial and effluent, which has major consequences for aquatic life, humans, plants, and the environment by disrupting photosynthesis and other physiochemical processes (Chadha et al., 2021). Due to their lower binding efficiency, studies have shown that up to 200,000 tonnes of dye are wasted in industrial discharge. Depending on their chemical structure, this amount can range from 2% to 50%. Due to their high thermal and photostability, which prevent biodegradation and promote persistence in the environment, dyes pose several environmental risks (Chadha et al., 2021). In particular, the release of dye-containing effluents into the water environment is undesirable because many of the released dyes and their breakdown products are toxic, hazardous, or mutagenic to life forms, primarily due to carcinogens, such as benzidine, naphthalene, and other aromatic compounds. These dyes have the potential to linger in the environment for a very long time if not properly treated. For instance, hydrolysed reactive blue 19 has a half-life of approximately 46 years at pH 7 and 25 °C (Carmen & Daniela, 2012).

Most dyes are nonbiodegradable, have carcinogenic effects, or can trigger allergies, dermatitis, and skin irritation because they are synthetic and have mostly aromatic structures. Additionally, several azo dyes, primarily aromatic chemicals, exhibit both acute and chronic toxicity. Adsorption of azo dyes and their breakdown products (toxic amines) through the gastrointestinal system, skin, and lungs, as well as the creation of haemoglobin adducts and disruption of blood formation, poses a significant health risk (Carmen & Daniela, 2012). It can also change the pH and oxygen levels in water, which might disrupt aquatic ecology. Additionally, it may be poisonous and mutagenic to aquatic flora and fauna. The remaining dyestuff is also linked to several negative consequences on human health, including inflammation, respiratory issues, and immune system damage. At concentrations between 10 and 100 mg/L, azo dyes can contain hazardous substances such as pentachlorophenol, toxic metals, chlorine, biocides, and carcinogenic amines. Additionally, it has harmful effects on aquatic flora and animals, causing serious environmental problems worldwide. Additionally, azo dyes have a deleterious impact on total suspension solids (TSS), biological oxygen demand (BOD), and chemical oxygen demand (COD) (Kokeb et al., 2023). The colour generated during the dyeing process harms aquatic life by preventing sunlight from penetrating the water, which decreases photosynthesis and depletes the water's oxygen content. It has been reported that Remazol Red Brilliant and Indigo dye effluents alter the aquatic food cycle, resulting in an ecological imbalance. It has also been reported that dye toxicity kills some soil microorganisms, which reduces agricultural yields and fruit quality (Hassaan et al., 2017).

15.2 Bionanotechnology as an Innovative Approach for Dye Degradation

To find a less expensive method for synthesizing nanoparticles (NP), scientists use a variety of biological agents, including plant, fungal, and microbial enzymes. They are typically responsible for the incorporation of metal compounds into their respective nanoparticles because of their reducing characteristics (Roy et al., 2022). This technology is used to manipulate and control small particles for the sake of humanity. The common and unique characteristics of NPs make them far more valuable than bulk materials. The use of nanotechnology has seen an important rise in dye degradation applications due to its unique advantages and faster reactions than other methods. Numerous microorganisms and plants have been employed in biological synthesis. The manufacture of Zn, Ag, Au, Cu, Fe, Pb and iron nanostructures involves the use of microorganisms such as bacteria, fungi, yeast, and algae viruses and plants (Joshi et al., 2022; Al-Buriahi et al., 2022).

15.2.1 Key Components of Bionanotechnology

Bionanotechnology consists of two main methods, biological components and nanotechnologies, which are composed of nanosized catalytic particles such as metals or metal oxides. Biological substrates and metal oxides are the two main components used for the preparation of nanoparticles.

15.2.1.1 Metal and Metal Oxides

Metal oxides have the potential to be used in energy production, conversion, storage, and environmental remediation (Joshi et al., 2022). Metal and metal oxide nanoparticles such as gold, silver, iron, cobalt, zinc, titanium, CuO, MnO_2, SnO_2, TiO_2, ZnO, ZrO_2, and CeO_2 are effectively used for the treatment of dye-containing effluents (Table 15.1). Gold is a noble metal that has various uses in the production of nanomaterials, including minimal toxicity to biological systems and conformational flexibility. AgNPs can be produced both biologically and chemically (Roy et al., 2022). The environmentally friendly, low-cost, and efficient methods of creating metal oxide nanoparticles via biological or green synthetic processes have drawn much attention. Several catalytic processes have made use of transition metals and their oxides. Additionally, metal oxides have the potential to be used in energy production, conversion, storage, and environmental remediation. Numerous studies on the biogenic synthesis of metallic NPs, including Ag, Au, Zn, ZnO, Cu,

Table 15.1 Metal nanoparticles used for dye degradation

Metal nanoparticle	Dyes	Conditions	Degradation %	Time	References
ZnO 0.08 g	Azo dyes 5 ppm	pH 7, 40 °C	85% and 98.5%	80 min	Hamada and Jabal (2022)
Mn-ZnO NPs	Methylene blue dye		99%	120 min	Chanu et al. (2019)
Co nanoparticle 0.5 g/L	Methyl orange 100 mg/L	pH 2.5	99%	4 min	Sha et al. (2016)
$MnFe_2O_4$	Reactive Red 198, Reactive Red 120	25 °C	53%, 50%	60 min	Mahmoodi (2015)
ZnO nanoparticles	Methylene blue 20 mg/L		81% and 92.5%	3 h	Balcha et al. (2016)
$ZnFe_2O_4$	Reactive Red 198, Reactive Red 120, 100 mg/L	25 °C	40%, 37%	60 min	Mahmoodi (2013)
α-Fe_2O_3 nanoparticles	Methylene blue, Methyl orange, Bromo green, Methyl red		(95%), (94%), (94%), (76%)	5 min	Gupta et al. (2020)
ZnO nanoparticles	Methylene blue dye 10 ppm	pH-5, pH-7, pH-10, pH-12	41%, 83%, 99%, 97%	120 min	Chanu et al. (2019)
TiO_2 nanoparticles	Congo red	pH 6.3	75%	100 min	Guo et al. (2012)
ZnO nanocrystals	Rhodamine-B and methylene blue10 mg/L	–	33.8%	2 h	Liu et al. (2012)
ZnO	Direct Yellow 27, Acid Blue 129	–	84%, 66%	390 min	Umar et al. (2015)

CuO, Ce, CeO, and Pt, employing various microbes and plants (Table 15.2), have been published by Patil and team (2021). Various processes have been used to synthesise ZnO NPs using microbial resources such as fungi, bacteria, algae, and others (Al-Buriahi et al., 2022).

15.2.1.2 Bioresources

For many years, different microbes—including bacteria, algae, and microorganisms—have been used to create ecologically safe nanoparticles. Compared to traditional techniques, biosynthesised nanoparticles can be used quickly and without restrictions to treat and decolorise dyes (Roy et al., 2022). The use of bionanomaterials for the removal of synthetic dyes and the development of new functional materials such as metal-organic frameworks (MOFs) has received an increased amount of attention in recent years. MOFs bind dye molecules (anionic and cationic) through electrostatic contact. In a variety of sectors, including aeronautical engineering, materials science, bioremediation, pharmacology, and biosensors, bionanomaterials have drawn significant interest due to their remarkable capacities, biologically derived nature, and biocompatible characteristics despite their small size (Srivastava et al., 2022). AgNPs can be produced in different parts of microbial cells. As reducing and stabilising agents, pure polysaccharides from sources such as plants, animals, and microflora have been employed in the manufacture of nanoparticles (Saravanan et al., 2017). Biogenic NPs are NPs that are synthesised *via* biological processes. These NPs show distinct potential due to their unique size and shape and are nontoxic. Microorganisms are the best choice for synthesising metallic nanoparticles due to their high yield, facile purification, and mild requirements. Thus, microorganisms have recently been widely reported to synthesise nanoparticles from moniker "nanofactories" (Patil et al., 2021). NPs can be created from bacteria or other biological materials due to their many benefits over other living materials. In the course of the synthesis process, microbial metabolite extracts serve as stabilising and reducing agents while mixing with metal precursors under suitable conditions (Roy et al., 2022).

15.2.2 Microorganisms in Bionanotechnology

The term "bio nanomaterials" refers to a class of possible materials formed from a variety of biological components, including peptides, nucleic acids, bacteria, plants, and fungi (Srivastava et al., 2022). Bacteria have the intrinsic ability to convert metal ions into appropriate metallic nanoparticles, making them valuable prospective nanofactories. Proteins and other biomolecules in bacteria can regulate the formation of inorganic crystals during biomineralization processes. *Pseudomonas aeruginosa, Rhizopus oryzae, Zoogleacamera*, and many other bacterial species have been investigated recently for the production of metallic nanoparticles

Table 15.2 Photodegradation mechanism of dyes via redox reactions in the presence of solar radiation

Plants name	Dye	Activity	Metal	Percentage of degradation	Time	References
Emblica seed extract	Methylene blue dye	Photocatalytic activity	Iron oxide	80%	30 min	Ashraf et al. (2023)
Glycyrrhiza glabra	Methylene blue dye	Photocatalytic activity	Fe_2O_3NPs	92%	180 min	Geethamala et al. (2023)
Cyperus rotundus L	Congo red dye	Photocatalytic activity	α-Fe_2O_3 NPs	64.3%	25 min	Basavegowda et al. (2017)
Dicliptera Roxburghiana	Methylene blue	Photocatalytic degradation	CdS	87.12%	120 min	Ullah et al. (2021)
jujube fruit extract	Methylene blue eriochrome black-T	Photocatalytic degradation	ZnO NPs	92%, 86%	5 h	Golmohammadi et al. (2020)
Daphne mezereum	Methyl orange	Catalytic	Fe_2O_3 NPs	81%	6 h	Beheshtkhoo et al. (2018)
Solanum trilobatum extract	Methylene blue dye	Catalytic	RGO-Fe_3O_4	89.35%	60 min	Vinothkannan et al. (2015)
Solanum lycopersicum	Crystal violet	Photocatalytic activity	Copper oxide	97%	30 min	Vaidehi et al. (2018)
Suaeda japonica	Methylene blue	Photocatalytic degradation	Zinc oxide	54%		Shim et al. (2019)
Syzygium cumini	Methylene blue	Photocatalytic degradation	Zinc oxide	91.4%	pH 7, 180 min	Sadiq et al. (2021)
Ruellia tuberosa	Crystal violet	Photocatalytic degradation	Iron oxide	80%	150 min	Vasantharaj et al. (2019)
Punica granatum	Methylene blue	Photocatalytic degradation	Silver	89%	48–72 h	Joshi et al. (2018)
Prosopis farcta	Methylene blue	Photocatalytic activity	Silver	30 min, 70.2%	30 min	Miri et al. (2018)
Nigella arvensis	Methylene blue	Catalytic activity	Gold	44%	48 h	Chahardoli et al. (2018)
Nigella Sativa	Congo red	Photocatalytic activity	Silver	98.5%	24 h	Chand et al. (2021)
Ficus carica extract	Alizarin Yellow R dye 120 mg/l	Photocatalytic degradation	CuNPs	89.71%	40 min	Usman et al. (2019)
Foeniculum vulgare	Methylene blue	Catalytic activity	AuNPs	Complete disappearance	2 min	Choudhary et al. (2017)

Table 15.3 List of nanoparticles synthesized from microbe used for dye degradation

Microorganisms	Dye	Types of nanoparticles	Percentage of degradation	References
Bacillus marisflavi	Congo red and methylene blue	Gold nanoparticles	98%	Nadaf and Kanase (2019)
Escherichia sp.	Congo red, malachite green, direct blue-1, reactive black-5	Copper nanoparticles	97.07%, 90.55%, 88.42%, 83.61%	Noman et al. (2020)
Chlamydomonas reinhardtii	Methylene blue	Cadmium sulphide NP	99%	Rao and Pennathur (2017)
Macrocystis pyrifera	Reactive black 5	Zerovalent Iron	69–80%	Garcia et al. (2018)
Spirulina platensis	Malachite Green	Cadmium Sulphide	60%	Mandal et al. (2016)
Chlorella vulgaris	Methylene blue	AgNPs	24 h, 96.51%	Rajkumar et al. (2021)
Sargassum wighitii	Methylene blue	MgO NPs		Pugazhendhi et al. (2019)
Cordyceps militaris fungus	Methylene blue dye	Zinc Oxide	97%	Li et al. (2019)
Cystoseira trinodis	Methylene blue	CuO NPs	89% and 87%	Gu et al. (2018)
Caulerpa serrulata	Congo red	Ag NPs,20Extract conc. (v/v) %	99.0%, 06(min)	Aboelfetoh et al. (2017)

(Srivastava & Mukhopadhyay, 2014). The use of nanoparticles from living organisms, such as plants and microbes, has gained interest due to its simplicity, nontoxicity, and green chemical approach for the synthesis of nanoparticles as well as the creation of ecologically friendly and sustainable materials. The extraordinary reduction in performance is attributed to energy-dependent ion efflux from the bacterial cell via membrane protein and chemical detoxification processes. As a result, microorganisms are a top choice for the biosynthesis of nanomaterials (Table 15.3) due to their intriguing features (Srivastava & Mukhopadhyay, 2014). Recent reports of the successful biosynthesis of metallic nanosized rods, wires, flowers, tubes, triangles, and spherical, hexagonal, tetragonal, and pentagonal shapes have been reported. These biologically synthesised nanomaterials have potential uses in a variety of fields, including medical diagnostics, therapy, surgical nanodevices, creation, and commercial product production (Vijayaraghavan & Kumar, 2017). Microbial cells can grow rapidly in a variety of environments and require easy maintenance (Patil et al., 2021). Yeast, which is related to the fungal class Ascomycetes, showed significant promise in the generation of nanoparticles. When fungi are exposed to aqueous metal ions, they produce both intra- and extracellular nanoparticles. Compared with the intracellular pathway, the extracellular technique results in the production of significantly more synthesized nanoparticles, but this process is much faster (Vijayaraghavan & Kumar, 2017). Most bacteria use extracellular synthesis to generate nanoparticles. There have been reports of several fungal species, including yeasts, producing metallic nanoparticles. The production of nanoparticles by microbes is typically linked to an enzymatic mechanism. Two hydrogenases were directly implicated in the reduction of technetium (VII) in *Shewanella oneidensis* MR-1; the NiFe hydrogenase showed greater Tc reduction activity than the Fe-only hydrogenase (Pereira et al., 2015). Since they easily handle and grow at a high rate, bacteria are among the best options for the production of nanoparticles because of their special capacity to convert metallic ions into small particles. Actinomycetes have traditionally been employed in the production of secondary metabolites and extracellular enzymes. Their high protein content and unparalleled ability to produce a wide range of bioactive chemicals have led to their adoption for the biosynthesis of nanoparticles. Although actinomycetes can produce nanoparticles through extracellular reduction as well as through intracellular means, the latter route is more frequently used and has a wider range of industrial uses. Another easy and basic method that has been thoroughly investigated for the creation of nanoparticles is fungal biosynthesis. Yeast is capable of absorbing large amounts of harmful metal ions from the environment and accumulating in the environment. Yeast cells use a variety of detoxification mechanisms, including bioprecipitation, chelation, and intracellular sequestration, to adjust to metal toxicity conditions. Many researchers have made use of this characteristic of yeast cells. Viruses can be used to synthesize metal nanoparticle nanocomposites and nanoconjugates, which are crucial bioengineering materials for cancer therapy and drug delivery (Gahlawat & Choudhury, 2019).

The main nanoparticle synthesis methods are generally classified into three types: (1) liquid phase, (2) biological/green, and (3) vapor state-based methods. Biological extracts containing several secondary metabolites are generally obtained from microbial sources such as bacteria, algae, fungi, and yeast. These biological components are used as reducing, capping, or stabilizing agents during the preparation of various metal and metal oxide nanoparticles (Joshi et al., 2022). The green synthesis process generally consists of three major phases: selecting a biological reducing agent and selecting materials for stabilization of the nanoparticles. Since biobased procedures for the synthesis of nanomaterials are based on the principles of ecological chemistry and are easy, inexpensive, and easily scaled up for large-scale production, they are environmentally and economically friendly (Keat et al., 2015). Within 10 min, *Aspergillus fumigatus* produced monodispersed silver nanoparticles in the 5–25 nm range extracellularly. With chemical and physical methods of nanoparticle synthesis, fungi produce nanoparticles more quickly. Consequently, *A. fumigatus* may be employed in the large-scale synthesis of nanoparticles. Therefore, *A. fumigatus* could be used for large-scale production of nanoparticles. FTIR analysis revealed that biological agents or biomolecule groups bound to the particles account for the stability of the nanoparticles and that the biological agents or biomolecule groups act as stabilizing and capping agents. Many microorganisms generate inorganic minerals through intra- or extracellular processes. The cell wall of microorganisms is essential for the intracellular synthesis of nanoparticles. The positive and negative charges of the metal ions interact electrostatically with the negative charge of the cell wall as part of the mechanism (Hulkoti & Taranath, 2014). The process of biological nanoparticle synthesis is completed in these steps. In the first step, the microorganisms or their metabolites are extracted. In the second step, the microorganisms mix well with metals or metal oxides. In the third step, the nanoparticle is synthesized. Finally, the synthesized nanoparticles are separated, characterized, and applied.

15.2.3 Nanotechnology for the Treatment of Textile Dye Wastewater

Nanoparticles in the size range of 1–100 nm exhibit a variety of unique properties not observed in their bulk form. Dye molecules are efficient electron acceptors, and nanoscale zero-valent iron (NZVI) particles are good electron donors. To accomplish full degradation, the NZVI particles must also break the auxochrome link to decolorize the dye molecules, and the resultant intermediate organic compounds must be mineralized into CO_2, H_2O, and inorganic ions (Raman & Kanmani, 2016). One of the most effective methods for addressing environmental pollution problems is semiconductor photocatalysis. Compared with other photocatalysts, TiO_2 NPs are regarded as one of the most promising dye treatment photocatalysts because of their nontoxicity, potent oxidizing capability, and excellent stability (Sathiyan et al., 2020). The large surface-to-volume ratio provided by nanostructures makes the filtration of water a promising application for nanotechnology (Hamada & Jabal, 2022). To evaluate their prospective applicability, synthetic magnetic iron oxide nanoparticles (MNPs) were used to treat wastewater by degrading organic dyes. A more environmentally friendly partnership of chemical science with analytical science, nanotechnology, and material science is required for the development of MNPs for the degradation of rhodamine B (RhB) dye. MNP synthesis adheres to moral standards and sustainable practices, such as the use of effective, affordable, secure, and clean raw materials as well as reducing agents produced from plants (Sharma et al., 2019). The NF of the biological and UF treatment effluents allowed for even more COD, color, and salt removal, resulting in water that was acceptable for reuse inside the textile industry (Grilli et al., 2011). A recently developed environmentally friendly and cost-effective alternative for the large-scale treatment of water contaminated with dyes and other organic chemicals is photocatalysis, which uses nanosized semiconductor photocatalysts (Balcha et al., 2016). The fastest dye decolorization rate was obtained from methyl orange dye degradation using the same amount of hollow Co nanoparticles, while for the other three dyes, where surface adsorption predominated the dye removal process, the decolorization rate followed the order of amaranth > orange G > Congo red. Commercially available solid, spherical Co and Ni nanoparticles have been compared with hollow Co and Ni nanoparticles to better understand the influence of hollow structures on the degradation of methyl orange dye (Sha et al., 2016). Chemical processes involve the use of chemicals that are not biodegradable and have the potential to be hazardous to people and other biological systems, such as organic solvents, reducing agents, and stabilizers. These techniques are frequently expensive due to the high temperatures and complicated control requirements. The need for clean, safe, affordable, and environmentally friendly alternatives is obvious given the drawbacks of producing nanoparticles physically and chemically (Kouhbanani et al., 2019). Phytoconstituents may be used in the production of nanoparticles as both reducing and stabilizing agents. Since various plant extracts include various combinations of organic reducing and capping agents, the source of the plant extract is known to impact nanoparticle production (Kalaiselvi et al., 2015; Al-Hakkani et al., 2021). Metal NPs (MNPs) made from plant extracts are stable and effectively monodispersed by adjusting the pH, incubation time, mixing ratio, and temperature. Plants with high metal ion hyperaccumulation and reductive capacity have been employed to recover metallic NPs because of their potential to reduce metal ions both on their surface and in several organs and tissues far from the ion penetration point (Dutta & Das, 2021). The earliest evidence of silver nanoparticle manufacturing was found in alfalfa plants. After that, a wide range of plants were successfully evaluated for AgNP production and applications. The creation of a multilayer of dye molecules on the catalytic surface of nanoparticles might explain why dye degradation percentages are lower at greater concentrations. This multilayer creation reduces the amount of light necessary to start the photooxidation process, which promotes dye degradation (Jain et al., 2021).

15.2.4 Microbial Nanoparticles for the Treatment of Textile Dye Wastewater

Bacillus amyloliquefaciens isolated from effluent from the dairy sector is used to make TiO_2 nanoparticles. The photocatalytic effectiveness of biosynthesized titanium dioxide nanoparticles toward the breakdown of the sulphonated textile dye Reactive Red 31 was examined. For the degradation of methyl orange (62%), reactive blue (160, 76%), and rhodamine B (63%) dyes, previous research revealed improved photocatalytic activity of Zn-, Ag- and Pt-doped TiO_2 compared to undoped TiO_2 (Khan & Fulekar, 2016). Using *Bacillus marisflavi* YCIS MN 5, Nadaf and Kanase (2019) explained for the first time an easy and bacteriogenic method to create gold nanoparticles. These authors also suggested that these particles may catalyze the degradation of Congo red and methylene blue in the presence of sodium borohydride ($NaBH_4$). The biosynthesized gold nanoparticles demonstrated complete catalytic degradation of 88% methylene blue by $NaBH_4$ within the following minute (Nadaf & Kanase, 2019). The synthesis of CuNPs by the native copper-resistant bacterial strain *Escherichia sp.* SINT7 and evaluation of the photocatalytic activity of the biogenic CuNPs for congo red, malachite green, direct blue-1 and reactive black-5 azo dye degradation. Azo dye degradation assays indicated that CuNPs decolorised congo red (97.07%), malachite green (90.55%), direct blue-1 (88.42%), and reactive black-5 (83.61%) at a dye concentration of 25 mg/L after 5 h of sunlight exposure (Noman et al., 2020).

According to Narayanan and Park (2015), it took 18 min to catalytically reduce methylene blue using gold nanoparticles with an average size of 32 nm in the presence of $NaBH_4$. These biosynthesised gold nanoparticles, in contrast to existing publications, promote faster methylene blue degradation in the presence of $NaBH_4$, making them superior nanocatalysts (Narayanan & Park, 2015). A green method for the synthesis of silver nanoparticles by *Chlorella vulgaris* and its use in photocatalytic dye degradation was reported by Rajkumar et al. (2021). *Saccharomyces cerevisiae* has been extensively studied for its ability to biosynthesise a variety of nanoparticles, including silver and palladium nanoparticles, and for its use in photocatalytic dye degradation. Silver nanoparticles were synthesised by Mechouche et al. (2022) from the breakdown of methylene blue dye by the *Streptomyces tuirus* strain. Within 6 h of incubation, the AgNPs exhibited a 71.3% photocatalytic decolourization capability for MB dye (20 ppm) under sunlight irradiation, whereas only 11.25% of the MB dye decomposed under UV irradiation (Mechouche et al., 2022). Silver ions can be reduced to silver nanoparticles by some lactic acid bacteria (LAB), including *Lactobacillus species, Pediococcus pentosaceus,* and *Enterococcus faecium.* EPS-AgNPs (exopolysaccharide-stabilised AgNPs) were generated by a strain of *Leuconostoc lactis* that produces EPS. Saravanan et al. (2017) established that EPS-stabilised AgNPs effectively facilitated the degradation of commercial textile dyes, including azo dyes such as methyl orange (MO) and Congo red (CR). The bacterial degradation of azo dyes by *Lactobacillus acidophilus*, *Lactobacillus fermentum,* and *Halomonas spp*. was less efficient than the breakdown of azo dyes by EPS-AgNPs. It was demonstrated that EPS-stabilised AgNPs were effective in the degradation of azo dyes, including methyl orange (MO) and Congo red (CR) (Saravanan et al., 2017). Proteins and amino acids are thought to play a substantial role in the bacterial *E. herbicola*-mediated production of SnO_2 nanoparticles. Biologically produced tetragonal crystalline SnO_2 nanoparticles range in size from 10 to 42 nm (TEM) (Srivastava & Mukhopadhyay, 2014).

15.2.5 Combined Microbial and Nanotechnological Approaches for the Treatment of Textile Dye Wastewater

A single technology cannot be efficiently used to remove azo dye; thus, hybrid technologies are less expensive, eco-friendly and sustainable and are preferred for azo dye pretreatment and removal. By using combined physicochemical and biological techniques, azo dyes can be degraded along with their metabolites (Saba et al., 2021). Hybrid methods, which combine nanotechnology with other systems to improve the biological treatment process, are common methods for obtaining high performance. Biological/membrane treatment (Fig. 15.1) is one of the most widely used hybrid techniques. This hybrid method was established to be effective and

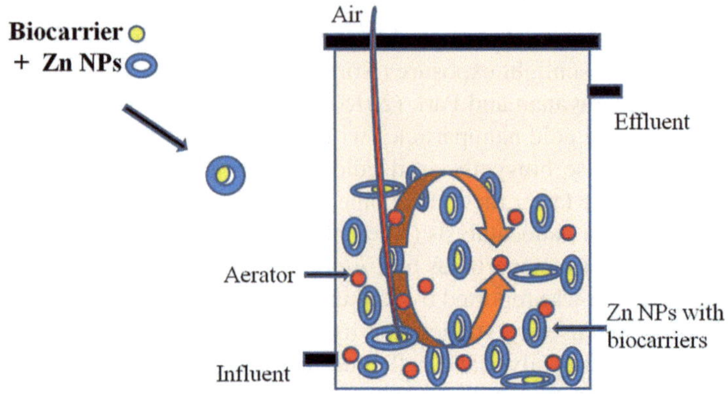

Fig. 15.1 Bionanoparticle integrated membrane bioreactor for the treatment of textile dye wastewater

beneficial due to its ability to remove contaminants from wastewater and produce less residual waste (Khosravi et al., 2020). However, biological methods such as sequencing batch reactors (SBRs), membrane bioreactors (MBRs) or upflow blanket bioreactors (UBRs) are more difficult to maintain and operate than physical processes. Compared with simple MFCs, hybrid systems have better performance.

15.3 Bionanoparticle Integrated Bioreactor for Textile Dye Wastewater

In such applications, biobased nanoparticles are integrated into membranes. These membranes are used for treatment purposes and are replacing typical polymeric membranes. In most cases, biological membranes are recognized for their high selectivity and permeability. Mixed matrix membranes, referred to as nanoparticle-blended or nanoparticle-entrapped membranes, are created when NPs are added to monomeric or polymeric membrane solutions (Fig. 15.2). One team developed PVDF hollow fiber membranes with TiO_2 using the phase inversion technique on a customised single-head spinning machine. Their membranes were subsequently tested in an algal membrane bioreactor (AMBR) for the treatment of wastewater (Sahu et al., 2019). Textile wastes were treated using a variety of reactor types, but the only ex-situ bioremediation method available was bioreactors. The reactors were divided into anaerobic and aerobic reactors. These materials are commonly used for the decolorization and biological deterioration of textile effluent. Many research papers on the combined, sequential, or integrated anaerobic–aerobic bioreactor treatment of azo-dye wastewater have been published (Ramachandran et al., 2013). The NF treatment was already relatively low because of the strong performance of the MBR system. Further COD, color, and salt removal were made

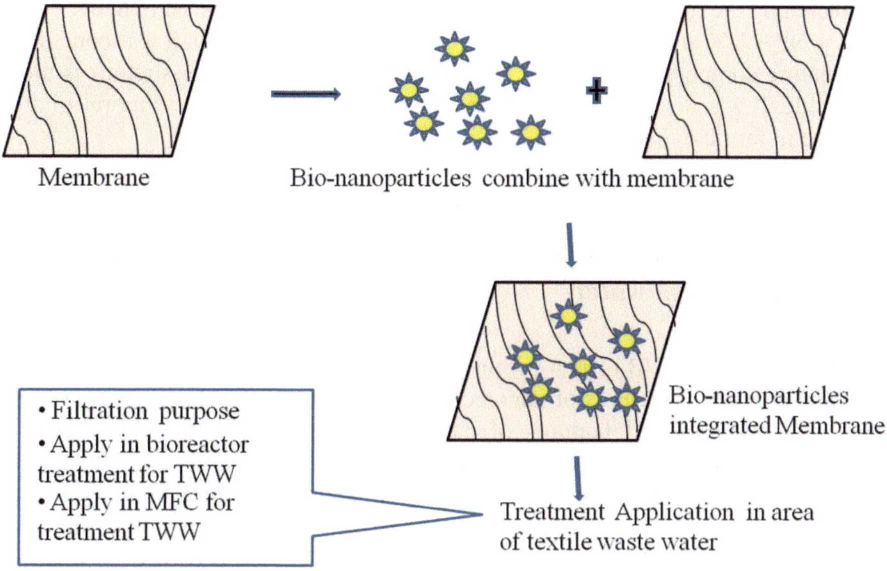

Fig. 15.2 Bionanoparticle-integrated membrane for the treatment of textile dye wastewater

possible by the NF of the biological and UF treatment effluent, which produced water with qualities that made it suitable for reuse inside the textile factory (Grilli et al., 2011). Membrane bioreactors (MBRs) are used when very high-quality effluents are needed, such as for water reuse. MBRs combine biological processes with membrane technologies. Additionally, MBRs have been regarded as the best pretreatment when additional filtration is needed. A series of filtration techniques, often comprising reverse osmosis, microfiltration (MF), ultrafiltration (UF), and nanofiltration (NF), have been used in conjunction with membranes to treat textile effluent (Grilli et al., 2011). According to Gomes et al. (2007), an upflow anaerobic sludge blanket (UASB) bioreactor was utilised in the successful application of anaerobic treatment for dye removal from nanofiltration. Biological processes led to considerable decolorization of the nanofiltration concentrate. Over 90% of dye removal occurred with both techniques. This was the result of anaerobic azo bond cleavage (Pazdzior et al., 2009). The color and COD removal performance of MBR processes using PVDF UF membranes, which ran at a concentration of 15,000 mg/L of MLSS and a hydraulic retention time (HRT) of 2 days, were reviewed. The average color and COD elimination were 72% and 97%, respectively. The effectiveness of the MBR + NF and MBR + BWRO procedures for treating industrial wastewater was studied. Using BWRO membranes, 96.7%, 97%, 74%, 90%, and 87% of the conductivity, salinity, COD, TOC, and colour, respectively, was removed. In dyeing tests using MBR effluent, sodium sulphate was not employed because the water already contained dissolved salts. Compared to softened process water, MBR + NF and MBR + BWRO effluents have comparable characteristics in terms of conductivity, turbidity, color, and hardness. The quality of the textile product was found to be unaffected by the recovered wastewater in the fabric dyeing tests using MBR + UV, MBR + NF + UV, and MBR + BWRO + UV effluents. In addition, it was concluded that treating composite wastewater instead of separate wastewater streams is more acceptable. It was discovered that MBR + BWRO + UV effluent was more suitable for reuse in fabric dying procedures (Cinperi et al., 2019).

15.3.1 Microbial Nanofiltration for the Treatment of Textile Dye Wastewater

Nanobiotechnology (NBT) is a recently developed area of technology that combines the prominent features of nanotechnology and biomolecules in living organisms. The concepts and features of nanotechnology can be used in biological processes to develop novel products. Owing to the many properties of living organisms, researchers have attempted to explore applications in which biological components canbe combined with nanoparticles. For such filtration purposes, biobased nanoparticles are combined into membranes. These membranes are utilized for wastewater treatment applications and can be used to replace traditional polymeric membranes. In many studies, biological membranes have been recognized for their excellent selectivity and permeability. The incorporation of NPs into monomeric or polymeric membrane solutions results in the formation of nanoparticle-blended membranes or nanoparticle-entrapped membranes, which are referred to as mixed matrix membranes (MMMs). NF is a biologically inspired membrane (Sahu et al., 2021). The large specific area of the filter nanoparticles indicates a high surface-to-volume ratio, which causes the polymer chain and the nanofiller to bind extensively. An organic-inorganic hybrid membrane based on biopolymers is a potential hybrid. A potential organic-inorganic hybrid for the development of a high-performance proton exchange membrane (PEM) is a biopolymer-based inorganic hybrid membrane (Terbish et al., 2023). A membrane-based filtration process involving reverse osmosis, ultrafiltration, nanofiltration, and microfiltration is an alternative technique for removing dyes from textile waste (Febrianto et al., 2019). The main advantage of membrane-based technology is that there is no need for the use of chemicals (Pandit et al., 2022). Globally, the combination of SBR and NF has current benefits in comparison with other technologies in textile wastewater treatment. NF is a technique that enables the production of sufficient quality wastewater. This membrane removed approximately 60% of the conductivity of the wastewater. The wastewater must undergo a prior biological process to be ready for the NF stage. An SBR can be used to remove the colour and biodegradable COD by alternating between the anaerobic and aerobic phases. Degradation of dyes also results in improved membrane performance because contamination is reduced (Zuriaga-Agusti et al., 2010). The membrane bioreactor-nanofiltration (MBR-NF) hybrid process is one of the most promising technologies studied in the fields of wastewater treatment and water reuse and has demonstrated high efficiency and stable performance in municipal wastewater treatment and reclamation (Li et al., 2020). This pretreatment made the membranes function more effectively and produced high-quality effluent. Wastewater may be effectively removed from SDS and RB21, which are highly soluble dyes, using the SBR/NF hybrid technique (Khosravi et al., 2020).

15.3.2 Nanocomposites and Microbial Fuel Cells in Textile Dye Wastewater Treatment

In recent decades, bioelectrochemical systems—most notably microbial fuel cells (MFCs)—have received increased attention as a viable means of producing clean energy while also offering the advantages of bioremediation. MFCs are capable of harvesting energy from organic wastes. Therefore, using wastewater as a substrate in MFCs to produce bioelectricity is considered "environmentally friendly" (Li et al., 2018). The development of microbial fuel cells (MFCs) as a sustainable bioenergy technology has made them extremely appealing because they provide power without burning. This biochemical system produces power and cleans waste by oxidising organic matter in wastewater through the action of microorganisms (Terbish et al., 2023). A novel Fe-Mn-Mg/CF composite cathode electrode was needed to remove refractory pollutants in a self-sustaining bioelectrochemical system, as the stable electrical energy output of the MFC®EF system is essential for its operation (Wang et al., 2022). Metal oxides combined with conducting polymers appear to be useful for the electrochemical catalytic activity of MFCs. After their preparation, conducting polymers exhibit great catalytic activity, reduced toxicity, good electroconductivity, environmental stability, rapid electrokinetics, and redox stability. Nanomaterials are recommended as an appropriate choice for MFC anode modifier catalysts. Because of their availability, low cost, and high catalytic activity, transition metal oxide (TMO) nanoparticles have been widely used as efficient catalysts in energy conversion and storage applications. The formation of diverse crystallographic forms in TMO nanoparticles endows them with a noteworthy role in rapid electrokinetic activity across a broad range of electrochemical applications. These novel, very effective nanosized preparations made from various metal oxides, including Mn, Fe, Co, Ni, and others, are among the most widely used and cost-effective supercatalytic materials. The use of these nanomaterials is also significantly influenced by their phase, size, and surface topography (Dessi & Tadesse, 2022). By incorporating living microorganisms as biocatalysts in the anode chamber, microbial fuel cells (MFCs) are a possible substitute. A proton exchange membrane (PEM) divides the anode in the anodic chamber and the cathode in the cathodic chamber of a standard two-chamber MFC system. The bacteria are deposited as a biofilm on the anode electrode, where they oxidise the substrate anaerobically to produce electrons. After that, these electrons are moved to the cathode chamber via an external circuit, where electron acceptors use them. In this manner, the anodic chamber treats wastewater while also producing energy. Methyl orange (MO) dye pollutant degradation in the cathodic compartment of a dual-chamber

MFC equipped with a bioanode and p-type Si nanowire photocathode modified by Pd nanoparticles. The dye removal efficiency and maximum power density under light illumination were found to be 84.5% and 119 mW m^{-2}, respectively, over a 36-h period, which was double that of a carbon paper cathode (Ahmadpour et al., 2020). In our previous study, researchers used a cathode material called Fe-Mn-Mg/CF, which linked Mg with Fe and Mn to form a three-metal electrode and produced great results. Although the industrial application of EF technology is limited by the cost of external power and the risk involved in H_2O_2 transportation, there are still some issues that must be resolved before field deployment in the degradation of azo dyes. The microelectricity utilisation rate of electro-Fenton and MFC-EF systems is also significantly influenced by biocenosis, bacteria in the anode, and the electrode materials, particularly the cathode. The essential parts of our previous study, such as the degradation performance of the azo dye, the application conditions, and thorough functional gene annotation, were not examined (Wang et al., 2022). Microbial fuel cells, a relatively new technology based on bioelectrochemical principles, are more advantageous because they can produce bioelectricity and are expensive to set up and maintain. In photocatalysis, a semiconductor—typically titanium dioxide (TiO_2)—is employed as a photocatalyst to produce electrons and whole pairs when exposed to light, such as sunlight or ultraviolet light. These materials can breakdown organic pollutants such as azo dyes by creating potent oxidants. The rapid recombination of electrons and holes at the photocatalyst surface is the main problem of this method. By delaying the recombination of electrons and holes, e.g., CB/h+ VB, the photocatalytic approach combines electrolytic and photocatalytic processes to overcome this drawback. This process leaves a positively charged whole (h+ VB) behind as the electron (e− CB) is ejected from the semiconductor's fully loaded valence band (VB) and into an empty conductive band (CB) in the presence of light irradiation. The increased power density in MFCs has been examined using surface-modified electrodes supplemented with additional materials, such as nanoparticles. Additional metals include Fe_2O_3, manganese oxide, hydrogenated NiO, and cobalt oxide (CoO). In one study, excellent green degradation was improved by using graphene oxide-coated electrodes in a two-chambered MFC. After only 7 days of treatment, more than 80% of the dye was removed. By increasing the total power output, graphene oxide adsorbs and degrades the dye. The transition metals RuO_2, NiO, MnO_2, and Fe_2O_3 are commonly used in MFCs (Saba et al., 2021).

The membrane used in MFCs can function as a physical separator, allowing only protons to cross and reach the cathode region. Additionally, the membrane prevents oxygen from diffusing from the cathode to the anode region. However, oxygen diffusion still occurs in MFCs using the commonly utilized Nafion 117 membrane since it is very permeable to oxygen. Furthermore, an MFC with a membrane has the potential to eventually result in biofouling, high internal resistance, and high operating costs (Thung et al., 2018). It has two electrodes, an electrolyte, and two redox reactions, which are the main processes involved in the conversion. As such, it is the most straightforward and economical electrochemical method for producing renewable energies. More modifications with other new and active nanomaterials, such as conducting polymers, are needed to increase the chemical, mechanical, and electrical conductivity of these materials (Dessi & Tadesse, 2022). The decolorization process in MFCs is the same, with minor differences on both membrane sides. The anodic chamber of a two-chamber system undergoes anaerobic biodegradation to achieve decolorization. Bacteria use the dye as a substrate (electron donor), breaking it down to release protons and electrons. As protons cross the PEM, the protons migrate in the direction of the anode and reach the cathode through the external circuit, completing the circuit and generating electricity. Protons, electrons, mediators, and enzymes are used at the cathode to decolorize the dye. Toxic intermediates such as amines are created in both situations. It has been demonstrated that the enzymes reductase, peroxidase, and laccase can break azo bonds. A white rot fungus produces laccase, which can catalyze the redox reaction that breaks down azo dye when copper ions are present. The combination of enzyme additive activity and biodegradation can improve dye degradation. Because the substrate uses some electrons for power generation and others for dye decolorization, power generation with simultaneous decolorization is a sluggish process (Saba et al., 2021).

15.3.3 Mechanism of Textile Dye Degradation by Nanoparticles

NPs function as catalysts and accelerate degradative processes. A source of light, such as UV or visible light, introduces nanoparticles to its rays. Usually, high-energy light sources are used directly to treat the surface of nanoparticles, or the photosensitization route is used for help (Roy et al., 2022). Photo and catalysis are the two components that make up the mechanism. In terms of photocatalytic processes, there are two categories: homogeneous and heterogeneous. In contrast to photocatalysts, which are used when ultraviolet (UV) and visible light are present, catalysts are often used to accelerate chemical reactions (Joshi et al., 2022). The first section of the article described photon absorption, the production of charge carriers, and dynamics. The generation of surface radicals and the reactivity of O_2, H_2O, and organic molecules at the surface are the topics of the second section (Aruna et al., 2022). The valence and conduction bands on the surface of metal oxide nanoparticles act as catalysts for photocatalysis (Fig. 15.3). The valence and conduction bands of holes and electrons were compared. In the presence of UV light, electrons are moved from the valence band to the conduction band through the appropriate band gap. Oxidation and reduction are the outcomes of the transfer of electrons. Reactive oxygen species (ROS) are ultimately formed as a result of this process. ROS frequently causes dye degradation. The oxidation and reduction processes that occur on the surface of any metal oxide and composite material are related to photocatalysis. Conduction bands (CBs), such as electrons ($e-$), and valence bands (VBs), such as holes ($h+$), which are produced by the absorption of ultraviolet light, play a role in this process. The generation of highly aggressive species, such as

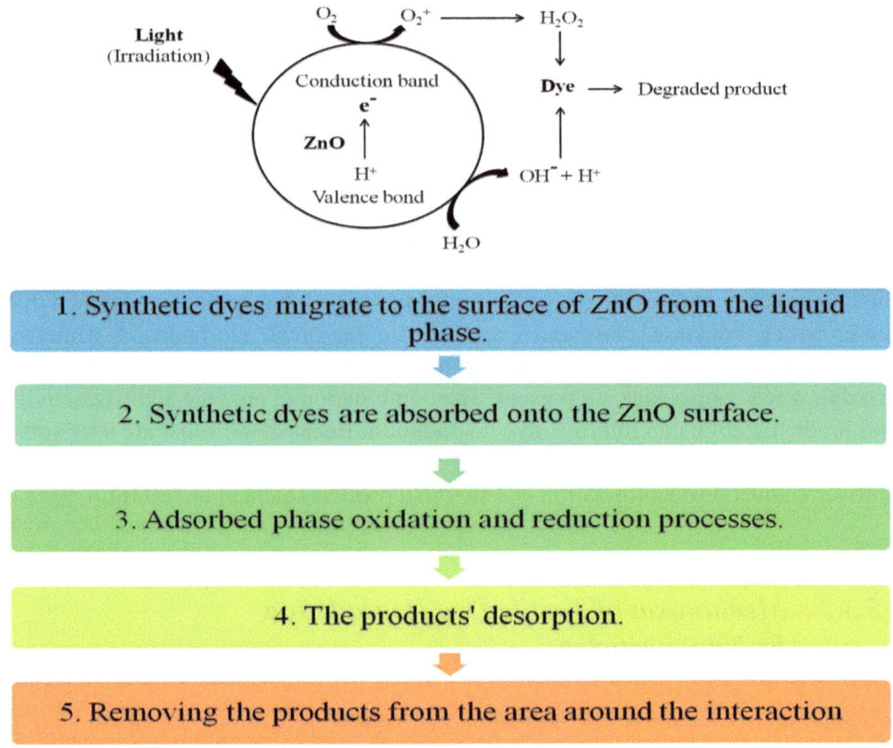

Fig. 15.3 Fundamental mechanism of zinc oxide photocatalysis. (Ong et al., 2018)

hydroxyl (OH) or superoxide radicals from moisture and ambient oxygen, is sped up by photogenerated pairs of h+ and e (Joshi et al., 2022; Aruna et al., 2022). The dye molecules that have been adsorbed on the photocatalyst can be degraded by these free radicals since they are highly reactive (Roy et al., 2022).

15.4 Advantages and Disadvantages of MAB for Textile Dye Wastewater Treatment

In the field of environmental remediation and biological sciences, nanotechnology has been widely used (Joshi et al., 2022). There are several ways to create nanoparticles, and photogenic synthesis has several benefits, including simpler management, high production, simple manufacturing, and a variety of applications (Hamada & Jabal, 2022). In comparison to conventional approaches, photocatalytic degradation by nanoparticles is preferable because of their fast oxidation, lack of polycyclic product formation, and oxidation of contaminants. It is a quick and efficient method for removing contaminants from wastewater (Kouhbanani et al., 2019). Compared to larger particles, nanoparticles have unique characteristics, such as size, shape, and dispersion, that give them extraordinary features. The catalytic, optical, magnetic, electrical, and thermal properties of nanoparticles are remarkable (Raina et al., 2020). In addition, textile industries are responsible for discharging harmful dyes that can be fatal to both people and aquatic life, such as effluent, into freshwater sources. Both promoting the need for clean drinking water and taking aggressive steps to clean up contaminated water are necessary for enhancing an individual's quality of life.

15.4.1 Environmental Friendliness

A remediation technique that uses microbes is beneficial for the environment. Biogenic materials are becoming more popular as affordable, safe, easy, and non-toxic biological solutions for these issues. Microorganism-assisted bionanoparticles are easy to use, can be produced on a large scale, and are simple to manipulate because of the use of natural reducing and capping agents (Rafique et al., 2019). The current evaluation is therefore focused on the fundamental analysis of several bio-synthesized eco-friendly nanoparticles for diverse industrially released dyes. Different biological approaches used for the synthesis of nanoparticles have been examined along with their effectiveness in reducing dyes (Thakare et al., 2022). An efficient, fast, and ecologically acceptable method of synthesizing nanoparticles is biological synthesis using microbes and their enzymes. Potential uses for synthesized nanoparticles in environmental cleanup.

15.4.2 Cost-Effectiveness

Cost-efficient, reliable, and environmentally friendly metal oxide-based photocatalysts should be developed for large-scale wastewater treatment operations. The suitability of a photocatalyst should be based on its cost, safety, band gap, ability to function under sunlight or other visible light, efficiency, and simple recovery and reusability. Nanoparticles made from plants work well for dye breakdown and are a good alternative for the cleanup procedure. Microorganisms assist bionanotechnology using natural reducing agents, so this is a cost-efficient technique.

15.4.3 Addressing Potential Drawbacks

NPs are very small molecules. Their size allows them to easily enter the cell membrane and may affect cell activities and organ function. Most nanoparticles affect nontarget samples adjacent to target cells. When they accumulate in the environment, NPs may cause adverse problems. NPs can have adverse effects on soil microbes and other soil flora. The nanoparticle synthesis process uses metals, which can cause health problems.

15.4.4 Challenges and Solutions

Researchers believe that producing metallic nanoparticles using plant extracts is the best option because plants are easier to cultivate than bacteria and fungi, which need complex environments to thrive and maintain properly. In addition, a variety of distinct main and minor metabolites that are found in plants can be used in a wide range of diverse ways. The best alternative among the different metal NPs produced by plants, such as Au, Ag, Cd, Zn, and Fe, is silver due to its antibacterial, insecticidal, water purification, nanomedicine, agricultural, and use as a coating agent for medical devices (Jain et al., 2020).

15.5 Comparison with Traditional Wastewater Treatment Methods

Several traditional physicochemical and biological treatments (Table 15.4) are used in the field of wastewater treatment, but these methods release large amounts of sludge and are not eco-friendly. As a result, there is an increasing need for the development of expensive, effective methods for removing dangerous chemicals from polluted water. Compared to other traditional technologies, nanotechnology is currently thought to be a promising solution for water purification. It can split complicated, long-chained chemical molecules that are typically toxic into simpler forms (Hamada & Jabal, 2022). Nanotechnology has been extensively used for many years in the bioremediation, wastewater purification, pharmaceutical, medical, and food processing sectors. The rapid advancement of nanobiotechnology has given manufacturers of nanostructure materials used to support immobilized enzymes a new advantage. Recently, the use of nanoparticles for the removal of dyes from wastewater has attracted the interest of many studies because of their unique properties, such as large specific surface areas, high adsorption capacities, and decreased diffusion resistance (Srivastava et al., 2022). Using nanoparticles, various technical developments have been made possible in the medical, catalysis, and environmental cleanup sectors. Silver and copper nanoparticles are among the noble metal nanoparticles that have been studied the most and recognised as possible nanoparticles for a long time. Copper and silver nanoparticles exhibited excellent catalytic characteristics. Silver is a common antimicrobial agent, and because of its high toxicity to a wide range of microorganisms, silver-based compounds have been widely used in numerous bactericidal applications (Raina et al., 2020). Several important

Table 15.4 Comparison of traditional bionanotechnology methods

Methods	Types	Advantages	Disadvantages
Physical	Coagulation/ flocculation	Easy, less expensive, small land occupation	Sludge generation, low efficiency
	Ion exchange	Easy recovery of solvent	Sludge generation is non-ineffective for all types of dyes
Chemical	Ozonation	Complete decolorisationation	Expensive
	Fenton reaction	Apply for soluble & insoluble dye	Sludge generation, long reaction time
Bioremediation	Microbial bioreactors	Environmentally friendly	Long degradation time
	Fungal degradation	High flexibility	lengthy and unstable growth phase
	Enzymatic degradation	Cost-effective, fast, nontoxic	Limited recovery
Nano remediation	Nanofiltration and ultrafiltration	Suitable for all dyes High separation capacity	High investment cost
	Photocatalytic	No sludge production, effective oxidation	Formation of byproducts
	Magnetic nanoparticles	Provide stability	In vitro cytotoxicity
Bio-Nanoremediation	Plant nanoparticles	Minimizing waste, reducing unnecessary derivatives, simplicity, no requirement of toxic chemicals	Affects the nontarget sample, causes health problems
	Microbial nanoparticles		
References	Joshi et al. (2022), Wong et al. (2019), Bhatia et al. (2017), and Srivastava et al. (2022)		

parameters, including catalyst dosage, pH, dye concentration, and reaction time, can impact the effectiveness of photocatalytic dye degradation (Mahmoodi et al., 2017).

15.6 Highlighting Notable Achievements in the Field

Bionanotechnology has increased in importance in the present day because of the development of many tools, materials, and techniques that are useful for solving major human and environmental issues, such as green agriculture, renewable energy production, diagnostic tools for detecting human and plant diseases at early stages, nanoparticles for the treatment of diseases, pollution checking through nanobiosensors, bioremediation of polluted sites, wastewater treatment and water purification. Hence, bionanotechnology has very high potential for combating current and future global problems (Rohela et al., 2019). Bacteria and microalgae are especially capable of synthesising distinctive nanomaterials such as exopolysaccharides, nanocellulose, and nanowires. These NPs have many uses, such as in the development of third-generation biosensors, biofilms, biosensors, cell imaging, sensoristic devices, and diagnostics. Additionally, bio fabricated NPs have demonstrated potential uses as antioxidants, antimicrobial agents, anticancer drugs, anticoagulants, antimigration agents, and antiproliferative agents (Patil et al., 2021). Nanochemistry is expected to have significant impacts on energy storage devices, air filtration, and wastewater treatment (Piracha et al., 2021). The different high-quality chemical sensor types used by nanotechnology allow sensors to detect small amounts of chemical vapours. These nanosensors are more stable for both commercial and industrial processes, and they are made using high-quality components to increase their sensitivity, operating capacity, and ultimate water chemical element detection. Nanotechnology-based sensors can employ a variety of sensing elements, including carbon nanotubes, zinc oxide nanowires, and palladium nanoparticles (Piracha et al., 2021). The multidisciplinary discipline of nanoscience is concerned with the engineering and design of functional systems at the molecular level. It is a field of applied science specializing in the creation, description, and usage of nanoscale materials and devices. In general, the art and science of modifying matter at the nanoscale to produce novel and distinctive materials is known as nanoscience (Vijayaraghavan & Kumar, 2017). The unique properties of silver nanoparticles make them particularly interesting for current nanotechnology research. These properties can be incorporated into a wide range of extensive applications, including food packaging, cosmetics, bioengineering, electrochemistry, catalysis, and antiseptic agents in the medical industry (Keat et al., 2015). The application of nanomaterials is currently being studied in many areas of human life. Carbon nanotubes, fluorescent nanocrystals, nanoporous silica, dendrimers, liposomes, graphene, and nanoparticles can all be included in the category of nanomaterials (Moritz & Moritz, 2013).

15.7 Future Prospects

Microbial nanoparticles will eventually be integrated with current water treatment systems as a result of advancements in these fields. To control costs, economic analysis is necessary. The fact that conventional disinfectants are affordable and readily available naturally is one of their benefits. The use of water treatment methods based on nanotechnology may be a potential solution for overcoming these issues, especially in light of the effects on the environment and human health. The parameters, including viability, scalability, safety, and cost of operating these systems, should be the main topics of future research. Innovations addressing these problems might overcome many of the current impediments and increase opportunities to revolutionise the treatment of drinking water distribution systems and wastewater reuse. The natural availability and low cost of conventional disinfectants are benefits. Only if the costs of the systems using NPs are reasonable, feasible, and effective can bionanotechnology-based methods of water treatment effectively compete with traditional technologies. These methods have several drawbacks, such as the need for expensive equipment, particular biological agents for dealing with various pollutants (organic chemicals, colours, nutrients), and ideal processing conditions for each microorganism.

15.8 Conclusion

Currently, environmental contamination is a severe issue that continues to worsen. Dye inclusion in wastewater and water treatment poses a major risk to human health and disrupts the environment. Traditional methods, including adsorption, ultrafiltration, chemical, and electrochemical procedures, as well as wastewater treatment systems, are commonly unable to eliminate them from wastewater. It is necessary to prevent the use of toxic dye substances, and this method must be efficient. Since nanotechnology uses small particles with special qualities that make it superior to bulk materials, the use of these materials is expanding extremely quickly. Green synthetic/biogenic strategies should be used for the development of nanoparticles to avoid the risks associated with the use of chemical-reducing agents and organic solvents. Biological agents are used as green solvents, reducing and stabilizing agents, and sources of renewable and affordable enzymes and metabolites. This will ensure the long-term viability of nanotechnology. Bioremediation should be a priority as remediation techniques that are both economical and environmentally friendly. Bionanoremediation is a novel method for dye degradation. When exposed to visible light, the biosynthesized NPs show very strong degradation activity. In general, our findings indicate that microbial nanoparticles manufactured from NPs can serve as helpful and appropriate nanoparticles for various scientific fields in the future, particularly for the treatment of organic dye-containing wastewater in the environment.

References

Aboelfetoh, E. F., El-Shenody, R. A., & Ghobara, M. M. (2017). Eco-friendly synthesis of silver nanoparticles using green algae (*Caulerpa serrulata*): reaction optimisation, catalytic and antibacterial activities. *Environmental Monitoring and Assessment, 189*, 1–5.

Ahmadpour, T., Aber, S., & Hosseini, M. G. (2020). Improved dye degradation and simultaneous electricity generation in a photoelectrocatalytic microbial fuel cell equipped with AgBr/CuO hybrid photocathode. *Journal of Power Sources, 474*, 228589.

Al-Buriahi, A. K., Al-Gheethi, A. A., Kumar, P. S., et al. (2022). Elimination of rhodamine B from textile wastewater using nanoparticle photocatalysts: A review for sustainable approaches. *Chemosphere, 287*, 132162.

Al-Hakkani, M. F., Gouda, G. A., & Hassan, S. H. (2021). A review of green methods for phytofabrication of hematite (α-Fe_2O_3) nanoparticles and their characterisation, properties, and applications. *Heliyon, 7*(1).

Aruna, V., Chada, N., Tejaswini Reddy, M., et al. (2022). Plant synthesised nanoparticles for dye degradation. In *Phytonanotechnology* (pp. 1–30). Springer.

Ashraf, I., Singh, N. B., & Agarwal, A. (2023). Green synthesis of iron oxide nanoparticles using Amla seed for methylene blue dye removal from water. *Materials Today Proceedings, 72*, 311–316.

Balcha, A., Yadav, O. P., & Dey, T. (2016). Photocatalytic degradation of methylene blue dye by zinc oxide nanoparticles obtained from precipitation and sol-gel methods. *Environmental Science and Pollution Research*, 25485–25493.

Basavegowda, N., Mishra, K., & Lee, Y. R. (2017). Synthesis, characterisation, and catalytic applications of hematite (α-Fe_2O_3) nanoparticles as reusable nanocatalyst. *Advances in Natural Sciences: Nanoscience and Nanotechnology, 8*(2), 025017. https://doi.org/10.1088/2043-6254/aa6885

Beheshtkhoo, N., Kouhbanani, M. A., Savardashtaki, A., et al. (2018). Green synthesis of iron oxide nanoparticles by aqueous leaf extract of Daphne mezereum as a novel dye removing material. *Applied Physics A, 124*, 1–7.

Bhatia, D., Sharma, N. R., Singh, J., & Kanwar, R. S. (2017). Biological methods for textile dye removal from wastewater: A review. *Critical Reviews in Environmental Science and Technology, 47*(19), 1836–1876.

Carmen, Z., & Daniela, S. (2012). *Textile organic dyes-characteristics, polluting effects and separation/elimination procedures from industrial effluents-A critical overview*. IntechOpen.

Chadha, P., Mehra, S., & Singh, M. (2021). Adverse impact of textile dyes on the aquatic environment as well as on human beings. *Toxicology International (Formerly Indian Journal of Toxicology), 28*(2), 165–176. https://doi.org/10.18311/ti/2021/v28i2/26798

Chahardoli, A., Karimi, N., Sadeghi, F., & Fattahi, A. (2018). Green approach for synthesis of gold nanoparticles from Nigella arvensis leaf extract and evaluation of their antibacterial, antioxidant, cytotoxicity and catalytic activities. *Artificial Cells, Nanomedicine, and Biotechnology, 46*(3), 579–588.

Chand, K., Jiao, C., Lakhan, M. N., et al. (2021). Green synthesis, characterisation and photocatalytic activity of silver nanoparticles synthesised with Nigella Sativa seed extract. *Chemical Physics Letters, 763*, 138218.

Chanu, L. A., Singh, W. J., Singh, K. J., & Devi, K. N. (2019). Effect of operational parameters on the photocatalytic degradation of Methylene blue dye solution using manganese doped ZnO nanoparticles. *The Results in Physics, 12*, 1230–1237.

Choudhary, M. K., Kataria, J., & Sharma, S. (2017). A biomimetic synthesis of stable gold nanoparticles derived from aqueous extract of Foeniculum vulgare seeds and evaluation of their catalytic activity. *Applied Nanoscience, 7*, 439–447.

Cinperi, N. C., Ozturk, E., Yigit, N. O., & Kitis, M. (2019). Treatment of woolen textile wastewater using membrane bioreactor, nanofiltration and reverse osmosis for reuse in production processes. *Journal of Cleaner Production, 223*, 837–848.

Dessie, Y., & Tadesse, S. (2022). Advancements in bioelectricity generation through nanomaterial-modified anode electrodes in microbial fuel cells. *Frontiers in Nanotechnology, 4*, 876014.

Dutta, D., & Das, B. M. (2021). Scope of green nanotechnology towards amalgamation of green chemistry for cleaner environment: A review on synthesis and applications of green nanoparticles. *Environmental Nanotechnology, Monitoring & Management, 15*, 100418.

Febrianto, G., Karisma, D., & Mangindaan, D. (2019). Polyetherimide nanofiltration membranes modified by interfacial polymerisation for treatment of textile dyes wastewater. *IOP Conference Series: Materials Science and Engineering, 622*, 1–012019.

Gahlawat, G., & Choudhury, A. R. (2019). A review on the biosynthesis of metal and metal salt nanoparticles by microbes. *RSC Advances, 23*, 12944–12967.

García, F. E., Plaza-Cazón, J., Montesinos, V. N., Donati, E. R., & Litter, M. I. (2018). Combined strategy for removal of Reactive Black 5 by biomass sorption on Macrocystis pyrifera and zerovalent iron nanoparticles. *Journal of Environmental Management, 207*, 70–79.

Geethamala, G. V., Poonkothai, M., & Swathilakshmi, A. V. (2023). Assessment on the photocatalytic and phytotoxic activities using ecobenevolently synthesised iron oxide nanoparticles from the root extracts of Glycyrrhiza glabra. *Environmental Science and Pollution Research*, 1–5.

Gomes, A. C., Gonçalves, I. C., de Pinho, M. N., & Porter, J. J. (2007). Integrated Nanofiltration and Upflow Anaerobic Sludge Blanket Treatment of Textile Wastewater for In-Plant Reuse. *Water environment research, 79*(5), 498–506.

Golmohammadi, M., Honarmand, M., & Ghanbari, S. (2020). A green approach to synthesis of ZnO nanoparticles using jujube fruit extract and their application in photocatalytic degradation of organic dyes. *Spectrochimica Acta Part A: Molecular and Biomolecular Spectroscopy, 229*, 117961.

Grilli, S., Piscitelli, D., Mattioli, D., et al. (2011). Textile wastewater treatment in a bench-scale anaerobic-biofilm anoxic-aerobic membrane bioreactor combined with nanofiltration. *Journal of Environmental Science and Health Part A, 46*(13), 1512–1518.

Gu, H., Chen, X., Chen, F., Zhou, X., & Parsaee, Z. (2018). Ultrasound-assisted biosynthesis of CuO-NPs using brown alga Cystoseira trinodis: Characterisation, photocatalytic AOP, DPPH scavenging and antibacterial investigations. *Ultrasonics Sonochemistry, 41*, 109–119.

Guo, H. X., Lin, K. L., Zheng, Z. S., et al. (2012). Sulphanilic acid-modified P25 TiO_2 nanoparticles with improved photocatalytic degradation on Congo red under visible light. *Dyes and Pigments, 92*(3), 1278–1284.

Gupta, N. K., Ghaffari, Y., Bae, J., & Kim, K. S. (2020). Synthesis of coral-like α-Fe_2O_3 nanoparticles for dye degradation at neutral pH. *Journal of Molecular Liquids, 301*, 112473.

Hamada, M. S., & Jabal, R. A. (2022). Doped (Ag) ZnO nanoparticles for removal of azo dyes from aqueous solutions. *International Journal Of Chemical And Biochemical Sciences, 21*, 210–217.

Hassaan, M. A., El Nemr, A., & Hassaan, A. (2017). Health and environmental impacts of dyes: Mini-review. *American Journal of Environmental Science and Engineering, 3*, 64–67.

Hulkoti, N. I., & Taranath, T. C. (2014). Biosynthesis of nanoparticles using microbes—A review. *Colloids and Surfaces B: Biointerfaces, 121*, 474–483.

Jain, A., Ahmad, F., Gola, D., et al. (2020). Multi dye degradation and antibacterial potential of Papaya leaf derived silver nanoparticles. *Environmental Nanotechnology, Monitoring & Management, 14*, 100337.

Jain, A. S., Pawar, P. S., Sarkar, A., et al. (2021). Bionanofactories for green synthesis of silver nanoparticles: Toward antimicrobial applications. *International Journal of Molecular Sciences, 22*(21), 11993.

Joshi, S. J., Geetha, S. J., Al-Mamari, S., & Al-Azkawi, A. (2018). Green synthesis of silver nanoparticles using pomegranate peel extracts and its application in photocatalytic degradation of methylene blue. *Jundishapur Journal of Natural Pharmaceutical Products, 13*(3).

Joshi, N. C., Gururani, P., & Gairola, S. P. (2022). Metal oxide nanoparticles and their nanocomposite-based materials as photocatalysts in the degradation of dyes. *Biointerface Research in Applied Chemistry*, 6557–6579.

Kalaiselvi, A., Roopan, S. M., Madhumitha, G., et al. (2015). Synthesis and characterisation of palladium nanoparticles using Catharanthus roseus leaf extract and its application in the photocatalytic degradation. *Spectrochimica Acta Part A: Molecular and Biomolecular Spectroscopy, 135*, 116–119.

Keat, C. L., Aziz, A., Eid, A. M., & Elmarzugi, N. A. (2015). Biosynthesis of nanoparticles and silver nanoparticles. *Bioresources and Bioprocessing, 2*(1), 1–11.

Khan, R., & Fulekar, M. H. (2016). Biosynthesis of titanium dioxide nanoparticles using Bacillus amyloliquefaciens culture and enhancement of its photocatalytic activity for the degradation of a sulfonated textile dye Reactive Red 31. *Journal of colloid and interface science, 475*, 184–191.

Khosravi, A., Karimi, M., Ebrahimi, H., & Fallah, N. (2020). Sequencing batch reactor/nanofiltration hybrid method for water recovery from textile wastewater contained phthalocyanine dye and anionic surfactant. *Journal of Environmental Chemical Engineering, 8*(2), 103701.

Kokeb, N., Tesfaye, T., Sharew, S., et al. (2023). Color adsorption performance of bone biocomponent for textile dyeing effluent treatment in Ethiopia. *Journal of Engineering, 2023*, 114. https://doi.org/10.1155/2023/4964194

Kouhbanani, M. A., Beheshtkhoo, N., Taghizadeh, S., et al. (2019). One-step green synthesis and characterisation of iron oxide nanoparticles using aqueous leaf extract of Teucrium polium and their catalytic application in dye degradation. *Advances in Natural Sciences: Nanoscience and Nanotechnology, 10*(1), 015007.

Li, M., Zhou, M., Tian, X., et al. (2018). Microbial fuel cell (MFC) power performance improvement through enhanced microbial electrogenicity. *Biotechnology Advances, 36*(4), 1316–1327.

Li, J. F., Rupa, E. J., Hurh, J., et al. (2019). Cordyceps militaris fungus mediated Zinc Oxide nanoparticles for the photocatalytic degradation of Methylene blue dye. *Optik, 183*, 691–697.

Li, K., Liu, Q., Fang, F., et al. (2020). Influence of nanofiltration concentrate recirculation on performance and economic feasibility of a pilot-scale membrane bioreactor-nanofiltration hybrid process for textile wastewater treatment with high water recovery. *Journal of Cleaner Production, 261*, 121067.

Liu, Y., Hu, Y., Zhou, M., et al. (2012). Microwave-assisted no aqueous route to deposit well-dispersed ZnO nanocrystals on reduced graphene oxide sheets with improved photoactivity for the decolorisationation of dyes under visible light. *Applied Catalysis B: Environmental, 125*, 425–431.

Mahmoodi, N. M. (2013). Zinc ferrite nanoparticle as a magnetic catalyst: Synthesis and dye degradation. *Materials Research Bulletin, 48*(10), 4255–4260.

Mahmoodi, N. M. (2015). Manganese ferrite nanoparticle: Synthesis, characterisation, and photocatalytic dye degradation ability. *Desalination and Water Treatment, 53*(1), 84–90.

Mahmoodi, N. M., Keshavarzi, S., & Ghezelbash, M. (2017). Synthesis of nanoparticle and modelling of its photocatalytic dye degradation ability from coloured wastewater. *Journal of Environmental Chemical Engineering, 5*(4), 3684–3689.

Mandal, R. P., Sekh, S., Sarkar, N. S., et al. (2016). Algae-mediated synthesis of cadmium sulphide nanoparticles and their application in bioremediation. *Materials Research Express, 3*(5), 055007.

Mechouche, M. S., Merouane, F., Messaad, C. E., et al. (2022). Biosynthesis, characterisation, and evaluation of antibacterial and photocatalytic methylene blue dye degradation activities of silver nanoparticles from Streptomyces tuirus strain. *Environmental Research, 204*, 112360.

Miri, A., Shahraki Vahed, H. O., & Sarani, M. (2018). Biosynthesis of silver nanoparticles and their role in photocatalytic degradation of methylene blue dye. *Research on Chemical Intermediates, 44*, 6907–6915.

Mondal, A., Adhikari, B., & Mukherjee, D. (2015). Room-temperature synthesis of air-stable cobalt nanoparticles and their use as catalyst for methyl orange dye degradation. *Colloids and Surfaces A: Physicochemical and Engineering Aspects, 482*, 248–257.

Moritz, M., & Geszke-Moritz, M. (2013). The newest achievements in synthesis, immobilisation and practical applications of antibacterial nanoparticles. *Chemical Engineering Journal, 228*, 596–613.

Nadaf, N. Y., & Kanase, S. S. (2019). Biosynthesis of gold nanoparticles by *Bacillus marisflavi* and its potential in catalytic dye degradation. *Arabian Journal of Chemistry, 12*(8), 4806–4814.

Narayanan, K. B., & Park, H. H. (2015). Homogeneous catalytic activity of gold nanoparticles synthesized using turnip (*Brassica rapa* L.) leaf extract in the reductive degradation of cationic azo dye. *Korean Journal of Chemical Engineering, 32*, 1273–1277.

Noman, M., Shahid, M., Ahmed, T., et al. (2020). Use of biogenic copper nanoparticles synthesised from a native Escherichia sp. as photocatalysts for azo dye degradation and treatment of textile effluents. *Environmental Pollution, 257*, 113514.

Ong, C. B., Ng, L. Y., & Mohammad, A. W. (2018). A review of ZnO nanoparticles as solar photocatalysts: Synthesis, mechanisms and applications. *Renewable and Sustainable Energy Reviews, 81*, 536–551.

Pandit, C., Roy, A., Ghotekar, S., et al. (2022). Biological agents for synthesis of nanoparticles and their applications. *Journal of King Saud University, Science, 34*(3), 101869.

Patil, N., Bhaskar, R. A., Vyavhare, V. I., et al. (2021). Overview on methods of synthesis of nanoparticles. *International Journal of Current Pharmaceutical Research, 13*(2), 11–16.

Paździor, K., Klepacz-Smółka, A., Ledakowicz, S., et al. (2009). Integration of nanofiltration and biological degradation of textile wastewater containing azo dye. *Chemosphere, 75*(2), 250–255.

Pereira, L., Mehboob, F., Stams, A. J., et al. (2015). Metallic nanoparticles: Microbial synthesis and unique properties for biotechnological applications, bioavailability and biotransformation. *Critical Reviews in Biotechnology, 35*(1), 114–128.

Piracha, S., Saleem, S., Momil, G., et al. (2021). Nanoparticle: role in chemical industries, potential sources and chemical catalysis applications. *Scholary International Journal of Chemistry and Material Sciences, 4*(4), 40–45.

Pugazhendhi, A., Prabhu, R., Muruganantham, K., et al. (2019). Anticancer, antimicrobial and photocatalytic activities of green synthesised magnesium oxide nanoparticles (MgONPs) using aqueous extract of Sargassum wightii. *Journal of Photochemistry and Photobiology B: Biology, 190*, 86–97.

Rafique, M., Sadaf, I., Tahir, M. B., et al. (2019). Novel and facile synthesis of silver nanoparticles using Albizia procera leaf extract for dye degradation and antibacterial applications. *Materials Science and Engineering: C, 99*, 1313–1324.

Raina, S., Roy, A., & Bharadvaja, N. (2020). Degradation of dyes using biologically synthesised silver and copper nanoparticles. *Environmental Nanotechnology, Monitoring & Management, 13*, 100278.

Rajkumar, R., Ezhumalai, G., & Gnanadesigan, M. (2021). A green approach for the synthesis of silver nanoparticles by Chlorella vulgaris and its application in photocatalytic dye degradation activity. *Environmental Technology & Innovation, 21*, 101282.

Ramachandran, P., Sundharam, R., Palaniyappan, J., & Munusamy, A. P. (2013). Potential process implicated in bioremediation of textile effluents: A review. *Advances in Applied Science Research, 4*(1), 131–145.

Raman, C. D., & Kanmani, S. J. (2016). Textile dye degradation using nano zero-valent iron: A review. *Journal of Environmental Management, 177*, 341–355.

Rao, M. D., & Pennathur, G. (2017). Green synthesis and characterisation of cadmium sulphide nanoparticles from Chlamydomonas reinhardtii and their application as photocatalysts. *Materials Research Bulletin, 85*, 64–73.

Rohela, G. K., Srinivasulu, Y., & Rathore, M. S. (2019). A review paper on recent trends in bionanotechnology: Implications and potentials. *Nanoscience and Nanotechnology - Asia, 9*(1), 12–20.

Roy, A., Murthy, H. A., Ahmed, H. M., et al. (2022). Phytogenic synthesis of metal/metal oxide nanoparticles for degradation of dyes. *Journal of Renewable Materials, 7*, 1911.

Saba, B., Kjellerup, B. V., & Christy, A. D. (2021). Eco-friendly bioelectrodegradation of textile dyes wastewater. *Bioresource Technology Reports, 15*, 100734.

Sadiq, H., Sher, F., Sehar, S., et al. (2021). Green synthesis of ZnO nanoparticles from Syzygium Cumini leaves extract with robust photocatalysis applications. *Journal of Molecular Liquids, 335*, 116567.

Sahu, J. N., Zabed, H., Karri, R. R. et al. (2019). Applications of nanobiotechnology for sustainable water purification. In *Industrial applications of nanomaterials* (pp. 313–340). Elsevier.

Sahu, J. N., Karri, R. R., Zabed, H. M., et al. (2021). Current perspectives and future prospects of nanobiotechnology in wastewater treatment. *Separation and Purification Reviews, 50*(2), 139–158.

Saravanan, C., Rajesh, R., Kaviarasan, T., et al. (2017). Synthesis of silver nanoparticles using bacterial exopolysaccharide and its application for degradation of azo-dyes. *Biotechnology Reports, 15*, 33–40.

Sathiyan, K., Bar-Ziv, R., Mendelson, O., & Zidki, T. (2020). Controllable synthesis of TiO_2 nanoparticles and their photocatalytic activity in dye degradation. *Materials Research Bulletin, 126*, 110842.

Sha, Y., Mathew, I., Cui, Q., et al. (2016). Rapid degradation of azo dye methyl orange using hollow cobalt nanoparticles. *Chemosphere, 144*, 1530–1535.

Sharma, M., Singh, J., & Basu, S. (2019). Efficient metal ion adsorption and photodegradation of Rhodamine-B by hierarchical porous Fe-Ni@ SiO_2 monolith. *Microchemical Journal, 145*, 708–717.

Sharma, A., Pal, K., Saini, N. et al. (2023). Remediation of contaminants from wastewater using algal nanoparticles via green chemistry approach: An organised review. *Nanotechnology*.

Shim, Y. J., Soshnikova, V., Anandapadmanaban, G., et al. (2019). Zinc oxide nanoparticles synthesised by Suaeda japonica Makino and their photocatalytic degradation of methylene blue. *Optik, 182*, 1015–1020.

Srivastava, N., & Mukhopadhyay, M. (2014). Biosynthesis of SnO_2 nanoparticles using bacterium Erwinia herbicola and their photocatalytic activity for degradation of dyes. *Industrial & Engineering Chemistry Research, 53*(36), 13971–13979.

Srivastava, R., & Singh, N. (2019). Importance of natural dye over synthetic dye: A critical. *International Journal of Home Science, 5*(2), 148–150.

Srivastava, A., Rani, R. M., Patle, D. S., & Kumar, S. (2022). Emerging bioremediation technologies for the treatment of textile wastewater containing synthetic dyes: A comprehensive review. *Journal of Chemical Technology & Biotechnology, 97*(1), 26–41.

Terbish, N., Popuri, S. R., & Lee, C. H. (2023). Improved performance of organic–inorganic nanocomposite membrane for bioelectricity generation and wastewater treatment in microbial fuel cells. *Fuel, 332*, 126167.

Thakare, Y., Kore, S., Sharma, I., & Shah, M. (2022). A comprehensive review on sustainable greener nanoparticles for efficient dye degradation. *Environmental Science and Pollution Research, 29*(37), 55415–55436.

Thung, W. E., Ong, S. A., Ho, L. N., et al. (2018). Biodegradation of Acid Orange 7 in a combined anaerobic-aerobic upflow membrane-less microbial fuel cell: Mechanism of biodegradation and electron transfer. *Chemical Engineering Journal, 336*, 397–405.

Ullah, A., Rasheed, S., Ali, I., & Ullah, N. (2021). Plant Mediated Synthesis of CdS Nanoparticles, their characterisation and application for photocatalytic degradation of toxic organic dye. *Chemical Review Letters, 4*(2), 98–107.

Umar, K., Aris, A., Parveen, T., et al. (2015). Synthesis, characterisation of Mo and Mn-doped ZnO and their photocatalytic activity for the decolorisationation of two different chromophoric dyes. *Applied Catalysis A: General, 505*, 507–514.

Usman, M., Ahmed, A., Yu, B., et al. (2019). Photocatalytic potential of bioengineered copper nanoparticles synthesised from Ficus carica extract for the degradation of toxic organic dye from waste water: Growth mechanism and study of parameter affecting the degradation performance. *Materials Research Bulletin, 120*, 110583.

Vaidehi, D., Bhuvaneshwari, V., Bharathi, D., & Sheetal, B. P. (2018). Antibacterial and photocatalytic activity of copper oxide nanoparticles synthesised using *Solanum lycopersicum* leaf extract. *Materials Research Express, 5*(8), 085403.

Vasantharaj, S., Sathiyavimal, S., Senthilkumar, P., et al. (2019). Biosynthesis of iron oxide nanoparticles using leaf extract of Ruellia tuberosa: Antimicrobial properties and their applications in photocatalytic degradation. *Journal of Photochemistry and Photobiology B: Biology, 192*, 74–82.

Vijayaraghavan, K., & Ashok Kumar, T. (2017). Plant-mediated biosynthesis of metallic nanoparticles: A review of literature, factors affecting synthesis, characterisation techniques and applications. *Journal of Environmental Chemical Engineering, 5*(5), 4866–4883.

Vinothkannan, M., Karthikeyan, C., Kim, A. R., & Yoo, D. J. (2015). One-pot green synthesis of reduced graphene oxide (RGO)/Fe_3O_4 nanocomposites and its catalytic activity toward methylene blue dye degradation. *Spectrochimica Acta Part A: Molecular and Biomolecular Spectroscopy, 136*, 256–264.

Wang, W., Jiang, J., Zhang, C., et al. (2022). Microelectricity utilisation performance and microbial mechanism in microbial fuel cell powered electro-Fenton system for azo dye treatment. *Biochemical Engineering Journal, 186*, 108583.

Wong, J. K., Tan, H. K., Lau, S. Y., Yap, P. S., et al. (2019). Potential and challenges of enzyme incorporated nanotechnology in dye wastewater treatment: A review. *Journal of Environmental Chemical Engineering, 7*(4), 103261.

Zuriaga-Agusti, E., Iborra-Clar, M. I., Mendoza-Roca, J. A., et al. (2010). Sequencing batch reactor technology coupled with nanofiltration for textile wastewater reclamation. *Chemical Engineering Journal, 161*(1–2), 122–128.

Correction to: Bioremediation of Various Industrial Effluents Utilizing Microorganism-Assisted Nanotechnology Within a Circular Economy

Shareefraza J. Ukkund, Bhavna Alke, and Usman T. Syed

Correction to:
Chapter 14 in: Y. K. Mohanta et al. (eds.),
***Nano-microbiology for Sustainable Development*,**
https://doi.org/10.1007/978-3-031-78845-1_14

The original version of the chapter was inadvertently published with incorrect affiliation details for the author **Shareefraza J. Ukkund**. The correct affiliation details have been updated in the chapter.

The updated version of this chapter can be found at
https://doi.org/10.1007/978-3-031-78845-1_14

Y. K. Mohanta et al. (eds.), *Nano-microbiology for Sustainable Development*,
https://doi.org/10.1007/978-3-031-78845-1_16

MIX
Papier aus verantwortungsvollen Quellen
Paper from responsible sources
FSC® C105338

If you have any concerns about our products,
you can contact us on
ProductSafety@springernature.com

In case Publisher is established outside the EU,
the EU authorized representative is:
Springer Nature Customer Service Center GmbH
Europaplatz 3, 69115 Heidelberg, Germany

Printed by Libri Plureos GmbH
in Hamburg, Germany